Lecture Notes in Computer Science

Lecture Notes in Artificial Intelligence   14474

Founding Editor

Jörg Siekmann

The series Lecture Notes in Artificial Intelligence (LNAI) was established in 1988 as a topical subseries of LNCS devoted to artificial intelligence.

The series publishes state-of-the-art research results at a high level. As with the LNCS mother series, the mission of the series is to serve the international R & D community by providing an invaluable service, mainly focused on the publication of conference and workshop proceedings and postproceedings.

Lu Fang · Jian Pei · Guangtao Zhai ·
Ruiping Wang
Editors

Artificial Intelligence

Third CAAI International Conference, CICAI 2023
Fuzhou, China, July 22–23, 2023
Revised Selected Papers, Part II

 Springer

Editors
Lu Fang 🆔
Tsinghua University
Beijing, China

Jian Pei
Duke University
Durham, NC, USA

Guangtao Zhai 🆔
Shanghai Jiao Tong Univeristy
Shanghai, China

Ruiping Wang 🆔
Chinese Academy of Sciences
Beijing, China

ISSN 0302-9743 ISSN 1611-3349 (electronic)
Lecture Notes in Artificial Intelligence
ISBN 978-981-99-9118-1 ISBN 978-981-99-9119-8 (eBook)
https://doi.org/10.1007/978-981-99-9119-8

LNCS Sublibrary: SL7 – Artificial Intelligence

This Springer imprint is published by the registered company Springer Nature Singapore Pte Ltd.
The registered company address is: 152 Beach Road, #21-01/04 Gateway East, Singapore 189721, Singapore

Paper in this product is recyclable.

Preface

The present book includes extended and revised versions of papers selected from the second CAAI International Conference on Artificial Intelligence (CICAI 2023), held in Fuzhou, China, during July 22–23, 2023.

CICAI is a summit forum in the field of artificial intelligence and the 2023 forum was hosted by Chinese Association for Artificial Intelligence (CAAI). CICAI aims to establish a global platform for international academic exchange, promote advanced research in AI and its affiliated disciplines, and promote scientific exchanges among researchers, practitioners, scientists, students, and engineers in AI and its affiliated disciplines in order to provide interdisciplinary and regional opportunities for researchers around the world, enhance the depth and breadth of academic and industrial exchanges, inspire new ideas, cultivate new forces, implement new ideas, integrate into the new landscape, and join the new era. The conference program included invited talks delivered by four distinguished speakers, Chenghu Zhou, Zhihua Zhou, Marios M. Polycarpou, and Xuesong Liu, as well as 17 tutorials on 8 themes, followed by an oral session of 13 papers, a poster session of 72 papers, and a demo exhibition of 16 papers. Those papers were selected from 376 submissions using a double-blind review process, and on average each submission received 2.9 reviews. The topics covered by these selected high-quality papers span the fields of AI-generated content, computer vision, machine learning, nature language processing, application of AI, and data mining, amongst others.

These two volumes contain 100 papers selected and revised from the proceedings of CICAI 2023. We would like to thank the authors for contributing their novel ideas and visions that are recorded in this book.

The proceedings editors also wish to thank all reviewers for their contributions and Springer for their trust and for publishing the proceedings of CICAI 2023.

October 2023

Lu Fang
Jian Pei
Guangtao Zhai
Ruiping Wang

Organization

General Chairs

Lu Fang   Tsinghua University, China
Jian Pei   Duke University, USA
Guangtao Zhai   Shanghai Jiao Tong University, China

Program Chair

Ruiping Wang   Chinese Academy of Sciences, China

Publication Chairs

Xiaohui Chen   University of Illinois at Urbana-Champaign, USA
Mengqi Ji   Beihang University, China

Presentation Chairs

Xun Chen   University of Science and Technology of China,
            China
Jiantao Zhou   University of Macau, China

Demo Chairs

Jiangtao Gong   Tsinghua University, China
Can Liu   City University of Hong Kong, China

Tutorial Chairs

Jie Song   ETH Zurich, Switzerland
Tao Yu   Tsinghua University, China

Grand Challenge Chairs

David Brady   University of Arizona, USA
Haozhe Lin   Tsinghua University, China

Advisory Committee

C. L. Philip Chen   University of Macau, China
Xilin Chen   Institute of Computing Technology, Chinese
                                     Academy of Sciences, China
Yike Guo   Imperial College London, China
Ping Ji   City University of New York, USA
Licheng Jiao   Xidian University, China
Ming Li   University of Waterloo, Canada
Chenglin Liu   Institute of Automation, Chinese Academy of
                                     Sciences, China
Derong Liu   University of Illinois at Chicago, USA
Hong Liu   Peking University, China
Hengtao Shen   University of Electronic Science and Technology
                                     of China, China
Yuanchun Shi   Tsinghua University, China
Yongduan Song   Chongqing University, China
Fuchun Sun   Tsinghua University, China
Jianhua Tao   Institute of Automation, Chinese Academy of
                                     Sciences, China
Guoyin Wang   Chongqing University of Posts and
                                     Telecommunications, China
Weining Wang   Beijing University of Posts and
                                     Telecommunications, China
Xiaokang Yang   Shanghai Jiao Tong University, China
Changshui Zhang   Tsinghua University, China
Lihua Zhang   Fudan University, China
Song-Chun Zhu   Peking University, China
Wenwu Zhu   Tsinghua University, China
Yueting Zhuang   Zhejiang University, China

Program Committee

Boxin Shi   Peking University, China
Can Liu   City University of Hong Kong, China

Feng Xu	Tsinghua University, China
Fu Zhang	University of Hong Kong, China
Fuzhen Zhuang	Beihang University, China
Guangtao Zhai	Shanghai Jiao Tong University, China
Hao Zhao	Tsinghua University, China
Haozhe Lin	Tsinghua University, China
Hongnan Lin	Chinese Academy of Sciences, China
Jian Zhao	Institute of North Electronic Equipment, China
Jian Zhang	Peking University, China
Jiangtao Gong	Tsinghua University, China
Jiannan Li	Singapore Management University, Singapore
Jiantao Zhou	University of Macau, China
Jie Song	ETH Zurich, Switzerland
Jie Wang	University of Science and Technology of China, China
Jinshan Pan	Nanjing University of Science and Technology, China
Junchi Yan	Shanghai Jiao Tong University, China
Le Wu	Hefei University of Technology, China
Le Wang	Xi'an Jiaotong University, China
Lei Zhang	Chongqing University, China
Liang Li	Institute of Computing Technology, Chinese Academy of Sciences, China
Lijun Zhang	Nanjing University, China
Lu Fang	Tsinghua University, China
Meng Yang	Sun Yat-sen University, China
Meng Wang	BIGAI, China
Mengqi Ji	Beihang University, China
Nan Gao	Tsinghua University, China
Peng Cui	Tsinghua University, China
Qi Liu	University of Science and Technology of China, China
Qi Dai	Microsoft Research, China
Qi Ye	Zhejiang University, China
Qing Ling	Sun Yat-sen University, China
Risheng Liu	Dalian University of Technology, China
Ruiping Wang	Institute of Computing Technology, Chinese Academy of Sciences, China
Shuhui Wang	VIPL, ICT, Chinese Academic of Sciences, China
Si Liu	Beihang University, China
Tao Yu	Tsinghua University, China
Wu Liu	AI Research of JD.com, China

Xiaoyan Luo	Beihang University, China
Xiaoyun Yuan	Tsinghua University, China
Xiongkuo Min	Shanghai Jiao Tong University, China
Xun Chen	University of Science and Technology of China, China
Ying Fu	Beijing Institute of Technology, China
Yubo Chen	Institute of Automation, Chinese Academy of Sciences, China
Yuchen Guo	Tsinghua University, China
Yue Gao	Tsinghua University, China
Yue Deng	Beihang University, China
Yue Li	Xi'an Jiaotong-Liverpool University, China
Yuwang Wang	Tsinghua University, China

Contents – Part II

Robotics

Contents – Part I

AI Ethics, Privacy, Fairness and Security

FairDR: Ensuring Fairness in Mixed Data of Fairly and Unfairly Treated Instances

Yuxuan Liu(ID), Kun Kuang(✉)(ID), Fengda Zhang(ID), and Fei Wu(ID)

Zhejiang University, Hangzhou, China
{yuxuanliu,kunkuang,fdzhang,wufei}@zju.edu.cn

Abstract. Fairness has emerged as a crucial topic in data mining and machine learning applications, driven by ethical and legal considerations. It is important to recognize that not all samples are treated unfairly, resulting in data heterogeneity in fair machine learning. Existing fair models primarily focus on achieving fairness across all heterogeneous data, yet they often fall short in ensuring fairness within specific subgroups, such as fairly treated and unfairly treated data. This paper presents a novel problem of training a fair model on heterogeneous data, aiming to achieve fairness for both types of data, with a particular emphasis on the unfairly treated subset. To address this challenge, an effective approach is to recover the distribution of both fairly and unfairly treated data. In this study, we adopt the Structural Causal Model (SCM) to model the heterogeneous data as a mixture of causal structures. Leveraging the perspective of SCM, we propose a framework called FairDR, which utilizes the Hirschfeld-Gebelein-Rényi (HGR) correlation to accurately recover the distribution of both fairly and unfairly treated data. FairDR can serve as a pre-processing method for other fair machine learning models, providing protection for the unfairly treated members. Through empirical evaluation on synthetic and real-world datasets, we demonstrate that the presence of heterogeneous data can introduce unfairness in previous algorithms. However, FairDR successfully recovers the distribution of fairly and unfairly treated data, thus improving the fairness of downstream algorithms when dealing with heterogeneous data.

Keywords: Fairness · Causality · Heterogeneity

1 Introduction

Fairness has become an important topic in data mining and machine learning applications. Lack of fairness considerations, machine learning algorithms may cause discriminatory behaviors against certain groups (e.g., race, gender, etc.) in applications such as law, medicine, sociology, and policy science [6–8,22,23]. In practice, the unfair results would violate the interests of specific groups, which is unethical and illegal. Therefore, more and more researchers and organizations [10] have started to focus on fairness in machine learning algorithms in recent years. To design a fairness algorithm, there are three main lines of adjustment

L. Fang et al. (Eds.): CICAI 2023, LNAI 14474, pp. 3–14, 2024.
https://doi.org/10.1007/978-981-99-9119-8_1

methods, including pre-processing methods [14, 30], in-processing methods [4, 29] and post-processing methods [16, 18]. These methods achieve fairness via either generating fair data [30], disentangling sensitive attributes [12], adding regular terms [3, 15], causal intervention [9, 19, 24] and etc.

Despite the efforts on training a fair model on an overall data set, a critical issue of fair machine learning is whether the potentially unfairly treated members could get a more fair result. Here, we argue that the data we obtain in practice is always heterogeneous that consists of fairly treated samples and unfairly treated samples. For example, female interviewers may consider more of gender fairness, even though gender discrimination is common in job-hunting scenarios. Fair machine learning algorithms themselves also produce fairly treated data that will be mixed with existing unfairly treated data. Unfortunately, as in our studies, most of the previous fair models trained on heterogeneous data may not perform that well on a mixture of fairly and unfairly treated data. In Fig. 1, we demonstrate the results of three kinds of methods (i.e., FFVAE [12], Cfair [32], Rényi [3]) on two heterogeneous datasets (i.e., Adult[1] [2] and COMPAS[2] [1]) and their corresponding fairly and unfairly treated samples. The results demonstrate that even when these models could perform significantly fairly on heterogeneous data, they could still be discriminatory on both subpopulations. In some cases, the unfairly treated members get a more discriminatory result, and the fairly treated ones also suffer from the risk of algorithmic discrimination. More details will be provided in Sect. 4. Therefore, training a fair model on an overall heterogeneous dataset cannot guarantee its performance on the fairly and unfairly treated samples separately. How to achieve fairness in heterogeneous data, especially for those unfairly treated samples, is still an open problem.

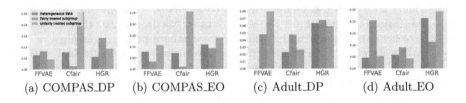

(a) COMPAS_DP   (b) COMPAS_EO   (c) Adult_DP   (d) Adult_EO

Fig. 1. Results of existing fair models on the heterogeneous data and its corresponding fairly and unfairly treated samples. Fairness on overall heterogeneous data cannot guarantee fairness on either subgroup of the data.

One possible way to address the data heterogeneity in fair machine learning is obtaining the fairly treated subgroup through sample selecting [26, 27]. But fairly treated samples are not always the majority of heterogeneous data. There are also researchers [5] trying to identify whether a sample is unfairly treated

---

[1] https://archive.ics.uci.edu/ml/datasets/Adult.
[2] https://www.propublica.org/datastore/dataset/compas-recidivism-risk-score-data-and-analysis.

and bridge the gap between two subpopulations in linear cases. How to solve the problem of data heterogeneity in fair machine learning more generally is still an open and challenging problem. At the same time, more attention has to be paid to protect the unfairly treated samples. As detecting whether a sample is unfairly treated is difficult, a more sufficient way is to recover the distribution of fairly and unfairly treated data in heterogeneous data. In this paper, we focus on heterogeneous data consisting of fairly treated samples and unfairly treated samples. We model the data heterogeneity in fair machine learning via Structural Causal Model (SCM) framework with theoretical analysis of why models trained on heterogeneous data do not guarantee fairness performance on fairly and unfairly treated data. Moreover, we propose a distribution recovery framework, named FairDR, to recover the data distribution for both fairly and unfairly treated data in heterogeneous data in continuous cases based on the HGR coefficient and SCM framework. Our approach first searches the adjacent nodes of sensitive attributes by independence test. In addition, we train two neural networks to fit the two nonlinear transformations in HGR correlation. At last, we recover the distribution of fairly and unfairly treated data with the individual HGR value. Theoretically, our framework can recover the two distributions well in different cases. Empirical evaluation on both synthetic data and real-world data is provided to prove: (1) data heterogeneity does have impacts on fair machine learning; (2) our framework can help downstream methods to get better results on both fairly and unfairly treated data. At the end of this paper, we discuss the open questions about data heterogeneity in fair machine learning.

Broadly, we categorize our contributions in the following points:

- We conceptualize the problem of data heterogeneity in fair machine learning via causal graphs, and analyze the possible impacts of data heterogeneity with theory and experiments.
- We propose a distribution recovery framework FairDR for continuous cases based on the Structural Causal Model framework and HGR correlation, which does not require prior knowledge of the causal graph.
- Extensive experiments demonstrate our method can recover the distribution of fairly and unfairly treated data. Also, our method can be used as a data pre-processing framework for other downstream fairness methods.

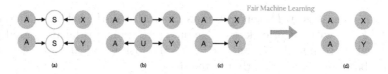

Fig. 2. Illustration of fair and unfair structure with SCM view. (a) Unfair structure based on selection bias (Collider). (b) Unfair structure based on confounders (Folk). (c) Unfair structure based on the cause (Chain). (d) Fair structure.

2   Analyses, Notations, and Problems

2.1   Heterogeneous Data on SCM Perspective

From the perspective of SCM, heterogeneity in the data means that data are generated from different causal structures. So we can use causal graphs to represent heterogeneity in data. According to [21], discrimination can be summarised as ***systemic discrimination*** and ***statistical discrimination***. As shown in Fig. 2(b) and Fig. 2(c), systemic discrimination can be modeled as a folk structure and chain structure. Similarly, statistical discrimination can be formed as selection bias in Fig. 2(a). To our observation, the mechanism of selection bias is usually complex social factors and not involved in the data (e.g., the selection bias between race and education level). In existing research on fairness, it is commonly assumed that the data only consists of biased samples. However, in real-world scenarios, not all samples are subjected to discrimination for various reasons. This heterogeneity of discrimination can manifest as variations in the causal structures within the data.

2.2   Notations and Problems

In heterogeneous data D^h, there are N fairly treated samples and M unfairly treated samples. For the fairly treated data, we have $D_f = \{(x_i^f, y_i^f, a_i^f)\}_{i=1}^N$, each sample $i \in \{1, 2, \ldots, N\}$ consists of feature vectors x_i^f, sensitive variable a_i^f and label y_i^f. Similarly, the unfairly treated data can be represented as $D_f = \{(x_j^u, y_j^u, a_j^u)\}_{j=1}^M$. The heterogeneous data $D^h = D^f \cup D^u$ where D^f, D^u, D^h are over $\mathcal{X} \times \mathcal{Y} \times \mathcal{A}$. We use $\boldsymbol{\theta}_h, \boldsymbol{\theta}_f, \boldsymbol{\theta}_u$ to denote parameters of a model trained on heterogeneous data, fairly treated data and unfairly treated data, respectively. So, as an example, $\hat{Y}_{\theta_h}^f$ means prediction on fairly treated data with parameters trained on heterogeneous data. Let θ_f and θ_g be the parameters of function f and g in HGR correlation. ρ_R denotes the HGR correlation coefficient and ρ_R denotes the individual HGR correlation value.

    As we described, fair models trained on heterogeneous data may not be guaranteed to be fair on both fairly and unfairly treated data. In this case, the unfairly treated members may not be protected from discrimination. Worse yet, those fairly treated ones may also suffer from algorithmic discrimination. Therefore, in this paper, we focus on the problem of fairness in heterogeneous data, which is defined as:

Problem 1. **Fairness in Heterogeneous Data.** Given the heterogeneous data $D^h = D^f \cup D^u$ without prior knowledge on which sample is fairly treated (i.e., D^f) and which is unfairly treated (i.e., D^u), the task is to learn a model which is fair on both fairly and unfairly treated data.

3   Fair Distribution Recovery from Heterogeneous Data

As we described in previous sections, we can recover the distribution of fairly and unfairly treated samples from heterogeneous data with the correlation between sensitive variable A and its adjacent nodes. In this section, we propose a fair distribution recovery (FairDR) framework based on a structural causal model perspective. To find the adjacent node of A, we choose independent test ways which are similar to [28]. We choose HGR correlation as an independence metric and use its individual value to recover the distribution. With our approach, we can reduce the negative effects of data heterogeneity in a number of fairness problems. A discussion of the difficulties of identifying data heterogeneity in fair machine learning will also be included in this section. Our distribution recovery framework FairDR consists of two parts:

1. Filtering neighboring nodes by independence test and conditional independence test.
2. Fitting the individual HGR value between each variable in adjacent nodes and the sensitive attribute A to recover the distribution.
3. Recovering the fair and unfair distributions

3.1   Conditional Independence Test for Distribution Recovery

The reason we filter the set of variables used for distribution recovery by independence tests is that we are concerned with the heterogeneous structure associated with sensitive attributes in fair machine learning. And due to the Markov property in the causal graph, a sensitive attribute is independent of all remaining variables given all its neighbors (in our setting, selection bias is some sampling mechanism instead of a variable). Therefore, the heterogeneous structure we need to identify exists only between sensitive attributes and their neighboring nodes in the causal graph. Consider the situation that a heterogeneous structure existing between A and Y and the unfairly treated data A is the cause of Y. In the fairly treated data, A and Y are independent, while in the unfairly treated data, A and Y are correlated. In the mixed data, the larger the proportion of the unfairly treated data, the stronger the correlation between A and Y. This means that when the correlation coefficient between A and Y is less than a threshold, there is no heterogeneous structure between A and Y (only A and Y independent data exist). For all variables associated with sensitive attributes, we can filter out all nodes that are not directly adjacent to A by the conditional independence test when the assumption of faithfulness and the Markov assumption is satisfied. With the independence test, we can exclude heterogeneous structures in the data that are not related to sensitive attributes. At the same time, the conditional independence test avoids the degradation of distribution recovery performance brought by a large number of invalid features when the number of features is large. We chose Kernel-based conditional independence (KCI) [31] as the independence test in our experiments, but other independence test methods (e.g., HSIC, mutual information, distance covariance) are also feasible. Algorithm 1 in Appendix summarizes our steps to filter the features.

3.2   Individual HGR Value Estimation

As in this paper we consider a heterogeneous structure containing only fairly and unfairly treated samples, we choose the distribution recovery method based on HGR correlation. The first reason we chose HGR is that it is a widely used and proven independence metric in

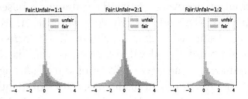

Fig. 3. Ideal distributions for fairly and unfairly treated data in the linear case.

fair machine learning. Another reason is that HGR correlation makes it easy to obtain individual values for each sample.

Definition 1. *HGR Correlation.*

$$\rho_R(A, B) = \sup_{f,g} \mathbb{E}[f(A)g(B)]$$

$$s.t. \quad \mathbb{E}[f(A)] = \mathbb{E}[g(B)] = 0, \quad \mathbb{E}\left[f^2(A)\right] = \mathbb{E}\left[g^2(B)\right] = 1 \tag{1}$$

With Definition 1, we can see that HGR correlation is actually calculating the Pearson correlation coefficient between $f(x)$ and $g(y)$. Here f and g are some kind of nonlinear transformations that make ρ_R take the maximum value. For the Pearson correlation coefficient, the individual values of two uncorrelated standard Gaussian variables will be the difference of two independent χ_1^2 distributions (As shown in 1). So it forms a K-form Bessel distribution with $\mu = 0$ and $\sigma^2 = 1$ [17]. When the n of chi-square distribution χ_n^2 is large, it converges to a Gaussian distribution [11]. So we approximate it as a Gaussian distribution in our method. In contrast, the Pearson individual values of two linearly correlated Gaussian variables form a χ_1^2 distribution with $n = 1$. Subject to $\mathbb{E}[A] = \mathbb{E}[B] = 0$, $\mathbb{E}\left[A^2\right] = \mathbb{E}\left[B^2\right] = 1$, Pearson correlation can be represented as $\text{Cov}(A, B)$. So in fairly treated data, Pearson individual values follow the Gaussian distribution. And in unfairly treated data, Pearson individual values follow a χ_1^2 distribution. As shown in Fig. 3, no matter how the ratio of two parts changes, we can get pure fair data D_f when the number is less than 0 in the perfect case (linear correlation without noise). Since 0 is both the mean and median of the standard normal distribution, we can get the sample size of the fair part of the data. According to the sample size of the fair part and the unfair part, we can reduce the distribution of each part.

Lemma 1. *Given 2 independent standard Gaussian variables A and B, $AB \sim \frac{1}{2}Q - \frac{1}{2}R$ where $Q, R \sim \chi_1^2$:*

Proof. See Appendix.

Since HGR correlation can be seen as a nonlinear generalization of Pearson's correlation coefficient, it is also feasible under nonlinear conditions when we fit f and g relatively well. Same as [15] we use neural networks to fit the functions f and g. The step of training f and g is described in Appendix A.

Lemma 2. *Given 2 variables* $A^h = A^f \cup A^u$ *and* $B^h = B^f \cup B^u$ *where* $A^f \perp B^f$ *and* $B^u \not\perp B^u$. $f(A^h), f(A^u), f(A^f), g(B^h), g(B^u), g(B^f)$ *form standard Gaussian distribution.* $J^h(\theta_f, \theta_g) = E[f(A^h) \cdot g(B^h)]$ *and* $J^u(\theta_f, \theta_g) = E[f(A^u) \cdot g(B^u)]$ *will converge to same* θ_f *and* θ_g.

Proof. See Appendix.

3.3 Fair and Unfair Distribution Recovery with HGR

In practice, because there is noise in the data, the χ^2 distribution we get is sometimes a little shifted. For this problem, we set the hyperparameter γ. We assume that fair for all samples with $\rho_R < \gamma$. As we know that the distribution of ρr for fairly treated data is a K-form Bessel distribution with $\mu = 0$ and $\sigma^2 = 1$, we can estimate the sample size of the fair part. So that we can know the proportion of the fair part and the unfair part in mixed data. For example, when $\gamma = -\frac{\sqrt{2}}{2}$, the fairly treated data will have about 16 percent of the data less than γ because $-\gamma$ is the standard deviation of the Gaussian distribution. With the proportion, we recover a Bessel function distribution and a shifted χ^2 distribution in the data. In this paper, we tried three different ways to recover the distribution:

- Discretization: We split the observed samples into 100 discrete bins, calculate the probabilities for samples from either distribution to fall in each bin, and samples in each bin according to the ratio of two probabilities to form our distribution.
- Gaussian KDE: We use a Gaussian kernel to estimate the probability density function of the mixed distribution, and assign each sample to the Gaussian part according to the ratio of p.d.f. value of a standard Gaussian to our KDE estimation at this certain sample point.
- χ^2 approximation: Alternative to kernel density estimation, we combine a χ_1^2 distribution with negative shift and a standard Gaussian distribution to approximate the distribution of ρ_R values from the unfairly treated data.

When we need to use sensitive attribute A to do distribution recovery with multiple variables $\{c_1, \ldots, c_l\}$, we repeat our algorithm to each pair of variables $\{(A, c_1), \ldots, (A, c_l)\}$ and we choose the pair of variables which is closest to the ideal Gaussian and χ_1^2 distribution. More discussion about FairDR will be discussed in Appendix.

4 Experiments

4.1 Experimental Setup

For each dataset, we first empirically reveal the limitations of the previous approaches, which only aim to train a model achieving fairness towards heterogeneous data. Then we use the proposed FairDR to recover the fair data distribution and unfair data distribution from the heterogeneous data. At last,

Table 1. Experiments results on synthetic data in terms of fairness on fairly treated samples and unfairly treated samples.

Method	Metric	Setting 1	Setting 2	Setting 3	Setting 4	Setting 5	Setting 6	Setting 7	Setting 8
Training w/o FairDR	Δ_{DP}: F	0.048±0.007	0.010±0.010	0.108±0.071	0.015±0.010	0.016±0.014	0.066±0.010	0.005±0.001	0.027±0.000
	Δ_{DP}: U	0.110±0.007	0.023±0.009	0.080±0.040	0.024±0.014	0.067±0.010	0.099±0.008	0.115±0.008	0.033±0.000
Training w/ FairDR (Ours)	Δ_{DP}: F	0.021±0.007	0.005±0.003	0.048±0.042	0.032±0.001	0.052±0.015	0.079±0.008	0.007±0.002	0.012±0.003
	Δ_{DP}: U	0.018±0.001	0.007±0.002	0.035±0.027	0.003±0.001	0.046±0.023	0.084±0.010	0.097±0.005	0.023±0.002
Training w/o FairDR	Δ_{EO}: F	0.051±0.004	0.035±0.004	0.032±0.008	0.061±0.029	0.022±0.013	0.073±0.015	0.008±0.005	0.064±0.005
	Δ_{EO}: U	0.021±0.008	0.169±0.004	0.045±0.016	0.120±0.026	0.011±0.009	0.017±0.015	0.029±0.003	0.026±0.002
Training w/ FairDR (Ours)	Δ_{EO}: F	0.033±0.019	0.071±0.008	0.019±0.010	0.059±0.039	0.008±0.005	0.086±0.007	0.006±0.000	0.004±0.003
	Δ_{EO}: U	0.030±0.019	0.140±0.005	0.024±0.014	0.092±0.005	0.019±0.011	0.010±0.006	0.017±0.005	0.006±0.004

we compare the performance of the algorithms with and without FairDR in terms of fairness on fairly treated samples and unfairly treated samples. In this paper, we adopt HGR regularization method [20] as the backbone algorithm, but we would like to emphasize that our proposed method can be adapted to any fairness algorithm.

Fairness Metrics. We restrict attention to statistical fairness, which asks for equality of some statistics evaluated over protected groups. We mainly consider the notions of **demographic parity** (DP) [13] and **equalized opportunity** (EO) [16] in this paper, but our method can be adapted to any type of statistical fairness. We use Δ_{DP} and Δ_{EO} to measure the degree of unfairness between the two groups. The greater the values of Δ_{DP} and Δ_{EO}, the more unfair the model is. For the case where both sensitive attributes and labels are continuous variables, we binarize them so that we can split the dataset into two groups and calculate the Δ_{DP} and Δ_{EO}.

4.2 Experiments on Synthetic Data

Dataset. We first use synthetic data to evaluate the fairness of models trained on heterogeneous data across subgroups. Specifically, we generate 6 different datasets based on causal structures in Fig. 2. Then followed the structure of the causal graph in the Appendix to create a data set with 8 variables.

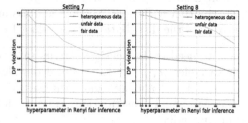

Fig. 4. Δ_{DP} using different hyperparameters in Setting 7 and Setting 8.

Details of how the synthetic data sets are generated will be described in Appendix B.

Impact of Data Heterogeneity. In this experiment, we first generate 6 different types of data heterogeneity with the causal structure shown in Fig. 1. Then we generate 2 causal graphs with 8 variables. We choose HGR regular term as our downstream algorithm. As shown in Fig. 4, fair models trained on heterogeneous data perform not that well on both fair and unfair parts. In some cases, the model performs lower on both fair and unfair components than on heterogeneous data. This proves the existence of possible reverse discrimination on fairly treated samples as we proposed. In Table 1, Setting 1 to Setting 6 correspond to six heterogeneous structures. Setting 7 and Setting 8 correspond to the two complex causal graphs. In all eight settings, we can observe that models trained on heterogeneous data fail to perform consistently on the fair and unfair parts. We found this situation to be particularly noticeable in the presence of variables in X between sensitive attributes A and label Y.

Distribution Recovery. In this experiment, we evaluate the accuracy of the HGR-based distribution recovery when unfair data based on different correlations are mixed with fair data. The proportion of fair and unfair data is 1 : 1. Here we choose four different types of correlations and add noise. We test three different ways to fit the distribution of the fair subgroup and unfair subgroup. In

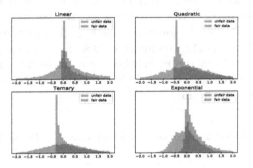

Fig. 5. Distribution HGR individual value on unfair and unfair subgroups.

Fig. 5, we can find that when the relationship between variables is nonlinear, the distribution of the fair part tends to be more Gaussian. The visualization of this experiment is attached in Appendix.

Fairness Results. In Table 1, when we choose DP as the fairness metric, models trained on our reduced distribution perform more consistently on both the fair and unfair components than those trained directly on heterogeneous data. In the first six settings, both models perform well when sensitive attribute A is adjacent to the label Y. In other cases our recovered distributions solve the problem posed by data heterogeneity well.

4.3   Experiments on Real-World Data

Dataset. We also verify the effectiveness of our method on the following real-world datasets: $Census^3$ and $Crime^4$ [25]. More details about these two datasets are described in Appendix.

Remark 1. Because we do not know whether the real-world data is heterogeneous or not, and the ground truth of the ratio of the two parts of the data, we try to

---

[3] https://www.kaggle.com/muonneutrino/us-census-demographic-data.
[4] https://archive.ics.uci.edu/ml/datasets/communities+and+crime.

deal with the data with a long-term fair machine learning process. Specifically, we first divide the dataset into three random equal parts. We train a fair machine learning model with the first part of the data and predict fair labels in the second part of the data. Finally, we mix the second and third parts of the data to form heterogeneous data. We expect to simulate data heterogeneity caused by fairness adjustment through the above process.

Impact of Data Heterogeneity. We first test the optimal fairness performance of the backbone algorithm trained based on mixed data on the mixed test set. We chose Census and Crime as the test dataset. Both methods can work well on mixed test sets. When we look at the performance of the fair and unfair parts of the mixed test set, the performance of the model becomes poor and unstable. In some cases such as Δ_{EO} on the Crime dataset, the convergence of Δ_{DP} and Δ_{EO} on mixed data seems to be simply due to its good performance on one of subgroups. In other cases, the models are all less fair on subgroups than on heterogeneous data. This may be due to the presence of reverse discrimination in two subgroups.

Table 2. Δ_{DP} on real-world data.

	census	crime
Unfair (w/o FairDR)	0.149(0.001)	0.141(0.037)
Fair (w/o FairDR)	0.070(0.002)	0.078(0.073)
Unfair (w/ FairDR)	0.040(0.002)	0.040(0.028)
Fair (w/ FairDR)	0.078(0.012)	0.076(0.048)

Table 3. Δ_{EO} on real-world data.

	census	crime
Unfair (w/o FairDR)	0.140(0.002)	0.141(0.029)
Fair (w/o FairDR)	0.028(0.006)	0.196(0.153)
Unfair (w/ FairDR)	0.096(0.002)	0.065(0.038)
Fair (w/ FairDR)	0.040(0.019)	0.185(0.167)

Fairness Results. In Table 2 and Table 3, models trained on the recovered distributions perform better than those trained on heterogeneous data. The models trained on our recovered distributions perform better on fair and unfair subgroups and have better overall fairness. As our fair data is generated with a fair algorithm, we have to point out that a real-world heterogeneous dataset with ground truth discrimination labels is needed for further study.

5   Conclusion and Future Work

In this work, we present the problem of data heterogeneity in fair machine learning. We find that not only accuracy but also fairness metrics can be affected by data heterogeneity. The fair model trained on heterogeneous data may still perform poorly on fair and unfairly treated data. To solve this problem, we propose FairDR, an SCM-based distribution recovery framework. With our framework, we can recover the distributions of the fair subgroup and the unfair subgroup well in many cases and improve the performance of downstream methods. As described in the paper, data heterogeneity is an essential and challenging problem for fair machine learning. Although our method performs well under continuous conditions, it remains a challenge to reduce the distribution in discrete or even binary cases. Also, data heterogeneity in fair machine learning is not just a mixture of fairly and unfairly treated data. There may also be different levels of discrimination that cannot be solved by our approach. We will put those issues in our future work.

Acknowledgement. This work was supported in part by Zhejiang Province Natural Science Foundation (LQ21F020020), National Natural Science Foundation of China (62006207, U20A20387), Young Elite Scientists Sponsorship Program by CAST (2021QNRC001), and the Fundamental Research Funds for the Central Universities (226-2022-00142, 226-2022-00051).

References

1. Angwin, J., Larson, J., Mattu, S., Kirchner, L.: Machine bias: there's software used across the country to predict future criminals and it's biased against blacks. ProPublica **23**, 77–91 (2016)
2. Asuncion, A., Newman, D.: Uci machine learning repository (2007)
3. Baharlouei, S., Nouiehed, M., Beirami, A., Razaviyayn, M.: R\'enyi fair inference. arXiv preprint arXiv:1906.12005 (2019)
4. Bahng, H., Chun, S., Yun, S., Choo, J., Oh, S.J.: Learning de-biased representations with biased representations. In: International Conference on Machine Learning, pp. 528–539. PMLR (2020)
5. Barik, A., Honorio, J.: Fair sparse regression with clustering: An invex relaxation for a combinatorial problem. In: Advances in Neural Information Processing Systems, vol. 34 (2021)
6. Berk, R.: Accuracy and fairness for juvenile justice risk assessments. J. Empir. Leg. Stud. **16**(1), 175–194 (2019)
7. Berk, R., Heidari, H., Jabbari, S., Kearns, M., Roth, A.: Fairness in criminal justice risk assessments: The state of the art. Sociological Methods Res. **50**(1), 3–44 (2021)
8. Bogen, M., Rieke, A.: Help wanted: An examination of hiring algorithms, equity, and bias (2018)
9. van Breugel, B., Kyono, T., Berrevoets, J., van der Schaar, M.: Decaf: Generating fair synthetic data using causally-aware generative networks. In: Advances in Neural Information Processing Systems, vol. 34 (2021)
10. Caton, S., Haas, C.: Fairness in machine learning: a survey. arXiv preprint arXiv:2010.04053 (2020)
11. Craig, C.C.: On the frequency function of xy. Ann. Math. Stat. **7**(1), 1–15 (1936)
12. Creager, E., et al.: Flexibly fair representation learning by disentanglement. In: International Conference on Machine Learning, pp. 1436–1445. PMLR (2019)
13. Dwork, C., Hardt, M., Pitassi, T., Reingold, O., Zemel, R.S.: Fairness through awareness. In: Goldwasser, S. (ed.) Innovations in Theoretical Computer Science 2012, Cambridge, MA, USA, January 8–10, 2012, pp. 214–226. ACM (2012). https://doi.org/10.1145/2090236.2090255
14. Dwork, C., Immorlica, N., Kalai, A.T., Leiserson, M.: Decoupled classifiers for group-fair and efficient machine learning. In: Conference on Fairness, Accountability and Transparency, pp. 119–133. PMLR (2018)
15. Grari, V., Ruf, B., Lamprier, S., Detyniecki, M.: Fairness-aware neural r\'eyni minimization for continuous features. arXiv preprint arXiv:1911.04929 (2019)

16. Hardt, M., Price, E., Srebro, N.: Equality of opportunity in supervised learning. In: Advances in Neural Information Processing Systems 29 (2016)
17. Johnson, N.L., Kotz, S., Balakrishnan, N.: Continuous univariate distributions, volume 2, vol. 289. John wiley & sons (1995)
18. Kim, M., Reingold, O., Rothblum, G.: Fairness through computationally-bounded awareness. In: Advances in Neural Information Processing Systems, vol. 31 (2018)
19. Kusner, M.J., Loftus, J., Russell, C., Silva, R.: Counterfactual fairness. In: Advances in Neural Information Processing Systems, vol. 30 (2017)
20. Mary, J., Calauzenes, C., El Karoui, N.: Fairness-aware learning for continuous attributes and treatments. In: International Conference on Machine Learning, pp. 4382–4391. PMLR (2019)
21. Mehrabi, N., Morstatter, F., Saxena, N., Lerman, K., Galstyan, A.: A survey on bias and fairness in machine learning. ACM Comput. Surv. (CSUR) **54**(6), 1–35 (2021)
22. Metz, C., Satariano, A.: An algorithm that grants freedom, or takes it away. The New York Times 6 (2020)
23. Mukerjee, A., Biswas, R., Deb, K., Mathur, A.P.: Multi-objective evolutionary algorithms for the risk-return trade-off in bank loan management. Int. Trans. Oper. Res. **9**(5), 583–597 (2002)
24. Pan, W., Cui, S., Bian, J., Zhang, C., Wang, F.: Explaining algorithmic fairness through fairness-aware causal path decomposition. In: Proceedings of the 27th ACM SIGKDD Conference on Knowledge Discovery & Data Mining, pp. 1287–1297 (2021)
25. Redmond, M.: Communities and crime unnormalized data set. UCI Machine Learning Repository, p. 66 (2011)
26. Roh, Y., Lee, K., Whang, S., Suh, C.: Sample selection for fair and robust training. In: Advances in Neural Information Processing Systems, vol. 34 (2021)
27. Roh, Y., Lee, K., Whang, S.E., Suh, C.: Fairbatch: Batch selection for model fairness. In: International Conference on Learning Representations (2020)
28. Spirtes, P., Glymour, C.: An algorithm for fast recovery of sparse causal graphs. Soc. Sci. Comput. Rev. **9**(1), 62–72 (1991)
29. Xu, D., Wu, Y., Yuan, S., Zhang, L., Wu, X.: Achieving causal fairness through generative adversarial networks. In: Proceedings of the Twenty-Eighth International Joint Conference on Artificial Intelligence (2019)
30. Xu, D., Yuan, S., Zhang, L., Wu, X.: Fairgan: Fairness-aware generative adversarial networks. In: 2018 IEEE International Conference on Big Data (Big Data)., pp. 570–575. IEEE (2018)
31. Zhang, K., Peters, J., Janzing, D., Schölkopf, B.: Kernel-based conditional independence test and application in causal discovery. arXiv preprint arXiv:1202.3775 (2012)
32. Zhao, H., Coston, A., Adel, T., Gordon, G.J.: Conditional learning of fair representations. arXiv preprint arXiv:1910.07162 (2019)

Blind Adversarial Training: Towards Comprehensively Robust Models Against Blind Adversarial Attacks

Haidong Xie[1](✉)(iD), Xueshuang Xiang[2](✉), Bin Dong[3,4,5], and Naijin Liu[2]

[1] China Aerospace Science and Technology Innovation Research Institute, Beijing, China
xiehaidong@aliyun.com
[2] Qian Xuesen Laboratory, China Academy of Space Technology, Beijing, China
xiangxs2@163.com
[3] Beijing International Center for Mathematical Research, Peking University, Beijing, China
[4] Center for Data Science, Peking University, Beijing, China
[5] Beijing Institute of Big Data Research, Beijing, China

Abstract. Adversarial training (AT) aims to improve models' robustness against adversarial attacks by mixing clean data and adversarial examples (AEs) into training. Most existing AT approaches can be grouped into restricted and unrestricted approaches. Restricted AT requires a prescribed uniform budget for AEs during training, with the obtained results showing high sensitivity to the budget. In contrast, unrestricted AT uses unconstrained AEs, and these overestimated AEs significantly lower the clean accuracy and robustness against small budget attacks. Thus, the existing AT approaches find it difficult to obtain a comprehensively robust model when confronting attacks with an unknown budget, which we name blind adversarial attacks. Considering this problem, this paper proposes a novel AT approach named blind adversarial training (BAT). The main idea is to use a cutoff-scale strategy to adaptively estimate a nonuniform budget to modify the AEs used in training, ensuring that the strengths of the AEs are dynamically located in a reasonable range and ultimately improving the comprehensive robustness of the AT model. We include a theoretical investigation on a toy classification problem to guarantee the improvement of BAT. The experimental results also demonstrate that BAT can achieve better comprehensive robustness than AT with several AEs.

Keywords: Adversarial Training · Comprehensive Robustness · Blind Adversarial Attacks

1 Introduction

Deep learning [9,15] has made great breakthroughs in many fields, such as computer vision [13], speech recognition [11,17], and natural language process-

L. Fang et al. (Eds.): CICAI 2023, LNAI 14474, pp. 15–26, 2024.
https://doi.org/10.1007/978-981-99-9119-8_2

ing [25]. Its weakness has attracted increasing attention, especially after adversarial examples (AEs) were introduced [10,26,29]. Many effective AE generation methods and defensive strategies have been proposed; see the review papers and references therein for details [1,28].

Adversarial training (AT) is a process of training a neural network on a mixture of clean data and AEs, in order to improve the robustness of the network against adversarial attacks; see [2,10,16,26] for white-box attacks and [14,23,27] for black-box attacks. These approaches focus on improving the generalization ability of AT by modifying the AE generation method or training loss. Furthermore, the limitation of incurring a minimal impact on clean accuracy requires the architecture of the network to be sufficiently expressive [1,28], as guaranteed by the universal approximation theorem [10,12] and regularization [22]; even so, it is still difficult to achieve improved robustness [10,19,27]. Thus, how to obtain a more robust model is still a hotspot issue to be studied. This paper discusses this issue from the point of view of the magnitude of the AE perturbation (budget) during training. Thus, most existing AT approaches can be divided into two categories: restricted and unrestricted AT. Next, we analyze the characteristics of these existing AT approaches.

Restricted AT approaches: that use norm-constrained AEs, such as FGSM-AT [10] and PGD-AT [16] with using FGSM and PGD AEs respectively, require a prescribed uniform training budget to constrain the magnitude of the AE perturbation during training and then evaluate the result on the AEs with the same budget attacks. The model obtained with the prescribed budget is only robust when confronting the attack with the same strength, while it is clearly weak for an attack stronger than the prescribed budget, and is overly defensive when confronting a small attack or encountering clean data. Madry et al. [16] found that a large budget is necessary to improve the AT effectiveness, but possibly lowering the accuracy on clean data. This result was also verified numerically by Song et al. [23]. It can be intuitively displayed through the experiment in Fig. 1 that the robustness of restricted AT is sensitive to the budget.

In contrast, the unrestricted AT approaches: are not affected by the prescribed budget and use unconstrained AEs [4] during training. The main goal of these AE approaches is to attack the models without constraints on AEs, such that the desired AEs should go beyond the decision boundary. To date, many types of unconstrained AEs have been examined; for example, DeepFool [18] attempts to obtain the smallest AEs aiming for the decision boundary, CW [5] can obtain AEs that balance the perturbation and confidence, and other innovative methods without a prescribed budget [3,24]. While we can directly apply these AEs to AT, the basic motivation of unconstrained AEs is to attack while aiming to fool the model by using the AEs beyond the decision boundary. As an improvement, DDN-AT [21] decouples the direction and norm of gradient-based attacks while inheriting the advantages of high computational efficiency. MMA training [7], focuses on maximizing the margins, which is an alternative approach for selecting the budget for each point individually. Other similar methods have the same characteristics [6,8,30]. It is clear that the introduction of the

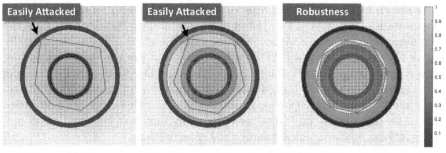

(a) Normal Training   (b) Adversarial Training   (c) Blind Adversarial Training

Fig. 1. Comparison of different training approaches on a two circles classification problem with the 1-hidden-layer 6-dim perceptron. The solid blue/red lines correspond to the datasets of two labels, the solid polygon lines represent the decision boundary of the classifiers, and the blue/red shadow zones in (b,c) show the manifold of AEs. (Color figure online)

unconstrained AEs into training actually changes the concerned original data manifold. Based on the obtained numerical experience, the loss function gradients are unstable in this case, leading to dramatic decision boundary fluctuations and giving rise to a very large number of AEs with too large a strength; this leads to the training process being unstable and severely decreases the robustness and lowers the accuracy on clean data.

From the above discussions on both restricted and unrestricted AT, the following may be summarized: if the attack budget is known exactly, we can definitely increase the model's robustness by using a restricted AT; moreover, if it is assumed that the attack budget is large, then the model's robustness can be increased by using an unrestricted AT. However, a more interesting situation is that of **practical application, one cannot expect to know the attack budget in advance**. This means that it is necessary to develop an AT approach when confronting attacks with an unknown budget, which we name **blind adversarial attacks**.

To address this problem, we propose a novel AT approach named **blind adversarial training (BAT)**, which uses a cutoff-scale (CoS) strategy to ameliorate the generation of AEs during training. See Algorithm 1 and Sect. 2.3 for details. Both strategies adaptively estimate a nonuniform budget and ensure that the AEs are located in a reasonable range in the blind case, such that the trained model will achieve a better comprehensive robustness. In the future, the CoS strategy can be combined with other methods as a plug-in to achieve better results. Moreover, considering that the existing robustness evaluation criterion cannot reflect the models' comprehensive robustness, we propose comprehensive robustness ($\mathcal{CR}$). $\mathcal{CR}$ can be considered the integration of a model's accuracies under the attack with varying budgets and clean data. See Sect. 2.4 for details.

As shown in Fig. 1 (c), BAT can generate a model with an exactly regular octagon decision boundary, i.e., the model with the best robustness within the given network architecture without the prescribed budget. We also theoretically analyze the improvement of BAT on a toy model and provide Propositions 1–3. Furthermore, compared with the FGSM-, PGD- and DeepFool-based ATs, using BAT to train LeNet-5 on Mnist, Fashion-Mnist and SVHN and to train FitNet-4 on Cifar10 and Cifar100, we can obtain models with better comprehensive robustness. In addition, on Mnist, we empirically find that both the cutoff and scale strategies make significant individual contributions.

2   Theory and Methodology

We consider a standard classification task with dataset $\{\mathbf{x}, \mathbf{y}\}$ and minimization objective $\min_\theta \mathbb{E}_{(\mathbf{x}, \mathbf{y})}[\mathcal{L}(\theta, \mathbf{x}, \mathbf{y})]$, where $\mathcal{L}$ is the loss function with weights θ. Adversarial training (AT) is also called brute-force AT and was first proposed by Szegedy et al. [26] and further developed by Goodfellow et al. [10]. The core idea of AT is to enhance the robustness by adding AEs to the training data; here, the total loss function can be written in a general form $\min_\theta \mathbb{E}_{(\mathbf{x}, \mathbf{y})}[\mathcal{L}(\theta, \mathbf{x}, \mathbf{y}) + \mathcal{L}(\theta, \mathbf{x} + \delta(\mathbf{x}), \mathbf{y})]$, where $\mathbf{x} + \delta(\mathbf{x})$ represents the AEs of data $\mathbf{x}$. AT will alternately generate AEs and optimize the network parameters until the levels of accuracy on clean data and AEs converge.

To theoretically compare the behavior of the ATs, we consider a toy classification problem (**TCP**), i.e., using a simplified 1-layer perceptron $y = \text{Sigmoid}(W\mathbf{x} + b)$ to classify two points $\mathbf{x}^1, \mathbf{x}^2$. It can be easily verified that the decision boundary of the model with the **best robustness** falls on the perpendicular bisector of the two points. All the proofs of the following propositions are provided in the appendix.

2.1   Restricted Adversarial Training

To generate AEs, Madry et al. [16] introduced a set of allowed perturbations $\mathcal{S}$ that formalize the manipulative power of the adversary, usually using the ℓ_∞-ball around $\mathbf{x}$ with the budget ε as $\mathcal{S}$ [10], so that the AT process can be reformulated as

$$\min_\theta \mathbb{E}_{(\mathbf{x}, \mathbf{y})} \max_{\delta(\mathbf{x}) \in \mathcal{S}} \mathcal{L}(\theta, \mathbf{x} + \delta(\mathbf{x}), \mathbf{y}) \tag{1}$$

This saddle point optimization problem specifies a clear goal of a robust classifier. The inner "max" (adversarial loss) aims to find an AE of the given data $\mathbf{x}$, while the outer "min" finds a model that minimizes the "adversarial loss". These AEs can be easily simplified to the widely accepted and used FGSM or PGD [16] AEs. We name this kind of AT **restricted AT**. Proposition 1 shows the improvement of restricted AT on NT.

Proposition 1. *For the TCP problem, both restricted AT and NT can obtain the model with the best robustness, and there exists a prescribed budget such that the restricted AT with this budget can accelerate the training process of NT.*

Proposition 1 demonstrates that the budget used in restricted AT can accelerate the training process. However, the optimal budget with maximum acceleration cannot be predicted, and the optimal choices vary between different datasets. This means that even for such a toy problem, one cannot expect to always obtain a better model given a prescribed budget.

2.2 Unrestricted Adversarial Training

To release the requirement of a prescribed budget in restricted AT, an intuitive approach is to use the unconstrained AEs during AT, i.e., **unrestricted AT**, formally

$$\min_{\theta} \mathbb{E}_{(\mathbf{x},\mathbf{y})} \max_{\delta(\mathbf{x})} \{ \mathcal{L}(\theta, \mathbf{x} + \delta(\mathbf{x}), \mathbf{y}) - \|\delta(\mathbf{x})\| \}, \tag{2}$$

where the $\mathcal{L}(\theta, \mathbf{x} + \delta(\mathbf{x}), \mathbf{y})$ term corresponds to guaranteeing the attacking ability, and the $\|\delta(\mathbf{x})\|$ term constrains the norm to be small. An example of an alternative approach to generate AEs is DeepFool [18], which minimizes the ℓ_2 norm of AEs slightly beyond the decision boundary of the current model. The AT with unconstrained AEs may obtain a more robust model than NT, but it cannot ensure better performance than restricted AT and will incur a heavy cost in terms of accuracy loss. Similarly, we state the following proposition.

Proposition 2. *For the TCP problem, unrestricted AT cannot be used to obtain the model with the best robustness.*

While simply using unconstrained AEs in unrestricted AT can avoid the shortcoming of choosing a prescribed budget, the AEs located on the decision boundary prevent the model from obtaining the best robustness. Thus, unlike restricted AT, for such a toy classification problem, unrestricted AT may output a bad model.

2.3 Blind Adversarial Training

As we claimed in Sect. 1, one cannot expect to know the attack budget in practice. Therefore, to alleviate the drawbacks of the restricted and unrestricted AT against blind adversarial attacks, we propose a cutoff-scale (CoS) strategy based on DeepFool-AT (DF-AT) in (2), and we name our approach blind adversarial training (BAT). To clarify the discussion, here, we only consider using the Deep-Fool AEs. We can similarly combine the CoS strategy with other approaches of generating unconstrained AEs, such as CW [5] or DDN [21], which will be the future plan. BAT can be formulated as

$$\min_{\theta} \mathbb{E}_{(\mathbf{x},\mathbf{y})} \max_{\delta(\mathbf{x})} \{ \mathcal{L}(\theta, \underbrace{\mathbf{x} + \rho\delta(\mathbf{x})}_{\text{Scale}}, \mathbf{y}) - \|\delta(\mathbf{x})\| - \underbrace{(\|\delta(\mathbf{x})\| - \varepsilon)_{+}}_{\text{Cutoff}} \}, \tag{3}$$

where we introduce two parameters ε and ρ to monitor the cutoff and the scale process, respectively. $\mathcal{L}(\theta, \mathbf{x} + \delta(\mathbf{x}), \mathbf{y})$ and $\|\delta(\mathbf{x})\|$ terms act in the same manner as unrestricted AT.

Algorithm 1. Blind Adversarial Training (BAT)

Input: Dataset $\{\mathbf{x}, \mathbf{y}\}$, scale factor ρ, learning rate α.
Output: Model θ.
  Initialize model θ.
  repeat
    $\mathcal{L}_{\mathrm{C}} = \mathcal{L}(\theta, \mathbf{x}, \mathbf{y})$,   ▷ Loss on clean data
    $\delta(\mathbf{x}) = \mathbf{x}_{\mathrm{adv}} - \mathbf{x}$,   ▷ $\mathbf{x}_{\mathrm{adv}}$: DeepFool AEs
    $\varepsilon = \mathbb{E}_{\mathbf{x}} \|\delta(\mathbf{x})\|_2$,   ▷ Adaptive Cutoff budget
    $\delta_{\mathrm{Co}}(\mathbf{x}) = \mathrm{cut}\{\delta(\mathbf{x}), \varepsilon\}$,   ▷ Cutoff AEs
    $\delta_{\mathrm{CoS}}(\mathbf{x}) = \rho \delta_{\mathrm{Co}}(\mathbf{x})$,   ▷ Scale AEs
    $\mathbf{x}_{\mathrm{CoS}} = \mathbf{x} + \delta_{\mathrm{CoS}}(\mathbf{x})$,   ▷ Get CoS AEs
    $\mathcal{L}_{\mathrm{AE}} = \mathcal{L}(\theta, \mathbf{x}_{\mathrm{CoS}}, \mathbf{y})$,   ▷ Loss on CoS AEs
    $\theta = \theta - \alpha(\nabla_\theta \mathcal{L}_{\mathrm{C}} + \nabla_\theta \mathcal{L}_{\mathrm{AE}})$,   ▷ Update with total loss
  until $\mathcal{CR}(0.5)$ converge.   ▷ Reach comprehensive robustness

---

The motivation of the cutoff and scale is to ensure that the AEs are dynamically located in a reasonable range such that the AT model can be robust when encountering an attack with varying strength. We use the cutoff to further penalize the AEs with a norm larger than the budget ε. Similar to the inner problem in (2), we can alternatively solve the inner problem in (3), followed by the DeepFool AEs. Due to the dramatic fluctuations in the decision boundary during training, the cutoff can avoid AEs with a large or even unreasonable strength, e.g., some failure or overestimation AEs of DeepFool, limiting the AEs to lie within the perfect decision boundary. We use the scale to prevent the AEs from going over the decision boundary, i.e., preventing the AEs of different labels from touching each other.

The procedure of BAT is given in Algorithm 1. Starting with the AEs generated by DeepFool, which corresponds to only maximizing the first two terms in the objective of (3) (with $\rho = 1$), we simply calculate the third term (equivalent to minimizing $(\|\delta(\mathbf{x})\| - \varepsilon)_+$) by cutting off the perturbations of AEs with a norm larger than ε, defined by $\mathrm{cut}\{\delta(\mathbf{x}), \varepsilon\}$: if $\|\delta(\mathbf{x})\| > \varepsilon$, $\delta(\mathbf{x}) \leftarrow \delta(\mathbf{x})/\|\delta(\mathbf{x})\| \cdot \varepsilon$; otherwise, there is no change in $\delta(\mathbf{x})$. Then, we scale the new perturbations with weight ρ and add these CoS AEs into the training process, i.e., $\{\mathbf{x} + \rho\delta(\mathbf{x})\}$. We set $\varepsilon = \mathbb{E}\|\delta(\mathbf{x})\|$ and $\rho < 1$ (a predefined parameter), corresponding to adaptively estimating a nonuniform budget. $\mathrm{Cut}\{\delta(\mathbf{x}), \mathbb{E}\|\delta(\mathbf{x})\|\}$ implies that the budget $\varepsilon \leftarrow \mathbb{E}\|\delta(\mathbf{x})\|$ is computed prior to cutting off the perturbations $\delta(\mathbf{x})$.

Similar to the restricted and unrestricted AT, the following can be proven theoretically for BAT:

Proposition 3. *For the TCP problem, BAT can also obtain the model with the best robustness and has the same convergence property as the restricted AT with the best budget.*

Although these propositions are based on the toy model, the conclusion is also of guiding significance for common neural networks. Comparison with Propositions 1-2 shows that BAT not only avoids the difficulty of making the best choice for the budget but also provides an AT approach that dynamically adjusts a nonuniform budget, seeks to provide a path to potentially guess the perfect decision boundary, and finally reaches the model with the best robustness.

2.4 Comprehensive Robustness

The traditional evaluation criterion of a model's robustness is its accuracy under adversarial attack with a given budget [10,16]. This means that in fact, it considers the behavior of a model against adversarial attacks with a known attack budget. Since we instead consider the performance of an AT approach against blind adversarial attacks, the traditional evaluation criterion is unreasonable at this time. To address this problem, we propose a new evaluation criterion, named comprehensive robustness ($\mathcal{CR}$). First, we define the adversarial accuracy under a given budget ε as $\mathcal{AA}(\varepsilon) = \mathcal{A}(\mathbf{x} + \mathrm{cut}\{\delta(\mathbf{x}), \varepsilon\})$, where $\mathcal{A}(\cdot)$ is the accuracy of the given data and ε is a threshold used to cut the AEs. Then, we use the average adversarial accuracy to show the comprehensive robustness in the interval $[0, \Theta]$,

$$\mathcal{CR}(\Theta) = \frac{1}{\Theta} \int_0^\Theta \mathcal{AA}(\varepsilon) d\varepsilon. \tag{4}$$

It is clear that a larger $\mathcal{CR}(\Theta)$ indicates that the model has a better comprehensive robustness against adversarial attacks with varying attack budgets less than Θ. Clearly, clean data accuracy is a special case of adversarial accuracy, i.e., $\mathcal{AA}(0)$. It appears that one can instead use constrained AEs, such as FGSM [10] or PGD [16], to define the adversarial accuracy. However, we note that these AEs are only effective under a small budget. Moreover, the proposed adversarial accuracy is consistent with our cutoff strategy in optimizing the objective (3) such that the evaluation criterion in Algorithm 1 is efficient during our training process.

   Since we numerically scale the input data to $[0, 1]$ in the experiments, we simply set $\Theta = 0.5$ in Algorithm 1. We suggest that the parameter Θ should be selected at the point that $\mathcal{AA}(\Theta)$ falls to 0. At this time, the absolute value of $\mathcal{CR}$ is large as a whole and has strong discrimination. If Θ is further reduced, the robustness information near Θ will be lost. In contrary, if Θ is further increased, the robustness information will not increase, but the overall value of $\mathcal{CR}$ will decrease in an equal proportion, only causing the value to appear smaller.

3 Experiments for Benchmark Datasets

This section will experimentally evaluate the performance of BAT on various benchmark classification problems. We will compare BAT with normal training (NT), two restricted ATs (FGSM-AT/PGD-AT), and one unrestricted AT (DeepFool-AT). Here, we choose the most widely used prescribed budgets for FGSM-AT/PGD-AT in the literature [10,16,23]. The budgets for FGSM-AT and PGD-AT are 0.3 (Mnist), 0.1 (Fashion) and 6/255 (SVHN, Cifar10 and

Table 1. Comparison of the comprehensive robustness ($\mathcal{CR}(\Theta)$ in (4)) of normal training (NT), two restricted ATs (FGSM-AT/PGD-AT), unrestricted AT (DeepFool-AT), and our BAT (DeepFool AEs). We mark the highest and second-highest results in bold and underlined, respectively.

Dataset	Mnist		Fashion		SVHN		Cifar10		Cifar100	
Θ	0.13	0.07	0.08	0.04	0.004	0.002	0.03	0.02	0.007	0.003
NT	28.7	49.8	22.4	35.6	44.6	65.4	58.9	67.5	34.6	49.6
FGSM-AT	26.0	45.0	17.2	26.5	43.0	63.5	54.3	63.2	51.2	55.9
PGD-AT	92.2	96.3	69.3	<u>79.6</u>	75.8	79.8	57.3	65.2	51.4	55.6
DeepFool-AT	<u>84.8</u>	94.2	<u>69.4</u>	77.8	**77.7**	<u>82.5</u>	**62.9**	<u>68.3</u>	<u>52.2</u>	<u>57.0</u>
BAT (Ours)	<u>84.8</u>	<u>94.8</u>	**73.9**	**83.2**	<u>76.2</u>	**83.8**	<u>62.3</u>	**68.8**	**52.8**	**59.7**

Cifar100). We only consider the BAT based on DeepFool AEs, and we will implement the BAT with CW/DDN AEs in future work. The code for these experiments is based on the open source library Cleverhans [20].

3.1   Overall Results

Comprehensive Robustness. Comprehensive Robustness results of BAT compared with other ATs are shown in Table 1 firstly. BAT approach achieve good comprehensive robustness. For each classification problem, we consider the comprehensive robustness against blind adversarial attacks with two different budget ranges, i.e., different Θ. Since $\mathbb{E}\|\delta(\mathbf{x})\|$ can represent the average distance between the data manifold and the model's decision boundary, we set $\Theta = \mathbb{E}\|\delta(\mathbf{x})\|$ and take one-half of it, because in applications, the focus is more on the robustness facing the AEs that do not exceed decision boundaries. Thus, we simply choose $\Theta = 0.5$ in Algorithm 1 because it is hard to estimate a suitable Θ in advance. These results in Table 1 clearly show that the proposed BAT is **in the top two of all datasets and ranks first in most cases**, showing that our method can balance the clean accuracy and the comprehensive robustness of the model. Since DeepFool-AT is effective for defending against large attacks, its accuracies under large Θ values are slightly higher than those of BAT in some cases, e.g., SVHN and Cifar10. For Mnist, PGD-AT works much better than FGSM-AT and outperforms DeepFool-AT and BAT, especially under larger Θ. The reason may be that for the simplest dataset with high accuracy, such as Mnist, the introduction of randomness in PGD may help to obtain the optimal AEs, such that the corresponding AT has a better comprehensive robustness. In contrast, for complex datasets, this behavior cannot be expected.

Due to space limitations, we will include a more detailed comparison about comprehensive robustness, the results of different methods facing different adversarial attacks, and the results of black-box attacks in the appendix.

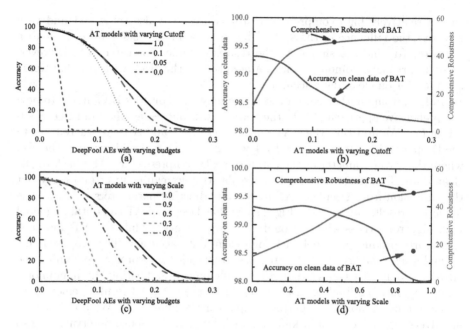

Fig. 2. An ablation study on cutoff and scale on Mnist. The results of AT models by Algorithm 1 with varying cutoff parameter ε and fixed $\rho = 1$ are shown in (a): the accuracies on DeepFool AEs with varying budgets and (b): the accuracy on clean data and comprehensive robustness $\mathcal{CR}(0.3)$. The corresponding results with varying scale parameter ρ and fixed $\varepsilon = 1$ are shown in (c) and (d). We plot the clean accuracy and comprehensive robustness $\mathcal{CR}(0.3)$ of BAT in (b) and (d). (Color figure online)

3.2   Ablation Study on Cutoff and Scale

We take Mnist as an example to perform an ablation study on cutoff and scale. For cutoff, we set the scale parameter $\rho = 1$ and strictly let the cutoff parameter ε be a predefined value rather than the adaptive value used in Algorithm 1. By setting $\varepsilon = 0, 0.05, 0.1, 1$, we obtain several ATs by Algorithm 1. Then, we test the performance of these AT models against the DeepFool attack with different budgets; the results are shown in Fig. 2(a). Figure 2(b) left (blue line) also shows the clean accuracy of these AT models. From these results, we find that the AT with larger ε has better performance against the attack with a larger budget, but with less clean accuracy. Figure 2(b) right (red line) also shows the comprehensive robustness $\mathcal{CR}(0.3)$ of the AT models with varying ε values. An examination of the results presented in Fig. 2(b) clearly indicates that there exists a trade-off between clean accuracy and comprehensive robustness. In practice, it is time-consuming to plot Fig. 2(b) for a given dataset, especially for large models. Thus, we are encouraged to adaptively choose a suitable ε. We plot the clean accuracy and comprehensive robustness of BAT, i.e., the AT with adaptive ε, in Fig. 2(b) with the blue and red points, respectively. Surprisingly, the results show that our BAT approach can adaptively estimate a good budget and obtain the model with a relatively high comprehensive robustness. We also set $\varepsilon = 1$ and vary ρ to test the influence of the scale. The corresponding results are presented in Figs. 2(c)–(d) and are similar to the results obtained for the cutoff. However, more surprisingly, combined with an adaptive ε, the BAT with $\rho = 0.9$ works much better on clean data accuracy.

In fact, from another point of view, we may also consider BAT in (3) to be a comprehensive approach of NT (the training without using AEs) and DeepFool-AT in (2). We find that the BAT with cutoff $\varepsilon = 1$ and scale $\rho = 1$ reduces to the DeepFool-AT. Moreover, the BAT with $\varepsilon = 0$ or $\rho = 0$ is identical to the NT, where the zero parameter of the cutoff or scale compresses the AEs back to the clean data. Therefore, the cutoff and scale strategy can connect and carry out the transition from DeepFool-AT to NT by adjusting parameters ε and ρ. The numerical results presented in Fig. 2 show that DeepFool-AT can obtain a model with high robustness when encountering large attacks, while the NT approach can clearly produce a model with high clean data accuracy. The proposed cutoff and scale strategy in BAT attempts to provide an approach combining the advantages of both DF-AT and NT by dynamically adjusting the nonuniform budget against blind adversarial attacks, and the obtained results demonstrate that the output model has better comprehensive robustness. We note that for an adversarial attack with a known attack budget, the above observations also encourage us to use the BAT with a prescribed cutoff and scale parameters for defense. This is an interesting future direction.

4   Conclusion

Both restricted and unrestricted AT approaches cannot obtain comprehensively robust models against blind adversarial attacks, i.e., attacks with an unknown budget. To address this problem, this paper proposed a blind adversarial training (BAT) approach by using the cutoff-scale strategy to adaptively estimate a nonuniform budget in the generation of each AE during adversarial training. The improvement of BAT was theoretically verified on a toy classification problem (TCP) and a two circles classification (TCC) problem. By using BAT to train the classification models on several benchmarks, we obtained models with better comprehensive robustness. The individual contributions of the cutoff and scale were also investigated in detail.

Acknowledgment. This work was supported by the National Natural Science Foundation of China (Grant No. 12004422) and by Beijing Nova Program of Science and Technology (Grant No. Z19110000111 9129).

References

1. Akhtar, N., Mian, A.: Threat of adversarial attacks on deep learning in computer vision: a survey. IEEE Access **6**, 14410–14430 (2018)

2. Athalye, A., Carlini, N., Wagner, D.: Obfuscated gradients give a false sense of security: circumventing defenses to adversarial examples. In: Proceedings of the 35th International Conference on Machine Learning, ICML 2018, July 2018. https://arxiv.org/abs/1802.00420

3. Bhattad, A., Chong, M.J., Liang, K., Li, B., Forsyth, D.A.: Unrestricted adversarial examples via semantic manipulation. arXiv:1904.06347 (2020)

4. Brown, T.B., Carlini, N., Zhang, C., Olsson, C., Goodfellow, I.: Unrestricted adversarial examples. arXiv:1809.08352 (2018)

5. Carlini, N., Wagner, D.: Towards evaluating the robustness of neural networks. In: 2017 IEEE Symposium on Security and Privacy (SP), pp. 39–57, May 2017

6. Croce, F., Hein, M.: Reliable evaluation of adversarial robustness with an ensemble of diverse parameter-free attacks. CoRR abs/2003.01690 (2020). https://arxiv.org/abs/2003.01690

7. Ding, G.W., Sharma, Y., Lui, K.Y.C., Huang, R.: MMA training: direct input space margin maximization through adversarial training. In: ICLR (2020)

8. Duan, R., Chen, Y., Niu, D., Yang, Y., Qin, A.K., He, Y.: Advdrop: adversarial attack to dnns by dropping information. CoRR abs/2108.09034 (2021), https://arxiv.org/abs/2108.09034

9. Goodfellow, I., Bengio, Y., Courville, A.: Deep Learning. MIT Press, Cambridge (2016)

10. Goodfellow, I.J., Shlens, J., Szegedy, C.: Explaining and harnessing adversarial examples. In: ICLR (2015)

11. Hinton, G., et al.: Deep neural networks for acoustic modeling in speech recognition: the shared views of four research groups. IEEE Signal Process. Mag. $29(6)$, 82–97 (2012)

12. Hornik, K., Stinchcombe, M., White, H.: Multilayer feedforward networks are universal approximators. Neural Netw. $2(5)$, 359–366 (1989)

13. Krizhevsky, A., Sutskever, I., Hinton, G.E.: Imagenet classification with deep convolutional neural networks. In: International Conference on Neural Information Processing Systems, pp. 1097–1105 (2012)

14. Kurakin, A., Goodfellow, I., Bengio, S.: Adversarial machine learning at scale. arXiv preprint arXiv:1611.01236 (2016)

15. LeCun, Y., Bengio, Y., Hinton, G.: Deep learning. Nature $521(7553)$, 436–444 (2015)

16. Madry, A., Makelov, A., Schmidt, L., Tsipras, D., Vladu, A.: Towards deep learning models resistant to adversarial attacks. arXiv:1706.06083 (2017)

17. Mikolov, T., Deoras, A., Povey, D., Burget, L., Černocký, J.: Strategies for training large scale neural network language models. In: 2011 IEEE Workshop on Automatic Speech Recognition and Understanding, pp. 196–201 (2011)

18. Moosavi-Dezfooli, S., Fawzi, A., Frossard, P.: Deepfool: a simple and accurate method to fool deep neural networks. In: CVPR, pp. 2574–2582 (2016)

19. Papernot, N., McDaniel, P., Wu, X., Jha, S., Swami, A.: Distillation as a defense to adversarial perturbations against deep neural networks. In: 2016 IEEE Symposium on Security and Privacy (SP), pp. 582–597, May 2016

20. Papernot, N., et al.: Technical report on the cleverhans v2.1.0 adversarial examples library. arXiv preprint arXiv:1610.00768 (2018)

21. Rony, J., Hafemann, L.G., Oliveira, L.S., Ayed, I.B., Sabourin, R., Granger, E.: Decoupling direction and norm for efficient gradient-based L2 adversarial attacks and defenses. arXiv:1811.09600 (2018)

22. Sankaranarayanan, S., Jain, A., Chellappa, R., Lim, S.N.: Regularizing deep networks using efficient layerwise adversarial training. In: arXiv preprint arXiv:1705.07819 (2017)
23. Song, C., He, K., Wang, L., Hopcroft, J.E.: Improving the generalization of adversarial training with domain adaptation. arXiv:1810.00740 (2018)
24. Song, Y., Shu, R., Kushman, N., Ermon, S.: Constructing unrestricted adversarial examples with generative models. arXiv:1805.07894 (2018)
25. Sutskever, I., Vinyals, O., Le, Q.V.: Sequence to sequence learning with neural networks. In: Advances in Neural Information Processing Systems, pp. 3104–3112 (2014)
26. Szegedy, C., et al.: Intriguing properties of neural networks. In: ICLR (2014)
27. Tramèr, F., Kurakin, A., Papernot, N., Goodfellow, I., Boneh, D., McDaniel, P.: Ensemble adversarial training: attacks and defenses. arXiv preprint arXiv:1705.07204 (2017)
28. Zhang, J., Jiang, X.: Adversarial examples: opportunities and challenges. arXiv preprint arXiv:1809.04790 (2018)
29. Zhang, L., Wang, X., Lu, K., Peng, S., Wang, X.: An efficient framework for generating robust adversarial examples. Int. J. Intell. Syst. 35(9), 1433–1449 (2020). https://doi.org/10.1002/int.22267, https://onlinelibrary.wiley.com/doi/abs/10.1002/int.22267
30. Zhao, Z., Liu, Z., Larson, M.A.: Towards large yet imperceptible adversarial image perturbations with perceptual color distance. CoRR abs/1911.02466 (2019), http://arxiv.org/abs/1911.02466

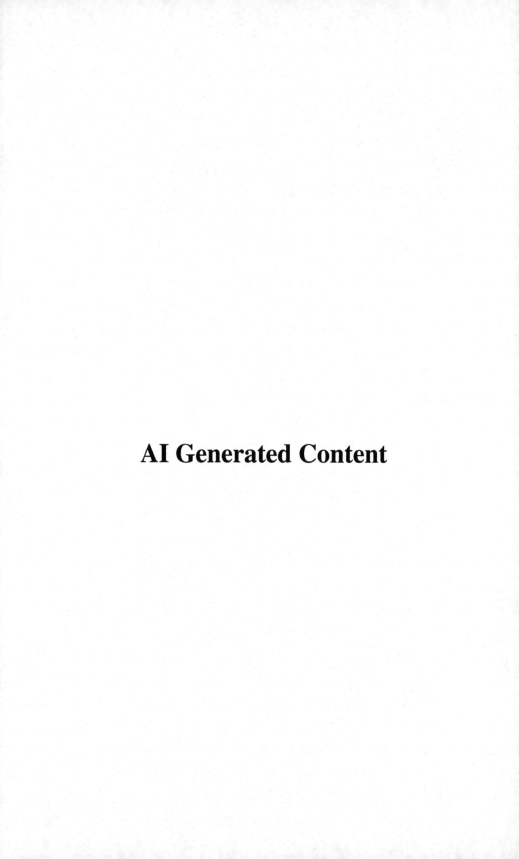

AI Generated Content

TOAC: Try-On Aligning Conformer for Image-Based Virtual Try-On Alignment

Yifei Wang[1], Wang Xiang[1], Shengjie Zhang[1], Dizhan Xue[2], and Shengsheng Qian[2(✉)]

[1] Zhengzhou University, Zhengzhou, China
[2] State Key Laboratory of Multimodal Artificial Intelligence Systems, Institute of Automation, Chinese Academy of Sciences, Beijing, China
shengsheng.qian@nlpr.ia.ac.cn

Abstract. Recently, *Image-based Virtual Try-on* has garnered increasing attention within the realm of online apparel e-commerce, which aims to virtually superimpose garments onto images of portraits. Image-based virtual try-on generally consists of two steps: Image-based Virtual Try-on Alignment and Image-based Virtual Try-on Generation. In this paper, we focus on Image-based Virtual Try-on Alignment (IVTA), which plays a pivotal role in virtual try-on and aligns the target garment with the portrait. Current approaches for IVTA mostly adopt Convolutional Neural Networks (CNN) to extract local detailed features of both garments and portraits, ignoring the significance of global extensive features. To address this problem, we propose a novel model named Try-On Aligning Conformer (TOAC) to effectively aligns the target garment with the portrait and improve virtual try-on. Firstly, we integrate both Swin Transformer and CNN to comprehensively extract both global patterns and local details. Secondly, we propose a robust learned perceptual loss between generative reconstructed garment images and the ground truth to alleviate the overlap problem. Extensive experiments demonstrate the superiority of our proposed model compared to the state-of-the-art methods for virtual try-on alignment.

Keywords: Virtual Try-on Alignment · Image Synthesis · Transformer

1 Introduction

Recently, Image-based Virtual Try-on [4,7,11,15] has garnered significant attention within the realm of online apparel e-commerce. This emerging technology seeks to virtually superimpose garments onto images of portraits, allowing users to visualize how a particular garment would appear on them without physically trying it on. The application of Image-based Virtual Try-on has the potential to revolutionize the online shopping experience by providing customers with a more realistic and interactive way to explore and evaluate garment items. Generally, image-based virtual try-on comprises two stages [16]: image-based virtual try-on alignment and image-based virtual try-on generation. As shown in Fig. 1, the virtual try-on alignment stage focuses on deforming the target garment to fit the portrait by inputting the features of the garment and the

Supported by National Natural Science Foundation of China (No. 62036012, 62276257).

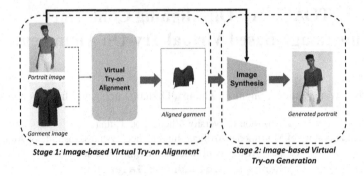

Stage 1: Image-based Virtual Try-on Alignment   Stage 2: Image-based Virtual
                                                              Try-on Generation

Fig. 1. The workflow of Image-based Virtual Try-on: Stage 1 focuses on deforming the target garment to fit the specific portrait, and Stage 2 synthesizes an image of the specific person wearing the target garment.

portrait. Afterward, the virtual try-on generation stage synthesizes an image of the specific person wearing the properly fitted garment by inputting the aligned garment and the portrait. In this paper, we focus on Image-based Virtual Try-on Alignment (IVTA), which can be the most crucial stage of image-based virtual try-on.

Though some existing image-based virtual try-on methods [2, 12, 29] have achieved compelling performance in warping garments, there are still notable deficiencies that hinder their practical application in real-world scenarios. Previous work [12] exploits Thin Plate Splines (TPS) transformation to deform the target garments. HR-VION [2] models the geometric changes of garments by building on two feature pyramid networks with Convolutional Neural Networks (CNN) [19,25,31]. However, the above work only adopts CNN to extract detailed features of garments and body information, which ignores global information in the virtual try-on alignment stage. As shown in Fig. 2(a), we utilize Grad-CAM [28] to demonstrate heat maps of regions focused by the state-of-the-art method HR-VITON [2], where hotter color means HR-VITON putting more attention on the corresponding regions. In the virtual try-on alignment stage, HR-VITON fails to capture the essential global interaction information which is necessary for accurately warping target garments to fit portraits. Instead, HR-VITON primarily focuses on specific garment details, such as sleeves and necklines. Therefore, we have to address **Challenge 1:** How to design an effective feature extraction model for IVTA that can integrate global interactions and local details?

Furthermore, VITON-HD [2] recognizes the significance of the portrait segmentation map in providing the model with essential knowledge about the body layout. This information aids in effectively separating the generated and preserved regions, thereby addressing the problem of body occlusion. However, despite this improvement, misalignment issues with the segmentation map still persisted, indicating the need for further enhancements in the alignment process. As we demonstrate an example shown in Fig. 2(b), HR-VITON generates adhesive artifacts and misaligns regions between the warped garment and the corresponding segmentation map. Therefore, we have to address **Challenge 2:** How to effectively alleviate the misalignment regions while deforming the target garment for better virtual try-on alignment?

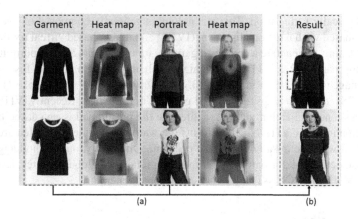

Fig. 2. (a) shows the heat maps of HR-VITON, (b) shows the generated results by HR-VITON.

Motivated by the above observations, we propose a novel Try-On Aligning Conformer (TOAC) for virtual try-on alignment to integrate both global interactions and local details and achieve a more robust garment deformation. For **Challenge 1**, we combine CNN and Swin Transformer [20] to effectively capture both global information and local representations. Specifically, we exploit CNN and Swin Transformer to represent both portraits and garments, which can alleviate the noise in feature extraction. For **Challenge 2**, we develop a Masked Autoencoders-based (MAE-based) learned perceptual loss to learn garment warp effectively. Inspired by the effectiveness of MAE [13] in image reconstruction, we develop Learned Perceptual Image Patch Similarity (LPIPS) [32] by comparing extracted MAE features of the synthesized image with the ground truth. Comprehensive experiments demonstrate that our proposed method achieves superior performance compared with the state-of-the-art methods for virtual try-on alignment.

In brief, the contributions of this paper are summarized as follows:

– We propose a novel Try-On Aligning Conformer (TOAC) for virtual try-on alignment, which can effectively learn features for portraits and garments and deform target garments while alleviating the misalignment regions.
– We adopt a hybrid architecture to leverage the strengths of CNN and Transformer to effectively capture both global information and local representations. Our hybrid network can extract comprehensive features to improve virtual try-on alignment.
– We develop an MAE-based learned perceptual loss for the conditional generation network, which improves the warping module and reduces misalignment regions.
– Extensive experiments demonstrate the superiority of our proposed model compared with the state-of-the-art methods for virtual try-on alignment.

2 Related Work

2.1 Image-Based Virtual Try-On Alignment

Image-based virtual try-on [2,3,11,12,29] aims to superimpose garments onto images of portraits, allowing users to visualize how a particular garment would appear on them without physically trying it on. This technology is generally composed of two stages, *i.e.*, image-based virtual try-on alignment (IVTA) and image-based virtual try-on generation. IVTA aims to generate a deformed garment image that fits a given portrait image. Previous approaches [2,2,3,11,12,29] mainly explicit warping module to fuse the input garment image with a given portrait. VITON [12] and CP-VTON [29] adopt a Thin Plate Splines (TPS) transformation to warp the garment deformation. Recently, VITON-HD [2] proposes a normalization method to alleviate the misaligned issue caused by complex deformation. HR-VITON [2] performs the warping module with a feature fusion block to deal with the misalignment-free problem. Differently, our method aggregates both diverse detailed deformation fields and comprehensive global features, leveraging the power of a hybrid CNN-Transformer conditional generator.

2.2   Image Synthesis

Generative adversarial networks (GANs) [9,26,27], especially the StyleGAN-based models [17,18] have recently achieved significant improvement in realistic image synthesis. Aiming to model the real image distribution by forcing the generated samples to be indistinguishable from the real images, in the field of human synthesis, most of the existing methods [5,6] utilize StyleGAN-based architecture to get high performance. StyleGAN-Human [6] depends on three momentous factors to find the key to improving high-quality human synthesis. InsetGAN [5] exploits various pretrained GANs to generate various parts of the human body (such as the face, hands, etc.). LaDI-VTON [22] utilizes a latent diffusion model and introduces an innovative autoencoder module that incorporates learnable skip connections. In this paper, our work focus on image-based alignment, which conducts aligned and discriminative generator to produce a realistic portrait image with a target garment.

3   Notation and Problem Definition

Given a guideline image of a portrait $P \in \mathbb{R}^{3 \times H \times W}$ and an image of a target garment $G \in \mathbb{R}^{3 \times H \times W}$ as inputs, where H, W denote the height and width of the given images. The target of virtual try-on alignment is to synthesize an image $\bar{P}_g \in \mathbb{R}^{3 \times H \times W}$ of an aligned garment that fits the portrait. We train the model to reconstruct P by inputting garment-agnostic portrait image P_a and the garment G that the persons are wearing. The garment-agnostic portrait image eliminates the garment region in P, and allows the model to generalize during the test when an arbitrary garment image is given.

4   Method

The overall architecture of our method is shown in Fig. 3.

4.1   Data Pre-processing

In the data pre-processing, we obtain a segmentation feature map $S \in \mathbb{L}^{H \times W}$ of a portrait, a garment mask g_m, and a densepose map $D \in \mathbb{R}^{3 \times H \times W}$ which remove the

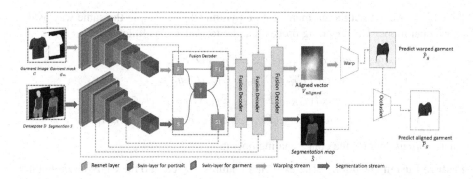

Fig. 3. Architecture of our Try-On Aligning Conformer (TOAC).: (1) The inputs of TOAC consist of a segmentation feature map $S \in \mathbb{L}^{H \times W}$ of a specific person, a garment mask g_m, a garment image G, and a densepose map $D \in \mathbb{R}^{3 \times H \times W}$. (2) TOAC deforms a target garment G to obtain an aligned garment $\bar{P}_g$ that fits a specific person.

garment, where $\mathbb{L}$ denotes a set of integers indicating the semantic labels. The Pixel-level Generation Network (PGN) [8] is employed for semantic segmentation, including edge detection and instance parsing. By processing the input image P, we obtain the segmented image S. Densepose [10] specializes in analyzing segmentation maps of portrait, enabling us to obtain the densepose D. Then, we utilize an open-source network named CarveKit [2] to extract the mask of the garment in the image. By inputting the garment image G into CarveKit, we obtain the corresponding mask g_m, which represents the garment region.

4.2   Conditional Generative Network

The conditional generative network aims to generate an image of an aligned garment $\bar{P}_g$ which is deformed to fit the body of a specific person.

Feature Extract Encoder.   In contrast to existing methods [2, 2, 3, 12, 29] that mostly rely on CNN blocks and only focus on local information, our feature extraction method takes into account both local and global information, allowing for a more comprehensive representation of the garments. Our conditional generation network consists of four feature extraction encoders, $i.e.$ a portrait ResNet [14, 24] encoder $E_{Rp} = \{E_{Rp_k}\}_{k=0}^{4}$ which focuses on local portrait information, a portrait Swin Transformer encoder $E_{Tp} = \{E_{Tp_k}\}_{k=0}^{4}$ which focuses on global portrait information, a garment ResNet encoder $E_{Rg} = \{E_{Rg_k}\}_{k=0}^{4}$ which focuses on local garment information, and a garment Swin-transformer encoder $E_{Tg} = \{E_{Tg_k}\}_{k=0}^{4}$ which focuses on global garment information, where $E_{.,\cdot_k}, k \in [0, 4]$ denotes k-th layer of four encoders, respectively. Given a target garment image G, a parallel garment mask g_m, a segmentation feature map S of a specific portrait and a corresponding densepose pose map D, we first introduce local portrait feature pyramid $\{F_{Rp_k}\}_{k=0}^{4}$ by the portrait ResNet encoder E_{Rp}, global portrait feature pyramid $\{F_{Tp_k}\}_{k=0}^{4}$ by the portrait Swin Transformer encoder E_{Tp}. Considering the robustness, we concatenate $\{F_{Rp_k}\}_{k=0}^{4}$ and $\{F_{Tp_k}\}_{k=0}^{4}$ then get

$\{F_{P_k}\}_{k=0}^4$. We extract the garment feature pyramid $\{F_{G_k}\}_{k=0}^4$ in the same way as the above operations. The encoding process can be formulated as follows:

$$
\begin{aligned}
F_{P_0} &= concat(E_{Rp_0}(D,S), E_{Tp_0}(D,S)), \\
F_{G_0} &= concat(E_{Rc_0}(G,g_m), E_{Tc_0}(G,g_m)), \\
F_{P_k} &= concat(E_{Rp_k}(F_{P_{k-1}}), E_{Tp_k}(F_{P_{k-1}})), k \in [1,4], \\
F_{G_k} &= concat(E_{Rc_k}(F_{G_{k-1}}), E_{Tc_k}(F_{G_{k-1}})), k \in [1,4],
\end{aligned}
\tag{1}
$$

where $concat$ denotes the concatenating operation.

Feature Fusion Decoder.   The extracted features are fed into the CNN-based feature fusion decoder $D = \{d_k\}_{k=0}^3$, where the feature maps obtained from the two different feature pyramids are fused to predict the segmentation map and the appearance aligned vectors for warping the garment image. The fusion decoder layer d_k generates V_k and $\hat{s}_k$ by splicing the results of the previous fusion decoder layer V_{k-1} and $\hat{s}_{k-1}$ with the features of the corresponding feature pyramid $F_{C_{3-k}}$ and $F_{P_{3-k}}$, and then uses the convolution layer to mix the features. Finally, the aligned vector $V_{aligned} = V_3$ and the segmentation map $\hat{S} = \hat{s}_3$ for the final fusion decoder are given, where $V_{aligned}$ denotes the Aligned vector and $\hat{S}$ denotes the segmentation map. The $warp$ module is to sample the original garment image G based on the aligned vector $\{v_3\}$ to obtain the distorted garment $\hat{P}_g$. The $occlusion$ module is based on the segmentation map $\hat{S}$, removing the part of the distorted garment $\hat{P}_g$ that will obscure the limbs and get the final image of the garment $\bar{P}_g$. The decoding process can be formulated as follows:

$$
\begin{aligned}
V_0, \hat{S}_0 &= d_0(concat(F_{G_4}, F_{G_3}), concat(F_{P_4}, F_{P_3})), \\
V_k, \hat{S}_k &= d_k(concat(V_{k-1}, F_{G_{3-k}}), concat(\hat{S}_{k-1}, F_{P_{3-k}})), k \in [1,3], \\
\hat{P}_g &= warp(V_{aligned}, G), k \in [1,3], \\
\bar{P}_g &= occlusion(\hat{S}, \hat{P}_g),
\end{aligned}
\tag{2}
$$

where $concat$ denotes the concatenating operation, $V_{aligned} = V_3$, $\hat{S} = \hat{s}_3$, $warp$ denotes the warp module, and $occlusion$ denotes the occlusion module. These two streams exchange information with each other, enabling the joint estimation of the aligned vectors and segmentation map and enhancing the effectiveness of alignment.

4.3   Loss Functions

We employ the pixel-wise cross-entropy loss, denoted as $\mathcal{L}_{CE}$, and the L-1 loss $\mathcal{L}_{L1}$ to incentivize the network to appropriately deform the garment in order to align with the portrait. Specifically, $\mathcal{L}_{CE}$ and $\mathcal{L}_{L1}$ can be formulated as follows:

$$
\mathcal{L}_{CE} = -\frac{1}{N} \sum_{i=1}^N \sum_{j=1}^M S_{ij} \log(\hat{S}_{ij}) + (1 - S_{ij}) \log(1 - \hat{S}_{ij}),
$$

$$
\mathcal{L}_{L1} = \sum_{i=0}^3 \omega_i \cdot \| W(g_m, v_i) - S_g \|_1 + \left\| \hat{S}_g - S_g \right\|_1,
\tag{3}
$$

where N denotes the number of pixels in an image, M denotes the number of categories, $\hat{S}$ denotes the predicted segmentation map, and S denotes the ground truth segmentation map. In $\mathcal{L}_{L1}$, ω_i is a trade-off factor, v_i denotes the aligned vectors, $W(g_m, v_i)$ denotes a distortion with (g_m, v_i), $\hat{S}_g$ and S_g denote the predicted segmentation map and the ground truth segmentation map corresponding to the garment mask g_m, respectively.

MAE [13] developed a simple but strong architecture to represent highly redundant information by exploiting masks. Inspired by Learned Perceptual Image Patch Similarity (LPIPS) [32] and the effectiveness of MAE, we exploit MAE-based learned perceptual loss to adjust the gap between the generated garment image after alignment vector warping and the image of the garment on the human body. The MAE-based learned perceptual loss can be formulated as follows:

$$\varphi(I_{pred}, I_{label}) = \|MAE(I_{pred}) - MAE(I_{label})\|_1 \,,$$
$$\mathcal{L}_{MAE} = \sum_{i=0}^{3} \rho_i \cdot \varphi\left(W\left(G, v_i\right), P_g\right) + \varphi\left(\bar{P}_g, P_g\right), \tag{4}$$

where ρ_i is a trade-off factor, $W(\cdot, \cdot)$ denotes the distortion of the garment map based on the alignment vector, φ denotes the gap of extracted image features using the pre-trained MAE, P_g and $\bar{P}_g$ denote the garment on the portrait and a final aligned garment.

L_{tv} is a total-variation loss that enhances the smoothness of the alignment vector [2]. Recall $V_{aligned}$ has a length of H pixels and a width of W pixels. We use $\triangledown$ to calculate the distance of the vector V_{up} from vertical 0 to $H - 1$ pixels and the vector V_{down} from vertical 1 to H pixels to obtain the vertical loss $\sigma(V_{up}, V_{down})$. A similar method is used to obtain the lateral loss $\sigma(V_{left}, V_{right})$. The longitudinal and transverse losses are then summed to obtain L_{tv} as follows:

$$\sigma(v_a, v_b) = |V_a - V_b| \cdot e^{-150|V_a - V_b|},$$
$$\triangledown(V_{aligned}) = \sigma(V_{up}, V_{down}) + \sigma(V_{left}, V_{right}),$$
$$\mathcal{L}_{tv} = \sum_{i=0}^{3} \triangledown(V_{align,i}). \tag{5}$$

The overall objective of optimization is expressed as follows:

$$\mathcal{L}_{TOCG} = L_{CE} + \lambda_{GAN} L_{GAN} + \lambda_{L1} L_{L1} + \lambda_{MAE} L_{MAE} + \lambda_{tv} L_{tv}, \tag{6}$$

where λ_{GAN}, λ_{L1}, λ_{MAE}, and λ_{tv} are hyperparameters to balance loss terms. L_{GAN} [21] loss uses a discriminant network to calculate the loss of the predicted segmentation.

5   Experiments

5.1   Training

In experiments, we use the high-resolution virtual try-on dataset introduced by VITON-HD [2], which contains 13,679 pairs of the model wearing garments image and the garment image. Specifically, the dataset is divided into a training set and a test set of

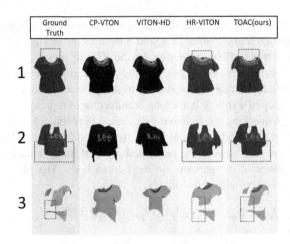

Fig. 4. The qualitative comparison among TOAC and baselines on VITON-HD dataset.

11,647 and 2,032 pairs, respectively. The original resolution of the images is 1024×768, and the images are bicubically downsampled to the desired resolution when needed.

We employ three widely used metrics (*i.e.*, Structural SIMilarity index (SSIM) [30], Perceptual distance (LPIPS) [32], and Fréchet Inception Distance (FID) [23]) to evaluate the similarity between synthesized and real images, in which SSIM and LPIPS are used for paired setting and FID are used for unpaired setting.

5.2   Qualitative Results

We compare our method with several state-of-the-art baselines, including CP-VTON [12], VITON-HD [2], and HR-VITON [2]. Figure 4 shows that our conditional generation network can generate alignment vectors that fit the portrait better than previous methods, resulting in a more sensory-aligned image of the garments: (1) In the first example, for the collar of the garment, the result of our network is much closer to the real treatment compared to HR-VITON. (2) In the second example, HR-VITON has an unconventional bulge at the edge of the garment, which does not exist in our method. (3) As can be seen in the third example, the segmentation result of our network is more correct when generating the segmentation map, which leads to more correct final processing of the occlusion than HR-VITON. Because our work mixes global and local information, the alignment process allows the network to understand the relationship of the garment as a whole. Moreover, we introduce a novel MAE-based loss function L_{MAE} that effectively addresses the issue of misaligned regions.

5.3   Quantitative Results

As reported in Table 1, we show the quantitative results within the default and no-occlusion settings by comparing the quality of the generated garments with and without occlusion, respectively. Based on the results, we have the following observations:

Table 1. The quantitative results of our TOAC and compared baselines on VITON-HD dataset. Bold denotes the best results.

Method	CP-VITON	VITON-HD	HR-VITON		TOAC (ours)	
Setting	default	default	no-occlusion	default	no-occlusion	default
SSIM ↑	0.7590	0.8875	0.8770	0.8823	0.8882	**0.8947**
FID ↓	42.1431	34.6652	**29.3391**	32.8603	38.4994	30.6294
LPIPS ↓	0.3745	0.1924	0.1950	0.1998	0.1864	**0.1515**

(1) Our proposed TOAC outperforms all baselines on both the default setting and the no-occlusion setting on VITON-HD dataset. Compared with the state-of-the-art HR-VITON, our TOAC achieves SSIM improvements of 0.0151 on the default setting and 0.0112 on the no-occlusion setting, respectively. This indicates that our method can diminish discrimination between garments and portraits, and effectively extract features. (2) Our method TOAC obtains a comparable low FID score with HR-VITON on the no-occlusion setting and the default setting. This is mainly because our TOAC contributes more to generative occluded segmentation to deform garments. (3) Our TOAC significantly surpasses the baselines in terms of LPIPS. Compared with the state-of-the-art HR-VITON, our method achieves LPIPS improvement of 0.0483 on the no-occlusion setting. This shows TOAC can improve the warping and reduce misalignment regions.

5.4 Ablation Study

In order to further investigate the impact of each loss component, we conduct the ablation experiments with different losses removed from the complete loss function as shown in Table 2. We add two additional metrics for comparing the generated images: **Mean Square Error (MSE)** and **Inception Score (IS)** [1]. It can be found that the impact of L_{L1} on the results is huge. Losing the loss component between the comparison masks, all five evaluation metrics drop rapidly, and IS_{mean} even all tend to infinity. From another perspective, the current dominant evaluation metrics have a relatively high weight on whether the shape of the image matches the ground truth. In addition, L_{tv} has little effect on other metrics, but has a greater effect on the FID. It is seen that a stable change in the alignment vector is more favorable to predict the statistical distribution of the image close to the actual distribution, thus reducing FID. Finally, L_{MAE} has little effect on several other metrics, but a large boost for LPIPS, which also reflects the data level that L_{MAE} effectively mitigates the problem of misaligned areas.

5.5 Attention Comparison

We conduct an attention analysis between our approach and HR-VITON [2] at the attention level. We leverage the final activation layer outputs of the encoder in HR-VITON to identify the active regions. In contrast, our approach incorporates attention layers within E_{Tp} and E_{Tc}, enabling direct visualization of the network's active regions. From

Table 2. Ablation study on loss components of our proposed TOAC. w/o means without.

Method	Variant	SSIM↑	FID↓	LPIPS↓	MSE↓	IS_{mean} ↓
TOAC (ours)	w/o L_{CE}	0.8989	35.0501	0.1401	0.0153	4.9142
	w/o L_{GAN}	0.8975	34.2817	0.1499	0.0149	4.2532
	w/o L_{L1}	0.5932	66.4024	0.3205	0.0790	Nan
	w/o L_{tv}	0.8916	83.4413	0.1569	0.0156	4.9376
	w/o L_{MAE}	0.8923	39.2897	0.1756	0.0177	4.2254
	All losses	0.8947	30.6295	0.1513	0.0161	4.1645

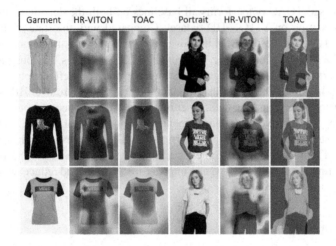

Fig. 5. The visualization of feature attention maps of the input garments and portraits learned by our TOAC and HR-VITON on VITON-HD dataset.

Fig. 5, we have the following observations: (1) HR-VITON exhibits a relatively small and concentrated active area. Conversely, our method can focus on the overall context while still attending to the garment and human body regions with clarity. (2) Regarding the human poses, HR-VITON focuses on the surrounding information of the human body (see the third example). Conversely, our TOAC prioritizes the human body, along with the overall pose. By visualizing the attention maps, it becomes apparent that our TOAC captures a more comprehensive range of information.

6   Conclusion

In this work, we propose a novel Try-On Aligning Conformer (TOAC) model for image-based virtual try-on alignment, which is capable of generating semantic-correct and photo-realistic warped garments. Specifically, to make garment warping robust to intricate inputs, TOAC adopts a hybrid feature network to integrate both Swin Transformer and CNN to effectively extract global information and grasp local representation respectively. Besides, to alleviate the overlap problem in existing methods, we propose an

MAE-based learned perceptual loss to capture reconstruction information. Extensive experiments illustrate the superiority of our proposed TOAC over existing methods.

References

1. Barratt, S., Sharma, R.: A note on the inception score. arXiv preprint arXiv:1801.01973 (2018)
2. Choi, S., Park, S., Lee, M., Choo, J.: VITON-HD: high-resolution virtual try-on via misalignment-aware normalization. In: IEEE Conference on Computer Vision and Pattern Recognition, CVPR 2021, virtual, 19–25 June 2021, pp. 14131–14140 (2021)
3. Chopra, A., Jain, R., Hemani, M., Krishnamurthy, B.: Zflow: gated appearance flow-based virtual try-on with 3d priors. In: 2021 IEEE/CVF International Conference on Computer Vision, ICCV 2021, Montreal, QC, Canada, 10–17 October 2021, pp. 5413–5422 (2021)
4. Dong, X., et al.: Dressing in the wild by watching dance videos. In: IEEE/CVF Conference on Computer Vision and Pattern Recognition, CVPR 2022, New Orleans, LA, USA, 18–24 June 2022, pp. 3470–3479 (2022)
5. Frühstück, A., Singh, K.K., Shechtman, E., Mitra, N.J., Wonka, P., Lu, J.: Insetgan for full-body image generation. In: IEEE/CVF Conference on Computer Vision and Pattern Recognition, CVPR 2022, New Orleans, LA, USA, 18–24 June 2022, pp. 7713–7722 (2022)
6. Fu, J. et al.: StyleGAN-human: a data-centric odyssey of human generation. In: Avidan, S., Brostow, G., Cissé, M., Farinella, G.M., Hassner, T. (eds.) Computer Vision - ECCV 2022, ECCV 2022, LNCS, Part XVI, vol. 13676, pp. 1–19. Springer, Cham (2022). https://doi.org/10.1007/978-3-031-19787-1_1
7. Ge, Y., Song, Y., Zhang, R., Ge, C., Liu, W., Luo, P.: Parser-free virtual try-on via distilling appearance flows. In: IEEE Conference on Computer Vision and Pattern Recognition, CVPR 2021, virtual, 19–25 June 2021, pp. 8485–8493 (2021)
8. Gong, K., Liang, X., Li, Y., Chen, Y., Yang, M., Lin, L.: Instance-level human parsing via part grouping network. In: Computer Vision - ECCV 2018–15th European Conference, Munich, Germany, 8–14 September 2018, Proceedings, Part IV, pp. 805–822 (2018)
9. Goodfellow, I.J., et al.: Generative adversarial nets. In: Advances in Neural Information Processing Systems 27: Annual Conference on Neural Information Processing Systems 2014, 8–13 December 2014, Montreal, Quebec, Canada, pp. 2672–2680 (2014)
10. Güler, R.A., Neverova, N., Kokkinos, I.: Densepose: dense human pose estimation in the wild. In: 2018 IEEE Conference on Computer Vision and Pattern Recognition, CVPR 2018, Salt Lake City, UT, USA, 18–22 June 2018, pp. 7297–7306 (2018)
11. Han, X., Huang, W., Hu, X., Scott, M.R.: Clothflow: a flow-based model for clothed person generation. In: 2019 IEEE/CVF International Conference on Computer Vision, ICCV 2019, Seoul, Korea (South), October 27 - November 2, 2019, pp. 10470–10479 (2019)
12. Han, X., Wu, Z., Wu, Z., Yu, R., Davis, L.S.: VITON: an image-based virtual try-on network. In: 2018 IEEE Conference on Computer Vision and Pattern Recognition, CVPR 2018, Salt Lake City, UT, USA, 18–22 June 2018, pp. 7543–7552 (2018)
13. He, K., Chen, X., Xie, S., Li, Y., Dollár, P., Girshick, R.B.: Masked autoencoders are scalable vision learners. In: IEEE/CVF Conference on Computer Vision and Pattern Recognition, CVPR 2022, New Orleans, LA, USA, 18–24 June 2022, pp. 15979–15988 (2022)
14. He, K., Zhang, X., Ren, S., Sun, J.: Deep residual learning for image recognition. In: Proceedings of the IEEE Conference on Computer Vision and Pattern Recognition, pp. 770–778 (2016)
15. He, S., Song, Y., Xiang, T.: Style-based global appearance flow for virtual try-on. In: IEEE/CVF Conference on Computer Vision and Pattern Recognition, CVPR 2022, New Orleans, LA, USA, 18–24 June 2022, pp. 3460–3469 (2022)

16. Huang, Z., Li, H., Xie, Z., Kampffmeyer, M., Cai, Q., Liang, X.: Towards hard-pose virtual try-on via 3d-aware global correspondence learning. arXiv preprint arXiv:2211.14052 (2022)
17. Karras, T., Aittala, M., Hellsten, J., Laine, S., Lehtinen, J., Aila, T.: Training generative adversarial networks with limited data. In: Advances in Neural Information Processing Systems, pp. 12104–12114 (2020)
18. Karras, T., et al.: Alias-free generative adversarial networks. In: Advances in Neural Information Processing Systems 34: Annual Conference on Neural Information Processing Systems 2021, NeurIPS 2021, 6–14 December 2021, virtual, pp. 852–863 (2021)
19. Krizhevsky, A., Sutskever, I., Hinton, G.E.: Imagenet classification with deep convolutional neural networks. In: Advances in Neural Information Processing Systems 25: 26th Annual Conference on Neural Information Processing Systems 2012, Proceedings of a meeting held 3–6 December 2012, Lake Tahoe, Nevada, United States, pp. 1106–1114 (2012)
20. Liu, Z., et al.: Swin transformer V2: scaling up capacity and resolution. In: IEEE/CVF Conference on Computer Vision and Pattern Recognition, CVPR 2022, New Orleans, LA, USA, 18–24 June 2022, pp. 11999–12009 (2022)
21. Mao, X., Li, Q., Xie, H., Lau, R.Y.K., Wang, Z., Smolley, S.P.: Least squares generative adversarial networks. In: IEEE International Conference on Computer Vision, ICCV 2017, Venice, Italy, 22–29 October 2017, pp. 2813–2821 (2017)
22. Morelli, D., Baldrati, A., Cartella, G., Cornia, M., Bertini, M., Cucchiara, R.: LaDI-VTON: latent diffusion textual-inversion enhanced virtual try-on. arXiv preprint arXiv:2305.13501 (2023)
23. Parmar, G., Zhang, R., Zhu, J.: On aliased resizing and surprising subtleties in GAN evaluation. In: IEEE/CVF Conference on Computer Vision and Pattern Recognition, CVPR 2022, New Orleans, LA, USA, 18–24 June 2022, pp. 11400–11410 (2022)
24. Qian, S., Chen, H., Xue, D., Fang, Q., Xu, C.: Open-world social event classification. In: Proceedings of the ACM Web Conference 2023, pp. 1562–1571 (2023)
25. Qian, S., Xue, D., Fang, Q., Xu, C.: Adaptive label-aware graph convolutional networks for cross-modal retrieval. IEEE Trans. Multimedia **24**, 3520–3532 (2021)
26. Qian, S., Xue, D., Fang, Q., Xu, C.: Integrating multi-label contrastive learning with dual adversarial graph neural networks for cross-modal retrieval. IEEE Trans. Pattern Anal. Mach. Intell. **45**(4), 4794–4811 (2022)
27. Qian, S., Xue, D., Zhang, H., Fang, Q., Xu, C.: Dual adversarial graph neural networks for multi-label cross-modal retrieval. In: Proceedings of the AAAI Conference on Artificial Intelligence, vol. 35, pp. 2440–2448 (2021)
28. Selvaraju, R.R., Cogswell, M., Das, A., Vedantam, R., Parikh, D., Batra, D.: Grad-cam: Visual explanations from deep networks via gradient-based localization. In: IEEE International Conference on Computer Vision, ICCV 2017, Venice, Italy, 22–29 October 2017, pp. 618–626 (2017)
29. Wang, B., Zheng, H., Liang, X., Chen, Y., Lin, L., Yang, M.: Toward characteristic-preserving image-based virtual try-on network. In: Computer Vision - ECCV 2018–15th European Conference, Munich, Germany, 8–14 September 2018, Proceedings, Part XIII, pp. 607–623 (2018)
30. Wang, Z., Bovik, A.C., Sheikh, H.R., Simoncelli, E.P.: Image quality assessment: from error visibility to structural similarity. IEEE Trans. Image Process. **13**(4), 600–612 (2004)
31. Xue, D., Qian, S., Fang, Q., Xu, C.: Mmt: Image-guided story ending generation with multimodal memory transformer. In: Proceedings of the 30th ACM International Conference on Multimedia, pp. 750–758 (2022)
32. Zhang, R., Isola, P., Efros, A.A., Shechtman, E., Wang, O.: The unreasonable effectiveness of deep features as a perceptual metric. In: 2018 IEEE Conference on Computer Vision and Pattern Recognition, CVPR 2018, Salt Lake City, UT, USA, 18–22 June 2018, pp. 586–595 (2018)

GIST: Transforming Overwhelming Information into Structured Knowledge with Large Language Models

Meng Wu, Xinyu Zhou, Gang Ma, Zhangwei Lu, Liuxin Zhang, and Yu Zhang[✉]

Lenovo Research, Beijing, China
Zhangyu29@lenovo.com

Abstract. This paper introduces GIST (Generative Information Synthesis Task-force), a novel personal knowledge management system that utilizes large-scale online language models to analyze and organize the information, generating structured results, including summaries, key points, and questions and answers. The system also utilizes a multimodal information processing approach to enhance comprehension of the content. As the user's knowledge base grows, GIST becomes a personal knowledge database and provides the necessary information at the right moment. GIST can be accessed on any device, serving as the brain and soul of the user's devices, and empowering them to effectively manage their personal knowledge. Our demo video is at https://youtu.be/ImtduHMQKFQ.

Keywords: Personal Knowledge Database · Multimodal Information Processing with LLM · Structured Knowledge

1 Introduction

The evolution of societal requirements and interactions has undergone significant changes in different eras. The era of computation saw the widespread availability of computational resources, making general computing a commodity that was affordable and accessible to more people. This led to the emergence of task processing for information digitization, which was pioneered by Microsoft. The primary interaction methods during this era were command sets and WIMP (Windows, Icons, Menus, Pointer) [1]. In the internet era, the ubiquitous accessibility of information resources made general information a commodity, leading to services for information acquisition, sharing, and consumption that constitute the entire internet ecosystem. The interaction methods mainly revolved around WIMP, touch, and recommendations, delivering information-related services anytime and anywhere. The interaction principle here is mobile, intuitive, invisible [2]. In the era of large-scale language models (LLM), the universalization of knowledge resources prompted the demand and business model for refined, customized knowledge management. Interaction emphasis was placed on spontaneous interaction and implicit interaction.

As we transition from the internet era to the age of large models, we process a wealth of information daily, ranging from work reports, meetings, books, videos, to our own

L. Fang et al. (Eds.): CICAI 2023, LNAI 14474, pp. 41–45, 2024.
https://doi.org/10.1007/978-981-99-9119-8_4

thoughts and ideas. However, this information is scattered across various devices and apps, making it difficult to find when we need it. For instance, we might bookmark webpages on our computers, like videos on TikTok on our mobile phones, or take notes during an open course on our tablets. Gist utilizes a multimodal information processing method with LLM to consolidate these fragmented pieces of information into structured knowledge. This system can analyze and generate structured results from audio, video, or text data, regardless of the mode of interaction. As the system accumulates knowledge and becomes familiar with the user's cognitive habits, it can provide the necessary knowledge at the right time, even becoming the user's digital alter ego. This assistant can reside in any device the user uses, operating both online and offline, thus creating a private, personalized knowledge base.

2 System Architecture

The system comprises several modules, including a vector database, an Automatic Speech Recognition (ASR) model, a large-scale language model, a module for processing multimodal information, a module for prompt engineering, a module for question-answering flow, and a user interface. The system workflow is presented as follows (Fig. 1).

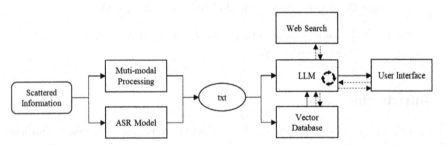

Fig. 1. The workflow of the system is illustrated by the solid line, depicting the automatic generation of summaries and themes by the system. The dotted line represents the flow of user and system interactions in the free chat feature.

A vector database is a specialized database that stores data in the form of vectors, which are mathematical objects possessing both magnitude and direction [3]. In a vector database, each data item is represented as a vector, enabling the storage of information such as position, velocity, or orientation. By encoding personal information as vectors, with the magnitude of each vector conveying the data and the direction providing additional context, the vector database can quickly and easily retrieve information, such as title or author, without requiring searches through traditional databases or spreadsheets. Furthermore, new information can be readily incorporated into the database by simply adding new vectors, making it a suitable method for managing personal collections of movies and books [4].

An ASR model is a type of machine learning model that is used to recognize spoken language in a variety of applications, such as voice assistants, dictation software, and call centers. The model takes as input an audio recording of speech, and outputs a transcription of the speech in written form. Besides, distinguish between different speakers [5].

A large-scale language model is a type of machine learning model that is used to generate text. LLMs are particularly useful for understanding the meaning of text because they are trained on large amounts of data and can learn to recognize patterns and relationships within language. This allows them to make connections between words and phrases, and to identify the underlying structure and organization of a document, even if it is complex and contains multiple ideas and themes [6]. LLMs provide the summary, theme, key point, Q&A, ToDo list according to the prompt.

Multimodal information processing refers to the ability of a system to process and analyze information that is presented in multiple forms, or modalities. This can include text, images, audio, video, and other types of data. The goal of multimodal information processing is to combine and integrate information from different modalities to gain a more complete and accurate understanding of a given situation or task [7]. For example, a system that uses multimodal information processing might be able to recognize and transcribe spoken words, while also analyzing the facial expressions and body language of the speaker to gain additional insights into their meaning and intent [8].

Prompt engineering is the process of designing and crafting input prompts for machine learning models, with the goal of obtaining the desired output or behavior from the model. A prompt can take many forms, including a natural language utterance, a visual or auditory cue, or a specific set of inputs or parameters. Prompt engineering is an important aspect of our system, as the quality and effectiveness of the prompt can have a significant impact on the accuracy and efficiency of the model's output. To guarantee the quality of the output, the system's prompts must encompass language limits, task requirements, further specific requirements, and the output format and presentation [9].

Question-answering flow refers to the process of determining whether a given question can be answered by a large pre-trained language model or whether additional information needs to be obtained from external sources, such as the internet. The basic idea behind question-answering flow is to use the LLM to determine whether the question can be answered based on the information contained within the original text, or whether additional information needs to be obtained from external sources. The goal of question-answering flow is to provide a seamless and efficient way to obtain answers to questions, whether they can be answered by the model or not. This can help to improve the user experience and enhance the overall effectiveness of the system.

The user interface of the system is designed to preserve the original content of the text and to highlight and structure the knowledge present in the text. To this end, the interface displays the original text or transcript of the conversation, along with a mind map that indexes the main ideas and topics discussed. Users can also access an editable summary of the entire conversation, which allows them to review and modify the machine-generated summary to ensure its accuracy. Additionally, the user interface provides a free dialog box where users can type in their own questions or concerns, which the system will then attempt to address using the structured knowledge it has learned from the conversation. This user-friendly interface allows users to easily navigate and interact with the system, making it a valuable tool for conducting natural language conversations.

3 Demo Procedure and Key Features

In the current stage, we have completed the first step of extracting structured knowledge from information. We have built a demo using online LLMs and ASR models. Users can upload audio or video files, and the system will automatically generate transcripts, perform content analysis, and output summaries, themes, key points, Q&A, and ToDo lists. If the generated content is not satisfactory, users can modify the content by adding or deleting parts and then regenerate it. They can also modify the generated content or trace it back to the original text (Fig. 2).

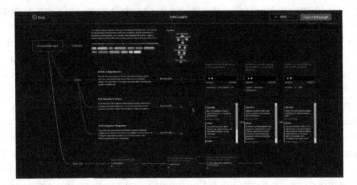

Fig. 2. Analysis user interface include summaries, themes, key points, Q&A, ToDo lists, etc.

To sum up, the key features of GIST are as followed:

1. Efficiently extract structured knowledge from scattered information.
2. Be compatible with multiple devices including PCs, smartphones, tablets, and wearable devices.
3. Support multiple modalities such as text, images, audio, and video.
4. Support multiple scenarios and provide a visualized and interactive experience.

4 Conclusion

The Gist framework successfully organizes fragmented information into structured knowledge, enabling users to browse, understand, and memorize the information while also facilitating direct use of structured knowledge with traceback from knowledge to information. Building upon the successful demonstration of extracting structured knowledge from multimodal/multiple-source fragmented information, we aim to systematically integrate personal and general knowledge, and continue to improve task perception and inference in implicit interaction scenarios based on multimodal information input. Our ultimate goal is to establish a personalized knowledge platform ecosystem that leverages the capabilities of Gist to provide users with a seamless and efficient way to access and use structured knowledge.

References

1. Myers, B.A.: A brief history of human-computer interaction technology. Interactions, **5**(2), 44–54 (1998)
2. Piccolo, L.S.G., De Menezes, E.M., De Campos Buccolo, B.: Developing an accessible interaction model for touch screen mobile devices: preliminary results. Presented at the Proceedings of the 10th Brazilian Symposium on Human Factors in Computing Systems and the 5th Latin American Conference on Human-Computer Interaction, pp. 222–226 (2011)
3. Stata, R., Bharat, K., Maghoul, F.: The term vector database: fast access to indexing terms for web pages. Comput. Netw. **33**(1–6), 247–255 (2000)
4. Lobentanzer, S., Saez-Rodriguez, J.: A platform for the biomedical application of large language models. arXiv preprint arXiv:2305.06488 (2023)
5. Gudepu, P.R., et al.: Whisper augmented end-to-end/hybrid speech recognition system-CycleGAN approach. Presented at the INTERSPEECH, pp. 2302–2306 (2020)
6. OpenAI, "GPT-4 Technical Report." https://cdn.openai.com/papers/gpt-4.pdf. Accessed 28 June 2023
7. Sarter, N.B.: Multimodal information presentation: Design guidance and research challenges. Int. J. Ind. Ergon. **36**(5), 439–445 (2006)
8. Khullar, A., Arora, U.: MAST: multimodal abstractive summarization with trimodal hierarchical attention. arXiv preprint arXiv:2010.08021 (2020)
9. Ekin, S.: Prompt Engineering for ChatGPT: A Quick Guide to Techniques, Tips, and Best Practices (2023). https://doi.org/10.36227/techrxiv.22683919

AIGCIQA2023: A Large-Scale Image Quality Assessment Database for AI Generated Images: From the Perspectives of Quality, Authenticity and Correspondence

Jiarui Wang[1], Huiyu Duan[1], Jing Liu[3], Shi Chen[4], Xiongkuo Min[1(✉)], and Guangtao Zhai[1,2]

[1] Institute of Image Communication and Network Engineering, Shanghai, China
{wangjiarui,huiyuduan,minxiongkuo,zhaiguangtao}@sjtu.edu.cn
[2] MoE Key Lab of Artificial Intelligence, AI Institute,
Shanghai Jiao Tong University, Shanghai, China
[3] Tianjin University, Tianjin, China
jliu_tju@tju.edu.cn
[4] Shanghai Polytechnic University, Shanghai, China
chenshi@sspu.edu.cn

Abstract. Recent years have witnessed a rapid growth of Artificial Intelligence Generated Content (AIGC), among which with the development of text-to-image techniques, AI-based image generation has been applied to various fields. However, AI Generated Images (AIGIs) may have some unique distortions compared to natural images, thus many generated images are not qualified for real-world applications. Consequently, it is important and significant to study subjective and objective Image Quality Assessment (IQA) methodologies for AIGIs. In this paper, in order to get a better understanding of the human visual preferences for AIGIs, a large-scale IQA database for AIGC is established, which is named as AIGCIQA2023. We first generate over 2000 images based on 6 state-of-the-art text-to-image generation models using 100 prompts. Based on these images, a well-organized subjective experiment is conducted to assess the human visual preferences for each image from three perspectives including *quality*, *authenticity* and *correspondence*. Finally, based on this large-scale database, we conduct a benchmark experiment to evaluate the performance of several state-of-the-art IQA metrics on our constructed database. The AIGCIQA2023 database and benchmark will be released to facilitate future research on https://github.com/wangjiarui153/AIGCIQA2023

Keywords: AI generated content (AIGC) · text-to-image generation · image quality assessment · human visual preference

1 Introduction

Artificial Intelligence Generated Content (AIGC) refers to the content, including texts, images, audios, or videos, *etc.*, that is created or generated with the

L. Fang et al. (Eds.): CICAI 2023, LNAI 14474, pp. 46–57, 2024.
https://doi.org/10.1007/978-981-99-9119-8_5

assistance of AI technology. Many impressive AIGC models have been developed in recent years, such as ChatGPT and DALLE [26], which have been utilized in various application scenarios. As an important part of AIGC, AI Generated Images (AIGIs) have also gained significant attention in recent years due to advancement in generative models including Generative Adversarial Network (GAN) [9], Variational Autoencoder (VAE) [14], diffusion models [27], *etc.*, and language-image pre-training techniques including CLIP [25], BLIP [18], *etc.*

However, the development of AIGI models also raises new problems and challenges. One significant challenge is that not all generated images are qualified for real-world applications, which often require to be processed, adjusted, refined or filtered out before being applied to practical scenes. However, unlike common image content, such as Natural Scene Images (NSIs) [7,8], screen content images [3,20], graphic images [5,20], *etc.*, which generally encounters some common distortions including noise, blur, compression, *etc.* [4,6], AIGIs may suffer from some unique degradations such as unreal structures, unreasonable combinations, *etc.* Moreover, the generated images may not correspond to the semantics of the text prompts [15,17,29]. Therefore, it is important to study the human visual preferences for AIGIs and design corresponding objective Image Quality Assessment (IQA) metrics for these images.

Many subjective IQA studies have been conducted for human captured or created images, and many objective IQA models have also been developed. However, these models are designed for assessing low-level distortions, while AIGIs generally contain both low-level artifacts and high-level semantic degradations. Some quantitative evaluation metrics such as Inception Score (IS) [10] and Fréchet Inception Distance (FID) [12] have been proposed to assess the performance of generative models and have been widely used to evaluate the authenticity of the generated images. However, these methods cannot evaluate the authenticity of a single generated image, and cannot measure the correspondence between the generated images and the text-prompts. As a new type of image content, previous IQA methods may fail to assess the image quality of AIGIs and cannot align well with human preferences due to the irregular distortions.

To gain a better understanding of human visual preferences for AIGIs and guide the design process of corresponding objective IQA models, in this paper, we conduct a comprehensive subjective and objective IQA study for AIGIs. We first establish a large-scale IQA database for AIGIs termed AIGCIQA2023, which contains 2,400 diverse images generated by 6 state-of-the-art AIGI models based on 100 various text prompts. Based on these images, a well-organized subjective experiment is conducted to assess the human visual preferences for each individual generated image from three perspectives including *quality*, *authenticity*, and *correspondence*. Based on the constructed AIGCIQA2023 database, we evaluate the performance of several state-of-the-art IQA models and establish a new benchmark. Experimental results demonstrate that current IQA methods cannot well align with human visual preferences for AIGIs, and more efforts should be made in this research field in the future. The main contributions of this paper are summarized as follows:

- We propose to disentangle the human visual experience for AIGIs into three perspectives including *quality*, *authenticity*, and *correspondence*.
- Based on the above theory, we establish a novel large-scale database, *i.e.*, AIGCIQA2023, to better understand the human visual preferences for AIGIs and guide the design of objective IQA models.
- We conduct a benchmark experiment to evaluate the performance of several current state-of-the-art IQA algorithms in measuring the quality, authenticity, and text-image correspondence of AIGIs.

The rest of the paper is organized as follows. In Sect. 2 we introduce the details of our constructed AIGCIQA2023 database, including the generation of AIGIs and the subjective quality assessment methodology and procedures. In Sect. 3 we present the benchmark experiment for current state-of-the-art IQA algorithms based on the established database. Section 4 concludes the whole paper and we discuss possible future research that can be conducted with the database.

2   Database Construction and Analysis

In order to get a better understanding of human visual preferences for AI-generated images based on text prompts, we construct a novel IQA database for AIGIs, termed AIGCIQA2023, which is a collection of generated images derived from six state-of-the-art deep generative models based on 100 text prompts, and corresponding subjective quality ratings from three different perspectives. Then we further analyze the human visual preferences for AIGIs based on the constructed database.

2.1   AIGI Collection

We adopt six latest text-to-image generative models, including Glide [24], Lafite [34], DALLE [26], Stable-diffusion [27], Unidiffuser [1], Controlnet [33], to produce AIGIs by using open source code and default weights. To ensure content diversity and catch up with the practical application requirements, we collect diverse texts from the PartiPrompts website [32] as the prompts for AIGI generation. The text prompts can be simple, allowing generative models to produce imaginative results. They can also be complex, which raises the challenge for generative models. We select 10 scene categories from the prompt set, and each scene contains 10 challenge categories. Overall, we collect 100 text prompts (10 scene categories × 10 challenge categories) from PartiPrompts [32]. The distribution of the selected scene and challenge categories is displayed in pie chart of Fig. 1. It can be observed that the dataset exhibits a high level of scene diversity, with images generated covering a broad range of challenges. Then we perform the text-to-image generation based on these models and prompts. Specifically, for each prompt, we generate 4 various images randomly for each generative model. Therefore, the constructed AIGCIQA2023 database totally contains 2400 AIGIs (4 images × 6 models × 100 prompts) corresponding to 100 prompts (Fig. 2).

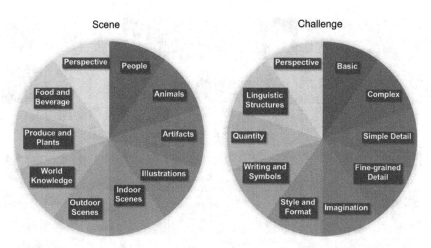

Fig. 1. Pie Chart of the ten challenge categories and ten scene categories selected from PartiPrompts [32].

2.2 Subjective Experiment Setup

Subjective IQA is the most reliable way to evaluate the visual quality of digital images perceived by the users. It is generally used to construct image quality datasets and served as the ground truth to optimize or evaluate the performance of objective quality assessment metrics. Due to the unnatural property of AIGIs and different text prompts having different target image spaces, it is unreasonable to just use one score, *i.e.,* "quality" to represent human visual preferences. In this paper, we propose to measure the human visual preferences of AIGIs from three perspectives including *quality*, *authenticity*, and text-image *correspondence*. For an image, these three visual perception perspectives are related but different.

The first dimension of AIGI evaluation is "quality" evaluation, *i.e.,* evaluating an AIGI from its clarity, color, lightness, contrast, *etc.,* which is similar to the assessment of NSIs. During the experiment procedure, subjects are instructed to evaluate whether the image outline is clear, whether the content can be distinguished, and the richness of details, *etc.* Fig. 3(a) shows 10 high quality examples and 10 low quality examples of the images generated by the prompt of "a corgi".

Considering the generation nature of AIGIs, an important problem of these images is that they may not look real compared to NSIs. Therefore, we introduce a second dimension of evaluation metrics for the generated images, *i.e.,* "authenticity" evaluation. For this dimension, subjects are instructed to assess the image from the authenticity aspect, *i.e.,* whether it looks real or whether they can distinguish that the image is AI-generated or not. Figure 3(b) shows 10 high authenticity and 10 low authenticity examples of images generated by the prompt of "a girl".

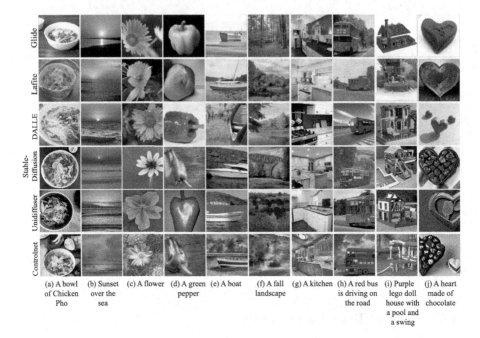

Fig. 2. Sample images from the AIGCIQA2023 database generated by six different generative models (Glide [24], Lafite [34], DALLE [26], Stable-diffusion [27], Unidiffuser [1], Controlnet [33].)

Since an AIGI is generated from a text, it is also important to evaluate its correspondence with the original prompt, *i.e.*, the third dimension, text-image "correspondence". For this purpose, subjects are instructed to consider textual information provided with the image and then give the correspondence score from 0 to 5 to assess the relevance between the generated image and its prompt. Figure 3(c) shows 10 high text-image correspondence and 10 low correspondence examples of images generated by the prompt of "a grandmother reading a book to her grandson and granddaughter".

2.3   Subjective Experiment Procedure

To evaluate the quality of the images in the AIGCIQA2023 and obtain Mean Opinion Scores (MOSs), a subjective experiment is conducted following the guidelines of ITU-R BT.500-14 [3]. The subjects are asked to rate their visual preference degree of exhibited AIGIs from the quality, authenticity and text-image correspondence. The AIGIs are presented in a random order on an iMac monitor with a resolution of up to 4096 × 2304, using an interface designed with Python Tkinter, as shown in Fig. 4. The interface allows viewers to browse the previous and next AIGIs and rate them using a quality scale that ranges from 0 to 5, with a minimum interval of 0.01. A total of 28 graduate students (14 males

(a) "a corgi"

(b) "a girl"

(c) "a grandmother reading a book to her grandson and granddaughter"

Fig. 3. Illustration of the images from the perspectives of quality, authenticity, and text-image correspondence. (a) 10 high quality examples and 10 low quality examples of the images generated by the prompt of "a corgi". (b) 10 high authenticity and 10 low authenticity examples of images generated by the prompt of "a girl". (c) 10 high text-image correspondence and 10 low correspondence examples of images generated by the prompt of "a grandmother reading a book to her grandson and granddaughter".

and 14 females) participate in the experiment, and they are seated at a distance of around 60 cm in a laboratory environment with normal indoor lighting.

2.4 Subjective Data Processing

We follow the suggestions recommended by ITU to conduct the outlier detection and subject rejection. The score rejection rate is 2%. In order to obtain the MOS for an AIGI, we first convert the raw ratings into Z-scores, then linearly scale them to the range $[0, 100]$ as follows:

$$z_{ij} = \frac{r_{ij} - \mu_{ij}}{\sigma_i}, \quad z'_{ij} = \frac{100(z_{ij} + 3)}{6},$$

$$\mu_i = \frac{1}{N_i} \sum_{j=1}^{N_i} r_{ij}, \quad \sigma_i = \sqrt{\frac{1}{N_i - 1} \sum_{j=1}^{N_i} (r_{ij} - \mu_{ij})^2}$$

where r_{ij} is the raw ratings given by the i-th subject to the j-th image. N_i is the number of images judged by subject i.

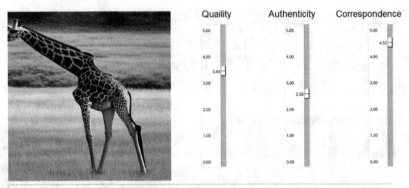

A giraffe walking through a green grass covered field

Fig. 4. An example of the subjective assessment interface. The subject can evaluate the quality of AIGIs and record the quality, authenticity, correspondence scores with the scroll bar on the right.

Next, the mean opinion score (MOS) of the image j is computed by averaging the rescaled z-scores as follows:

$$MOS_j = \frac{1}{M} \sum_{i=1}^{M} z'_{ij}$$

where MOS_j indicates the MOS for the j-th AIGI, M is the number of valid subjects, and z'_{ij} are the rescaled z-scores.

2.5   AIGI Analysis from Three Perspectives

To further illustrate the differences of the three perspectives, we demonstrate several example images and their corresponding subjective ratings from three aspects in Fig. 5. For each subfigure, it can be noticed that the right AIGI outperforms the left AIGI on two evaluation dimensions but is much worse than the left AIGI on another dimension, which demonstrates that each evaluation perspective (quality, authenticity, or text-image correspondence) has its own unique perspective and value.

Figure 6 demonstrates the MOS and score distribution for quality evaluation, authenticity evaluation, and text-image correspondence evaluation, respectively, which demonstrate the images in AIGCIQA 2023 cover a wide range of perceptual quality.

3   Experiment

3.1   Benchmark Models

Since the AIGIs in the proposed AIGCIQA2023 database are generated based on text prompts and have no pristine reference images, they can only be evaluated

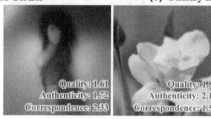

(a) the word 'START'   (b) a library filled with kids reading books

(c) A girl

Fig. 5. Comparison of the differences between three evaluation perspectives. (a) Left image has better quality, but worse authenticity and correspondence. (b) Left image has better authenticity, but worse quality and correspondence. (c) Left image has better correspondence, but worse quality and authenticity.

by no-reference (NR) IQA metrics. In this paper, we select fifteen state-of-the-art IQA models for comparison. The selected models can be classified into two groups:

- **Handcrafted-based** models, including: NIQE [23], BMPRI [21], BPRI [19], BRISQUE [22], HOSA [30], BPRI-LSSn [19], BPRI-LSSs [19], BPRI-PSS [19], QAC [31], HIGRADE-1 and HIGRADE-2 [16].

  These models extract handcrafted features based on prior knowledge about image quality.
- **Deep learning-based** models, including: CNNIQA [13], WaDIQaM-NR [2], VGG (VGG-16 and VGG-19) [28] and ResNet (ResNet-18 and ResNet-34) [11].

  These models characterize quality-aware information by training deep neural networks from labeled data.

3.2 Evaluation Criteria

In this study, we utilize the following four performance evaluation criteria to evaluate the consistency between the predicted scores and the corresponding ground-truth MOSs, including Spearman Rank Correlation Coefficient (SRCC), Pearson Linear Correlation Coefficient (PLCC), Kendall's Rank Correlation Coefficient (KRCC), and Root Mean Squared Error (RMSE).

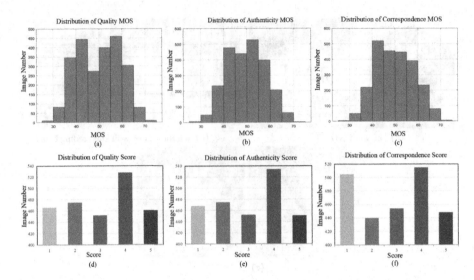

Fig. 6. (a) MOSs distribution of quality score. (b) MOSs distribution of authenticity score. (c) MOSs distribution of correspondence score. (d) Distribution of the quality score. (e) Distribution of the authenticity score. (f) Distribution of the correspondence score.

3.3 Experimental Setup

All the benchmark models are validated on the proposed AIGCIQA2023 database. For traditional handcrafted-based models, they are directly evaluated based on the database. For deep trainable models, we first randomly split the database into an 4:1 ratio for training/testing while ensuring the image with the same prompt label falls into the same set. The partitioning and evaluation process is repeated several times for a fair comparison while considering the computational complexity, and the average result is reported as the final performance. For deep learning-based models, we applied CNNIQA [13], WaDIQaM-NR [2], VGG (VGG-16 and VGG-19) [28] and ResNet (ResNet-18 and ResNet-34) [11] to predict the MOS of image quality. The repeating time is 10, the training epochs are 50 with an initial learning rate of 0.0001 and batch size of 4.

3.4 Performance Discussion

The performance results of the state-of-the-art IQA models mentioned above on the proposed AIGCIQA2023 database are exhibited in Table 1, from which we can make several conclusions:

- The handcrafted-based methods achieve poor performance on the whole database, which indicates the extracted handcrafted features are not effective for modeling the quality representation of AIGIs. This is because most employed handcrafted features of these methods are based on the prior knowledge learned from NSIs, which are not effective for evaluating AIGIs.

Table 1. Performance comparisons of the state-of-the-art IQA methods on the AIG-CIQA2023 database. The best performance results are marked in RED and the second-best performance results are marked in BLUE.

Method	Quality				Authenticity				Correspondence			
	SRCC	KRCC	PLCC	RMSE	SRCC	KRCC	PLCC	RMSE	SRCC	KRCC	PLCC	RMSE
NIQE [23]	0.5060	0.3420	0.5218	7.9461	0.3715	0.2453	0.3954	7.3999	0.3659	0.2460	0.3485	7.7721
QAC [31]	0.5328	0.3644	0.5991	6.3062	0.4009	0.2673	0.4428	7.2236	0.3526	0.2414	0.4062	7.5768
BRISQUE [22]	0.6239	0.4291	0.6389	7.1655	0.4705	0.3142	0.4796	7.0695	0.4219	0.2865	0.4280	7.4941
PRI-PSS [19]	0.3556	0.2373	0.4183	8.4605	0.2409	0.1583	0.2625	7.7739	0.2670	0.1794	0.2960	7.9203
PRI-LSSs [19]	0.5141	0.3512	0.5618	7.7054	0.3721	0.2460	0.3998	7.3845	0.3230	0.2160	0.3473	7.7756
PRI-LSSn [19]	0.5245	0.3523	0.5935	7.4964	0.3838	0.2528	0.5465	6.7467	0.3655	0.2474	0.4594	7.3653
BPRI [19]	0.6301	0.4307	0.6889	6.7517	0.4740	0.3144	0.5207	6.8783	0.3946	0.2657	0.4346	7.4680
HOSA [30]	0.6317	0.4311	0.6561	7.0297	0.4716	0.3101	0.4985	6.9841	0.4101	0.2765	0.4252	7.5051
BMPRI [21]	0.6732	0.4661	0.7492	6.1693	0.5273	0.3554	0.5756	6.5878	0.4419	0.3014	0.4827	7.2619
Higrade-1 [16]	0.4849	0.3220	0.4966	8.0847	0.4175	0.2791	0.4181	7.3183	0.3319	0.2207	0.3379	7.8041
Higrade-2 [16]	0.2344	0.1568	0.3189	8.8282	0.2654	0.1742	0.3106	7.6579	0.1756	0.1170	0.2144	8.0990
WaDIQaM-NR [2]	0.4447	0.3036	0.4996	8.7400	0.3936	0.2715	0.3906	7.4627	0.3027	0.2057	0.2810	6.0477
CNNIQA [13]	0.7160	0.4955	0.7937	5.8816	0.5958	0.4085	0.5734	6.7231	0.4758	0.3313	0.4937	7.3839
VGG16 [28]	0.7961	0.5843	0.7973	6.2143	0.6660	0.4813	0.6807	6.0273	0.6580	0.4548	0.6417	6.9292
VGG19 [28]	0.7733	0.5376	0.8402	5.0860	0.6674	0.4843	0.6565	6.1705	0.5799	0.4090	0.5670	6.9851
Resnet18 [11]	0.7583	0.5360	0.7763	6.9897	0.6701	0.4740	0.6528	6.4597	0.5979	0.4165	0.5564	7.0957
Resnet34 [11]	0.7229	0.4835	0.7578	6.4806	0.5998	0.4325	0.6285	6.5344	0.7058	0.5111	0.7153	6.7605

- The deep learning-based methods achieve relatively more competitive performance results on three evaluation perspectives. However, they are still far away from satisfactory.
- Most of the IQA models achieve better performance on quality evaluation and worse on text-image correspondence score assessment. The reason is that the text prompts for image generation are not utilized for the IQA model training. This makes it more challenging for the IQA models to extract relation features from AIGIs, which inevitably leads to performance drops.

4   Conclusion and Future Work

In this paper, we study the human visual preference problem for AIGIs. We first construct a new IQA database for AIGIs, termed AIGCIQA2023, which includes 2400 AIGIs generated based on 100 various text-prompts, and corresponding subjective MOSs evaluated from three perspectives (*i.e., quality, authenticity, and text-image correspondence*). Experimental analysis demonstrates that these three dimensions can reflect different aspects of human visual preferences on AIGIs, which further manifests that the evaluation of Quality of Experience (QoE) for AIGIs should be considered from multiple dimensions. Based on the constructed database, we evaluate the performance of several state-of-the-art IQA models and establish a new benchmark to facilitate future research.

In future work, we will further explore the human visual perception for AIGIs and develop corresponding objective evaluation models for better assessing the quality of AIGIs from the three perspectives proposed in this paper.

Acknowledgement. This work is supported by National Key R&D Project of China (2021YFE0206700), NSFC (61831015, 62101325, 62271312, 62225112), Shanghai Pujiang Program (22PJ1407400), Shanghai Municipal Science and Technology Major Project (2021SHZDZX0102), STCSM (22DZ2229005).

References

1. Bao, F., et al.: One transformer fits all distributions in multi-modal diffusion at scale. ArXiv abs/2303.06555 (2023)
2. Bosse, S., Maniry, D., Müller, K.R., Wiegand, T., Samek, W.: Deep neural networks for no-reference and full-reference image quality assessment. IEEE Trans. Image Process. (TIP) **27**(1), 206–219 (2017)
3. Duan, H., Min, X., Zhu, Y., Zhai, G., Yang, X., Le Callet, P.: Confusing image quality assessment: toward better augmented reality experience. IEEE Trans. Image Process. (TIP) **31**, 7206–7221 (2022)
4. Duan, H., et al.: Develop then rival: A human vision-inspired framework for super-imposed image decomposition. IEEE Trans. Multimed. (TMM) (2022)
5. Duan, H., Shen, W., Min, X., Tu, D., Li, J., Zhai, G.: Saliency in augmented reality. In: Proceedings of the ACM International Conference on Multimedia (ACM MM), pp. 6549–6558 (2022)
6. Duan, H., et al.: Masked autoencoders as image processors. arXiv preprint arXiv:2303.17316 (2023)
7. Duan, H., Zhai, G., Min, X., Zhu, Y., Fang, Y., Yang, X.: Perceptual quality assessment of omnidirectional images. In: Proceedings of the IEEE International Symposium on Circuits and Systems (ISCAS), pp. 1–5 (2018)
8. Duan, H., Zhai, G., Yang, X., Li, D., Zhu, W.: Ivqad 2017: An immersive video quality assessment database. In: Proceedings of the IEEE International Conference on Systems, Signals and Image Processing (IWSSIP), pp. 1–5. IEEE (2017)
9. Goodfellow, I., et al.: Generative adversarial networks. Commun. ACM **63**(11), 139–144 (2020)
10. Gulrajani, I., Ahmed, F., Arjovsky, M., Dumoulin, V., Courville, A.C.: Improved training of wasserstein gans. In: Proceedings of the Advances in Neural Information Processing Systems (NeurIPS) 30 (2017)
11. He, K., Zhang, X., Ren, S., Sun, J.: Deep residual learning for image recognition. In: Proceedings of the IEEE Conference on Computer Vision and Pattern Recognition (CVPR), pp. 770–778 (2016)
12. Heusel, M., Ramsauer, H., Unterthiner, T., Nessler, B., Hochreiter, S.: Gans trained by a two time-scale update rule converge to a local nash equilibrium. In: Proceedings of the Advances in Neural Information Processing Systems (NeurIPS) 30 (2017)
13. Kang, L., Ye, P., Li, Y., Doermann, D.: Convolutional neural networks for no-reference image quality assessment. In: Proceedings of the IEEE Conference on Computer Vision and Pattern Recognition (CVPR), pp. 1733–1740 (2014)
14. Kingma, D.P., Welling, M.: Auto-encoding variational bayes. arXiv preprint arXiv:1312.6114 (2013)
15. Kirstain, Y., Polyak, A., Singer, U., Matiana, S., Penna, J., Levy, O.: Pick-a-pic: an open dataset of user preferences for text-to-image generation. arXiv preprint arXiv:2305.01569 (2023)

16. Kundu, D., Ghadiyaram, D., Bovik, A.C., Evans, B.L.: Large-scale crowdsourced study for tone-mapped hdr pictures. IEEE Trans. Image Process. (TIP) **26**(10), 4725–4740 (2017)

17. Lee, K., et al.: Aligning text-to-image models using human feedback. arXiv preprint arXiv:2302.12192 (2023)

18. Li, J., Li, D., Xiong, C., Hoi, S.: Blip: bootstrapping language-image pre-training for unified vision-language understanding and generation. In: International Conference on Machine Learning, pp. 12888–12900. PMLR (2022)

19. Min, X., Gu, K., Zhai, G., Liu, J., Yang, X., Chen, C.W.: Blind quality assessment based on pseudo-reference image. IEEE Trans. Multimed. (TMM) **20**(8), 2049–2062 (2017)

20. Min, X., Ma, K., Gu, K., Zhai, G., Wang, Z., Lin, W.: Unified blind quality assessment of compressed natural, graphic, and screen content images. IEEE Trans. Image Process. (TIP) **26**(11), 5462–5474 (2017)

21. Min, X., Zhai, G., Gu, K., Liu, Y., Yang, X.: Blind image quality estimation via distortion aggravation. IEEE Trans. Broadcast. **64**(2), 508–517 (2018)

22. Mittal, A., Moorthy, A.K., Bovik, A.C.: No-reference image quality assessment in the spatial domain. IEEE Trans. Image Process. (TIP) **21**(12), 4695–4708 (2012)

23. Mittal, A., Soundararajan, R., Bovik, A.C.: Making a "completely blind" image quality analyzer. IEEE Signal Process. Lett. **20**(3), 209–212 (2012)

24. Nichol, A., et al.: Glide: towards photorealistic image generation and editing with text-guided diffusion models, pp. 16784–16804 (2021)

25. Radford, A., et al.: Learning transferable visual models from natural language supervision. In: International Conference on Machine Learning, pp. 8748–8763. PMLR (2021)

26. Ramesh, A., Dhariwal, P., Nichol, A., Chu, C., Chen, M.: Hierarchical text-conditional image generation with clip latents. arXiv preprint arXiv:2204.06125 (2022)

27. Rombach, R., Blattmann, A., Lorenz, D., Esser, P., Ommer, B.: High-resolution image synthesis with latent diffusion models. Proceedings of the IEEE Conference on Computer Vision and Pattern Recognition (CVPR), pp. 10674–10685 (2021)

28. Simonyan, K., Zisserman, A.: Very deep convolutional networks for large-scale image recognition. arXiv preprint arXiv:1409.1556 (2014)

29. Xu, J., et al.: Imagereward: Learning and evaluating human preferences for text-to-image generation. arXiv preprint arXiv:2304.05977 (2023)

30. Xu, J., Ye, P., Li, Q., Du, H., Liu, Y., Doermann, D.: Blind image quality assessment based on high order statistics aggregation. IEEE Trans. Image Process. (TIP) **25**(9), 4444–4457 (2016)

31. Xue, W., Zhang, L., Mou, X.: Learning without human scores for blind image quality assessment. In: Proceedings of the IEEE Conference on Computer Vision and Pattern Recognition (CVPR), pp. 995–1002 (2013)

32. Yu, J., et al.: Scaling autoregressive models for content-rich text-to-image generation. arXiv preprint arXiv:2206.10789 (2022)

33. Zhang, L., Agrawala, M.: Adding conditional control to text-to-image diffusion models. ArXiv abs/2302.05543 (2023)

34. Zhou, Y., et al.: Towards language-free training for text-to-image generation. In: Proceedings of the IEEE/CVF Conference on Computer Vision and Pattern Recognition (CVPR), pp. 17907–17917, June 2022

Applications of Artificial Intelligence

Space Brain: An AI Autonomous Spatial Decision System

Jiachen Du, Boyang Jia, and Xinyi Fu[✉]

The Future Laboratory, Tsinghua University, Beijing, China
dujiachen1226@163.com, Boyangjia@outlook.com, fuxy@mail.tsinghua.edu.cn

Abstract. Ubiquitous computing has proposed the idea of seamless integration of computing devices into smart spaces earlier. The proliferation of Internet of Things (IoT) technologies offers users an expanding array of device control options, including voice activation and home automation. However, these technologies necessitate user intervention for adjusting device state, relying on information derived from the environment and events within the smart space. This article introduces an AI autonomous decision system designed to address the reliance on user intervention in adjusting device states within IoT-enabled smart space. The proposed system consists of a sensing layer, transmission layer, decision layer, and execution layer, responsible for information sensing, transmission, decision generation, and device control, respectively. The decision layer incorporates a large language model (LLM) and accompanying modules to facilitate real-time decision making. This study validates the core functionalities of the system within an unmanned smart home and examines the advantages, disadvantages, potential security risks, and future development directions of the AI autonomous decision.

Keywords: Smart Space · Large Language Model · Autonomous Systems

1 Introduction

The control methods in smart spaces, such as smart homes, have evolved from manual physical switch control to encompass diverse modes including voice control and remote control [12]. Technology advancements have facilitated users to implement trigger-action programming (TAP) configuration [10] according to their living habits, enabling devices within a space to automatically change their functional states in response to different trigger conditions. The evolution of these control methods gradually relieves the complex action behavior of user-controlled devices. Among the three processes of sensing, thinking and executing level in smart space [13], existing smart systems are dedicated to helping users to complete the sensing and executing part. For example, human body sensors and light sensors help the whole system to perceive the environmental state and event in smart space. While smart lights and other smart appliances can help users to perform certain switching actions automatically. However, less research has focused on the thinking level of smart space, and the control strategy of core devices still relies mainly on the user's judgment now [7].

The purpose of this study is to develop an AI autonomous decision system for Internet of Things (IoT) devices in smart space at the thinking level, so that each device can react to the action according to the real-time decision of AI and free the users from the tedious work of condition judgment, switching action and home automation configuration.

In this study, large language model (LLM) is chosen as the basis of the underlying smart space decision brain. LLM processed the capability of logical reasoning and context awareness that can produce corresponding action instructions based on spatial information and judge the necessity of following actions in conjunction with the previous states of devices, then help users to intervene in the process of devices execution.

The contributions of this study are 1) Propose and validate a autonomous spatial decision system based on LLM; 2) Identify the advantages, disadvantages and security risks of spatial autonomous decision system; 3) Provide reference suggestions for future research and technology development related to LLM based smart space design.

2   Related Work

Mark Weiser [11] proposed the idea of seamless integration of computers into the world, that is, to minimize the presence of computers as input and output devices, so that users can use computing devices without thinking, and psychologically focus on new goals beyond the devices, which developing into the concept of ubiquitous computing. With the development of technology, ideas such as wireless communication and miniaturization of computing devices have been realized, but the ideal of seamless computer access in IoT and smart space is still far away.

Smart spaces originated and evolved in the home environment, and as technology advanced, these automation technologies were used in more scenarios, such as public office spaces, warehouse spaces, production spaces, and mobile spaces [5]. In this process, smart spaces and space automation arise. The first domotic technology [9] was developed in 1975 and the concept of home automation emerged as a communication protocol and module. Then, as a stepping stone to the emergence of smart home, home automation was gradually developed and promoted. However, this technology had not been widely used by users for a long time [2]. In early 2000, R.J.C. Nunes et al. proposed the use of easy programming user interfaces for home automation configuration in IoT systems [8]. Home automation was given a new concept, i.e., a user-programmed interactive system for device connection and trigger-action programming. Home automation survived as an important function and interaction method for smart home.

Some of the current studies had applied LLM to the smart home system to assist users in device control and automation configuration using fuzzy com-

mands [4]. To some extent, it simplified the operational process of home automation [1], but did not diminish the strong input-output computer primitive feature
of smart home. In the existing system framework, users need to view, predict
and understand spatial information in real time or ahead of time, and then give
commands to change the functional state of devices [3]. In this process, thinking
function of smart space is always done by the users.

3   System Architecture

(See Fig. 1)

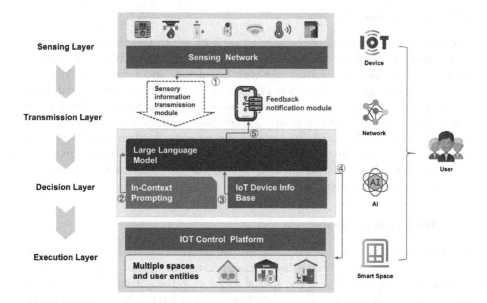

Fig. 1. Architecture diagram of the spatial autonomous system. 1) Sensing network can
sense environment and event information in the space, the transmission layer captures
this information and passes it to the decision layer. 2) The In-Context Prompting
module outputs task prompts to the LLM based on changes in sensing information. 3)
Information such as function, status and device number from the device information
base is input to the LLM. 4) After generating decision instructions, the decision layer
inputs the instructions into the IOT control platform to create a dynamically adjusted
and more appropriate space for the user or other entities. 5) While generating the
instructions, the decision layer sends a notification to the user describing the expected
effect of the decision instructions after they are issued.

3.1   Overview

The whole system framework is divided into sensing layer, transmission layer,
decision layer and execution layer. The sensing network in smart space consist of

the sensing layer. The transmission layer is responsible for transmitting sensor data and other environmental information in space to the decision layer, as well as informing users of the control commands issued by the decision layer. The core part of the whole system is LLM based decision layer. In order to make LLM capable of logical reasoning and instruction generation for specific IoT devices every time the spatial conditions change, we design the automatic prompt language module and the device function library module to ensure the integrity and continuity of the system functions. The automation commands then will be passed to the execution layer to control the smart appliances using IoT control platform in smart space.

3.2   Sensing Layer

The sensing layer consists of the sensing network which is capable of detecting environmental and event information in space in real time. When the spatial information changes, the sensing network can detect and record the type and value of the changes. For example, changes in temperature, humidity, sunlight intensity, wind speed, etc. in the space recorded by the sensing network can reflect the environmental conditions in the space, and changes in the state of human passing by, door and window switches or other equipment switches can also reflect the event information in the space. All these information are recorded and then captured by the transmission layer as the base information for decision generation at the decision layer.

3.3   Transmission Layer

The transmission layer is responsible for the information transfer between the decision layer and the sensing layer. Each change of spatial information and environmental information of the spatial location in the sensing-execution layer is captured by the sensing information transmission module and transmitted to the decision layer as the basis for the decision of the decision layer to adjust the equipment status. Each time the decision layer issues a control command to adjust the state of the equipment, the result of the execution of the command is also sent back to the user.

3.4   Decision Layer

The decision layer is the core part of the whole system and is the key to the system's ability to adjust the device state without the user's manual judgment. We adopt LLM technology with the capability of logical reasoning and context awareness. These capabilities can help the decision layer to generate decisions and output commands based on the spatial change information delivered by the above-mentioned modules. In this process, the LLM is fully autonomous in judging the environmental information, event information and other perceptual information in space according to its relatively reasoning process. By combining all kinds of information, the LLM autonomously decides whether it needs to execute actions, which devices to issue control commands to, and describes the overall execution results and the spatial changes brought by the devices after the control commands are issued in natural language. Due to the passive nature of the LLM, if cannot autonomously carry out the task of spatial decision making, and the ideal smart space should monitor the environmental changes in real time for dynamic adjustment. We design the prompt engineering execution module, which dispatches tasks for spatial decision making to the LLM at a specific time, and when the sensing layer reports changes in spatial information, the decision tasks will be input to the LLM along with the state sensing information, giving the LLM the ability to make real-time decisions about the smart space. The device list and functional state information required by the LLM in this process are provided by the device function library, and this module avoids device forgetting by the LLM, which is a drawback caused by the forgetting feature of the LLM. With the cooperation of the LLM and other modules in the decision layer, the system has the ability to make spatial decisions without human intervention.

3.5 Execution Layer

The execution layer consists of IoT control platform and IoT devices. Under the unified control of the IoT control platform, IoT devices can create a more suitable spatial environment for the entities served by the space through switch combinations and functional changes. For example, living environment, office environment, storage environment, production environment, etc. The decision layer is able to issue real-time or delayed control commands as needed, and the execution layer receives the commands to control the devices according to the corresponding control strategy. The environment in which the system is designed and experimented is a real unmanned space, and the system can complete a series of smart appliances control in the space.

4   Conclusion

This study identified the problem of user decision dependence in the existing framework of smart space and IoT control systems, and proposed a spatial autonomous decision system to help users in spatial information judgment and decision command generation, so that the devices in space can operate autonomously without user intervention. In terms of system design, we divided the whole system into sensing layer, transmission layer, decision layer and execution layer. We fully considered the features and defects of LLM, and encapsulated them in the decision system with the modular combination method, so that LLM can generated device control commands in real time according to certain strategies. In order to introduce the autonomous decision-making capability into the smart space and meet the realistic needs of the users and service objects in the smart space, we introduced other modules such as transmission module into the system, so that the decision system can accomplish spatial autonomy in the smart space in a suitable way. Our system activated the capability of autonomous decision making of smart space, alleviated the users' cost of using IoT devices, and relieved various problems in the process of home automation configuration.

At the same time, spatially autonomous systems may also have problems [6] such as failure to meet users' individual needs, possible implementation deviations and environmental control risks. The future developing directions of spatially autonomous systems may concern more user intervention, personalized service needs, security, and scenario applications, so that smart space can better meet users' needs. Related work in this area may bring new research directions for smart home and introduce new development paths for IoT control technologies.

Demo video: https://ispacedemo.github.io/CICAIdemo/

References

1. Asadullah, M., Raza, A.: An overview of home automation systems. In: 2016 2nd International Conference on Robotics and Artificial Intelligence (ICRAI), pp. 27–31. IEEE (2016)
2. Brush, A.B., Lee, B., Mahajan, R., Agarwal, S., Saroiu, S., Dixon, C.: Home automation in the wild: challenges and opportunities. In: Proceedings of the SIGCHI Conference on Human Factors in Computing Systems, pp. 2115–2124 (2011)
3. Jabbar, W.A., et al.: Design and fabrication of smart home with internet of things enabled automation system. IEEE Access **7**, 144059–144074 (2019)
4. King, E., Yu, H., Lee, S., Julien, C.: Sasha: creative goal-oriented reasoning in smart homes with large language models. arXiv preprint arXiv:2305.09802 (2023)
5. Korzun, D.G., Balandin, S.I., Gurtov, A.V.: Deployment of smart spaces in internet of things: Overview of the design challenges. In: Internet of Things, Smart Spaces, and Next Generation Networking: 13th International Conference, NEW2AN 2013 and 6th Conference, ruSMART 2013, St. Petersburg, Russia, August 28–30, 2013. Proceedings. pp. 48–59. Springer (2013)
6. Lei, L., Tan, Y., Zheng, K., Liu, S., Zhang, K., Shen, X.: Deep reinforcement learning for autonomous internet of things: Model, applications and challenges. IEEE Commun. Surv. Tutorials **22**(3), 1722–1760 (2020)
7. Mekuria, D.N., Sernani, P., Falcionelli, N., Dragoni, A.F.: Smart home reasoning systems: a systematic literature review. J. Ambient. Intell. Humaniz. Comput. **12**, 4485–4502 (2021)
8. Nunes, R.J., Delgado, J.C.: An internet application for home automation. In: 2000 10th Mediterranean Electrotechnical Conference. Information Technology and Electrotechnology for the Mediterranean Countries. Proceedings. MeleCon 2000 (Cat. No. 00CH37099), vol. 1, pp. 298–301. IEEE (2000)
9. Rye, D.: My life at x10 (1999). https://web.archive.org/web/20161015080410/http://www.hometoys.com/content.php?url=/htinews/oct99/articles/rye/rye.htm

10. Ur, B., McManus, E., Pak Yong Ho, M., Littman, M.L.: Practical trigger-action programming in the smart home. In: Proceedings of the SIGCHI Conference on Human Factors in Computing Systems, pp. 803–812 (2014)
11. Weiser, M.: The computer for the 21st century. ACM SIGMOBILE Mob. Comput. Commun. Rev. **3**(3), 3–11 (1999)
12. Xinyi, F., He, Z., Cheng, X., Tongxin, S.: A review of the frontier research on future smart home. Sci. Technol. Rev. **41**, 36–52 (2023)
13. Xinyi, F., He, Z., Cheng, X., Xinyang, L., Zhe, S., Yingqing, X.: Design research and application practice of integrated experimental platform for smart home. Packaging Eng. **43**, 50–58 (2022)

TST: Time-Sparse Transducer for Automatic Speech Recognition

Xiaohui Zhang[1,2], Mangui Liang[1(✉)], Zhengkun Tian[2], Jiangyan Yi[2], and Jianhua Tao[3]

[1] School of Computer and Information Technology, Beijing Jiaotong University, Beijing, China
{21120320,mgliang}@bjtu.edu.cn
[2] Institute of Automation, Chinese Academy of Science, Beijing, China
[3] Department of Automation, Tsinghua University, Beijing, China

Abstract. End-to-end model, especially Recurrent Neural Network Transducer (RNN-T), has achieved great success in speech recognition. However, transducer requires a great memory footprint and computing time when processing a long decoding sequence. To solve this problem, we propose a model named time-sparse transducer, which introduces a time-sparse mechanism into transducer. In this mechanism, we obtain the intermediate representations by reducing the time resolution of the hidden states. Then the weighted average algorithm is used to combine these representations into sparse hidden states followed by the decoder. All the experiments are conducted on a Mandarin dataset AISHELL-1. Compared with RNN-T, the character error rate of the time-sparse transducer is close to RNN-T and the real-time factor is 50.00% of the original. By adjusting the time resolution, the time-sparse transducer can also reduce the real-time factor to 16.54% of the original at the expense of a 4.94% loss of precision.

Keywords: speech recognition · human-computer interaction · computational paralinguistics

1 Introduction

In recent years, significant advancements have been made in end-to-end speech recognition models, including the connectionist temporal classification (CTC) [1–3], attention-based sequence-to-sequence models (AED) [4–7], and recurrent neural network transducer (RNN-T) [8–13]. The CTC algorithm, employed by many models, performs frame-level decoding by converting speech sequences to corresponding label sequences. However, this method relies on the assumption of conditional independence among speech frames, making it unable to effectively model the dependencies between outputs. On the other hand, the RNN-T leverages its recurrent structure to overcome the conditional independence assumption and optimizes acoustic and language components jointly through

L. Fang et al. (Eds.): CICAI 2023, LNAI 14474, pp. 68–80, 2024.
https://doi.org/10.1007/978-981-99-9119-8_7

the introduction of language and joint networks. Consequently, the RNN-T has found success in online automatic speech recognition (ASR) systems [14,15].

Despite its advantages, the RNN-T imposes a higher memory demand compared to AED and CTC methods [10,16]. During the forward-backward pass of the RNN-T, posteriors are calculated at each point within the grid composed of the encoder and prediction network. Computing a long decoding sequence in this grid consumes more memory and time than the aforementioned methods, making the vocabulary less dependent on training/inference speech and more reliant on sequence length [17–19]. Therefore, reducing memory consumption and improving computing speed are crucial for deploying RNN-T models on low-resource devices [16,20,21].

This paper proposes a model, named the time-sparse transducer (TST), designed to address the memory cost and computing time consumption of the RNN-T. Our approach consists of a convolutional front end, an acoustic encoder, a time-sparse mechanism, and a decoder. The encoder maps input acoustic frames into high-level representations, while the decoder, analogous to a conventional language model, combines these representations to produce a distribution over the output target through a softmax layer. The prediction network and joint network collectively form the decoder. The time-sparse mechanism reduces the time resolution by decomposing the encoder's hidden states into intermediate encoded representations using a sliding pooling window. These representations are then combined into sparse hidden states using a weighted average algorithm, and subsequently fed to the joint network. As a result, the sequence length of the sparse hidden states outputted by the time-sparse mechanism is significantly smaller than that of the encoder. This compression in the length of hidden states effectively reduces the GPU memory footprint and computing time. Additionally, introducing an attention mechanism [5] during the combination of intermediate encoded representations enhances the coefficients of representations with valuable information and suppresses noisy representations. Furthermore, the attention coefficients contribute to a lower character error rate (CER) for our model compared to the RNN-T baseline, as demonstrated through experiments conducted on the AISHELL-1 dataset.

The remaining sections of this paper are organized as follows: Sect. 2 provides an overview of the RNN-T method, while Sect. 3 describes the structure of the time-sparse mechanism and highlights key considerations during the generation of sparse representations. This section also presents the strategies for generating the weighted average coefficients. In Sect. 4, we detail the experiments conducted and their respective results. Finally, in Sect. 5, we present our conclusions, summarizing the key aspects and effects of the TST model.

2   Background

Our proposed approach is based on the Recurrent Neural Network Transducer (RNN-T) model. In this section, we provide an overview of the RNN-T structure, training strategy, and decoding process [22].

The RNN-T consists of two distinct networks: the acoustic encoder and the prediction network, which are connected through the joint network. The acoustic encoder maps an input frame $\mathbf{x}_t$ to a hidden state vector $\mathbf{h}_t$. The linguistic state vector $\mathbf{g}_u$ is generated by appending the prediction "non-blank" symbol from the previous time step to the prediction network. The joint network is a feed-forward network that combines the hidden state vector $\mathbf{h}_t$ and the linguistic state vector $\mathbf{g}_u$ as follows:

$$\mathbf{z}_{t,u} = f_{Activate}(W\mathbf{h}_t + V\mathbf{g}_u + b) \tag{1}$$

Here, W and V are weight matrices, b is a bias vector, and $f_{Activate}$ represents an activation function such as Tanh or ReLU. The output $\mathbf{z}_{t,u}$ is then linearly transformed:

$$\mathbf{m}_{t,u} = \text{Linear}(\mathbf{z}_{t,u}) \tag{2}$$

To obtain the posterior probability distribution $p(k|t, u)$ over the next output symbol, a softmax function is applied to $\mathbf{m}_{t,u}$:

$$p(k|t, u) = \text{Softmax}(\mathbf{m}_{t,u}) \tag{3}$$

The probability distribution is computed at each point in the grid formed by the prediction network and the acoustic encoder. The RNN-T model employs the forward-backward algorithm to sum the probabilities of all possible paths. However, this processing approach results in significant memory requirements [10]. The loss function for RNN-T is defined as the negative log-likelihood of the target sequence $y*$:

$$\mathcal{L}_{RNN-T} = -\ln(P(y * | x)) \tag{4}$$

In terms of inference, the RNN-T performs frame-by-frame computation, which can be slow when processing long sequences. The decoder employs beam search and greedy search methods to identify the most likely sequence as the output of the network [8].

Overall, the RNN-T model exhibits a distinctive architecture, comprising an acoustic encoder, prediction network, and joint network. The forward-backward algorithm is employed to compute probabilities, while the negative log-likelihood serves as the loss function. The inference process can be time-consuming for long sequences, and decoding methods like beam search and greedy search are employed to obtain the output sequence (Fig. 1).

3   Methodology

Our proposed method, Time-Sparse Transformer (TST), is based on the decomposition of encoded hidden states using a time-sparse mechanism. The time-sparse mechanism consists of two components: a decomposition of hidden states based on a sliding pooling window and a combination process of sparse hidden states based on the weighted average algorithm. In this section, we provide detailed explanations of each component.

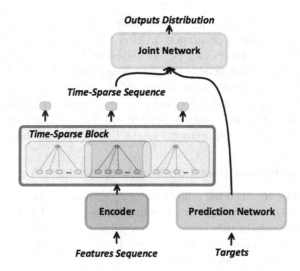

Fig. 1. Illustration of the structure of our proposed model time-sparse transducer. Compared with the RNN-T, a sparse block in time is introduced between the encoder and the joint network as shown in the figure. All intermediate representations are generated by sliding a window, which is shown in the Time-Sparse Block of this figure, on hidden states with two hyperparameters: window length and stride. Both of them can be set before the training process and would make observed disparity on the reasoning output. After that, the time-sparse block combines the encoded representations through the weighted average algorithm and feeds them into the joint network. The sequence length of the sparse hidden state is reduced after this process.

3.1 The Decomposition Based on Sliding Pooling Window

We first reduce the time resolution by decomposing the encoded hidden states into intermediate encoded representations by the TST algorithm. In the time-sparse mechanism, the intermediate representations are generated by sliding a window on hidden states with fixed window length and stride. Through this process, the information carried by the output of the encoder can be spread over various encoded representations r_k with smaller time resolution than hidden states as Eq. 5:

$$[\mathbf{r_1}, \mathbf{r_2}, \mathbf{r_3}, ... \mathbf{r_k}, ... \mathbf{r_n}] = f_{win}(\mathbf{x}, input, length, stride) \qquad (5)$$

where the $\mathbf{x}$ and $\mathbf{r_k}$ are the input and intermediate encoded representation output by decomposition respectively and n is the number of intermediate representations. The window can overlap partly during sliding to ensure continuous information between each representation. The pooling process is primarily affected by two factors, namely, the window length affecting the size of the sliding window, and the window stride affecting the size of the overlap between windows. The time resolution of the intermediate encoded representation decreases as the window length and stride size increase. So a large window length and stride size will cause the loss of detailed information but retain more global information when the encoder output is decomposed. After the decomposition, the intermediate encoded representations can be calculated with less computational effort in a shorter time due to the smaller sequence length than encoded hidden states. Conversely, setting a smaller window length and stride size can retain more detailed information, but also generate representations with greater sequence length than setting smaller sequence length and size, thus increasing the processing time and memory requirements of the time-sparse mechanism.

3.2   The Combination Based on Weighted Average Algorithm

After decomposing the encoded hidden states by sliding window, we combine the encoded representations through the weighted average algorithm and feed them into the joint network. In this process, the weighted average algorithm does not need to change the sequence length of the input and output, so that the time resolution of the sparse hidden state fed into the joint network by TST is much smaller than that of the hidden state fed into the joint network by RNN-T.

Combine with Absolute Average. In weighted average combination, an intuitive way is that all weight coefficients are initialized to be identical, as shown in Eq. 6.

$$h_t' = \sum_{k=1}^{n} \frac{1}{n} \cdot r_k \qquad (6)$$

This method reduces the time it consumes to calculate the coefficient but ignores the difference in the information carried by intermediate encoded representations.

Combine with Learnable Coefficients. In addition to the absolute average, the weight coefficient can also be initialized with a set of random coefficients and jointly optimized with other parameters through the RNN-T loss function during model training [23]. This process is illuminated as

$$h_t' = \sum_{k=1}^{n} w_k \cdot r_k \qquad (7)$$

where w_k is the learnable coefficient. The precision of the TST can be improved by increasing the coefficient of the intermediate encoded representation with the information that has a positive impact on the prediction and suppressing the coefficient of the representation that carries noise information.

Combine with Attention Mechanism. Inspired by the self-attention mechanism [5], we introduce the attention mechanism to calculate the weighted average coefficients. The attention weights α_k are computed as follows:

$$e_k = \text{Linear}(r) \qquad (8)$$

$$\alpha_k = \text{Softmax}(e_k) \tag{9}$$

$$\mathbf{h}_t^{'} = \sum_{k=1}^{n} \alpha_k \cdot \mathbf{r}_k \tag{10}$$

In this case, the attention mechanism allows the model to focus more on positive information. By calculating the attention weights, TST ensures that the intermediate encoded representations with significant contributions are given higher coefficients in the combination process.

In conclusion, the decomposition of the encoded hidden states and the combination of intermediate encoded representations occur only between the encoder and the joint network. From the decoder's perspective, the features inputted to the decoder in TST are indistinguishable from those in RNN-T. Therefore, any decoder used for RNN-T can be employed for TST.

4   Experiments and Results

4.1   Experimental Setup

All of our experiments are conducted on a public Mandarin speech corpus AISHELL-1 [24]. We use 80-dimension Mel-filter Bank coefficients (FBank) features with 3-dim pitch features computed on 25 ms window with 10 ms shift, which is known to be effective in Mandarin speech recognition. The 4234 characters (including a padding symbol <PAD>, an unknown token <UNK>, a begin-of-sequence token <BOS> and an end-of-sequence token <EOS>) are chosen as modeling units.

For the baseline RNN-T model, the front-end convolutional block followed by the encoder consists of two 2D-Convolution layers with a ReLU activation, stride size 2, channels 384, kernel size 3, and output size 384. The acoustic encoder consists of 12 transformer blocks with 4 heads in multi-head attention. The feed-forward size of the encoder is 384 and the hidden size is 768. We utilize three types of decoders. The first is a transformer decoder with 6 blocks and 4 heads in multi-head attention [6,25,26]; The second is a state-less decoder [27], and the third is an RNN decoder with 2-layer Long Short-Term Memory (LSTM) model. The configuration of TST is the same as RNN-T, except for introducing a time-sparse mechanism between the RNN-T encoder and joint network. Experiments are conducted on the sliding pooling window length from 1 to 10 with fixed window stride 1 and stride from 1 to 10 with window length 10. The generation strategies of intermediate encoded representation coefficients are absolute equality (AE), optimization of the random initialization coefficients through the RNN-T loss function (LC), and calculation through the self-attention mechanism (SA). We utilize the CER to evaluate the accuracy of different models and the real-time factor (RTF) to evaluate the inference speed.

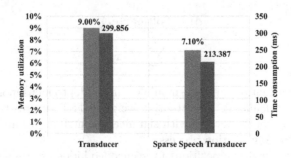

Fig. 2. Illumination of the GPU memory utilization and time consumption of RNN-T and TST on batch data. Both experiments are conducted on a Tesla K80 GPU.

Table 1. Comparison of our TST with the RNN-T after the downsampling process along time. To illustrate the disparity between before and after time sparing, we also show the performance of the commonly used RNN-T as the baseline. The time resolutions of both TST and downsampling RNN-T are reduced to 1/10 that of the RNN-T.

Model	RNN-T	Downsampling	TST		
			AE	LC	SA
CER (%)	7.824	43.926	15.784	13.411	12.760
RTF	0.122	0.019	0.021	0.021	0.021

4.2 Results

The Comparison of GPU Consumption Between RNN-T and TST. We conducted a comparative analysis of GPU memory utilization and time consumption between RNN-T and TST, as depicted in Fig. 2. The primary objective was to evaluate the impact of the time-sparse mechanism on these performance metrics. Our results revealed that the utilization of GPU memory decreased from 9% to 7.1% when utilizing TST instead of RNN-T. Additionally, the time consumption exhibited a noticeable improvement, reducing from 299.856 ms to 213.387 ms when processing batch data. These findings clearly demonstrate that TST achieves lower GPU occupancy and faster computation speed compared to RNN-T. By incorporating the time-sparse mechanism into RNN-T, we effectively reduce the sequence length of the hidden states received by the prediction network. Consequently, this reduction in sequence length leads to decreased GPU occupancy and significantly reduces the time required for subsequent computations.

The Influence of Downsampling in the Encoded Hidden State of TST and RNN-T. In this section, we evaluate the performance of TST and RNN-T when the time resolution is reduced. Table 1 presents the results, highlighting that there is no significant difference in terms of Real-Time Factor (RTF) between TST and RNN-T. Both models achieve a similar RTF reduction, approx-

Table 2. Comparison of TST with different sliding pooling window lengths. The window stride of all experiments is 1.

Window Length	AE		LC		SA	
	CER (%)	RTF	CER (%)	RTF	CER (%)	RTF
10	9.454	0.160	9.213	0.161	8.714	0.158
8	9.225	0.162	8.521	0.163	8.231	0.159
6	9.013	0.160	8.244	0.163	7.942	0.160
4	8.562	0.161	7.685	0.162	**7.418**	0.161
2	8.259	0.162	7.812	0.164	7.533	0.159
1	8.205	0.162	8.139	0.163	**7.824**	0.159

Table 3. Comparison of TST with different sliding pooling window strides. The window length of all experiments is 10.

Window Stride	AE		LC		SA	
	CER (%)	RTF	CER (%)	RTF	CER (%)	RTF
10	15.784	0.021	13.411	0.022	12.760	**0.021**
8	13.598	0.039	11.531	0.039	11.320	0.039
6	11.347	0.050	10.338	0.050	9.672	0.049
4	10.442	0.061	9.866	0.061	8.831	0.060
2	10.215	0.093	9.734	0.093	8.799	0.093
1	9.454	0.160	9.213	0.161	8.714	0.158

imately to 1/6 of the original value, when the time resolution is decreased to 1/10. However, it is important to note that downsampling the data along the time axis leads to a substantial loss in accuracy, amounting to 36.102%.

Regarding our TST, it incorporates a time-sparse mechanism, enhanced by a self-attention block, which effectively mitigates the accuracy loss. Specifically, TST achieves a remarkable reduction in accuracy loss to 4.936%. The introduction of the time-sparse mechanism, with its self-attention component, enables TST to better preserve the relevant information in the sparse representations, thus significantly minimizing the adverse effects of downsampling on accuracy.

The Influence of the Sliding Pooling Window Length and Stride on the Model Performance. This section presents a comparative analysis of models employing different sliding pooling window lengths and stride sizes. The experimental results, as depicted in Tables 2 and 3, clearly demonstrate that reducing the window length and stride size improves the accuracy of the models. When a large window with a large step size slides over the encoded hidden state, it leads to the loss of more detailed information. Conversely, a small window with a small step size preserves more information but results in a longer sequence

Table 4. The Result of RNN-T and TST with different decoders. Both window length and window stride are 4. The window type is SA.

Model	CER (%)	RTF
RNNT-T	7.979	0.508
TST-T (Ours)	**7.744**	**0.285**
RNNT-S	7.824	0.122
TST-S (Ours)	**7.528**	**0.061**
RNNT-R	11.454	0.397
TST-R (Ours)	**10.193**	**0.222**

length. The outcomes reveal that, for TST, the weighted average combination approach effectively reduces the CER when the window length is less than 4, accomplishing this by increasing the weight assigned to positive information while suppressing the weight assigned to noise. Moreover, employing the SA yields the lowest CER among the tested approaches. However, it should be noted that an increase in window length diminishes the information content, thereby decreasing the accuracy of TST. Notably, the tables demonstrate that the RTF remains largely unaffected by changes in window length but decreases as the stride size increases. This observation highlights that the inference speed of TST primarily depends on the decoding sequence length rather than the scale of intermediate encoded representation.

The Experimental Results of Different Window Types. This section primarily investigates the impact of different strategies employed for generating weighted average coefficients within the Time-Sparse Transformer (TST) framework. Notably, these strategies have no bearing on the decoding sequence length, thereby ensuring consistent computational speed for TST. The experimental findings presented in Tables 1, 2, and 3 provide compelling evidence regarding the efficacy of the SA strategy, which yields the lowest CER on the AISHELL-1 dataset. Comparing the LC approach with the SA strategy, it becomes evident that the inclusion of the attention mechanism significantly aids the model in learning the weighted average coefficients. Specifically, the self-attention mechanism strengthens the coefficient associated with intermediate encoded representations that convey pertinent information while simultaneously diminishing the coefficient assigned to representations containing noise. Consequently, the resulting sparse hidden state exhibits enhanced recognition capabilities pertinent to the target task. These results shed light on the discriminative power and adaptability provided by the self-attention mechanism within the TST framework, reinforcing its effectiveness for optimizing speech recognition performance.

Table 5. The results of RNN-T and TST with different decoders. Both window length and window stride are 10. The window type is SA.

Model	CER (%)	RTF
RNNT-T	7.979	0.508
TST-T (Ours)	12.934	**0.084**
RNNT-S	7.824	0.122
TST-S (Ours)	12.760	**0.021**
RNNT-R	11.454	0.397
TST-R (Ours)	15.968	**0.089**

The Experimental Results of Sliding Pooling Window with Different Decoders. This section focuses on comparing the performance of TST and RNN-T with different decoders, namely RNN-T with Transformer decoder (RNNT-T), RNN-T with State-Less decoder (RNNT-S), RNN-T with RNN decoder (RNNT-R), TST with Transformer decoder (TST-T), TST with State-Less decoder (TST-S), and TST with RNN decoder (TST-R). Experiments were conducted to evaluate their performance in terms of CER and recognition speed.

Table 4 presents the results, indicating that TST consistently achieves a lower CER and higher recognition speed compared to RNN-T in all three types of decoder. Specifically, when comparing TST-S with RNNT-S, it was found that the CER of TST-S is very close to that of RNN-T, while the RTF is reduced to 50% of RNNT-S. These findings highlight that TST-S achieves comparable accuracy to RNN-T while significantly improving computational efficiency.

Furthermore, Table 5 provides insights when the time resolution is reduced to 1/10. The CERs of models with different decoders are observed to increase, but the RTFs are reduced to 16.535% to 22.418% of the original values. Notably, among all the experiments, TST-S achieves the highest accuracy and its RTF is reduced to 17.213% of that of RNNT-S. These results demonstrate that TST offers the potential to improve both recognition accuracy and computational efficiency across various decoder types. Moreover, when the time resolution is decreased (see Table 4), TST maintains superior performance in terms of accuracy and computational efficiency compared to RNN-T.

5 Conclusion

In this study, we have introduced a novel model called the time-sparse transducer, which incorporates a time-sparse mechanism between the recurrent neural network (RNN) transducer encoder and the prediction network. Our proposed model offers several advantages compared to conventional transducers, including reduced memory consumption and accelerated computation through the reduction of time resolution in the encoded hidden state. Through our experimentation on the AISHELL-1 dataset, we have observed that the incorporation of the time-sparse mechanism in the prediction phase leads to a notable decrease in the character error rate. This finding underscores the efficacy of the time-sparse transducer in enhancing recognition accuracy and reducing inference time consumption when employed in conjunction with various decoders. While this study primarily focused on the development and evaluation of the time-sparse transducer, there are avenues for future research that warrant exploration. Specifically, we intend to investigate optimizing strategies for weighted average coefficients on diverse speech corpora. Such exploration will enable us to better understand the implications and generalized ability of our proposed model across different linguistic contexts.

References

1. Graves, A., Fernández, S., Gomez, F., Schmidhuber, J.: Connectionist temporal classification: labelling unsegmented sequence data with recurrent neural networks. In: Proceedings of the 23rd International Conference on Machine Learning, pp. 369–376 (2006)
2. Graves, A., Jaitly, N.: Towards end-to-end speech recognition with recurrent neural networks. In: International Conference on Machine Learning, pp. 1764–1772. PMLR (2014)
3. Amodei, D., et al.: Deep speech 2: end-to-end speech recognition in English and mandarin. In: International Conference on Machine Learning, pp. 173–182. PMLR (2016)
4. Bahdanau, D., Cho, K., Bengio, Y.: Neural machine translation by jointly learning to align and translate. arXiv preprint arXiv:1409.0473 (2014)
5. Vaswani, A., et al.: Attention is all you need. Advances in neural information processing systems 30 (2017)
6. Dong, L., Xu, S., Xu, B.: Speech-transformer: a no-recurrence sequence-to-sequence model for speech recognition. In: 2018 IEEE International Conference on Acoustics, Speech and Signal Processing (ICASSP), pp. 5884–5888. IEEE (2018)
7. Kim, S., Hori, T., Watanabe, S.: Joint ctc-attention based end-to-end speech recognition using multi-task learning. In: 2017 IEEE International Conference on Acoustics, Speech and Signal Processing (ICASSP), pp. 4835–4839. IEEE (2017)
8. Graves, A.: Sequence transduction with recurrent neural networks. arXiv preprint arXiv:1211.3711 (2012)
9. Graves, A., Mohamed, A.r., Hinton, G.: Speech recognition with deep recurrent neural networks. In: 2013 IEEE International Conference on Acoustics, Speech and Signal Processing, pp. 6645–6649. IEEE (2013)
10. Rao, K., Sak, H., Prabhavalkar, R.: Exploring architectures, data and units for streaming end-to-end speech recognition with rnn-transducer. In: 2017 IEEE Automatic Speech Recognition and Understanding Workshop (ASRU), pp. 193–199. IEEE (2017)
11. He, Y., et al.: Streaming end-to-end speech recognition for mobile devices. In: ICASSP 2019–2019 IEEE International Conference on Acoustics, Speech and Signal Processing (ICASSP), pp. 6381–6385. IEEE (2019)
12. Han, W., et al.: Contextnet: improving convolutional neural networks for automatic speech recognition with global context. ArXiv abs/2005.03191 (2020)

13. Zhang, Q., et al.: Transformer transducer: A streamable speech recognition model with transformer encoders and rnn-t loss. In: ICASSP 2020–2020 IEEE International Conference on Acoustics, Speech and Signal Processing (ICASSP), pp. 7829–7833 (2020). https://doi.org/10.1109/ICASSP40776.2020.9053896

14. Kannan, A., et al.: Large-scale multilingual speech recognition with a streaming end-to-end model. arXiv preprint arXiv:1909.05330 (2019)

15. Variani, E., Rybach, D., Allauzen, C., Riley, M.: Hybrid autoregressive transducer (hat). In: ICASSP 2020–2020 IEEE International Conference on Acoustics, Speech and Signal Processing (ICASSP), pp. 6139–6143 (2020). https://doi.org/10.1109/ICASSP40776.2020.9053600

16. Li, J., Zhao, R., Hu, H., Gong, Y.: Improving RNN transducer modeling for end-to-end speech recognition. In: 2019 IEEE Automatic Speech Recognition and Understanding Workshop (ASRU), pp. 114–121. IEEE (2019)

17. Venkatesh, G., et al.: Memory-efficient speech recognition on smart devices. In: ICASSP 2021–2021 IEEE International Conference on Acoustics, Speech and Signal Processing (ICASSP), pp. 8368–8372 (2021). https://doi.org/10.1109/ICASSP39728.2021.9414502

18. Han, Y., Zhang, C., Li, X., Liu, Y., Wu, X.: Query-based composition for large-scale language model in lvcsr. In: 2014 IEEE International Conference on Acoustics, Speech and Signal Processing (ICASSP), pp. 4898–4902 (2014). https://doi.org/10.1109/ICASSP.2014.6854533

19. Zhang, Y., Sun, S., Ma, L.: Tiny transducer: a highly-efficient speech recognition model on edge devices. In: ICASSP 2021–2021 IEEE International Conference on Acoustics, Speech and Signal Processing (ICASSP), pp. 6024–6028 (2021). https://doi.org/10.1109/ICASSP39728.2021.9413854

20. Chen, X., Wu, Y., Wang, Z., Liu, S., Li, J.: Developing real-time streaming transformer transducer for speech recognition on large-scale dataset. In: ICASSP 2021–2021 IEEE International Conference on Acoustics, Speech and Signal Processing (ICASSP). pp. 5904–5908 (2021). https://doi.org/10.1109/ICASSP39728.2021.9413535

21. Kim, C., et al.: A review of on-device fully neural end-to-end automatic speech recognition algorithms. In: 2020 54th Asilomar Conference on Signals, Systems, and Computers, pp. 277–283 (2020). https://doi.org/10.1109/IEEECONF51394.2020.9443456

22. Tian, Z., Yi, J., Tao, J., Zhang, S., Wen, Z.: Hybrid autoregressive and non-autoregressive transformer models for speech recognition. IEEE Signal Process. Lett. **29**, 762–766 (2022)

23. Ostmeyer, J., Cowell, L.: Machine learning on sequential data using a recurrent weighted average. Neurocomputing **331**, 281–288 (2019)

24. Bu, H., Du, J., Na, X., Wu, B., Zheng, H.: Aishell-1: an open-source mandarin speech corpus and a speech recognition baseline. In: 2017 20th Conference of the Oriental Chapter of the International Coordinating Committee on Speech Databases and Speech I/O Systems and Assessment (O-COCOSDA), pp. 1–5. IEEE (2017)

25. Moritz, N., Hori, T., Le, J.: Streaming automatic speech recognition with the transformer model. In: ICASSP 2020–2020 IEEE International Conference on Acoustics, Speech and Signal Processing (ICASSP), pp. 6074–6078. IEEE (2020)

26. Tian, Z., Yi, J., Tao, J., Bai, Y., Zhang, S., Wen, Z.: Spike-triggered non-autoregressive transformer for end-to-end speech recognition. arXiv preprint arXiv:2005.07903 (2020)
27. Ghodsi, M., Liu, X., Apfel, J., Cabrera, R., Weinstein, E.: Rnn-transducer with stateless prediction network. In: ICASSP 2020–2020 IEEE International Conference on Acoustics, Speech and Signal Processing (ICASSP), pp. 7049–7053. IEEE (2020)

Enhancing Daily Life Through an Interactive Desktop Robotics System

Yuhang Zheng[1], Qiyao Wang[2], Chengliang Zhong[2,3](✉), He Liang[2], Zhengxiao Han[4], and Yupeng Zheng[5]

[1] Beihang University, Beijing, China
[2] Tsinghua University, Beijing, China
zhongcl19@mails.tsinghua.edu.cn
[3] Xi'an High-Tech Research Institution, Xi'an, China
[4] Beijing University of Chemical Technology, Beijing, China
[5] Institute of Automation, Chinese Academy of Sciences, Beijing, China

Abstract. In this demo, we develop an intelligent desktop operating robot designed to assist humans in their daily lives by comprehending natural language with large language models and performing a variety of desktop-related tasks. The robot's capabilities include organizing cluttered objects on tables, such as dining tables or office desks, placing them into storage cabinets, as well as retrieving specific items from drawers upon request. This paper provides the design, development, and functionality of our robotics system, highlighting its advanced language understanding capabilities, perception algorithms, and manipulation techniques. Through real-world experiments and user evaluations, we demonstrate the effectiveness and practicality of our robotic companion in assisting individuals with everyday desktop tasks.

Keywords: Table organization · Natural language processing · Robotic perception and manipulation

1 Introduction

In recent years, there has been an increasing interest in the development of robotic systems aimed at providing assistance to humans in their daily lives, thereby enhancing convenience, productivity, and overall well-being [1–4]. In this demonstration, our focus lies on the task of table rearrangement: an intelligent robot designed to assist individuals in moving every object on the table to its specific location. This concise paper offers a comprehensive overview of the design, development, and functionality of our robotic companion, aiming to

---

Y. Zheng and Q. Wang—Contribute equally to this work.

Supplementary Information The online version contains supplementary material available at https://doi.org/10.1007/978-981-99-9119-8_8.

L. Fang et al. (Eds.): CICAI 2023, LNAI 14474, pp. 81–86, 2024.
https://doi.org/10.1007/978-981-99-9119-8_8

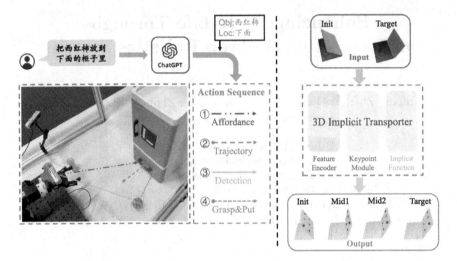

Fig. 1. The architecture of our robotics system. Our speech recognition module analyzes input speech to identify the objects to be manipulated and their target places. Using this information, our robot arm executes a sequence of actions to complete the organization task. The most challenging aspect lies in predicting the trajectory of opening the cabinet door, for which we introduce a 3D Implicit Transporter network to effectively manipulate the articulated object, as depicted in the right panel.

revolutionize how we interact with our immediate workspace and elevate the quality of our lives.

One of the key considerations in designing this robotic system is ensuring its user-friendliness. Humans can effortlessly communicate with the robot, issuing commands and requesting specific actions to be performed. Additionally, the inherent clutter often encountered on desks, be it in home or office settings, presents considerable challenges in terms of automated organization and task management [5]. Furthermore, the system is required to rapidly adapt to novel scenarios, such as encountering new tables and unfamiliar objects.

Our robot has been meticulously engineered to tackle these challenges, offering a reliable and intuitive solution to handle a wide range of desktop-related tasks. Through harnessing advancements in natural language processing, computer vision, and robotic manipulation, this innovative robot is capable of seamlessly understanding user commands, accurately identifying and categorizing objects, and proficiently executing specified actions to enhance desktop organization. We evaluate our system quantitatively on a real-world task and find that our robot achieves a success rate of 87.3% on grasp and correctly places 90.0% of objects.

2   System Architecture

As shown in Fig. 1, the robot system consists of two parts: (1) speech recognition and (2) perception and manipulation. Detailed information on each component is described below.

2.1   Speech Recognition

Taking human speech as input, the speech recognition module first converts voice to text and then outputs the names of organized objects and the parts or locations of the cabinets. In the speech recognition module, we choose Microsoft Azure's speech recognition and speech synthesis cloud services and select Chat-GPT [6] as the dialogue interaction robot to achieve the integration of the intelligent interaction system. To enable the robot to flexibly answer various questions related to organizing objects on the desktop, it is necessary to send a human-written prompt to ChatGPT for online training after initializing the interaction system, and then start the dialogue and Q&A. The content of the prompt mainly includes **1)** the robot's responsibilities and main tasks to be completed; **2)** names of objects and the parts or locations of the cabinets; **3)** several example scenarios that illustrate how the robot answers human questions.

2.2   Perception and Manipulation

Affordance Prediction. Affordance refers to the potential uses or actions that a particular object or feature in the environment can offer, such as a door handle affording the action of grasping. The concept of affordance provides valuable semantic information for robot agents, as it enables them to understand how they can interact with their surroundings to perform various tasks. In particular, we leverage the notion of affordance to guide our robot arm's grasping actions. To achieve this, we employ AffCorrs, a one-shot transfer method [7], to generate a grasping affordance map based on a source image containing labeled regions for grasping and an image sequence to be labeled. AffCorrs outputs the grasping region for each frame, which we then use in conjunction with aligned depth maps and camera intrinsic parameters to reconstruct a point cloud representing the location of the grasped object. This approach enables our robot arm to accurately locate and grasp objects in its environment.

Grasp Pose Planning. This function determines the best way for the gripper to hold and lift an object. To grasp the object, we need to obtain its position and estimate the grasping pose of the gripper. Here, we adopt the method of object detection. The system takes RGBD images captured by Realsense D455 as input and a YOLO-v7 network [8], trained with self-collected data, is utilized for detection. For grasping, we combined the detection results with the depth frames got previously to obtain the grasping pose.

Articulated Object Manipulation. Articulated objects play a ubiquitous role in our everyday lives, serving as storage units for various items [9]. Manipulating such objects presents a significant challenge due to the inherent shape variations and dynamic changes in their topology over time. In order to tackle this challenge, we have developed a novel network that leverages temporally consistent keypoints to infer the kinematic structure and movement of different parts. This network builds upon our previous work, denoted as 3D Implicit Transporter.

Our network operates on two input point clouds that capture the dynamic movement of object parts. The primary objective of our method is to reconstruct the shape of the target state by transporting explicit feature grids from the source state based on the predicted 3D keypoints. This process is carried out in a self-supervised manner. By leveraging temporally aligned keypoints, we are able to predict the direction of movement for the articulated object. In contrast to employing a single-step action to reach the desired target, our approach generates a series of sequential actions over an extended timeframe, gradually transforming the articulation state. The effectiveness of these long-horizon sequential actions is demonstrated in our accompanying demo.

In our system, we utilize MoveIt [10], the most popular motion planning framework, to control the movement of the robotic arm. To minimize any potential planning failures, we ensure that the robotic arm returns to its preset home position before every manipulation.

3   Results

We demonstrate the effectiveness of our system in both perception and manipulation tasks. For the object detection module, we achieve 98.2% map@50% on the collected dataset. Additionally, the average relative repeatability of the keypoint prediction is 83.1%. The robot can compute the moving direction with the temporally aligned keypoints, which enables it to perform the correct action to manipulate articulated objects and achieve a success rate of 87% in the Pybullet emulator.

The qualitative results are shown in Fig. 2. We show the key action of our robot arm after parsing the input speech. In addition, (a) and (b) shows the process of our robotic system searching for the target object and the grasping position respectively.

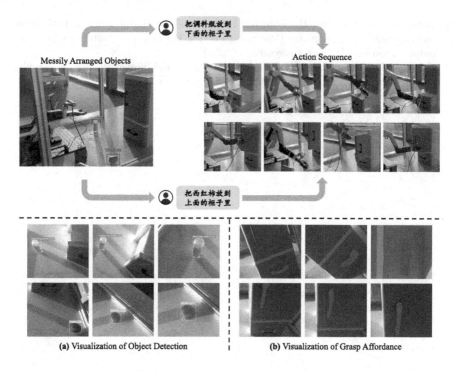

(a) Visualization of Object Detection (b) Visualization of Grasp Affordance

Fig. 2. The results of our robotics system. (a) and (b) visualize the process of our robotic system searching for the target object and the grasping position respectively.

References

1. Billard, A., Kragic, D.: Trends and challenges in robot manipulation. Science **364**(6446), eaat8414 (2019)
2. Shridhar, M., Manuelli, L., Fox, D.: Cliport: what and where pathways for robotic manipulation. In: Conference on Robot Learning. PMLR, pp. 894–906 (2022)
3. Wu, J., Antonova, R., Kan, A., et al.: Tidybot: personalized robot assistance with large language models. arXiv preprint arXiv:2305.05658 (2023)
4. Driess, D., Xia, F., Sajjadi, M.S.M., et al.: Palm-e: an embodied multimodal language model. arXiv preprint arXiv:2303.03378 (2023)
5. Liu, Z., Liu, W., Qin, Y., et al.: Ocrtoc: a cloud-based competition and benchmark for robotic grasping and manipulation. IEEE Robot. Autom. Lett. **7**(1), 486–493 (2021)
6. Ouyang, L., Wu, J., Jiang, X., et al.: Training language models to follow instructions with human feedback. Adv. Neural. Inf. Process. Syst. **35**, 27730–27744 (2022)
7. Hadjivelichkov, D., Zwane, S., et al.: One-shot transfer of affordance regions? affcorrs! In: Conference on Robot Learning. PMLR, pp. 550–560 (2023)
8. Wang, C.Y., Bochkovskiy, A., Liao, H.Y.M.: YOLOv7: trainable bag-of-freebies sets new state-of-the-art for real-time object detectors. In: Proceedings of the IEEE/CVF Conference on Computer Vision and Pattern Recognition, pp. 7464–7475 (2023)

9. Jiang, Z., Cheng-Chun, H., Zhu, Y.: Ditto: building digital twins of articulated objects from interaction. In: Proceedings of the IEEE/CVF Conference on Computer Vision and Pattern Recognition (2022)
10. Coleman, D., Sucan, I., Chitta, S., et al.: Reducing the barrier to entry of complex robotic software: a moveit! case study. arXiv preprint arXiv:1404.3785 (2014)

Model Distillation for Lane Detection on Car-Level Chips

Zixiong Wei[1], Zerun Wang[2], Hui Chen[2], Tianyu Shao[1], Lihong Huang[3], Xiaoyun Kang[4], and Xiang Tian[5(✉)]

[1] Zhuoxi Institute of Brain and Intelligence, Hangzhou, China
[2] Tsinghua University, Beijing, China
[3] HoloMatic Technology, Beijing, China
[4] Beijing Normal University, Beijing, China
[5] Zhejiang University, Hangzhou, China
tianx@zju.edu.cn

Abstract. As one of the most essential perception modules in the autonomous driving system, the lane detection module must accurately and efficiently detect each traffic lane's location, color and type to secure the vehicle's safety. State-of-the-art lane detection models have already shown great performance using deep and complex architecture. However, since the embedded system of car-level chips such as TDA4 supports limited computing resources and operators, these models could not be deployed directly. Besides, simply compressing models using a smaller backbone usually yields unsatisfied performance. In this work, to tackle these problems, we propose a lightweight and practical lane detection model based on model distillation. Specifically, our model learns the knowledge of a pre-trained teacher model using spatial softmax. We also re-train our model using the method of quantization aware training to further compress the size of model. Compared with the original PINet [1] model, experimental results on TDA4 demonstrate that our proposed method merely consumes a quarter of the parameters while enjoying a close detection precision of 0.7524 and an IoU score of 0.7621.

Keywords: Model distillation · Lane detection · Car-level Chips

1 Introduction

Lane detection is a challenging task in the autonomous driving system. It has broad application values in the real world including route planning, cruise control, lane-keeping, auto emergency, etc. To cope with various circumstances, such as different lanes, environments, cracks on the roads and lights, a robust and real-time lane detection method is intensively needed for safe driving.

Traditional lane detection methods extract features such as color, edges and shapes by using hand-crafted operators [2–4], and then apply statistics methods

Supported by Zhejiang Provincial Natural Science Foundation of China under Grant No. LDT23F01013F01.

such as Hough Transformation [5], Random Sampling Consensus [6] and Kalman filter [7] to fit the lane. Although these methods are simple and fast, they fail to maintain robustness in various traffic scenarios with different lighting and obstacles. Deep-learning-based lane detection methods have made great progress in both accuracy and effectiveness because of their powerful capacity to capture both the high-level and low-level features of lanes from the image [8,9]. For example, anchor-based [10,11] and parameter-based methods [12] have obtained outstanding detection accuracy with a fast inference speed. Recently, with the help of pixel-wise labeled images, segmentation-based methods [13,14] further enhance the detection performance by making compromises on efficiency. However, there still exist many challenges for detecting lanes accurately as well as deploying a deep model on car-level chips.

Since traffic lanes usually share similar features such as color and shape, it is not easy to distinguish them into different instances. A common solution is to find the key points of each lanes [13–16] and then apply some post-processing techniques [4,13] to cluster these points into different lane instances. However, these methods demand the number of lanes to be pre-defined and struggle in occasions when multiple lanes mix with each other. Another problem is that some complex and fancy methods [11,12] require too many parameters to be deployed on car-level chips such as TDA4 [19]. Even when some large models can be compressed into feasible sizes, they cannot be deployed due to unsupported operators and kernels.

In this work, we propose a lightweight lane detection model, dubbed Mixed Distilled Point Instance Network (MD-PINet), aiming to derive a lightweight model via knowledge transferred from an effective but heavy teacher model, i.e. PINet [1]. Specifically, we have a teacher network PINet with four hourglass modules as the backbone to fully utilize low-level, mid-level and high-level features. We choose PINet as our teacher network because most car-level chips support its simple operators and its stacked hourglass architecture allows self distillation. Unlike the common PINet, our MD-PINet is designed with only one hourglass module as its backbone, which is merely one-forth of the parameters of the larger teacher network. Once the teacher network is trained, both the four distillation (attention) layers of the four hourglass modules and the outputs of the last hourglass modules can be used as knowledge to supervise the training of MD-PINet. Therefore, we introduce a mixture of feature distillation and confidence distillation to enhance the performance of the student network with knowledge transferred from the teacher network. Moreover, to deploy our MD-PINet on the car-level chip, we further apply a quantization aware method to compress parameters from 32-bits to 8-bits. We show that our MD-PINet can be successfully deployed on TDA4 chip with little drop in performance after quantization.

The contributions of this study are summarized as follows:

- We propose a novel Mixed Distilled Point Instance Network dubbed MD-PINet in which a mixture of feature distillation and confidence distillation are introduced to transfer the knowledge from a teacher network to a student one. Thanks to the mixed distillation, we can obtain a lightweight but strong lane detection model for the autonomous driving system.

– To meet the requirement of model inference efficiency on the car-level chip, i.e., TDA4, we further accelerate the inference speed of MD-PINet by the quantization-aware training method.

– We systematically verify the effectiveness of MD-PINet and its quantized version on a real-world traffic lane dataset. We show that our MD-PINet can obtain comparable performance as the teacher network while enjoying much low inference speed, which forcefully demonstrates its effectiveness and superiority.

2 Related Work

2.1 Lane Detection

Existing traffic lane detection methods can be categorized into three different types of approaches: segmentation-based methods, parameter-based methods and anchor-based methods.

Segmentation-based lane detection methods [9,13–16] are most common and perform well. One typical model is Spatial Convolution Neural Network (SCNN) [9], in which each pixel manages to receive messages from every other pixel. This characteristic enables SCNN to build a spatial relationships between different lanes and hence solve the problem of the no-visual-evidence problem [17]. Another typical model, CurveLaneNAS [16], applies a neural network search method as a way to find the optimal network to detect different features. However, both models are impractical because they are cumbersome and too slow to meet the real-time inference requirement on car-level chips.

Unlike segmentation methods, parameter-based methods regress the lane curve equation directly without predicting key points of the lane. For example, LSTR model [12] applies transformer [18] to extract the high-level features and regress the equation of each lane. Even though the inference speed is fast, their performance is worse than segmentation-based approaches.

Anchor-based methods [10,11] apply anchor boxes to resolve no visual evidence problem. CondLaneNet [11] introduces the conditional convolution to predict the starting points of each lane. It then performs a row-wise anchor-based detection method to retrieve the whole lane from the starting point. One problem with the anchor-based method is that the use of anchors limits the shape of the lane, which might cause inaccuracy in special cases.

2.2 Knowledge Distillation

Knowledge distillation is a method to transfer knowledge from complex and powerful models to a small and weak model. Since the knowledge types play a crucial role in student learning, we discuss three types of knowledge: response-based knowledge, feature-based knowledge and relation-based knowledge.

Response-based distillation methods involve the response of the output layer of the teacher network. In this way, the student network can directly mimic the result [22] of the teacher network. A temperature method [22] is further adapted to enable the student network to learn the soft target with lower risks of over-fitting. Another category, feature-based distillation methods learn the intermediate feature map [23,24] as well as the response of the output layer of the teacher network. By learning hints from the feature maps, student network no longer merely mimics the result. The last category, relation-based distillation methods do not learn the outputs from any layer of teacher models. They learn the relationship between different feature maps using the flow of solution process (FSP) [25], which calculates the inner products between feature maps of different layers.

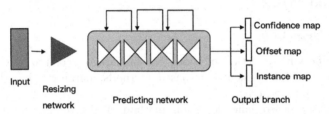

Fig. 1. Framework of the PINet. The input image is resized and fed into a backbone with four hourglass. Each hourglass module outputs a confidence map, an offset map and an instance map. The confidence map of previous hourglass module acts as the input of the next hourglass.

3   Method

Both our teacher and student networks are modified from PINet [1] which can be roughly demonstrated by Fig. 1. Section 3.1 describes the architecture of PINet in details. Section 3.2 proposes a mixed-distilled PINet (MD-PINet). Section 3.3 states the overall supervisions for MD-PINet. Section 3.4 demonstrates the model verification on car-level chips.

3.1   PINet

As shown in Fig. 1, PINet consists of three components, including a resizing network, a predicting network and an output branch. Any input image is fed into the resizing network and compressed to a size of 64×32 using three consecutive convolution layers. Each layer has a filter size of 3×3 and stride 2. Each convolution is equipped with a non-linear activation function, i.e. PReLU, and Batch Normalization. This procedure greatly saves memory and reduce the training time [1].

The predicting network contains four stacked hourglass modules. An hourglass module includes three parts, i.e., four down layers as the encoder, four up layers as the decoder and four distillation layers as the skip-connection.

Thanks to its encoder-decoder architecture, the hourglass module can effectively extracting both high-level and low-level features while comprehensively capturing the latent correspondence between them [20].

The output branch generates three 64×32 maps: a confidence map, an offset map and an instance map. Each cell of the maps represents a specific region of the original images. The confidence map predicts the probability that a key lane point exists in a cell. The offset map indicates exact locations of key points of traffic lanes. Finally, the instance map stores the embedding features of each lane instance.

3.2   Mixed Distilled PINet

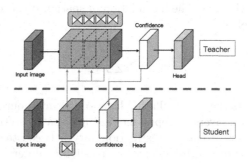

Fig. 2. Framework of MD-PINet. The green lines indicate the feature distillation while the blue line represents the confidence distillation (Color figure online)

The incentive of MD-PINet arises from the nature of the stacked hourglass modules of PINet. Since deep hourglass modules usually have more capacity to capture more informative features, they can enable self-distillation by transferring knowledge to those shallow hourglass ones. In other words, the performance of a lightweight one-hourglass student network can be improved by knowledge transferred from a deeper teacher network. Specifically, the four distillation layers in the deeper hourglass modules can be used as hints to supervise the feature activation of the shallow hourglass modules. However, since the teacher network learns more high-level features while the student network learns more low-level features, enforcing the student network to directly mimic the feature map of the teacher network can result in under-fitting. Besides, if directly regularizing the confidence map of the student network with the confidence map of the teacher network, it might cause over-fitting due to the failure of capturing intermediate-level supervision in the student network. Consequently, we propose to a mixed distillation strategy based on both the feature maps and the confidence map. We show that a mix of feature distillation and confidence distillation can be applied together to yield better detection.

As shown in Fig. 2, in the proposed MD-PINet, we stack four hourglass modules in the teacher network to effectively capture different levels of features. For the student network, we employ only one hourglass module to obtain a simple and lightweight model. The proposed feature distillation and the confidence distillation are performed between different hourglass modules accordingly.

The feature distillation provides instructive hints from the distillation layers of the teacher network for the student network. Since enforcing the student network to imitate the intermediate feature maps of the teacher network might lead to severe under-fitting, we need to use a loss function which provides hints rather than the feature map itself. Sufficient experiments have validated that spatial softmax [28] is an effective method to distill features between different layers of a hourglass modules [27]. For each channel of the feature map in the distilled layer, spatial softmax calculates the softmax over the spatial scope and then calculates the exact two-dimensional location of points, yielding the maximal activation for each channel. The formulation of spatial softmax in one hourglass module is shown below:

$$F(A_m) = S(\sum_{i=1}^{N} |A_{mi}|^2) \tag{1}$$

where A_{mi} represents the ith distillation layer of the mth hourglass and N is the number of distillation layers. S represents a single spatial softmax operation.

Since the purposes of feature distillation vary between teacher and student networks, their feature distillation losses are slightly different. In the teacher network, the purpose of self-distilled is to make the network deeper and hence improve the performance of the fourth hourglass [29]. In the student network, it only learns the last hourglass of the teacher network to yield better accuracy. Then, the feature distillation losses are calculated using the following equations:

$$L_{dist\_teacher} = \sum_{m}^{4} D(F(B_4) - F(B_m)) \tag{2}$$

$$L_{dist\_student} = D(F(B_4) - F(C_1)) \tag{3}$$

where B_m represents the mth hourglass of the teacher network and C_1 represents the only hourglass of the student network. D is the sum of square.

The confidence distillation requires the student network to mimic the confidence map of the teacher network. Since the value of confidence map indicates the probability that a key point of lane exists in the cell, we can treat it as a response-based distillation hints. The confidence distillation loss is calculated as follows:

$$L_{conf\_dist\_student} = \sum_{i=1}^{64} \sum_{j=1}^{32} |A_{ij} - B_{ij}|^2 \tag{4}$$

where A_{ij} represents the cell at column i and row j of the confidence map of the fourth hourglass of the teacher network. B_{ij} represents the cell at column i and row j of the confidence map of the student network.

3.3   Overall Supervisions for MD-PINet

The overall loss function contains four main opponents for teacher network and five main opponents for student networks. The loss function is inspired from YOLO [26] and modified from PINet [1]. The loss function of the teacher network is the same as PINet, while the loss function of the student network contains confidence loss, instance loss, offset loss, feature distillation loss and output distillation loss.

The confidence loss aims to penalizes the prediction error for cells with and without a lane key point simultaneously. The confidence loss is defined as follows:

$$L_{confidence} = \frac{1}{N_e} \sum_{C_c \in G_e} (C_c^* - C_c)^2 + \frac{1}{N_n} \sum_{C_c \in G_n, C_c > 0.01} (C_c^* - C_c)^2 \quad (5)$$

where N_e indicates the number of cell that contains a traffic lane key point. N_n indicates the number of cell without a traffic lane key point. G_e represents cells that contain a traffic lane key point and G_n represents cells without a traffic lane key point. C_c denotes to the predicted confidence value of each cell and C^* denotes the ground truth values which is 1 for key point cells and 0 for background cells.

The instance loss is essential to cluster key points into distinct lane instances. It is trained to make cells of the same instance have close embedding features. The instance loss is shown below:

$$L_{instance} = \frac{1}{N_e^2} \sum_i^{N_e} \sum_j^{N_e} l(i,j) \quad (6)$$

$$l(i,j) = \begin{cases} |F_i - F_j|_2, & if\ I_{ij} = 1 \\ max(0, 1 - |F_i - F_j|_2) & if\ I_{ij} = 0 \end{cases} \quad (7)$$

where F_i represents the predicted feature of a cell that contains a lane key point and F_j represents the predicted feature of cell that does not contain a lane key point. I_{ij} indicates whether cell i and cell j belongs to the same lane instances. When two cells belong to the same instance, the instance loss function calculates the difference of their features, otherwise this loss function increases their feature difference.

The offset loss helps find the exact location of key points in the cell. The offset losses on both the x-axis and the y-axis are calculated.

$$L_{offset} = \frac{1}{N_e} \sum_{C_x \in G_e} (C_x^* - C_x)^2 + \frac{1}{N_e} \sum_{C_y \in G_e} (C_y^* - C_y)^2 \quad (8)$$

Since our MD-PINet is also designed to detect attributes such as color and types, cross entropy losses for color and type are added accordingly.

The final loss function for the teacher network is:

$$L_{teacher} = 1.0 * L_{confidence} + 0.5 * L_{offset} + 0.5 * L_{instance}$$
$$+ 0.1 * L_{dist\_teacher} + 0.5 * L_{color} + 0.5 * L_{type}. \quad (9)$$

The final loss function for the student network is:

$$L_{student} = 1.0 * L_{confidence} + 0.5 * L_{offset} + 0.5 * L_{instance} +$$
$$0.2 * L_{dist\_student} + 0.05 * L_{conf\_dist\_student} + 0.5 * L_{color} + 0.5 * L_{type}. \quad (10)$$

We choose a larger weight for the confidence loss and smaller weights for the distillation losses to avoid the over-fitting.

3.4   Model Verification on Car-Level Chips

Since the primary goal of our model is to perform effectively and accurately on car-level chips, we build a lane detection verification system on TDA4. Once our MD-PINet has finished training on GPU, it first undergoes quantization aware training to compress model size. Then its operators and architecture are checked for its adaptability on TDA4. Finally it would be deployed and tested on TDA4.

4   Experiments

4.1   Dataset

The dataset is recorded and labelled by our team. The structure and labels of data are similar to CULane [9] and TuSimple [33]. Even though these public datasets have plenty of data, attributes such as color and lane type are not labelled. In order to fully utilize the detection capacity of our model, we collect and label our private dataset. The total number of data is 91820 with scenarios such as rainy, night, crowds, no lanes and cracked roads. An input image has a resolution of $1920 \times 1080 \times 3$ with key points and each pixel labelled. The type and color of each lane are also specified.

4.2   Evaluation Metrics

A positive lane detection label is judged by its IoU with the ground true lane. It needs a confidence score greater than 0.95 and a length longer than 5 pixels. Once a lane is predicted, we treat it as a 25-pixel wide polygon and set it as a positive detection if its IoU with ground truth is higher than 0.5. Then common lane detection evaluations such as precision, recall and IoU are calculated. We also calculate the pixel error using the following equation to examine the practicability of models:

$$distance\_error = \frac{\sum_{i=1}^{n} \sqrt{(A_{ix} - B_{ix})^2 + (A_{iy} - B_{iy})^2}}{n} \quad (11)$$

where A_{ix} indicates the x value of prediction i and B_{iy} indicates the y value of ground truth point that is closest to A_i.

4.3   Implementation Details

All input images are resized to 512×256, and RGB values are normalized between 0 and 1 before images are fed into the network. Then images are augmented using flipping, shadowing and Gaussian-noise as previous works [10,11]. We use one Tesla V100 (32G) GPU to train our model. During training, each batch contains 24 images. The learning rate is initialized as 1e-3 and updated by the Adam optimizer. Step learning rate decay [30] and early stop are applied to accelerate the training process. We also compress the proposed MD-PINet by the quantization aware method. The built-in function torch.quantization [31] can be applied to compress each parameter from 32-bit to 8-bit.

Table 1. Lane detection performance.

Method	Recall	Precision	IoU	Distance Error (pixel)	Parameter (M)
PINet (4H)	**0.6711**	**0.7592**	**0.7821**	**15.23**	5.31
PINet (1H)	0.6611	0.7440	0.7259	16.28	1.42
PINet (1H no distill)	0.6421	0.7303	0.7404	18.01	1.08
MD-PINet	**0.6694**	**0.7524**	**0.7621**	**15.52**	1.42

5   Results

Three Baselines to Compare. The first baseline, PINet(4H), is the original PINet with four hourglass modules. Then, we clip the first hourglass of the fully trained PINet(4H) to obtain the second baseline PINet(1H). Finally, we train a PINet with only one hourglass module without self-distillation as our last baseline PINet(1H no distill).

Quantitative Results. Quantitative results are shown in Table 1. Compared with the PINet(1H no distill), our MD-PINet greatly improves recall by 2.73%, precision by 2.21%, IoU by 2.17% and drops pixel error by 2.49 with merely 0.34M more parameters. Compared with the PINet(4H), our network compresses three quarters of parameters in compensation for barely a 0.17% drop in recall, a 0.68% drop in precision, a 2.0% drop in IoU and a 0.29 increase in pixel error. These results well manifest the effectiveness of our method.

Table 2. Lane attribute performance.

Method	Type	Color
PINet (1H with self-distillation)	0.8403	0.8441
MD-PINet (1H)	**0.8494**	**0.8767**

As information about the type an color of lanes is also vital in the autonomous driving system, it is notable that PINet can be easily trained to detect attributes of the lane using the same hourglass backbone by adding new heads. The experiment results in Table 2 indicates that MD-PINet can also achieve an obvious improvement in attributes as it does in lane detection. Our MD-PINet network remarkably improves the accuracy of lane color by 3.26% and lane type by 0.91% (Table 3).

Table 3. Impact of quantization on MD-PINet.

Method	Recall	Precision	IoU	Distance error (pixel)
MD-PINet	0.6694	0.7524	0.7621	15.52
MD-PINet(quantized)	0.6643	0.7271	0.6772	16.51
Changes	−0.0051	−0.0253	−0.0849	+0.99

Effects of Quantization. In order to be deployed on car-level chips, MD-PINet requires quantization to compress its model size. The results in Table refquant indicate that the effects of quantization on recall, precision and distance error are limited. Even though the IoU score decreases by 0.0849, its influence on the overall detection performance is negligible. Besides, we investigate the efficiency of the MD-PINet before and after quantization. As shown in Table 4, remarkably, quantization speeds up the post-processing time by 1.38, the inference time by 1.85 and the total time by 1.51. These results forcefully demonstrate that quantization greatly accelerates the operation speed of our network while maintaining a satisfactory performance.

Table 4. Inference time and post-processing time.

Procedure	Inference	Post-processing	Total
Before Quant (ms)	13.5	28.5	42.0
After Quant (ms)	7.3	20.6	27.9
Time save (ms)	6.2	7.9	14.1
Speedup	1.85	1.38	1.51

Visualization Analysis. The visualization results of MD-PINet and the original PINet in Fig. 3 show that our model manages to detect different line instances accurately. Both MD-PINet and PINet are able to detect lanes that are partially covered by obstacles. In a situation when most of a lane is covered (the third column), MD-PINet also performs extremely close to PINet(4H).

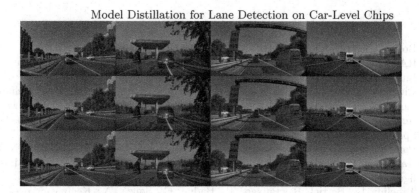

Fig. 3. Visualize results of the ground truth (first row), the MD-PINet (second row), and the original PINet (last row).

6   Conclusion

In this work, we propose MD-PINet, an efficient and effective lane detection method that is able to be deployed on car-level chips. We successfully transfer knowledge from a large and powerful teacher network to a small and weak student network by combining hint feature distillation and confidence map distillation. With the help of quantization aware training, we manage to deploy our model on TDA4 under strict memory and operator restrictions, achieving a sufficient detection speed. Experiment results show that our MD-PINet can consistently surpass the baseline model in terms of both the detection result and the inference speed, well demonstrating its effectiveness and possibility of being broadly deployed on large-scale business autonomous vehicles.

References

1. Ko, Y., Lee, Y., Azam, S.: Key points estimation and point instance segmentation approach for lane detection. IEEE Trans. Intell. Transp. Syst. (2020). IEEE
2. Jiang, Y., Gao, F., Xu, G.: Computer vision based multiple-lane detection on straight road and in a curve. In: International Conference on Image Analysis and Signal Processing, pp. 114–117 (2010)
3. Zhou, S., Jiang, Y., Xi, J., Gong, J.: A novel lane detection based on geometrical model and Gabor filter. In: IEEE Intelligent Vehicles Symposium (IV), pp. 59–64 (2010)
4. Kim, Z.: Robust lane detection and tracking in challenging scenarios. IEEE Trans. Intell. Transp. Syst. **9**(1), 16–26 (2008)
5. Liu, G., Worg, F., Markeli, I.: Combining statistical Hough transform and particle filter for robust lane detection and tracking. In: IEEE Intelligent Vehicles Symposium (IV), pp. 993–997 (2010)
6. Jiang, R., Klette, R., Vaudrey, T., Wang, S.: New lane model and distance transform for lane detection and tracking. In: International Conference on Computer Analysis of Images and Patterns, pp. 1044–1052 (2009)

7. Borkar, A., Hayes, M., Smith, M.T.: Robust lane detection and tracking with RANSAC and Kalman filter. In: 2009 16th IEEE International Conference on Image Processing, pp. 3261–3264, IEEE (2009)
8. Chougule, S., Koznek, N., Ismail, A., Adam, G., Narayan, V., Schulze, M.: Reliable multilane detection and classification by utilizing CNN as a regression network. In: Leal-Taixé, L., Roth, S. (eds.) ECCV 2018. LNCS, vol. 11133, pp. 740–752. Springer, Cham (2019). https://doi.org/10.1007/978-3-030-11021-5_46
9. Pan, X., Shi, J., Luo, P., Wang, X., Tang, X.: Spatial as deep: Spatial CNN for traffic scene understanding. In: Thirty-Second AAAI Conference on Artificial Intelligence (2018)
10. Tabelini, L., Berriel, R., Paixao, T., Badue, C., Souza, A.: Keep your eyes on the lane: real-time attention-guided lane detection. In: IEEE/CVF Conference on Computer Vision and Pattern Recognition, pp. 294–302 (2021)
11. Liu, L., Chen, X., Zhu, S., Tan, P.: CondLaneNet: a top-to-down lane detection framework based on conditional convolution. In: ICCV 2021 International Conference on Computer Vision (2021)
12. Liu, R., Yuan, Z.: End-to-end lane shape prediction with transformers. In: IEEE/CVF Winter Conference on Applications of Computer Vision, pp. 3694–3702 (2021)
13. Neven, D., Brabandere, B., Georgoulis, S., Proesmans, M., Gool, L.: Towards end-to-end lane detection: an instance segmentation approach. In: IEEE Intelligent Vehicles Symposium (IV), pp. 286–291 (2018)
14. Hou, Y., Ma, Z., Liu, C., Hui, T., Loy, C.: Inter-region affinity distillation for road marking segmentation. In: IEEE Conference on Computer Vision and Pattern Recognition, pp. 12486–12495 (2020)
15. Yoo, S., et al.: End-to-end lane marker detection via row-wise classification. In: IEEE Conference on Computer Vision and Pattern Recognition Workshops, pp. 1006–1007 (2020)
16. Xu, H., Wang, S., Cai, X., Zhang, W., Liang, X., Li, Z.: CurveLane-NAS: unifying lane-sensitive architecture search and adaptive point blending. In: Vedaldi, A., Bischof, H., Brox, T., Frahm, J.-M. (eds.) ECCV 2020. LNCS, vol. 12360, pp. 689–704. Springer, Cham (2020). https://doi.org/10.1007/978-3-030-58555-6_41
17. Tabelini, L., Berriel, R., Paixao, T., Badue, C., Souza, A., Santos, T.: Keep your eyes on the lane: attention-guided lane detection, arXiv preprint arXiv:2010.12035 (2020)
18. Vaswani, A., et al.: Attention is all you need. In: Advances in Neural Information Processing Systems (2017)
19. TI Deep Learning Library User Guide. https://software-dl.ti.com/jacinto7/esd/processor-sdk-rtos-jacinto7/06_02_00_21/exports/docs/tidl_j7_01_01_00_10/ti_dl.html. Accessed 10 Apr 2022
20. Newell, A., Yang, K., Deng, J.: Stacked hourglass networks for human pose estimation. In: Leibe, B., Matas, J., Sebe, N., Welling, M. (eds.) ECCV 2016. LNCS, vol. 9912, pp. 483–499. Springer, Cham (2016). https://doi.org/10.1007/978-3-319-46484-8_29
21. Yang, W., Li, S., Ouyang, W., Li, H., Wang, X.: Learning feature pyramids for human pose estimation. In: IEEE International Conference on Computer Vision, pp. 1281–1290 (2017)
22. Hinton, G., Vinyals, O., Dean, J.: Distilling the knowledge in a neural network. arXiv preprint arXiv:1503.02531 (2015)
23. Romero, A., Ballas, N., Kahou, S., Chassang, A., Gatta, C., Bengio, Y.: Fitnets: hints for thin deep nets. Computing Research Repository, arXiv:1412.6550 (2014)

24. Budnik, M., Avrithis, Y., Asymmetric metric learning for knowledge transfer, arXiv preprint arXiv:2006.16331 (2020)
25. Yim, J., Joo, D., Bae, J., Kim, J.: A gift from knowledge distillation: fast optimization, network minimization and transfer learning. In: 2017 IEEE Conference on Computer Vision and Pattern Recognition (CVPR), pp. 7130–7138 (2017). https://doi.org/10.1109/CVPR.2017.754
26. Redmon, J., Divvala, S.: You only look once: unified, real-time object detection. In: IEEE Conference on Computer Vision and Pattern Recognition, pp. 779–788 (2016)
27. Zagoruyko, S., Komodakis, N.: Paying more attention to attention: improving the performance of convolutional neural networks via attention transfer, arXiv preprint arXiv:1612.03928 (2016)
28. Levine, S., Finn, C., Darrell, T.: End-to-end training of deep visuomotor policies. J. Mach. Learn. Res. **17**, 1–40 (2016)
29. Hou, Y., Ma, Z.: Learning lightweight lane detection CNNs by self attention distillation. In: IEEE International Conference on Computer Vision, pp. 1013–1021 (2019)
30. Loshchilov, I., Hutter, F.: Decoupled weight decay regularization. In: International Conference on Learning Representations (ICLR) (2019)
31. Pytorch code for Torch.QUANTIZATION. https://pytorch.org/docs/stable/modules/torch/quantization.html. Accessed 10 Apr 2022
32. Wei, Y., Pan, X., Qin, H., Ouyang, W., Yan, J.: Quantization mimic: towards very tiny CNN for object detection. In: Ferrari, V., Hebert, M., Sminchisescu, C., Weiss, Y. (eds.) ECCV 2018. LNCS, vol. 11212, pp. 274–290. Springer, Cham (2018). https://doi.org/10.1007/978-3-030-01237-3_17
33. The tusimple lane challenge. http://benchmark.tusimple.ai/. Accessed 10 Apr 2022

A Weakly Supervised Learning Method for Recognizing Childhood Tic Disorders

Ruizhe Zhang[1], Xiaojing Xu[2], Zihao Bo[1], Junfeng Lyu[1], Yuchen Guo[3], and Feng Xu[1(✉)]

[1] School of Software and BNRist, Tsinghua University, Beijing, China
feng-xu@tsinghua.edu.cn
[2] China-Japan Friendship Hospital, Beijing, China
[3] BNRist, Tsinghua University, Beijing, China

Abstract. So far, the number of individuals with Tic disorder worldwide has reached 59 million, and the prevalence of the disorder is rapidly increasing globally. In this work, we focus on weakly supervised learning methods for recognizing childhood tic disorders. In situations with limited data availability, we design a relative probability metric based on the characteristics of the data and a multi-phase learning algorithm is proposed based on relative probability in order to efficiently utilize coarse-labeled data in a "from easy to difficult" manner. Furthermore, the effectiveness of our method is validated through ablation experiments. Through extensive experiments on the test dataset, we demonstrate that our method behaves extraordinarily compared to baseline approaches, improving AUC by 3.0%, and facilitating expedited diagnostic assessment for medical practitioners.

Keywords: tic disorders · facial data processing · weakly supervised learning

1 Introduction

Tic disorder [1, 2] is a motor or vocal muscle spasm characterized by symptoms such as frequent eye blinking, head jerking, facial distortions, repetitive coughing, and throat clearing. Diagnosing Tic disorder in clinical settings is typically a complex process, further complicated by the fact that the majority of affected individuals are children, who often have low cooperation, leading to diagnostic challenges. Research [10, 11] has primarily focused on pathology and clinical aspects over the past few decades, with limited studies on the identification and detection of tic disorder symptoms in patients.

In recent years, machine learning has been widely applied to medical problems, particularly in the areas of disease diagnosis and classification. Some studies have employed video-based action recognition to diagnose diseases. The mainstream approach for video action recognition is based on Convolutional Neural Networks (CNNs) [3, 5]. One popular approach is the two-stream architecture [16–18]. Another approach is the use of 3D CNNs [19–24] that can directly capture spatiotemporal information from video sequences. Furthermore, attention mechanisms [4] allow the model to allocate more attention to relevant parts of the video, improving both accuracy and efficiency. These

L. Fang et al. (Eds.): CICAI 2023, LNAI 14474, pp. 100–112, 2024.
https://doi.org/10.1007/978-981-99-9119-8_10

networks above need fully-supervised data but for the problem of recognizing tic disorder, data annotation requires professional doctors, which incurs high manpower costs and poses challenges in annotation. Moreover, our available labeled data is limited. Therefore, the fully supervised methods are not suitable for our research problem.

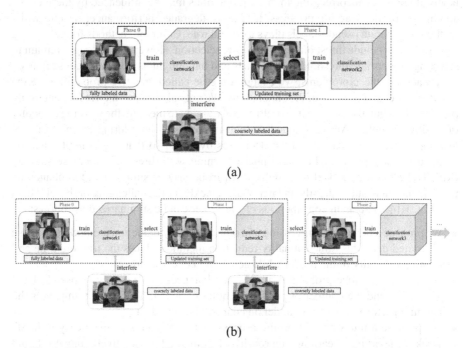

Fig. 1. Frameworks of one-phase training (a) and our multi-phase training method (b).

Weakly supervised learning [25–36] (WSL) is the method to solve this issue, which aims at improving the performance of models by exploiting many unlabeled data. Among various techniques in weakly supervised learning, pseudo-labeling methods have gained significant attention due to their effectiveness in leveraging unlabeled data. Pseudo-labeling is a technique that assigns labels to unlabeled data based on the predictions of a trained model. These assigned labels are considered "pseudo-labels" and are then used to augment the training set for further model refinement. Typically, a one-phase learning scheme in Fig. 1(a) is adopted.

However, it is insufficient for knowledge excavation to exploit the unlabeled data only once, so multi-phase learning comes out further enhances the performance of weakly supervised learning methods. Multi-phase learning, divides the weakly supervised learning process into multiple stages, each with a specific objective or set of labeled and unlabeled data. In each phase, the model is trained and pseudo-labels are generated based on the current phase's predictions. These pseudo-labels are then used as training data for the next stage, enabling the model to learn progressively and capture more complex patterns over successive stages.

But issues arise as a result of these methods. Firstly, As the model is trained with these pseudo-labels, it becomes biased towards making predictions that align with the labels generated in the previous phase, in which way, increasing the number of phases becomes meaningless for model performance improvement. Moreover, if the pseudo-labels are noisy or incorrect, the model's predictions may be influenced by these errors and hinder further learning progress, leading to degraded performance. Secondly, the method to generate pseudo-labels plays a crucial role. Common methods for generating pseudo-labels include thresholding and Top-K selection. However, challenges remain in selecting appropriate thresholds or K values and handling noisy or uncertain samples.

To address these problems, we design a metric called "relative probability" (RPr) based on the characteristics of the annotations. We not only use this metric for generating pseudo-labels but also involve it in multi-phase training to measure the learning difficulty of positive samples. An RPr-guided method is proposed, and during the multi-phase learning process, the RPr threshold decreases by phase so that easy samples can be selected in early phases. This multi-phase learning with thresholds decrease strategy (MPLTD) allows the model to initially learn from easy or simple data to enhance its performance, and subsequently, in later phases, tackle more challenging or difficult data (see Fig. 1(b)). The main contributions of this paper can be summarized as follows:

- We propose a facial data processing and dimensionality reduction method. In the case of limited data, this dimensionality reduction method reduces the training time and training difficulty of the model while achieving better accuracy.
- We design a relative probability metric that balances the accuracy of pseudo-label generation and the number of positive samples obtained. It effectively improves the learning performance of the model on coarsely labeled data.
- We propose a multi-phase learning process that implements a "from easy to hard" weakly supervised learning approach. This method is relatively universal and applicable.

2   Related Work

Tic disorder diagnosis has no great progress made in this area until the 2010s. In 2010, Bernabei et al. [10] conducted a study using wearable devices with accelerometers to detect twitching movements in the limbs and trunks of Tourette syndrome patients, achieving an accuracy of 80.5%. In 2016, Shute et al. [11] conducted research based on brain electrical stimulation and observed low-frequency central medial-prefrontal (CM-PF) activity to detect tic symptoms in patients.

Facial landmark detection is the process of automatically locating and identifying key points or landmarks on a human face. Cootes et al. [12] proposed the Active Appearance Models (AAM) which model the shape and texture of the face as random variables and estimate them through optimization methods, thus achieving facial landmark detection. Kazemi et al. [13] introduced a fast and accurate method for facial landmark detection based on ensemble models of regression trees, enabling rapid detection of facial landmarks. In 2015, Yang et al. [14] proposed a cascaded regression approach for robust facial landmark tracking. This method progressively improves the localization accuracy of facial landmarks by training a series of regressors in a cascade. Bulat et al. [15]

provided a review of 2D and 3D facial landmark detection problems and presented a large-scale 3D facial landmark dataset.

Weakly supervised learning focuses on developing algorithms and techniques to address the challenges of training machine learning models with limited or noisy supervision. The most important line of research in WSL explores methods for generating pseudo-labels [31, 33], which are inferred labels assigned to unlabeled data based on some heuristics or assumptions. These pseudo-labels are used to train the model in a semi-supervised [32–36] or self-supervised manner. Many works are devoted to semi-supervised these years, such as self-training, label propagation [29], and so on.

3 Method

In this section, we first define the problem of tic disorder recognition and classification. To address privacy concerns, we employ dimensionality reduction techniques to convert facial images into facial landmark points, thus preserving the privacy of the patients. Firstly, we train an initial model with fully labeled data. Then we use this model to generate pseudo-labels for the coarse labeled data, selecting reliable positive samples to be added to the training set. We retrain the model and repeat this process iteratively. In the pseudo-label generation step, we introduce the concept of relative probability, which ensures that the selected positive samples exhibit similar features to the most prominent movements in the long segments. For the iterative part, we propose a method to gradually decrease the threshold value so that the model initially learns from simple samples to improve accuracy and then focuses on difficult samples to enhance generalization.

3.1 Data Description

We collected a total of 129 videos from children with tic disorders. Based on the level of annotation detail, we divided all the videos into two categories: fully labeled videos (42 videos) and coarsely labeled videos (87 videos). The fully labeled videos consist of short segments, where each annotated segment has a length of 2 s. On the other hand, the coarsely labeled videos consist of long segments, where each segment has a length ranging from 3 to 10 s.

3.2 Tic Disorder Recognition Problem Definition

Let $\aleph$ be the set of all videos, where each video $X \in \aleph$ consists of several short segments $x_1, x_2, x_3, ..., x_N \in X$. Each short segment x_i is composed of several frames $a_1, a_2, a_3, ..., a_N \in x_i$ (usually 48 frames). Each frame a_i is the basic unit of our data processing, but not the basic unit for model prediction and tic recognition. The smallest unit of tics is the short segment x_i. In a video, each short segment can be one of the following: eye tic, mouth tic, nose tic, or normal. Among them, the first three can occur simultaneously, while the normal class can only occur alone.

We define the tic recognition task to determine whether a short segment x is a tic segment. In this task, we combine the tics in the eye, mouth, and nose regions as the positive class for binary classification, while the normal segments are the negative class. For convenience, we refer to it as "face binary classification" in the subsequent tables. Additionally, we define three tic disorder classification tasks to differentiate the tic regions. In each task, the tic region of interest is considered the positive class, while normal segments are the negative class. For example, in the eye tic disorder classification task, the positive class is eye tics, and the negative class is normal actions.

In summary, we define four binary classification tasks, where the positive and negative class samples are composed of multiple short segments. The labels for these samples are $y_i \in \{0, 1\}$, where 0 represents the negative class and 1 represents the positive class. Our goal is to achieve high classification accuracy (ACC) and area under the ROC curve (AUC) for these tasks.

3.3 Feature Extraction in Facial Data

In the context of limited data, to enhance the generalization of the algorithm, we perform feature point extraction, facial segmentation, and face alignment on each frame of the video segments. The overall process is in Algorithm 1 and visualized results are presented in Fig. 2.

Specifically, in step 3 of the algorithm, the method for calculating the rotation matrix is as follows: first, calculate the center coordinates of the left and right eyes ($centerX$, $centerY$). Then calculate the angle between the line connecting the left and right eyes and the horizontal line. This angle represents the rotation angle θ. Finally, we can calculate the rotation matrix M as follows:

$$M = \begin{bmatrix} \cos\theta & -\sin\theta & (1-\cos\theta) \times centerX + \sin\theta \times centerY \\ \sin\theta & \cos\theta & (1-\cos\theta) \times centerY - \sin\theta \times centerX \\ 0 & 0 & 1 \end{bmatrix} \quad (1)$$

By utilizing the rotation matrix M, we can transform the coordinates of any point (x, y) in the original image into the coordinates (x', y') of the corresponding point in the new image. The transformation relationship between them is given by:

$$\begin{bmatrix} x' \\ y' \\ 1 \end{bmatrix} = M \cdot \begin{bmatrix} x \\ y \\ 1 \end{bmatrix} \quad (2)$$

Algorithm 1 Facial Landmark Alignment

Input:

Single frame

Output:

Aligned single frame with facial landmark coordinates

1. Apply a face detector to detect facial landmark.

in the video.

3. Calculate the rotation matrix based on the coordinates of the left and right eyes.

4. Perform an affine transformation on the image to obtain the rotation-aligned image and its corresponding landmark.

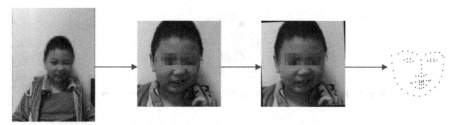

Fig. 2. Face alignment algorithm flow. Our method ultimately compresses the (1080, 1920) image into a facial landmark sequence of size (68, 2).

3.4 Relative Probability Guided Multi-phase Learning

Relative Probability (RPr). To proceed with our method, we propose the concept of Relative Probability. For each long segment, where PR_{cut} is our defined relative probability indicator, it represents the model's confidence in predicting the current segment. PR_{max} is the maximum value of confidence scores among all short segments cut in the long segment, and PR_{min} is the minimum. The calculation formula for relative probability is as follows:

$$PR_{relative} = \frac{PR_{cut} - PR_{min}}{PR_{max} - PR_{min}} \qquad (3)$$

Two thresholds $thsd_1$ and $thsd_2$ are set in advance and a short segment is marked as a positive sample in the following condition:

$$PR_{relative} > thsd_1 \,\&\&\, PR_{cut} > thsd_2 \qquad (4)$$

Through this approach, we effectively exploit the prior information inherent in the coarse annotations, assuming that the short segment with the highest confidence score corresponds to the most salient movement within the given long segment. Considering the inherent similarity of movement patterns within each long segment, our objective is to select positive samples that not only surpass the confidence threshold but also exhibit a high degree of resemblance to the most prominent movement feature present in the segment. This strategic selection process aims to mitigate the risk of false positives, thereby enhancing the reliability and precision of our approach.

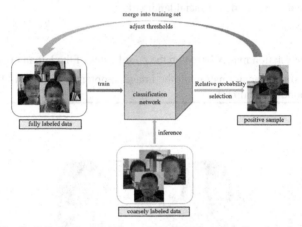

Fig. 3. Relative Probability guided multi-phase learning flowchart

Multi-phase Learning with Thresholds Decrease. Although we have introduced the concept of relative probability to improve the accuracy of generating pseudo-labels, it is inevitable that erroneous pseudo-label noise may still occur, potentially misleading the model during training. Additionally, the integration of rough labeled data into the training set requires careful consideration of techniques and strategies. If we simply incorporate all positively labeled samples into the training set, in the subsequent iterations, the model may tend to assign high confidence scores to these selected positive samples, resulting in pseudo-labels that are nearly identical to those of the previous round. Consequently, this iterative process can become stagnant, hindering any improvement in model accuracy.

Considering these problems, we propose a multi-phase learning algorithm with a threshold decrease. The overall process is shown in Algorithm 2.

We have designed multiple expressions to update the threshold with respect to the number of phases and we discover that the easiest and most effective method is a linear decay strategy (see Eq. (5)). And after sufficient experiments, we find that when $step_1 = step_2 = 0.05$ and thresholds stop decreasing in 4th phase, the proposed method get the best result.

$$thsd_i = thsd_i - step_i \quad i = 1, 2 \tag{5}$$

Algorithm 2 Multi-phase learning with threshold decrease

Input:

Video data (including coarse and fine annotations), the number of training phases N

Output:

Trained model

Training:

For j=0; j<=N; j++ **do**

1. Train the model on the training set (initially fully labeled) until convergence, and record the model's accuracy.

2. Perform inference on all coarse annotated data using the model.

3. Apply equation (4) to generate pseudo-labels based on the current thresholds.

4. Incorporate all selected positive samples that are not already in the training set.

5. Decrease the thresholds by equation (5).

End for

Testing:

Feed the test dataset into the trained model to obtain test results.

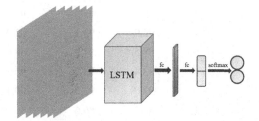

Fig. 4. The architecture of our classification network

As for the classification network in Fig. 1 and Fig. 3, we find that Long Short-Term Memory (LSTM) networks can capture the temporal motion features of facial landmarks, leading to superior classification performance.

4 Experiments

In this section, we train our models on our train dataset (which contains 2436 short segments) and coarse labeled data (87 videos). We evaluate the proposed method on our independent test dataset (which contains 833 short segments). For the methods that use the sequence of facial landmark points as input, we calculate the displacement between consecutive frames, resulting in a sequence of displacement vectors representing the motion of facial landmarks. For the methods that use original images as input, we perform facial segmentation to isolate the face region and apply grayscale normalization to enhance the consistency of the input data. All models are trained in RTX 3090.

4.1 Comparisons to Existing Methods

To the best of our knowledge, the field of tic disorder recognition lacks publicly available datasets, and there is a scarcity of relevant research with no established state-of-the-art (SOTA) method. Considering this, we conduct experiments using ResNet-3D and I3D, which are widely adopted methods in the domain of action recognition. We employ these models to tackle the task of movement disorder recognition, aiming to assess their performance and suitability. Considering the limited amount of data, we encounter challenges in evaluating transformer-based methods. For our facial feature extraction and privacy preservation method, which generates facial landmark points as input, we explore the performance of traditional machine learning methods.

As shown in Table 1, after face alignment, our method achieves an average AUC of 95.1%, 1.9% higher than ResNet-3D. The methods that use original images as input behave poorly even though they have much more parameters. I believe that the insufficient training data is one of the reasons. Additionally, simple LSTM or MLP models are already sufficient to capture the features of tic behaviors and our facial feature extraction method not only preserves privacy but also leads to better classification results.

Table 1. Our facial feature extraction and multi-phase learning based LSTM method vs. current method for video action detection. The numbers in the table represent AUC (%).

method	face	eye	mouth	nose	avg
ResNet-3D	94.7	92.9	93.1	92.0	93.2
I3D	92.6	88.5	87.9	88.6	89.4
MLP	95.0	93.4	93.2	90.9	93.1
MLP w/o alignment	93.3	92.0	92.5	90.1	92.0
RF [7, 8]	89.1	88.3	86.6	86.9	87.7
RF w/o alignment	88.0	88.1	85.6	86.7	87.1
LSTM (ours)	**97.0**	**96.2**	**93.5**	**93.8**	**95.1**
LSTM w/o alignment	95.7	94.9	92.9	92.8	94.1

For our best method LSTM in Table 1, we conduct complete experiment and find that while other methods may have higher AUC in the first phase, our method gradually surpasses them in subsequent stages and converges around four phases, demonstrating clear advantages compared to one-phase methods (Fig. 5).

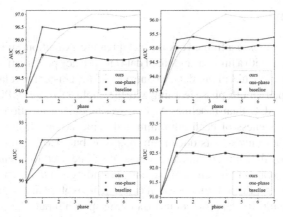

Fig. 5. AUC's change curve with phase on four binary classification tasks. From left to right, they are: face, eyes, mouth, nose. Baseline refers to the one-phase method with Top2 selection. One-phase changes Top2 selection into RPr selection.

4.2   Ablation Study

In this subsection, we evaluate the effect of proposed RPr selection and MPLTD algorithm. We conduct our experiment on ResNet-3D and LSTM. The results are in Table 2.

For LSTM, our RPr and MPLTD methods achieve the average AUC of 95.1%, outperforming baseline by 3%. For ResNet-3D, our method achieves the average AUC of 93.2%, 4.2% higher than baseline. Moreover, we discover that both RPr and MLTD methods lead to a significant increase in AUC. Through these proposed methods, we leverage both coarse-labeled and fine-labeled data in a comprehensive manner and achieve a more robust and effective training process for our model.

Table 2. Quantitive evaluation of our proposed method RPr and MLTD. The effectiveness is tested both in LSTM and ResNet-3D.

model	RPr	MPLTD	avg. AUC (%)
LSTM			92.1
	✓		94.3
		✓	93.9
	✓	✓	**95.1**
ResNet-3D			89.0
	✓		90.5
		✓	92.1
	✓	✓	**93.2**

5 Conclusion

In this work, we proposed a framework for facial feature extraction, suitable for weakly supervised learning with a limited amount of data and privacy preservation. Furthermore, based on the characteristics of our data, we introduced the concept of relative probability (RPr) and developed a multi-phase learning with threshold decrease (MPLTD) algorithm, achieving higher AUC than baseline. At last, we conducted ablation experiments to validate the effectiveness of each algorithm and achieved an ideal result. Our high-accuracy model not only assists doctors in diagnosis but also has the potential to be applied throughout the entire treatment process. It can be used to monitor and analyze the recovery and treatment progress of patients, providing guidance on medication and treatment approaches. In the future, we will continue to explore the tic disorder in limbs and address the multi-modal problem incorporating speech input.

Acknowledgements. This work was supported by Beijing Natural Science Foundation (M22024).

References

1. Leckman, J.F., Bloch, M.H.: Tic disorders. In: Rutter's Child and Adolescent Psychiatry, pp. 757–773 (2015)
2. Cohen, S.C., Leckman, J.F., Bloch, M.H.: Clinical assessment of Tourette syndrome and tic disorders. Neurosci. Biobehav. Rev. **37**(6), 997–1007 (2013)
3. Szegedy, C., et al.: Going deeper with convolutions. In: Proceedings of the IEEE Conference on Computer Vision and Pattern Recognition (CVPR), pp. 1–9 (2015)
4. Vaswani, A., et al.: Attention is all you need. In: Advances in Neural Information Processing Systems, pp. 5998–6008 (2017)
5. Bishop, C.M.: Neural Networks for Pattern Recognition. Oxford University Press (1995)
6. Cortes, C., Vapnik, V.: Support-vector networks. Mach. Learn. **20**(3), 273–297 (1995)
7. Breiman, L.: Random forests. Mach. Learn. **45**(1), 5–32 (2001)
8. Liaw, A., Wiener, M.: Classification and regression by random forest. R News **2**(3), 18–22 (2002)
9. Hochreiter, S., Schmidhuber, J.: Long short-term memory. Neural Comput. **9**(8), 1735–1780 (1997)
10. Bernabei, M., et al.: Automatic detection of tic activity in the Tourette Syndrome. In: 2010 Annual International Conference of the IEEE Engineering in Medicine and Biology, pp. 422–425. IEEE, August 2010
11. Shute, J.B., et al.: Thalamocortical network activity enables chronic tic detection in humans with Tourette syndrome. NeuroImage Clin. **12**, 165–172 (2016)
12. Cootes, T.F., Edwards, G.J., Taylor, C.J.: Active appearance models. IEEE Trans. Pattern Anal. Mach. Intell. **23**(6), 681–685 (2001)
13. Kazemi, V., Sullivan, J.: One millisecond face alignment with an ensemble of regression trees. In: Proceedings of the IEEE Conference on Computer Vision and Pattern Recognition (CVPR), pp. 1867–1874 (2014)
14. Yang, H., Liu, H.: Cascaded regression based landmark localization for robust facial feature tracking. IEEE Trans. Image Process. **24**(8), 2479–2490 (2015)
15. Bulat, A., Tzimiropoulos, G.: How far are we from solving the 2D & 3D face alignment problem? (and a dataset of 230,000 3D facial landmarks). In: Proceedings of the IEEE International Conference on Computer Vision (ICCV), pp. 1021–1030 (2017)

16. Simonyan, K., Zisserman, A.: Two-stream convolutional networks for action recognition in videos. In: Advances in Neural Information Processing Systems, vol. 27 (2014)
17. Donahue, J., et al.: Long-term recurrent convolutional networks for visual recognition and description. In: Proceedings of the IEEE Conference on Computer Vision and Pattern Recognition, pp. 2625–2634(2015)
18. Wu, Z., Jiang, Y.G., Wang, X., Ye, H., Xue, X.: Multi-stream multi-class fusion of deep networks for video classification. In: Proceedings of the 24th ACM International Conference on Multimedia, pp. 791–800, October 2016
19. Tran, D., Bourdev, L., Fergus, R., Torresani, L., Paluri, M.: Learning spatiotemporal features with 3D convolutional networks. In: Proceedings of the IEEE International Conference on Computer Vision, pp. 4489–4497 (2015)
20. Simonyan, K., Zisserman, A.: Very deep convolutional networks for large-scale image recognition (2014). arXiv preprint arXiv:1409.1556
21. He, K., Zhang, X., Ren, S., Sun, J.: Deep residual learning for image recognition. In: Proceedings of the IEEE Conference on Computer Vision and Pattern Recognition, pp. 770–778 (2016)
22. Hara, K., Kataoka, H.: Can spatiotemporal 3D CNNs retrace the history of 2D CNNs and ImageNet? In: Proceedings of the IEEE Conference on Computer Vision and Pattern Recognition, pp. 6546–6555 (2018)
23. Jiang, B., Zhang, L., Zhang, D., Zhang, M., Yang, H., Guo, Y.: T3D: temporal 3D ConvNet for real-time action recognition. In: Proceedings of the AAAI Conference on Artificial Intelligence, vol. 34, no. 07, pp. 12309–12316 (2020)
24. Qiu, Z., Yao, T., Mei, T.: Learning spatio-temporal representation with pseudo-3D residual networks. In: Proceedings of the IEEE International Conference on Computer Vision, pp. 5534–5542 (2017)
25. Pathak, D., Krähenbühl, P., Darrell, T.: Constrained convolutional neural networks for weakly supervised segmentation. In: Proceedings of the IEEE International Conference on Computer Vision (ICCV), pp. 1796–1804 (2015)
26. Zhang, Z., Xu, J., Yang, L., Xiong, Y.: Deep learning based intervertebral disc segmentation from weakly labeled training data. J. Med. Syst. **42**(6), 100 (2018)
27. Durand, T., Mordan, T., Thome, N.: Weakly supervised object detection: a survey. Int. J. Comput. Vision **127**(9), 1191–1234 (2019)
28. Tarvainen, A., Valpola, H.: Mean teachers are better role models: weight-averaged consistency targets improve semi-supervised deep learning results. In: Advances in Neural Information Processing Systems, vol. 30 (2017)
29. ZhuΓ, X., GhahramaniΓн, Z.: Learning from labeled and unlabeled data with label propagation (2002)
30. Ma, X., et al.: Dimensionality-driven learning with noisy labels. In: International Conference on Machine Learning, pp. 3355–3364. PMLR, July 2018
31. Lee, D.H.: Pseudo-label: the simple and efficient semi-supervised learning method for deep neural networks. In: Workshop on Challenges in Representation Learning, ICML, vol. 3, no. 2, p. 896, June 2013
32. Li, X., Yu, L., Chen, H., Fu, C.W., Xing, L., Heng, P.A.: Transformation-consistent self-ensembling model for semisupervised medical image segmentation. IEEE Trans. Neural Netw. Learn. Syst. **32**(2), 523–534 (2020)
33. Wang, Z., Li, Y., Guo, Y., Fang, L., Wang, S.: Data-uncertainty guided multi-phase learning for semi-supervised object detection. In: Proceedings of the IEEE/CVF Conference on Computer Vision and Pattern Recognition, pp.4568–4577 (2021)
34. Yang, X., Song, Z., King, I., Xu, Z.: A survey on deep semi-supervised learning. IEEE Trans. Knowl. Data Eng. (2022)

35. Huynh, T., Nibali, A., He, Z.: Semi-supervised learning for medical image classification using imbalanced training data. In: Computer Methods and Programs in Biomedicine, p. 106628 (2022)
36. Zheng, M., You, S., Huang, L., Wang, F., Qian, C., Xu, C.: SimMatch: semi-supervised learning with similarity matching. In: Proceedings of the IEEE/CVF Conference on Computer Vision and Pattern Recognition, pp. 14471–14481 (2022)

Detecting Software Vulnerabilities Based on Hierarchical Graph Attention Network

Wenlin Xu[1,3]([✉]), Tong Li[2], Jinsong Wang[3], Tao Fu[3], and Yahui Tang[4]

[1] School of Information Science and Engineering, Yunnan University,
Kunming, China
wlxu1120@ynufe.edu.cn
[2] School of Big Data, Yunnan Agricultural University, Kunming, China
[3] Information Management Center, Yunnan University of Finance and Economics,
Kunming, China
[4] School of Software, Chongqing University of Posts and Telecommunications,
Chongqing, China

Abstract. Detecting software vulnerabilities is a crucial part of software security. At present, the most commonly used methods are to train supervised classification or regression models from the source code to detect vulnerabilities, which require lots of high-quality labeled vulnerabilities. However, high-quality labeled vulnerabilities are not easy to be obtained in practical applications. To alleviate this problem, we present an effective and unsupervised method to detect software vulnerabilities. We first propose a new source code representation that maintains both the source code's natural language information and high-level programming logic information, and then we effectively embed the software function into a compact and low-dimensional representation based on hierarchical graph attention network. Finally, we obtain vulnerabilities by applying an outlier detection algorithm on the low-dimensional representation. We carry out extensive experiments on six datasets and the effectiveness of our proposed method is demonstrated by the experimental results.

Keywords: Vulnerability detection · Source code representation ·
Hierarchical graph attention network · Outlier detection

1 Introduction

The software vulnerabilities, which might be exploited by attackers and bring about significant financial and social damage, are crucial for software security. Since quantity of software vulnerabilities is rising quickly, detecting software vulnerabilities has been studied by many researchers [1]. One of the successful vulnerability detection methods is to design features manually and apply machine learning techniques to detect vulnerabilities [2–4]. However, this method requires domain specialists and is often laborious.

Since deep learning achieves outstanding performances at dealing with big data and can automatically learn features from raw data, vulnerability detection

L. Fang et al. (Eds.): CICAI 2023, LNAI 14474, pp. 113–124, 2024.
https://doi.org/10.1007/978-981-99-9119-8_11

methods based on deep learning have attracted much attention lately [5]. Among these works, most are treat source code as natural languages [6–8]. Compared to natural languages, source code is actually more logical and structured. When source code is treated as natural languages, it is quite difficult to learn entire program structural information, which effects the effectiveness of vulnerability detection results [9]. Thus, some researchers treat source code as graphs and use Graph Neural Network to detect vulnerabilities [10–12]. Typically, to determine whether there are any vulnerabilities in the testing software functions, the process involves creating graphs from the source code and training classifiers. Although these methods can effectively detect vulnerable software functions, there are three problems need to be concerned: (1) in practical applications, the functions'vulnerable labels are not easily obtained, (2) how to create a representation of source code that maintains its structural and semantical information, and (3) how to train a graph neural network to learn effective features from the representation of the source code.

To address these concerns, in this study, we present a novel approach for detecting software vulnerabilities. For the first problem, we detect vulnerable functions in an unsupervised way which increases the practical use of the method by eliminating the need for users to provide the labeled vulnerable functions for model training. For the second problem, we propose an effective source code representation method to turn the source code into a token graph, where the tokens' vectors are generated by a pretraining language model BERT that preserves the natural language information of the code [13] and edges are constructed by concatenating the Program Dependence Graph (PDG), the Control Flow Graph (CFG) and the Abstract Syntax Tree (AST) to preserve the structural information of the code. For the last issue, we design a Hierarchical Graph Attention Network (HGAT) consisting of a token-level and a function-level GAT to extract software function features from the heterogeneous graph, which could be effectively trained with deep autoencoder [14].

We test the proposed approach on six real-world datasets by comparing it with three vulnerability detectors FlawFinder, Rats and Joern. As a result, our proposed method performs better than other approaches across all datasets, and our method's average improvements on F1 score are 11%, 75% and 80%, respectively. The following is a summary of our primary contributions:

- We propose a new and effective source code representation method to transform the source code into a token graph, which could maintain the source code's natural language and structural information.
- We design a hierarchical graph attention network to learn the useful software function features from the heterogeneous graph. Besides, by considering the dependency relations among functions and the use of vulnerable related APIs, we further improve the effectiveness of our detection results.
- We implement the proposed method and assess its effectiveness using datasets from C/C++ program. The outcomes demonstrate that our method outperforms other comparison vulnerability detectors.

The remainder parts are organized as follows: We present relevant work in Sect. 2. The mechanism of our proposed method is discussed in Sect. 3. Section 4 displays the specific experimental configurations and outcomes. After that, Sect. 5 discusses the conclusions and the future work.

2   Related Work

The effectiveness of deep learning has motivated academics to use it in more automated ways to detect vulnerabilities in source code [6–8]. The research that investigates the potential of natural language processing approaches in vulnerability detection, typically takes source code as flat sequences of natural language. For example, [6] presents a detection approach utilizing text mining and a deep neural network with self-attention, and [15] builts BLSTM model for vulnerability detection. In order to overcome the constraints of the aforementioned models on expressing the source code's structure and semantic information, several studies have tried to investigate more structural neural networks to detect software vulnerability [9–12]. [9] constructed a composite code graph based on ASTs and used the *Conv* module in gated graph neural network model to detect vulnerable functions. [10] defined a new code representation called Slice Property Graph (SPG) and offered a model called VulSPG to find potential vulnerability in SPG.

Different from these works, we detect software vulnerability in an unsupervised way. We construct a novel comprehensive source code representation by employing pretraining language model BERT and integrating the AST, CFG and PDG, then we build a hierarchical graph attention network to learn features of software functions from the source code representation. Besides, considering that a function is more likely to be vulnerable when calling another vulnerable function or vulnerable related APIs [16], our work extracts the call relations as dependency relations among functions when constructing the function graph, and assigns the weight of the tokens in vulnerable related APIs when training the hierarchical graph attention network, which further improve the effectiveness of our detection results.

3   Methodology

In this paper, we aim to detect function level vulnerabilities in source code. Given a function set $F = \{f_1, ..., f_m\}$ (m is the number of functions), our task is to determine whether a given function $f_i \in F$ is vulnerable or not. Figure 1 demonstrates our method's framework. There are three phases in it: (1) source code representation to transform source code into a token graph, (2) hierarchical graph attention network consisting of a token-level GAT and a function-level GAT to extract function features, and (3) outlier detection module to obtain vulnerable functions in an unsupervised way. We next describe our method in details.

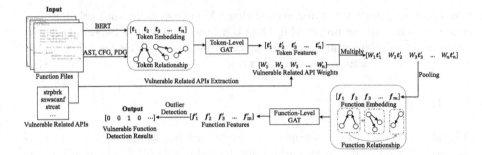

Fig. 1. Framework of our software vulnerability detection method

3.1 Source Code Representation

This subsection introduces how to transform source code into a token graph by code token embedding and code token relation extracting.

We employ a powerful pretraining language BERT to obtain the code token embedding T_v since BERT can acquire appropriate initialization for downstream tasks and has demonstrated impressive performance in numerous NLP tasks [17]. We construct the token graph by concatenating the AST, CFG and PDG of the source code, where the CFG explicitly describes the sequence of code statements and the requisite conditions for the execution of a particular path, the PDG explicitly represents dependencies among statements and predicates, and the AST faithfully expresses the organization of the statements and expressions [18]. By concatenating the AST, CFG and PDG, We maintain the code's structural information. Figure 2 demonstrates how the token graph is built.

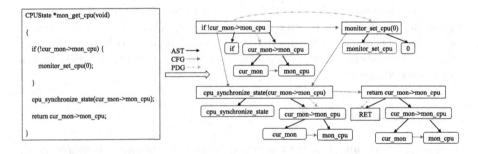

Fig. 2. The construction of the token graph

3.2 Hierarchical Graph Attention Network (HGAT)

Graph attention network is a network architecture that can aggregate the representations of the nearby nodes and operates on graph-structured data [19]. To detect the vulnerable software functions, we have to obtain the features of software functions. Thus, in this subsection, we design a hierarchical graph attention network to extract software function features from the heterogeneous graph. The HGAT includes a token-level GAT and a function-level GAT, where the token-level GAT is to obtain token features by aggregating the representations of the neighborhood tokens and the function-level GAT is to obtain the function features. The architecture of HGAT is shown in Fig. 3.

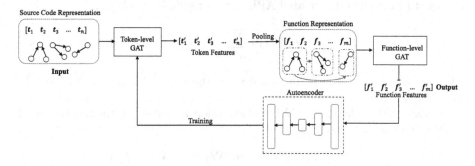

Fig. 3. The architecture of hierarchical graph attention network

Token-Level Graph Attention Network. Token features can be learned with Token-level graph attention network. The input is a set of token edges T_e and a set of token embedding $T_v = \{t_1, ..., t_n\}$, where n refers to the quantity of tokens, $t_i \in R^d$ refers to the embedding of ith token. The output is $T'_v = \{t'_1, ..., t'_n\}$, $t'_i \in R^{d'}$, where d' refers to the new feature length of tokens. The attention weight a_{ij} is calculated as Eq. (1),

$$\alpha_{ij} = \frac{exp(LeakyReLU(\mathbf{a}^T[\mathbf{W} \times \boldsymbol{t}_i || \mathbf{W} \times \boldsymbol{t}_j]))}{\sum_{r \in N_i} exp(LeakyReLU(\mathbf{a}^T[\mathbf{W} \times \boldsymbol{t}_i || \mathbf{W} \times \boldsymbol{t}_r]))} \tag{1}$$

where $\mathbf{W}$ refers to the weight matrix, $\mathbf{a}$ refers to the weight vector, and *LeakyReLU* is used as the activation function inspired by [19]. The ith token feature t'_i is calculated by Eq. (2),

$$t'_i = \sigma(\sum_{j \in N_i} \alpha_{ij} \mathbf{W} \times \boldsymbol{t}_j) \tag{2}$$

where σ refers to the activation function.

Function-Level Graph Attention Network. Function-level graph attention network is used to learn function features. The input is a set of edges of functions F_e and a set of function embedding $F_v = \{\boldsymbol{f}_1, ..., \boldsymbol{f}_m\}$, $\boldsymbol{f}_i \in R^{d'}$, where m refers to the functions' number and d' refers to the functions' feature length. The output is $F'_v = \{\boldsymbol{f}'_1, ..., \boldsymbol{f}'_m\}$, $\boldsymbol{f}'_i \in R^{d''}$, where d'' refers to the new feature length of functions.

We adopt the algorithm from [20] to build a directed function dependency graph as the relations among functions. We use $G = (F, E)$ describes the directed function dependency graph, where F refers to the function set and E refers to the edge set [20]. We employ mean pooling to obtain function embedding from the token features. Considering that the functions are more likely to be vulnerable when calling the related vulnerable APIs, we add the weight of the token which belongs to the vulnerable related APIs. The formula is as follows,

$$f_i = \frac{1}{|f_i|} \sum_{t_j \in f_i} t'_j \cdot \mathbf{w}_j \qquad (3)$$

where t'_j is the jth token feature, and $\mathbf{w}_j$ is the vulnerable related API weight for the jth token.

Similar to the token-level GAT, the attention weight α'_{ij} in the function-level GAT is calculated as Eq. (4),

$$\alpha'_{ij} = \frac{exp(LeakyReLU(\mathbf{a}_1^T[\mathbf{W}_1 \times f_i || \mathbf{W}_1 \times f_j]))}{\sum_{r \in N_i} exp(LeakyReLU(\mathbf{a}_1^T[\mathbf{W}_1 \times f_i || \mathbf{W}_1 \times f_r]))} \qquad (4)$$

where $\mathbf{W}_1$ refers to the weight matrix and $\mathbf{a}_1$ refers to the weight vector. The output of the ith function feature f'_i is calculated by Eq. (5),

$$f'_i = \sigma(\sum_{j \in N_i} \alpha'_{ij} \mathbf{W}_1 \times f_j) \qquad (5)$$

where σ refers to the activation function.

Training the Hierarchical Graph Attention Network. A deep autoencoder is a dimensionality reduction neural network that minimizes reconstruction error to learn an efficient and compact representation for given input [21]. We adopt autoencoder to train the hierarchical graph attention network in an unsupervised way. An encoder function $h = encoder(\mathbf{F}'_v)$ and a decoder function $\mathbf{F}''_v = decoder(h)$ are fundamental components of deep autoencoder. The training process is to minimize a Mean Square Error (MSE) loss function $L(\mathbf{F}'_v, \mathbf{F}''_v)$, which can be defined as Eq. 6,

$$L(\mathbf{F}'_v, \mathbf{F}''_v) = \sum_{i=1}^{m} ||f'_i - f''_i||^2 \qquad (6)$$

where $f'_i \in \mathbf{F}'_v$ and $f''_i \in \mathbf{F}''_v$.

3.3   Vulnerability Detection

Local Outlier Factor (LOF) are used as the outlier detection method to predict vulnerable functions. For an object $\mathbf{x}$, the LOF score indicates how far $\mathbf{x}$ is from the surrounding neighborhood, expressed by Eq. 7 [22]:

$$LOF_k(\mathbf{x}) = \frac{1}{\mid L_k(\mathbf{x}) \mid} \sum_{\mathbf{y} \in L_k(\mathbf{x})} \frac{A_k(\mathbf{x})}{A_k(\mathbf{y})} \tag{7}$$

where k refers to the quantity of closest neighbors, $L_k(\mathbf{x})$ refers to the points of $\mathbf{x}$'s k-nearest neighbors and $A_k(\mathbf{x})$ refers to the average reachability distances of $\mathbf{x}$ and $L_k(\mathbf{x})$. In this paper, each function's LOF score is calculated, and the functions with the greatest LOF scores are selected as vulnerable.

The proposed method's specific steps are displayed in Algorithm 1. Since the time complexity of HGAT and autoencoder is not affected by the amount of the input data, and the time complexity of constructing FDDG and LOF are $O(m^2 l)$ and $O(m^2)$, respectively. Accordingly, our method's time complexity is $O(m^2 l)$, where m refers to the quantity of functions and l refers to the quantity of dependency functions in each function.

Algorithm 1. USVD

Input: the set of function files F, the k closest neighbors in LOF.
Output: the set of detected vulnerable functions O
1: $T_v \leftarrow$ generating token embedding by BERT
2: $T_e \leftarrow$ generating token dependency graph based on AST, CFG and PDG
3: training token-level GAT(T_v, T_e) with MSE loss
4: $T_v' \leftarrow$ extracting token features from GAT(T_v,T_e) by equation 1 and 2
5: $F_v \leftarrow$ generating function embedding by pooling token features T_v'
6: $F_e \leftarrow$ generating function dependency graph by FDDG
7: training function-level GAT(F_v,F_e) with MSE loss
8: $F_v' \leftarrow$ extracting function features from GAT(F_v,F_e) by equation 4 and 5
9: $O \leftarrow$ detecting vulnerable functions by LOF_k on F_v'
10: **return** O

---

4   Experimental Study

This part is dedicated to evaluating the performance of the proposed approach in detecting software functions that are vulnerable. We carry out three series of tests to evaluate: (1) the effectiveness of the method proposed in our study, (2) the impacts of the source code representations for the detection results, and (3) the impacts of the dependency features and vulnerable related APIs for the detection results.

4.1   Experimental Settings

Datasets. We test our method with datasets from (1) Software Assurance Reference Dataset (SARD) [23], which has numerous production, synthesis, and scholarly functions that are known to be vulnerable, (2) three real-world open-source projects including Chrome [24], FFMPeg [9] and Qemu [9], and (3) the U.S. government's repository for data on vulnerability management based on standards NVD [25]. Table 1 provides an overview of the datasets' statistics.

Table 1. The statistics for the dataset

Datasets	Functions	Vul Functions	Non-Vul Functions	Vul rate
SARD^C	9,052	2,578	6,474	28%
SARD^{C++}	8,745	2,812	5,933	32%
Chrome	3,579	294	3,285	8%
FFMPeg	5,632	969	4,663	17%
Qemu	10,653	1,933	8,720	18%
NVD	1,146	206	940	18%

Comparison Methods. Three relevant methods are compared to our method: (1) **FF** (FlawFinder), a simple yet efficient and rapid source code scanning tool for C/C++ program, which flags calls relevant to typical vulnerable library functions [26], (2) **Rats**, a tool that scans source code in C, C++, Perl, PHP, and Python to identify common security-related programming mistakes [27], and (3) **Joern**, a helpful tool for discovering vulnerabilities through statically analysis [28].

Metrics for Evaluation. We assess the effectiveness of our approach using three commonly used measures: (1) **Precision**, the ratio of functions that are properly identified as vulnerable to those flagged as vulnerable, (2) **Recall**, the percentage of vulnerable functions that were appropriately identified relative to all vulnerable functions, and (3) **F1**, the precision and recall harmonic mean, which evaluates the entire impact by taking both into account.

Implementation. We used a hierarchical graph attention network to implement our approach to detect vulnerable functions. All trials were carried out on computers running 64-bit Windows 11 professional system, equipped with an AMD R9 16-Core 3.40 GHz CPU, an Nvidia RTX4090 GPU, and 128 GB of RAM.

4.2   Experimental Results

(I) Effectiveness of Our Method. We compared our suggested method's effectiveness with FF, Rats, and Joern in the first round of testing. The outcomes are displayed in Table 2. According to the results, our method performs better than other methods on all datasets. Across all datasets, our approach yields the highest F1 and raises F1 by (2%, 4%, 5%, 38%, 10%, 5%), (29%, 48%, 40%, 202%, 72%, 59%) and (129%, 245%, 5%, 1%, 26%, 72%) over FF, Rats and Joern on (SARD^C, SARD^{C++},Chrome, FFMPeg, Qemu, NVD), respectively. Moreover, our method also has the highest precision and recall compared with other methods on almost all datasets. It needs to mention that users do not need to provide the labeled vulnerabilities for training in our method. This demonstrates the effectiveness of our proposed approach.

Table 2. The performance of different unsupervised approaches

Method	SARD^C			SARD^{C++}			Chrome		
	Pre.	Rec.	F1	Pre.	Rec.	F1	Pre.	Rec.	F1
FF	36.58	56.87	44.52	44.90	49.00	46.86	13.26	12.59	12.91
Rats	33.86	36.81	35.27	37.68	29.41	33.03	**15.11**	07.14	09.70
Joern	**41.57**	13.11	19.94	35.17	08.85	14.15	11.99	13.95	12.89
Our	37.57	**58.03**	**45.61**	**46.80**	**50.92**	**48.77**	12.92	**14.29**	**13.57**

Method	FFMPeg			Qemu			NVD		
	Pre.	Rec.	F1	Pre.	Rec.	F1	Pre.	Rec.	F1
FF	17.37	18.37	17.85	20.36	11.17	14.43	20.30	26.70	23.06
Rats	14.71	05.68	08.19	17.88	06.21	09.22	20.33	12.14	15.20
Joern	19.55	32.30	24.36	17.91	09.67	12.56	18.85	11.17	14.02
Our	**19.90**	**32.61**	**24.72**	**22.25**	**12.29**	**15.84**	**21.09**	**28.16**	**24.12**

(II) Impacts of Source Code Representations. In the following set of experiments, we assessed the effects of different kinds of source code representations. We compare our source code representation with the AST, CFG and PDG. The outcomes are displayed in Table 3. According to the results, our source code representation provides better precision, recall and F1 score compared to the AST, CFG and PDG. Additionally, our method improves F1 score by (16%, 12%, 44%, 7%, 5%, 7%), (9%, 22%, 61%, 10%, 8%, 9%) and (9%, 6%, 67%, 5%, 1%, 18%) over AST, CFG and PDG on (SARD^C, SARD^{C++},Chrome, FFMPeg, Qemu, NVD), respectively.

Table 3. Comparison with other source code representations

Method	SARD^C			SARD^{C++}			Chrome		
	Pre.	Rec.	F1	Pre.	Rec.	F1	Pre.	Rec.	F1
AST	32.32	49.92	39.24	41.67	45.34	43.43	09.01	09.86	09.42
CFG	34.48	53.26	41.86	38.40	41.67	39.97	08.07	08.84	08.44
PDG	34.58	53.41	41.98	44.02	47.90	45.88	07.76	08.50	08.12
Our	**37.57**	**58.03**	**45.61**	**46.80**	**50.92**	**48.77**	**12.92**	**14.29**	**13.57**

Method	FFMPeg			Qemu			NVD		
	Pre.	Rec.	F1	Pre.	Rec.	F1	Pre.	Rec.	F1
AST	18.51	30.34	23.00	21.22	11.72	15.10	19.64	26.21	22.45
CFG	18.14	29.72	22.53	20.56	11.36	14.63	19.27	25.73	22.04
PDG	18.89	30.96	23.46	21.97	12.14	15.64	17.82	23.79	20.37
Our	**19.90**	**32.61**	**24.72**	**22.25**	**12.29**	**15.84**	**21.09**	**28.16**	**24.12**

(III) Impacts of the Function Dependency Features and the Vulnerable Related APIs. In the third round of experiments, we assessed the impact of the function dependency features and vulnerable related APIs. In order to have a more comprehensive understanding of incorporating with the function dependency features and the vulnerable related APIs, we compare our method with the method without the function dependency features (denoted as NDF) and without vulnerable related APIs (denoted as NAPIs). The outcomes are displayed in Table 4. According to the results, our method improves precision, recall and F1 score when incorporating the function dependency features and the vulnerable related APIs. Additionally, our method improves F1 score by (15%, 20%, 62%, 10%, 7%, 26%) and (28%, 38%, 14%, 2%, 18%, 12%) over NDF and NAPIs on ($SARD^C$, $SARD^{C++}$, FFMPeg, Qemu, NVD) respectively.

Table 4. Comparison with NDF and NAPIs

Method	$SARD^C$			$SARD^{C++}$			Chrome		
	Pre.	Rec.	F1	Pre.	Rec.	F1	Pre.	Rec.	F1
NDF	32.80	50.66	39.82	38.92	42.35	40.57	08.00	08.84	08.40
NAPIs	29.36	45.35	35.64	34.02	37.02	35.46	11.38	12.52	11.95
Our	**37.57**	**58.03**	**45.61**	**46.80**	**50.92**	**48.77**	**12.92**	**14.29**	**13.57**

Method	FFMPeg			Qemu			NVD		
	Pre.	Rec.	F1	Pre.	Rec.	F1	Pre.	Rec.	F1
NDF	18.07	29.62	22.45	20.75	11.46	14.77	16.73	22.33	19.13
NAPIs	19.58	32.09	24.33	18.87	10.43	13.34	18.91	25.24	21.62
Our	**19.90**	**32.61**	**24.72**	**22.25**	**12.29**	**15.84**	**21.09**	**28.16**	**24.12**

Summary. Based on the results of experimentation, the following is discovered: (1) The method we proposed outperforms the existing unsupervised vulnerability detection methods across all datasets. Actually, the average improvements of our method in terms of F1 are 11%, 75% and 80% compared with FF, Rats and Joern, and (2) using combination of AST, CFG and PDG as source code representation enhances the effectiveness of our proposed method. Moreover, incorporating the function dependency features and the vulnerable related APIs also enhances the effectiveness of our proposed method.

5   Conclusion and Future Work

This research provides a novel unsupervised software vulnerability detection method using hierarchical graph attention network. We present a new source code representation and design a HGAT to extract function features from the source code representation. Moreover, we incorporate with the function dependency features and the vulnerable related APIs, which help to generate more powerful and vulnerable related features for vulnerability detection. We evaluate the proposed method and the results indicate that our proposed method significantly outperforms other comparison methods across evaluation metrics.

We are to modify the graph attention networks to further promote the effectiveness of detection result and to apply our method to detect vulnerable software functions written by other programming languages.

Acknowledgements. This work was supported by Program of Yunnan Key Laboratory of Intelligent Systems and Computing (202205AG070003) and Natural Science Foundation of Yunnan Provincial Department of Education (2021J0567).

References

1. Hin, D., Kan, A., Chen, H., Babar, M. A.: LineVD: statement-level vulnerability detection using graph neural networks. In: Proceedings of the 19th International Conference on Mining Software Repositories, pp. 596–607. ACM, Pittsburgh, PA, USA (2022)
2. Gupta, A., Suri, B., Kumar, V., Jain, P.: Extracting rules for vulnerabilities detection with static metrics using machine learning. Int. J. Syst. Assur. Eng. Manag. **12**(1), 65–76 (2021)
3. Kronjee, J., Hommersom, A., Vranken, H.: Discovering software vulnerabilities using data-flow analysis and machine learning. In: Proceedings of the 13th International Conference on Availability, Reliability and Security, pp. 6:1–6:10. Springer, Hamburg (2018)
4. Grieco, G., Grinblat, G. L., Uzal, L., Rawat, S., Feist, J., Mounier, L.: Toward Large-scale vulnerability discovery using machine learning. In: Proceedings of the Sixth ACM on Conference on Data and Application Security and Privacy, pp. 85–96. New Orleans, LA, USA (2016)
5. Liu, H., Lang, B.: Machine learning and deep learning methods for intrusion detection systems: a survey. Appl. Sci. **9**(20), 4396 (2019)
6. Vishnu, P.R., Vinod, P., Yerima, S.Y.: A deep learning approach for classifying vulnerability descriptions using self attention based neural network. J. Netw. Syst. Manag. **30**(1), 1–27 (2022)
7. Wartschinski, L., Noller, Y., Vogel, T., Kehrer, T., Grunske, L.: VUDENC: vulnerability detection with deep learning on a natural codebase for Python. Inf. Softw. Technol. **144**, 106809 (2022)
8. Thapa, C., Jang, S. I., Ahmed, M. E., Camtepe, S., Pieprzyk, J., Nepal, S.: Transformer-based language models for software vulnerability detection. In: Proceedings of the 38th Annual Computer Security Applications Conference, pp. 481–496. Austin, TX, USA (2022)
9. Zhou, Y., Liu, S., Siow, J., Du, X., Liu, Y.: Devign: effective vulnerability identification by learning comprehensive program semantics via graph neural networks. In: Advances in Neural Information Processing Systems, vol. 32 (2019)
10. Zheng, W., Jiang, Y., Su, X.: VulSPG: vulnerability detection based on slice property graph representation learning. In: 2021 IEEE 32nd International Symposium on Software Reliability Engineering (ISSRE), pp. 457–467. IEEE, Vancouver, BC, Canada (2021)

11. Cheng, X., Wang, H., Hua, J., Xu, G., Sui, Y.: Deepwukong: statically detecting software vulnerabilities using deep graph neural network. ACM Trans. Softw. Eng. Methodol. (TOSEM) 30(3) (2021)

12. Nguyen, V.A., Nguyen, D.Q., Nguyen, V., Le, T., Tran, Q.H., Phung, D.: ReGVD: revisiting graph neural networks for vulnerability detection. In: Proceedings of the ACM/IEEE 44th International Conference on Software Engineering: Companion Proceedings, pp. 178–182. ACM/IEEE, Pittsburgh, PA, USA (2022)

13. Devlin, J., Chang, M., Lee, K., Toutanova, K.: BERT: pre-training of deep bidirectional transformers for language understanding. arXiv preprint arXiv:1810.04805, (2018)

14. Zeng, J., Liu, T., Jia, W., Zhou, J.: Fine-grained question-answer sentiment classification with hierarchical graph attention network. Neurocomputing 457 (2021)

15. Li, Z., et al.: Vuldeepecker: a deep learning-based system for vulnerability detection. In: 25th Annual Network and Distributed System Security Symposium (NDSS), San Diego, CA, USA (2018)

16. Zou, D., Wang, S., Xu, S., Li, Z., Jin, H.:μVulDeePecker: a deep learning-based system for multiclass vulnerability detection. IEEE Trans. Depend. Secure Comput. 18(5) (2019)

17. Hao, Y., Dong, Li., Wei, F., Xu, K.: Visualizing and understanding the effectiveness of BERT. In: EMNLP-IJCNLP 2019, pp. 4141–4150. Hong Kong, China (2019)

18. Yamaguchi, F., Golde, N., Arp, D., Rieck, K.: Modeling and discovering vulnerabilities with code property graphs. In: 2014 IEEE Symposium on Security and Privacy, pp. 590–604. IEEE, Berkeley, California, USA (2014)

19. Veličković, P., Cucurull, G., Casanova, A., Romero, A., Lio, P., Bengio, Y.: Graph attention networks. In: 6th International Conference on Learning Representations (ICLR), Vancouver, BC, Canada (2018)

20. Xu. W., Li, T., Wang, J., Tang, Y.: Detecting vulnerable software functions via text and dependency features. Soft Comput. 27(9), (2023)

21. Zhang, S., Yao, Y., Hu, J., Zhao, Y., Li, S., Hu, J.: Deep autoencoder neural networks for short-term traffic congestion prediction of transportation networks, 19(10) (2019)

22. Breunig, M.M., Kriegel, H., Ng, R.T., Sander, J.: LOF: identifying density-based local outliers. In: Proceedings of the 2000 ACM SIGMOD International Conference on Management of Data, pp. 93–104. ACM, Dallas, Texas, USA (2000)

23. SARD https://samate.nist.gov/SRD/

24. Chakraborty, S., Krishna, R., Ding, Y., Ray, B.: Deep learning based vulnerability detection: are we there yet. IEEE Trans. Softw. Eng. 48(9) (2021)

25. NVD https://nvd.nist.gov/

26. FlawFinder https://dwheeler.com/flawfinder/

27. Rats. https://code.google.com/archive/p/rough-auditing-tool-for-security/

28. Joern. https://joern.io/

Domain Specific Pre-training Methods for Traditional Chinese Medicine Prescription Recommendation

Wei Li[1], Zheng Yang[2], and Yanqiu Shao[1](✉)

[1] School of Information Science, Beijing Language and Culture University,
Beijing 100081, China
{liweitj47,shaoyanqiu}@blcu.edu.cn
[2] School of Traditional Chinese Medicine, Beijing University of Chinese Medicine,
Beijing 100029, China
yangzheng@bucm.edu.cn

Abstract. Traditional Chinese Medicine (TCM) is an important constituent of medical treatment. During the development history of TCM, there have been a large number of medical records accumulated, which embody the experiential judgement of the TCM practitioners. There are usually the symptoms observed by the practitioner and the according treatment methods within the records. In the treatment procedure, TCM practitioners often refer to the classical records and the prescriptions within them, which makes recommending prescriptions from the records based on the observation of the symptoms valuable in practice. Based on these observations, we propose to model this problem as a matching based recommendation task. To precisely model the relation between symptoms and prescriptions, inspired by the success of pre-trained language models, we propose a TCM domain specific hybrid input construction method and multi-grained negative sampling methods and training objectives. To verify the effectiveness of the proposed method, we conduct extensive experiments on the symptom-prescription dataset. The experiment results show that our proposed method can accurately recommend suitable prescriptions with more abundant candidates for the reference of TCM practitioners, making it more valuable in practice.

Keywords: Prescription Recommendation · Traditional Chinese Medicine · Pre-trained Language Model

1 Introduction

In recent years, with the development of deep learning and natural language processing, artificial intelligence (AI) has been applied in numerous domains. Among them, the integration of AI and healthcare is considered one of the most promising directions. Current research on AI and healthcare primarily focuses on

---

W. Li and Z. Yang—Equal Contribution.

L. Fang et al. (Eds.): CICAI 2023, LNAI 14474, pp. 125–135, 2024.
https://doi.org/10.1007/978-981-99-9119-8_12

modern medical fields, while lacking attention to traditional Chinese medicine (TCM). Leveraging deep learning and natural language processing techniques to explore and utilize the rich knowledge inherited from the historical practices of TCM has significant theoretical and practical implications. Particularly, recommending suitable herbal formulas based on the diagnosis and symptom descriptions provided by TCM practitioners is an important application scenario with practical and theoretical significance.

Previous studies have explored the use of machine learning and deep learning methods for recommending herbal prescriptions based on diagnostic information. [4] initially proposed the use of a sequence-to-sequence model with an improved objective function to generate the herbal components of prescriptions based on textual symptom descriptions. [6,7], and [3] respectively suggested leveraging expert knowledge, attention models to learn the associations between symptoms and herbs, as well as associations between different herbs, and incorporating external herbal knowledge to assist in prescription generation. [8] applied transfer learning using a pre-trained bidirectional encoder, known as BERT (Bidirectional Encoder Representations from Transformers), to the task of generating traditional Chinese medicine prescriptions. These works primarily focus on recommendation through a generative approach. However, generative methods possess certain inherent limitations that are challenging to overcome, such as limited interpretability, difficulty in providing recommendation justifications, and relatively fixed patterns. In actual clinical practice, high reliability is crucial, and these limitations restrict the practical utility of generative methods in assisting traditional Chinese medicine practitioners during the diagnosis and treatment process.

Inspired by the application of next sentence prediction in prompt tuning [2,5] based on pre-trained language models [9], we propose using the next sentence classification objective to match diagnostic texts with herbal prescription components. To effectively leverage the information in herb names, we suggest incorporating both the textual representation and the ID identifier of the herbs as inputs to the model. This approach not only allows the model to capture the intrinsic characteristics of the herbs but also facilitates modeling of herbs that are difficult to automatically identify by considering their textual descriptions. Furthermore, for the constructed diagnostic-prescription inputs, we propose adapting the model through masked language modeling, enabling the establishment of associations at a finer-grained level, including the relationships between diagnosis-herb, herb-herb, and herb name-herb.

Considering the characteristics of traditional Chinese medicine record studies, in order to better train the matching between symptom descriptions and prescription compositions, we introduce two granularity levels of negative sample construction. This involves randomly replacing the original prescription at the prescription level and herb level, respectively, and requires the model to detect the substitutions at different granularity levels, enabling it to differentiate between differences in granularity among prescriptions.

We conducted extensive experiments on a dataset specifically transformed for recommendation scenarios. The results demonstrate that our proposed method achieves more accurate herbal prescription recommendations compared to generative methods. Additionally, the recommended prescriptions exhibit better diversity, thereby providing better assistance to traditional Chinese medicine practitioners in practical diagnosis and treatment processes. Furthermore, the experimental results indicate that directly utilizing the next sentence prediction objective from pre-trained language models for training and prediction does not yield satisfactory matching performance. However, by incorporating model designs that consider the characteristics of traditional Chinese medicine record studies, we achieve significant improvements in matching effectiveness.

The main contributions of this paper can be summarized as follows:

- We propose modeling the objective of recommending herbal prescriptions based on diagnosis as a retrieval-based recommendation task. We introduce the utilization of the next sentence prediction method, based on pre-trained language models, to match symptom descriptions with herbal prescriptions.
- Addressing the characteristics of traditional Chinese medicine prescription recommendation, we propose a hybrid model input construction pattern and a multi-granularity negative sampling method, as well as matching training objectives that align with tasks in the field of traditional Chinese medicine.
- We conducted extensive experiments and analysis on a diagnostic-prescription dataset to validate the effectiveness of the proposed approach.

2 Approach

In this section, we describe how we construct the inputs for the model based on the symptom-formula pairs, as well as how we create training examples and training objectives for matching training.

2.1 Input Construction for Pre-trained Language Model

Based on observations on the characteristics of traditional Chinese medicine (TCM) record data, we propose modeling the correspondence between symptom descriptions and herbal formulations as a next sentence prediction relationship. In other words, if there is a correspondence between the symptom description and the herbal formulation, they form a sentence pair relationship; otherwise, they do not form a sentence pair relationship.

Since the composition of a prescription consists of herbal medicine, we initially consider using the entire herb entity as the input unit. However, due to the nature of Chinese herbal medicine, which is derived from various natural sources, and taking into account the presence of non-standardized herb descriptions in ancient medical texts, directly using the entire herb entity as input would render these herbs out of vocabulary (OOV), thereby reducing the availability of effective context. Additionally, many herb names exhibit certain similarities with their corresponding standardized herb names, with the only difference being the use of different names or the inclusion of preparation methods, places of origin, and other information. For example, "生地" actually refers to the same medicinal substance as "生地黄". In such cases, the textual descriptions of herbs themselves provide valuable information. Based on these observations, we propose combining the entire herb entity with the textual herb name as the input for the composition of a prescription.

Taking into account the considerations mentioned above and drawing inspiration from the input format of the next sentence prediction task in pre-trained language models, we propose constructing the symptom description-prescription pairs that require judgment in the following form:

$$[CLS]X[SEP]Y_s Y_m[SEP]$$

Here, Y_s and Y_m respectively refer to the textual representation of the herb name and the ID identifier of the herb as a whole. The special symbol [CLS] is used to learn the representation at the sample level for the pair, while [SEP] is used to separate the symptom description and the prescription composition and marks the end of the input. Additionally, to differentiate the roles of the symptom description and the prescription, we follow the approach of BERT and incorporate token types in the model input. The symptom description is marked with 0, while the herb portion of the prescription is marked with 1.

2.2 Training Data Sampling

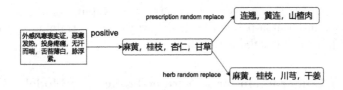

Fig. 1. Example of negative sample construction strategies.

The symptom-prescription pairs in the data naturally form positive examples for training the matching relationship. To train the model's ability to judge whether there is a match, we also need to construct negative examples. Drawing inspiration from the method of constructing negative examples for next sentence prediction and considering the characteristics of TCM medical records, we propose two granularities of random replacement schemes (as shown in Fig. 1). The first scheme is to randomly replace the entire prescription with another prescription from the training set, which is a prescription-level random replacement. The second scheme is to randomly replace certain herbs in the corresponding prescription with another herb, which is a herb-level random replacement.

The first scheme replaces the prescription with a valid prescription, but it may differ significantly from the original paired prescription, resulting in insufficient discriminative ability learned by the model, especially at a finer granularity level such as the herb level. The second scheme constructs prescriptions that are closer to the original paired prescription, allowing the model to learn the distinction and grasp the local herb information within the prescription. However, the constructed prescriptions may not be feasible in practice, meaning that the compatibility of the herbs in the prescription may be compromised. In the actual construction of negative examples, we choose to replace less than half of the herbs in the prescription to maintain the original framework of the prescription as much as possible, and the number of replacements itself is determined by random sampling.

For prescriptions with a small number of herbs (less than 3 ingredients), we adopt the first scheme of replacing the entire prescription because replacing individual herbs in this case wouldn't have much significance. For other prescriptions, we randomly choose one of the two schemes with an equal probability, meaning there's a 50% chance of using the first scheme (replacing the entire prescription) and a 50% chance of using the second scheme (replacing individual herbs).

2.3   Training Objective

To enable the model to learn both coarse-grained and fine-grained alignment information, we propose training the model using Masked Language Modeling (MLM) objective, the Symptom-Prescription Matching (SPM) objective, and the Herb Replacement Detection (HRD) objective to train the model from different perspectives.

Masked Language Modeling Objective. Due to the lack of ID representations for the complete herb names in the pre-trained word vector parameters of the pre-trained models, and considering that the training corpus of the pre-trained language models consists of general domain data, it is necessary to adapt the training set data using the input construction method described in Sect. 2.1. This adaptation involves employing a masked language model to learn herb word vector representations and the associations between herb textual descriptions and herb whole IDs proposed in this paper. When randomly replacing input tokens, we drew inspiration from BERT's approach, but with a modification. We only replace non-special characters with the "[MASK]" token with a probability of 15%, allowing the model to learn associations between the masked words or herbs and their contexts. To better capture the associations between symptoms and herbs, we slightly deviate from the original BERT model's masking strategy. Specifically, in some instances, we mask only the symptoms or the formulations separately, while in other instances, we perform completely random masking.

Symptom-Prescription Matching Objective. To directly model the overall matching relationship between symptoms and the composition of the formula, we employ a training objective similar to that of the next sentence prediction task. We utilize the hidden vector representation of the special token [CLS], which is encoded by the BERT encoder, to predict whether the input pair is a match. If there is a match, a label of 1 is assigned; otherwise, a label of 0 is assigned. The loss function for the matching relationship is the cross-entropy between the predicted match and the ground truth label.

$$L_{match} = -\sum_{i=0}^{1} y_i log p_i \tag{1}$$

where p represents the predicted probability for the overall matching, y denotes the actual label indicating whether there is a match or not, and i takes the value of 0 or 1, indicating the match or non-match scenario, respectively.

Herb Replacement Detection Objective. In order to enable the model to differentiate more fine-grained matching information between the symptoms and individual herbs in the prescriptions (i.e., which parts of the herbs match the symptoms and which parts do not), we draw inspiration from the work of [1] and propose a method to train the model to detect specific mismatched herb information in non-matching prescriptions while simultaneously learning the overall matching relationship between symptoms and prescriptions. Specifically, for the negative examples constructed through the method of replacing local herbs mentioned in Sect. 2.2, we train the model to predict which herbs in the prescriptions are original (matching the symptoms) and which herbs are replaced (not matching the symptoms). We assign a label of 1 to the originally correctly matched herbs and a label of 0 to the replaced herbs. The logic behind label assignment is consistent with the coarse-grained labels, aiming to help the model learn the finer-grained reasons for mismatches. For the negative examples constructed through the method of randomly replacing prescriptions at the prescription level, we do not train the model to detect whether herbs are replaced, as the majority of herbs are replaced in this case. The loss function used in this context is similar to the cross-entropy used for coarse-grained labels but applied to each herb in the negative examples (for the method of replacing herbs):

$$L_{token} = -\frac{1}{L} \sum_{j} \sum_{i=0}^{1} y_i^j log p_i^j \tag{2}$$

The overall training loss is the sum of three components: the loss of the fine-grained masked language model, the loss of the coarse-grained symptom-prescription matching, and the loss of the replaced herb detection. For the fine-grained masked language model, the loss is calculated based on the predicted probability p of whether a herb is replaced, the actual label y indicating whether it is a replaced herb, the matching label i (0 or 1) indicating whether the symptom and prescription match, the position j in the sample, and the input sequence length L. The final training loss can be expressed as follows:

$$Loss = L_{mlm} + L_{match} + L_{token} \tag{3}$$

During the inference testing phase, we directly use the probability p_i of the coarse-grained matching judgment from Eq. 1 as the prediction probability. We then sort the probabilities in descending order and obtain the actual order of recommended prescriptions.

3 Experiment

In this chapter, we introduce the experimental setup, the data used in the experiments, the evaluation metrics employed, as well as the experimental results and analysis.

3.1 Setting

The BERT model used in this study is Guwen-bert (base)[1], which is pretrained on classical Chinese language corpus. The hidden layer size of the model is 768, with 12 layers and 12 heads in the multi-head attention mechanism. The masked language model was trained for 10 epochs on the symptom-prescription pairs data. The symptom-prescription matching objective was trained for 5 epochs. The model with the highest Macro-F1 score on the development set during training was selected as the test model. Regarding the size of the herb vocabulary, we selected the top 3000 herbs with the highest frequency of occurrence as the vocabulary for whole herbs when represented as characters. The remaining herbs (including noise that has not been cleaned) were represented in textual form. For each positive sample in the matching training, two negative samples were sampled. The batch size during training was set to 24 (limited by GPU memory). For the symptom description, the first 150 characters were extracted, and for the prescription, the first 50 herbs were extracted. The selection of hyper-parameters was based on the highest Macro F1 score obtained on the development set.

3.2 Data

Based on the Chinese medical record data used by [4], we transformed the data into a format suitable for the recommendation task. Using the Jaccard matching method, we first found the top 20 symptom-prescription pairs in the prescription database that were closest to the target symptom description (excluding the symptom-prescription pairs in the test set). These 20 identified prescriptions were mixed with the target prescription as negative examples, and the model was required to find the most suitable prescription for the target symptom from these 21 prescriptions. For the sake of comparison, we used the same test set as [4]. The test set was divided into two parts.

---

[1] GuwenBERT https://github.com/ethan-yt/guwenbert.

Table 1. Overall results on TextBook and Crawl test set. Precision@5, Recall@5 and F1@5 are provided after "/" for our proposed method. seq2seq and multi-label are the baselines applied in [4].

TextBook	MRR	MAP	MacroPrecision	MacroRecall	MacroF1
proposal	28.68	28.68	40.42/79.52	47.38/84.64	42.07/80.44
seq2seq	-	-	30.97/-	23.70/-	26.85/-
multi-label	-	-	13.51/-	40.49/-	20.26/-
Li and Yang [4]	-	-	38.22/-	30.18/-	33.73/-
Crawl					
proposal	21.17	21.17	24.07/54.53	24.73/55.60	23.21/52.97
seq2seq	-	-	26.03/-	13.52/-	17.80/-
multi-label	-	-	10.83/-	29.72/-	15.87/-
Li and Yang [4]	-	-	29.57/-	17.30/-	21.83/-

3.3 Evaluation Metrics

In this section, we introduce the evaluation metrics used in our experiments. To assess the performance of the model from different perspectives, we employ two types of evaluation metrics. The first type is commonly used in recommender systems, namely MRR (Mean Reciprocal Rank) and MAP (Mean Average Precision). These metrics focus on the relative ranking of the model's results, where higher scores are assigned when the correct answer is ranked higher by the model. Another type focuses on the degree of overlap between the herb composition of recommended prescriptions and the herb composition of standard answers, aiming for finer granularity. The higher the degree of overlap, the closer the recommended prescriptions are to the answers. This type of method includes Macro Precision, Macro Recall, and Macro F1.

3.4 Results

In Table 1, we present the experimental results of our approach in comparison to the results reported in previous work [4], which used the same dataset as ours. It can be observed that our proposed method achieved significant improvements in Macro F1 values compared to the results obtained by the previous generative models, particularly on the more accurate TextBook test set. The Macro F1@1 reached 42.07, a substantial improvement over the 33.73 achieved by Li and Yang's method [4].

Furthermore, our method achieved a Macro F1@5 of 80.44 on the TextBook test set and 52.97 on the Crawl dataset. A higher Macro F1@5 indicates that, in the context of prescription recommendation, providing the top 5 prescriptions that the model considers optimal as candidate recommendations can yield correct recommendations with a high probability, making our approach more practical compared to generative methods.

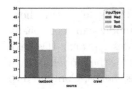 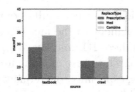 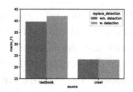

Fig. 2. Macro-F1 for different input construction methods. "Med" indicates only ID of herbs are used, "Text" indicates only textual names are used.

Fig. 3. Macro-F1 for different negative sampling methods. "Prescription" indicates prescription level negative sample, "Med" indicates herb level negative sampling.

Fig. 4. Macro-F1 for whether herb replacement detection objective is applied.

Although our method did not achieve high scores for retrieval-related evaluation metrics such as MRR and MAP, this is due to the characteristic of Traditional Chinese Medicine records, where similar medical conditions may have different treatment approaches and, therefore, different prescription solutions. Additionally, in the context of prescription recommendation, the absence of an exact match with the prescriptions in textbooks or medical records does not necessarily mean the answer is incorrect. Some discrepancies may arise from non-standardized herb terminologies, while others may result from variations in diagnostic details while still providing prescriptions that are similar to the answers but with some additions or omissions of herbs. These aspects can be reflected in the Macro F1 value. We also provide specific examples in Appendix to further illustrate this.

3.5 Analysis

In this section, we will analyze the effectiveness of our proposed method from several different aspects.

Input Construction Effect. In Fig. 2, we present the results of Macro F1@1 obtained from different input construction methods mentioned in Sect. 2.1 on two test sets (other metrics show a similar trend to Macro F1@1). From the results, we can observe that the performance is weakest when solely using herb text as input (labeled as "Text" in Fig. 2). We believe this is because the herb-related text encountered by the pre-trained language model in the pre-training corpus is sparse, which makes it difficult for the model to accurately differentiate and recognize different herbs based solely on their textual representations. Although the training of the masked language model in our proposed method involves herb-symptom pairs, where herbs are more densely present, the overall quantity is still insufficient to support the model in learning precise herb recognition and differentiation abilities. On the other hand, using herb identifiers (IDs) as input (labeled as "ID" in Fig. 2) yields better results compared to solely using herb text. We attribute this improvement to the fact that the model can more easily learn the associations between herb IDs and the symptom descriptions. IDs have a smaller semantic space compared to text, making it easier for the model to learn more accurate representations, especially for frequently used herbs. Compared to the two aforementioned individual herb input construction methods, the hybrid input construction method that combines herb text and herb IDs (labeled as "Both" in Fig. 2) provides richer information. It can capture the information of herb IDs for common herbs and the information of herb text representations for less common herbs. Additionally, it allows the model to learn the associations between herb text and herb IDs, leading to the best performance.

Negative Sampling Effect. In Fig. 3, we present the results of different negative sampling methods mentioned in Sect. 2.2 on Macro F1 (@1). It is important to note that the only difference here lies in the sampling methods, while the number or proportion of negative samples remains the same. From the graph, we can observe that both combined negative sampling methods proposed in this paper achieve the best performance on both test sets. On the TextBook test set, the effect of randomly replacing herbs at a finer-grained herb level is significantly better than randomly replacing herbs at a coarser-grained prescription level. On the Crawl test set, the two methods show similar performance. We believe this is because the TextBook test set has higher data quality, making it more sensitive to differences in the model's understanding of herb details. By combining negative samples at two different granularities, the model can better learn how to match symptom descriptions and prescriptions at different levels, resulting in the best matching performance.

Herb Replacement Detection Effect. In Fig. 4, we present the results of whether to use the replacement herb detection objective mentioned in Sect. 2.3 during training. From the results, we can observe that using this training objective brings some improvement on the TextBook test set, but the difference is not significant on the Crawl test set. We believe this phenomenon is due to the higher data quality of the TextBook test set, which better reflects the model's ability to grasp herb details. In fact, for the macro F1@5 metric (not shown in the graph), after adding the replacement herb detection objective, the macro F1@5 of TextBook improved from 76.89 to 80.44, and the macro F1@5 of Crawl improved from 52.18 to 52.97. This further confirms the effectiveness of this training objective from another perspective.

4   Conclusion

This article proposes a symptom-prescription matching method based on pre-trained language models for the task of recommending prescriptions based on symptom descriptions. In this method, we model the symptom-prescription matching as the next sentence prediction task in pre-trained language models. Considering the characteristics of TCM medical records, we propose a hybrid medication input construction method, a multi-granularity negative sampling method, and training objectives that are adapted to the task, allowing the model to learn the associations and matching relationships at different levels between symptom descriptions, prescriptions, and herbs. Extensive experiments and analysis demonstrate that our proposed method can provide more accurate prescription recommendations compared to generative methods and offer more diverse candidate answers, thereby enhancing the practical diagnostic process for TCM practitioners.

Acknowledgements. This research project is supported by National Key R&D Program of China (2020YFC2003100, 2020YFC2003102), Innovation Team and Talents Cultivation Program of National Administration of Traditional Chinese Medicine. (No: ZYYCXTD-C-202001), Science Foundation of Beijing Language and Culture University (supported by "the Fundamental Research Funds for the Central Universities") (No. 21YBB19)

References

1. Chuang, Y.S., et al.: Diffcse: difference-based contrastive learning for sentence embeddings. arXiv preprint arXiv:2204.10298 (2022)
2. Han, X., Zhao, W., Ding, N., Liu, Z., Sun, M.: PTR: prompt tuning with rules for text classification. arXiv preprint arXiv:2105.11259 (2021)
3. Li, C., Liu, D., Yang, K., Huang, X., Lv, J.: Herb-know: knowledge enhanced prescription generation for traditional Chinese medicine. In: 2020 IEEE International Conference on Bioinformatics and Biomedicine (BIBM), pp. 1560–1567. IEEE (2020)
4. Li, W., Yang, Z.: Exploration on generating traditional Chinese medicine prescriptions from symptoms with an end-to-end approach. In: Tang, J., Kan, M.-Y., Zhao, D., Li, S., Zan, H. (eds.) NLPCC 2019. LNCS (LNAI), vol. 11838, pp. 486–498. Springer, Cham (2019). https://doi.org/10.1007/978-3-030-32233-5_38
5. Liu, P., Yuan, W., Fu, J., Jiang, Z., Hayashi, H., Neubig, G.: Pre-train, prompt, and predict: a systematic survey of prompting methods in natural language processing. arXiv preprint arXiv:2107.13586 (2021)
6. Liu, Z., et al.: Attentiveherb: a novel method for traditional medicine prescription generation. IEEE Access **7**, 139069–139085 (2019)
7. Ruan, C., Luo, H., Wu, Y., Yang, Y.: TPGEN: prescription generation using knowledge-guided translator (2021)
8. Shi, Q.Y., Tan, L.Z., Seng, L.L., Wang, H.J., et al.: Intelligent prescription-generating models of traditional Chinese medicine based on deep learning. World J. Traditional Chin. Med. **7**(3), 361 (2021)
9. Sun, Y., Zheng, Y., Hao, C., Qiu, H.: NSP-BERT: a prompt-based zero-shot learner through an original pre-training task-next sentence prediction. arXiv abs/2109.03564 (2021)

Ensemble Learning with Time Accumulative Effect for Early Diagnosis of Alzheimer's Disease

Zhou Zhou, Hong Yu$^{(\boxtimes)}$, and Guoyin Wang

The School of Computer Science and Technology, Chongqing University of Posts and Telecommunications, Chongqing, China
yuhong@cqupt.edu.cn

Abstract. Alzheimer's disease (AD) is a neurodegenerative disorder. Early diagnosis of AD is critical for disease management and treatment options to slow progression. The existing early diagnosis algorithms for AD ignore the distinct time accumulative effect seen in chronic diseases and do not address the problem of adaptation of multi-source heterogeneous data to a single learner. We use the idea of ensemble learning to train multi-source heterogeneous data using different learners to solve the problem. The time accumulative operator is fusing while being trained. The outcomes of many learners are then combined using the decision fusion approach. Experimental results demonstrate that our algorithmic framework attains an average accuracy of 75.75% and the time accumulative effect also benefits our model.

Keywords: Alzheimer's disease · Ensemble learning · Machine learning · Time accumulative effect

1 Introduction

Alzheimer's disease (AD), a prevalent form of dementia, is a progressive and irreversible neurodegenerative condition. According to the Alzheimer's Association, the current global estimate for the prevalence of dementia exceeds 55 million individuals [7,11,13]. Despite advancements in clinical practices [8,16], the accurate diagnosis of AD pathology and disease progression remains below 50% [1] in practical medical scenarios. The first symptoms of AD are usually memory loss and disorganized speech. These symptoms come on suddenly for the patient, but the brain lesions that cause them may have started 20 years or more before the symptoms appear [14]. Therefore, the urgent requirement for accurate diagnosis and efficacious intervention has become a meaningful research [12].

Current research in the field of AD prediction algorithms primarily concentrates on identifying the cognitive impairment individuals who do not meet the clinical diagnostic criteria for AD. According to the Diagnostic and Statistical Manual of Mental Disorders (DSM-5), these individuals are considered to be in the mild cognitive impairment (MCI) stage. MCI serves as a crucial transitional stage for interventions aimed at preventing AD. Making interventions targeting the MCI phase potentially is the most efficacious strategy for delaying the onset of AD.

The time accumulative effect plays a crucial role in the medical field, particularly in the context of chronic diseases like AD. This phenomenon refers to the cumulative impact of repeated transient changes acting on the same receptor over time. Early diagnosis aims to leverage the precursor symptoms exhibited by individuals who have not yet developed the disease, and these symptoms accumulate gradually over time. Consequently, in real-world scenarios, physicians must conduct continuous long-term testing and refer to previous data to make accurate diagnoses.

Although some progress has been made in the studies, several challenges persist in applying AI algorithms to real-world medical scenarios. When dealing with multi-source heterogeneous data, training a single learner on the combined data may overlook important details. Furthermore, AI algorithms require the incorporation of medical knowledge to enhance correctness and interpretability. Yet combining medical knowledge with algorithmic models still presents difficulties.

The current stage of AD early diagnosis algorithm research ignores the problem of adaptability of different data to a single learner and the time accumulative effect. To address these issues, this paper proposes an algorithmic framework for early diagnosis of AD with the following contributions:

1. A time accumulative operator, which can integrate medical knowledge into mathematical model calculations, was designed by analyzing the conceptual features of the time accumulative effect.
2. An AD early diagnosis prediction algorithm M3TA is proposed, which can take into account multi-source heterogeneous data in real-world scenarios to provide decision support for clinicians.

2   Related Work

Most researchers today translate the early diagnosis of AD into a three-stage "CN-MCI-AD" multiclassification problem. Current algorithmic research in this domain can be broadly categorized into three main approaches: deep learning, traditional machine learning, and multi-task learning.

Deep learning has demonstrated immense potential for clinical decision support in AD [10,17]. One key advantage of DL lies in its capability to directly learn highly predictive features. However, it has been shown that depth models influence the results by combining weights at multiple layers [4], leading to their clinically meaningful non-interpretability.

Research on cross-sectional analysis of AD using traditional machine learning algorithms tends to be systematic and mature. Shen *et al.* [15] use support vector machines(SVM) to classify MCI patients on both auxiliary and predictive data. Wang *et al.* [20] regard structural connectomics as complementary information and use logistic regression to predict AD patients, verifying that structural connectomics has a contribution to AD prediction. Dimitriadis *et al.* [5] select multiple features from a subset of the entire dataset by random forest(RF) and use fusion methods for classification and obtain the integrated classification results by majority voting. Farouk *et al.* [6] extract texture features from the gray level co-occurrence matrix and voxel-based morphometric neuroimaging analysis for supervised machine learning classification of AD.

Multi-task Learning (MTL) is a machine learning framework that learns multiple interrelated tasks together to mine the valuable information present in them to improve the generalization performance of all tasks. Liu *et al.* [9] propose a multi-task learning method, combined with hyper-graphs to extract higher-order relationships between samples. Michele *et al.* [2] propose the SS-MTL method using a time window to capture the temporal correlation of the electronic health record data on for kidney disease prediction. Wang *et al.* [19] use different regularizations to capture the intrinsic associations of tasks across time and to select the most discriminative features for AD early diagnostic prediction. Brand *et al.* [3] improve on this basis for group-guided sparse group lasso regularization, performing both regression and classification tasks. Tang *et al.* [18] propose a novel feature-aware sparse-induced regularization that exploits correlations between brain regions to predict cognitive scores and identify stable biomarkers.

3   Materials and Methods

The purpose of this study is to develop an algorithmic framework for the early diagnosis of AD that can be employed in real-world medical scenarios. The structure of the proposed framework is illustrated in Fig. 1. In the first step, the multi-source heterogeneous data are separately utilized to train distinct learners. Next, the outputs of the different learners are combined using the time accumulative operator. Finally, a multi-source data weight assignor is trained to perform decision fusion on the multiple prediction results, leading to generate the output of the complete early diagnosis algorithm framework. The main mathematical notations are summarized in Table 1.

Table 1. Notations

Symbol	Description
n	# of sample
m	# of feature
t	# of follow-up visit
c	# of categories
ω	Between time periods influence the weighting
X, x_i	The matrix of input $X \in \mathbb{R}^{n \times t \times m}$ and input sample $x_i \in \mathbb{R}^{t \times m}$
W, w_j	The matrix of param $W \in \mathbb{R}^{c \times m \times t}$ and each category $w_j \in \mathbb{R}^{m \times t}$
Y, y_i	The matrix of labels $Y \in \mathbb{R}^{n \times t \times c}$ and the i_{th} label $y_i \in \mathbb{R}^{t \times c}$
T	The time accumulative operator $T \in \mathbb{R}^{t \times t}$

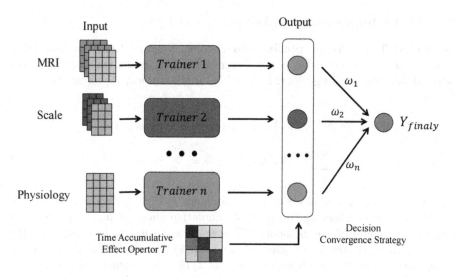

Fig. 1. Model architecture diagram.

3.1   Dataset

The data used in this study are obtained from the Alzheimer's Disease Neuroimaging Initiative (ADNI) database (ADNI.loni.ucla.edu). We categorize the data according to the source and form of collection into MRI image data, scale examination data, plasma measurement data, physical examination data, vital sign data, and symptom data. MRI image data, scale examination data, and plasma measurement data are follow-up data, which are sampled every six months, with three samples per patient. Physical examination data, vital sign data, and symptom data are baseline data. A total of 114 patients are selected, with a total of 342 follow-up data.

3.2   Pre-processing

MRI image data are image preprocessed in FreeSurFer (http://surfer.nmr.mgh.harvard.edu), and the remaining data has three types of missing data, missing samples in the follow-up data, missing continuous-value features of samples in the baseline data, and missing discrete-value features. Out of 114 patients, 57 samples are missing. For the missing samples, we fill them with the mean of the grouping. For the continuous missing values, the same mean of the same standard data is used to fill in. For the discrete missing values, the same labeled multi-valued data are used to fill in the missing values. All continuous values normalize with the mean-standard deviation method.

3.3 M3TA Implementation Details

Design of Time Accumulative Operators. Starting from the concept of time accumulative effect, this paper designs a time accumulative operator incorporated into the existing algorithm and its mathematical representation is as follows:

$$Y_t = \omega_0 Y_t' + \omega_1 Y_{t-1}' + \omega_2 Y_{t-2}' + \cdots + \omega_{t-1} Y'_1$$

$$s.t. \sum_{i=0}^{t-1} \omega_i = 1, \ \omega_0 > \omega_1 > \omega_2 > \cdots > \omega_{t-1} \quad (1)$$

where Y_t denotes the final result at the moment t, Y_t' denotes the result of the model without considering the time accumulative effect at moment t, and ω_t denotes the weight of the moments t. The weights are assigned according to the following rules: (1) the sum of the weights is 1; (2) the weights are decreasing in order of the time distance from the moment t; (3) the moments after t have no effect on moment t. Based on the above representation, the time accumulative operator is designed to factorize as follows:

$$T = \begin{pmatrix} \omega_{1,1} & & & \\ \omega_{2,1} & \omega_{2,2} & & \\ \vdots & \vdots & \ddots & \\ \omega_{t,1} & \omega_{t,2} & \cdots & \omega_{t,t} \end{pmatrix}$$

$$s.t. \ \omega_{i,.} = 1, \ \omega_{ij} \in [0,1], \ \omega_{i,1} < \omega_{i,2} < \cdots < \omega_{i,i} \quad (2)$$

where $\omega_{i,j}$ denotes the influence weight of period i on period j.

Follow-Up Data Learner Design. In this paper, we use a longitudinal multi-task learning model on the follow-up data. At the same time, we integrate the time accumulative operator into the longitudinal multi-task learning model. The output function of the integrated time accumulative operator is:

$$Y = \text{M3TA\_F}(X, W, T) \quad (3)$$

where $X = \{x_1, x_2, \cdots, x_n\} \in \mathbb{R}^{n \times t \times m}, x_i = \{x_i^1, x_i^2, \cdots, x_i^t\} \in \mathbb{R}^{t \times m}$ denotes the model input, $W = \{w_1, w_2, \cdots, w_c\} \in \mathbb{R}^{c \times m \times t}, w_j = \{w_j^1, w_j^2, \cdots, w_j^m\} \in \mathbb{R}^{m \times t}$ denotes the model parameters and $Y = \{y_1, y_2, \cdots, y_n\} \in \mathbb{R}^{n \times t \times c}, y_i = \{y_i^1, y_i^2, \cdots, y_i^t\} \in \mathbb{R}^{t \times c}$ denotes the model output.

The loss function of the longitudinal multi-task learning model is obtained as:

$$L(W) = \|\text{M3TA\_F}(X, W, T) - Y\|_F^2 \quad (4)$$

Based on existing longitudinal multi-task algorithms [2], the objective function M3TA that takes into account the time accumulative effect can be expressed as follows:

$$\min_{W} J\left(W\right) = L\left(W\right) + \lambda_1 \left\| W \right\|_1$$
$$+ \lambda_2 \left\| RW^T \right\|_1 + \lambda_3 \left\| W \right\|_{2,1} \tag{5}$$

where λ_1, λ_2 and λ_3 are the regularization parameters. $\left\| W \right\|_1$ is the Lasso coefficient regularization term, $\left\| W \right\|_{2,1} = \sum_{k=0}^{c} \sum_{i=0}^{m} \sqrt{\sum_{j=0}^{t} w_{kij}^2}$ is the group sparse regularization term, they allow for joint feature selection and task-specific feature selection for multiple tasks during model optimization. $\left\| RW^T \right\|_1$ is the temporal smoothing regularization term, which takes into account the temporal correlation between tasks, R is the temporal smoothing matrix, which takes values as:

$$R_{i,j} = \begin{cases} 1, & i = j \\ -1, & i = j - 1 \end{cases} \quad , R \in \mathbb{R}^{t-1 \times t} \tag{6}$$

Fusion of Multi-source Heterogeneous Prediction Results with Time Accumulative Operator. Unlike follow-up data, the baseline data trainer need to combine the prediction results with the time accumulative operator in the stage of fusion of the prediction results, and the combination steps are shown in Fig. 2:

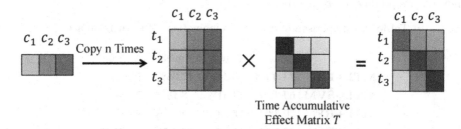

Fig. 2. Steps for combining multi-source heterogeneous prediction results with the time accumulative operator.

We replicate the baseline data predictions t times and then multiply them with the time accumulative operator to obtain the combined baseline data predictions.

$$Y' = YT \tag{7}$$

After obtaining all the multi-source heterogeneous prediction results considering the time accumulative effect, the weight assignor needs to be trained to weight and sum the multi-source heterogeneous prediction results to obtain the final prediction results.

$$Y = \beta_1 Y_1' + \beta_2 Y_2' + \cdots + \beta_s Y_s' \tag{8}$$

4 Results

4.1 Evaluation Metrics

We use 10-fold cross-validation (10-fold CV) to evaluate the predictive performance of this algorithmic framework. We encode the labels "CN", "MCI" and "AD" as one-hot coding for learning. We use accuracy (acc), precision (pre), specificity (spe), sensitivity (sen) and f1-score (f1) to evaluate the performance of the algorithmic framework.

4.2 Comparison of Different Combinations of Learners

We use longitudinal multi-task learning models to train the follow-up data. For the baseline data, we explore the best model combination by using LR, SVM, RF and decision tree(DT) algorithms in combination with longitudinal multi-task learning, respectively. The results are shown in Table 2 and Table 3, where Time1, Time2 and Time3 represent the current time, the next 6 months and the next year's prediction, respectively.

Table 2. Comparison of different combinations of learners on acc

	Time1	Time2	Time3	Average
MTL+LR	**71.11%**	75.34%	77.35%	74.60%
MTL+SVM	69.42%	74.51%	78.18%	74.04%
MTL+RF	70.38%	**75.60%**	74.62%	73.53%
MTL+DT	70.58%	75.00%	**81.67%**	**75.75%**

Based on the analysis of the experimental results, it shows that the model combination of MTL+DT performed the best on each time group with average precision, average specificity, average sensitivity, average f1-score and average accuracy of 76.83%, 88.39%, 79.53%, 36.75%, and 75.75%, respectively. For subsequent experimental setups, the MTL+DT model combination will be utilized as the M3TA algorithm framework.

4.3 Comparison with Single Learner

We compare the M3TA framework with a single learner. We input the first sampled data from the follow-up data as the baseline data and input the replication

Table 3. Comparison of different combinations of learners on precision, specificity, sensitivity and f1-score

CN-Time1	pre	spe	sen	f1
MTL+LR	**68.01%**	**78.88%**	98.25%	**38.53%**
MTL+SVM	66.04%	76.38%	98.00%	37.76%
MTL+RF	65.53%	77.62%	95.50%	37.54%
MTL+DT	65.22%	75.41%	98.00%	37.37%
CN-Time2	pre	spe	sen	f1
MTL+LR	76.33%	87.45%	98.00%	41.83%
MTL+SVM	74.05%	86.95%	93.00%	40.40%
MTL+RF	**80.67%**	**89.56%**	95.50%	**42.54%**
MTL+DT	76.33%	87.45%	**98.00%**	41.83%
CN-Time3	pre	spe	sen	f1
MTL+LR	76.33%	87.45%	98.00%	41.83%
MTL+SVM	78.00%	89.36%	94.67%	41.44%
MTL+RF	**84.33%**	**92.73%**	98.00%	**44.52%**
MTL+DT	75.81%	86.57%	**100.00%**	42.17%
MCI-Time1	pre	spe	sen	f1
MTL+LR	83.25%	90.62%	53.85%	**32.07%**
MTL+SVM	**84.98%**	**91.57%**	50.25%	30.88%
MTL+RF	82.27%	89.11%	53.57%	31.66%
MTL+DT	83.45%	90.70%	**54.09%**	32.01%
MCI-Time2	pre	spe	sen	f1
MTL+LR	83.32%	88.86%	**62.58%**	**35.22%**
MTL+SVM	**87.44%**	**92.40%**	56.03%	33.66%
MTL+RF	87.15%	90.02%	58.83%	34.54%
MTL+DT	86.44%	90.86%	57.40%	33.65%
MCI-Time3	pre	spe	sen	f1
MTL+LR	97.50%	98.00%	57.17%	34.69%
MTL+SVM	94.17%	96.00%	60.40%	36.05%
MTL+RF	91.66%	96.57%	50.98%	31.77%
MTL+DT	**98.00%**	**98.00%**	**64.43%**	**38.13%**
AD-Time1	pre	spe	sen	f1
MTL+LR	64.30%	88.68%	74.72%	33.05%
MTL+SVM	64.33%	87.36%	**76.67%**	33.07%
MTL+RF	67.33%	89.44%	75.00%	34.02%
MTL+DT	**74.44%**	**91.24%**	73.89%	**34.57%**
AD-Time2	pre	spe	sen	f1
MTL+LR	**63.03%**	**86.75%**	73.75%	32.86%
MTL+SVM	59.48%	84.18%	**82.92%**	32.95%
MTL+RF	58.81%	84.83%	82.50%	**33.11%**
MTL+DT	60.45%	85.22%	80.83%	32.89%
AD-Time3	pre	spe	sen	f1
MTL+LR	65.05%	84.13%	89.17%	36.09%
MTL+SVM	66.33%	84.13%	**94.17%**	37.22%
MTL+RF	54.88%	76.29%	92.50%	33.16%
MTL+DT	**71.33%**	**90.10%**	89.17%	**38.15%**

of the baseline data as the follow-up data. We compare single models on accuracy and the results are shown in Table 4.

Table 4. Comparison of single learners on acc

	Time1	Time2	Time3	Average
MTL	59.70%	66.70%	76.29%	67.56%
LR	61.72%	64.85%	72.73%	66.73%
SVM	33.33%	32.35%	20.91%	28.86%
RF	43.01%	46.86%	47.27%	45.71%
DT	36.36%	36.89%	30.00%	34.41%
OURS	**70.58%**	**75.00%**	**81.67%**	**75.75%**

Based on the analysis of the experimental results, our algorithmic framework surpasses all single learner models in terms of average accuracy for all time groups. Among the single learner models, the longitudinal multi-task learning model exhibits the highest performance. This outcome can be attributed to the substantial presence of follow-up data, which accounts for approximately 89% of the dataset used in the experiment.

4.4 Validity of the Time Accumulative Operator

To verify the validity of time accumulative operator, we conduct ablation experiments for all models as well as M3TA. The experimental results are shown in Table 5.

Table 5. Comparison of time accumulative operator on various models

	Time1	Time2	Time3	Average
MTL (noTA)	55.90%	62.31%	74.55%	64.25%
MTL	59.70%	66.70%	76.29%	67.56%
LR (noTA)	50.30%	47.73%	39.39%	45.81%
LR	61.72%	64.85%	72.73%	66.43%
SVM (noTA)	33.33%	32.35%	20.91%	28.86%
SVM	33.33%	32.35%	20.91%	28.86%
RF (noTA)	33.64%	32.80%	21.82%	29.42%
RF	43.01%	46.86%	47.27%	45.71%
DT (noTA)	33.33%	32.35%	20.91%	28.86%
DT	36.36%	36.89%	30.00%	34.42%
OURS (noTA)	70.28%	74.38%	80.76%	75.14%
OURS	**70.58%**	**75.00%**	**81.67%**	**75.75%**

Based on the experimental results, the use of time accumulative operator leads to improvements in average accuracy across each time groups. Notably, SVM is insensitive to time accumulative operator. Furthermore, our algorithmic framework is improved, confirming the effectiveness of the time accumulative operator in enhancing the predictive performance.

5 Conclusion

We propose an algorithmic framework for early diagnosis of AD based on ensemble learning and time accumulative effects. The framework is designed to address the challenges posed by different data in real-world scenarios and fulfill the requirements of predicting disease progression and integrating medical knowledge. This approach contributes to the mathematicalization of medical knowledge and promotes interdisciplinary research in medicine and computer science. Our algorithm framework has some practical applications for various real-world scenarios and helps doctors and patients to take preventive measures in advance.

Although our algorithmic framework has a good performance in the prediction of disease evolution trends, there is still much room for improvement. In the subsequent research, the time accumulative operator can be learned dynamically during the training of the model. Our algorithm is only validated on the ADNI dataset, and subsequent studies will be conducted on larger samples, more complex dataset, and even real dataset.

Acknowledgment. The investigators within the ADNI are thanked for contributing to the design and implementation of ADNI and/or provided data, but do not participate in analysis or writing of this study. A complete listing of ADNI investigators can be found at: https://adni.loni.usc.edu/wp-content/uploads/how_to_apply/ADNI_Acknowledgement_List.pdf.

This work was jointly supported by the National Key R&D Program of China under grant number 2021YFF0704103, and the National Natural Science Foundation of China under grant number 62136002.

References

1. Alzheimer's Association: 2016 Alzheimer's disease facts and figures. Alzheimer's Dementia **12**(4), 459–509 (2016)
2. Bernardini, M., Romeo, L., Frontoni, E., Amini, M.R.: A semi-supervised multi-task learning approach for predicting short-term kidney disease evolution. IEEE J. Biomed. Health Inform. **25**(10), 3983–3994 (2021)
3. Brand, L., Nichols, K., Wang, H., Shen, L., Huang, H.: Joint multi-modal longitudinal regression and classification for Alzheimer's disease prediction. IEEE Trans. Med. Imaging **39**(6), 1845–1855 (2019)
4. Che, Z., Purushotham, S., Khemani, R., Liu, Y.: Distilling knowledge from deep networks with applications to healthcare domain (2015)

5. Dimitriadis, S.I., Liparas, D., Tsolaki, M.N., Initiative, A.D.N., et al.: Random forest feature selection, fusion and ensemble strategy: combining multiple morphological MRI measures to discriminate among healthy elderly, MCI, CMCI and Alzheimer's disease patients: from the Alzheimer's disease neuroimaging initiative (ADNI) database. J. Neurosci. Methods **302**, 14–23 (2018)
6. Farouk, Y., Rady, S.: Supervised classification techniques for identifying Alzheimer's disease. In: Hassanien, A.E., Tolba, M.F., Shaalan, K., Azar, A.T. (eds.) AISI 2018. AISC, vol. 845, pp. 189–197. Springer, Cham (2019). https://doi.org/10.1007/978-3-319-99010-1_17
7. Gustavsson, A., et al.: Global estimates on the number of persons across the Alzheimer's disease continuum. Alzheimer's Dementia **19**(2), 658–670 (2023)
8. Kumar, A., Fontana, I.C., Nordberg, A.: Reactive astrogliosis: a friend or foe in the pathogenesis of Alzheimer's disease. J. Neurochem. **164**(3), 309–324 (2023)
9. Liu, M., Gao, Y., Yap, P.T., Shen, D.: Multi-Hypergraph learning for incomplete multimodality data. IEEE J. Biomed. Health Inform. **22**(4), 1197–1208 (2017)
10. Liu, S., et al.: Multimodal neuroimaging feature learning for multiclass diagnosis of Alzheimer's disease. IEEE Trans. Biomed. Eng. **62**(4), 1132–1140 (2014)
11. Nichols, E., et al.: Estimation of the global prevalence of dementia in 2019 and forecasted prevalence in 2050: an analysis for the global burden of disease study 2019. Lancet Publ. Health **7**(2), e105–e125 (2022)
12. Pearson, R.K., Kingan, R.J., Hochberg, A.: Disease progression modeling from historical clinical databases. In: Proceedings of the Eleventh ACM SIGKDD International Conference on Knowledge Discovery in Data Mining. KDD '05, pp. 788–793. Association for Computing Machinery, New York, NY, USA (2005). https://doi.org/10.1145/1081870.1081974
13. Prince, M., Wimo, A., Guerchet, M., Ali, G.C., Wu, Y.T., Prina, M.: World Alzheimer report 2015. The global impact of dementia: an analysis of prevalence, incidence, cost and trends. Research report, Alzheimer's Disease International (2015). https://unilim.hal.science/hal-03495438
14. Quiroz, Y.T., et al.: Plasma neurofilament light chain in the Presenilin 1 e280a autosomal dominant Alzheimer's disease kindred: a cross-sectional and longitudinal cohort study. Lancet Neurol. **19**(6), 513–521 (2020)
15. Shen, H.T., et al.: Heterogeneous data fusion for predicting mild cognitive impairment conversion. Inf. Fusion **66**, 54–63 (2021)
16. Stevenson-Hoare, J., et al.: Plasma biomarkers and genetics in the diagnosis and prediction of Alzheimer's disease. Brain **146**(2), 690–699 (2023)
17. Suk, H.I., Lee, S.W., Shen, D., Initiative, A.D.N.: Deep sparse multi-task learning for feature selection in Alzheimer's disease diagnosis. Brain Struct. Funct. **221**, 2569–2587 (2016)
18. Tang, S., Cao, P., Huang, M., Liu, X., Zaiane, O.: Dual feature correlation guided multi-task learning for Alzheimer's disease prediction. Comput. Biol. Med. **140**, 105090 (2021)
19. Wang, M., Zhang, D., Shen, D., Liu, M.: Multi-task exclusive relationship learning for Alzheimer's disease progression prediction with longitudinal data. Med. Image Anal. **53**, 111–122 (2019)
20. Wang, Y., et al.: Diagnosis and prognosis of Alzheimer's disease using brain morphometry and white matter connectomes. NeuroImage Clin. **23**, 101859 (2019)

LTUNet: A Lightweight Transformer-Based UNet with Multi-scale Mechanism for Skin Lesion Segmentation

Huike Guo🆔, Han Zhang🆔, Minghe Li🆔, and Xiongwen Quan(✉)🆔

College of Artificial Intelligence, Nankai University, Tianjin, China
{2120210394,liminghe}@mail.nankai.edu.cn, {zhanghan,quanxw}@nankai.edu.cn

Abstract. Medical image segmentation separates target structures or tissues within medical images to promote precise diagnoses. Automated image segmentation algorithms can help dermatologists to diagnose skin cancer by identifying skin lesions. Many popular image segmentation algorithms combine UNet and Transformer, but cannot fully utilize the global information between different scales and also have a huge number of parameters. To this end, this paper proposes a lightweight Transformer-based UNet (LTUNet) method for medical image segmentation, which designs an effective approach to extract and fully use multi-scale features. Firstly, the multi-scale feature maps of images are extracted by the inverted residual blocks of lightweight UNet encoder. Then, the feature maps are concatenated as the input of the Transformer's encoder blocks to compute intra- and inter-scale attention scores, and the scores are used to enhance the feature map of each scale. Finally, we fuse the upsampled results of all scales on UNet to improve the performance of segmentation. Our method achieves 0.9432, 0.8948, 0.9348 for mDice, mIoU and mACC on the ISIC2016 dataset, and 0.9058, 0.8138, 0.8968 on the ISIC2018 dataset respectively, which outperforms state-of-the-art methods. Besides, our network has a smaller number of parameters and converges faster.

Keywords: Medical image segmentation · UNet · Transformer · Multi-scale · Skin lesion

1 Introduction

Medical image segmentation is the process to identify and separate target structures or tissues within medical images, which enables physicians to make precise diagnoses and formulate effective treatment plans [28]. As an important application field, skin cancer has grown to be a serious disease that has an impact on people's health as a result of incorrect ultraviolet radiation, persistent stimulation, and other causes [17]. The dermatologist diagnoses skin cancer by observing

This research is supported by Major Program of the National Social Science Foundation of China (Grant No. 21&ZD102).

L. Fang et al. (Eds.): CICAI 2023, LNAI 14474, pp. 147–158, 2024.
https://doi.org/10.1007/978-981-99-9119-8_14

skin lesions on dermoscopic images. Automated image segmentation algorithms can identify and quantify the size, shape, and color of skin lesions with high precision and accuracy so as to improve the efficiency of diagnosis and treatment. Thus, it is crucial to develop automated image segmentation algorithms for skin lesion early diagnosis.

Current deep learning models often mine the data pattern from huge scale datasets e.g. ImageNet [6]. Due to the expensive labeling cost and privacy protection, medical images are available in limited quantities. Moreover, sophisticated models may overfit on relatively small datasets, whereas lightweight models sometimes fit well. In addition, multi-scale techniques can enhance the accuracy and robustness of medical image segmentation by addressing these issues, such as scale variation, noise perturbation, and variation in object size [16]. Combined with the attention mechanism, the multi-scale network can obtain intra- and inter-scale scores by converting similarity to weight.

Based on above discuss, in this paper, we propose LTUNet by combining Transformer [27] and UNet [21] for medical image segmentation, which designs a novel mechanism based on a combined framework to extract multi-scale features and achieves a lightweight network. Firstly, the multi-scale feature maps of images are extracted by the inverted residual blocks of LTUNet's encoder. Then, the feature maps are concatenated as the input of the Transformer's encoder blocks to compute intra- and inter-scale attention scores, which are used to enhance the feature map of each scale. Finally, we fuse the upsampled results of all scales to improve the performance of segmentation. Our contributions can be summarized in the following three parts:

- We propose a lightweight network to segment skin lesion images, which combine the UNet structure with Transformer to fuse local features and global features effectively.
- We use inverted residual blocks to extract features of different scales and reconstruct the features after attention and feature fusion respectively, where the features of each scale with global spatial attention are fused for more effective segmentation.
- LTUNet achieves competitive mDice, mIoU, mACC on ISIC2016 as well as ISIC2018 datasets and fast convergence by fewer parameters compared with other methods.

2   Related Works

2.1   Medical Image Segmentation Networks

Due to a variety of medical equipment and medical image modalities, the task of medical image segmentation is of great significance for assisting doctors in diagnosis [23]. UNet uses a U-shaped structure, which combines shallow features with deep features. The operation of downsampling and upsampling can compress and reconstruct the image, and the segmented image with the same size as the original image can be obtained. It is very effective for biomedical image segmentation.

In recent research, many variant UNets have been proposed. Attention UNet [19] adds the Attention Gate to control the feature input of the encoder part and then reconstructs the image through the decoder. Inception U-Net [20] draws on the idea of Inception [25], introduces multiple different convolution kernels, and finally connects the outputs of different branches in the channel dimension, which can make full use of large convolution kernels and small convolution kernels. Considering that the accuracy will become worse as the network deepens, ResUNet [7] applies the skip connection structure of ResNet [12]. Each module in the encoder and decoder contains multiple residual connections, which can effectively alleviate the vanishing gradient problem of deep networks. UNet++ [31] is a neural network architecture that enhances the original UNet structure by combining its 1–4 layers together, forming both long and short connections between them. Additionally, UNet++ incorporates deep supervision to further improve its effectiveness. However, CNN limits the receptive field to the convolution kernel and lacks the ability to model the image globally.

2.2   Model Lightweight

Due to the limitation of storage space and computing resources, the storage and computation of neural network models on mobile devices and embedded devices is still a huge challenge. Many techniques are used to simplify the model and reduce the number of model parameters. SqueezeNet [15] stacks the fire module with the squeeze and expand parts, drastically reducing the parameters. With depth-wise separable convolution, including depth-wise and point-wise, MobileNet [14] replaces conventional convolution. It can typically be accelerated by 8–9 times assuming a 3×3 convolution kernel. In order to produce more feature maps, GhostNet [10] replaces partial convolution calculations with a sequence of efficient linear operations. The number of model parameters is significantly reduced by such methods. In addition to network structure design, Hinton et al. [13] propose to use knowledge distillation to compress the model, and a complex teacher network can be used to guide a lightweight student network. Additionally, several weight-sharing methods [11,26] are applied to lighten models.

2.3   Transformer on Image Segmentation

Transformer is not limited by convolutional kernels and has multi-head attention to enhance globality. It has been widely used in image analysis recently. Based on Vision Transformer (ViT) [8], many methods have been proposed to achieve image segmentation. Segmenter [24] proposed a semantic segmentation model using only Transformer, and all performances exceed the most advanced convolution method on the ADE20K dataset. TransUNet [2] uses a CNN-based encoder to extract features from the input image, which are then fed into a Transformer-based decoder to generate the final segmentation map. SwinUNet [1] replaces the convolutional blocks in TransUNet with Transformer blocks, making global semantic feature learning more advantageous. In fact, although

CNN lacks globality, this structure can bring inductive bias (i.e. a part of prior knowledge), such as translation invariance and local relationship information. However, Transformer requires a large amount of data for training to learn this prior knowledge. Based on this observation, we combine the advantage of CNN and Transformer for medical image segmentation.

3   Method

Our LTUNet structure is shown in Fig. 1. To be specific, the inverted residual blocks of the various layers in the UNet encoders extract the multi-scale feature maps of the images. Then, the feature maps are concatenated along the channel dimension as the input of the Transformer's encoder blocks to compute intra- and inter-scale attention scores, which are used to enhance the feature map of each scale. Finally, we reconstruct the features to the original size and fuse the upsampled results of all scales to improve the performance of segmentation. After upsampling and reconstruction to the original image size, the segmentation result is obtained by the segmentation head. The details are described in the following subsections.

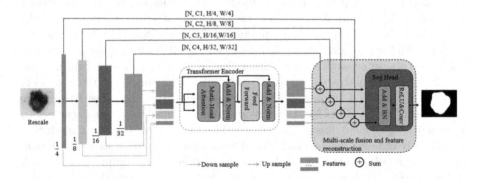

Fig. 1. The Structure of LTUNet, which consists of feature extraction and downsampling, Transformer encoder blocks, multi-scale fusion and segmentation three parts.

3.1   Feature Extraction and Downsample

To obtain multi-scale features as the input of the vision Transformer, we use CNN to extract different scale features of medical images. Considering that, the dimension of the feature map is lower, the complexity of the convolution and the parameter amount of the entire network is smaller. But low-dimensional features contain less information and may drop the accuracy of the model. Inspired by MobileNetv2 [22] we first use a 1×1 convolution kernel to expand the dimension, and then use a 3×3 depth separable convolution kernel to extract features.

Afterward, another 1×1 convolution kernel is used to match the initial number of channels. The output is then added to the input features to realize the residual connection. This type of inverted residual block can keep output low-dimensional characteristics and reduce complexity when extracting information from higher-dimensional features. When the stride is 1, the block is shown in Eq. (1), and when the stride is 2, the block is shown in Eq. (2).

$$out1 = x + conv1(\sigma(dconv3(\sigma(conv1(x))))) \tag{1}$$

$$out2 = conv1(\sigma(dconv3(\sigma(conv1(x))))) \tag{2}$$

where x is the input tensor, $conv1$ represents the normal 1×1 convolution operation, $dconv3$ represents the 3×3 depth-wise convolution operation, and σ represents the ReLU6 activation function. Compared with the UNet encoder, our network has fewer channels to reduce parameters. The extracted feature maps have different scales and numbers of channels. In order to fuse the features of different scales later, we downsample the features of different scales to the same size and then concatenate them along channel dimension.

3.2   Transformer Blocks Extract Multi-scale Semantic Information

Traditionally, ViT divides the input image into several patches, and each patch obtains a vector representation through a linear transformation. These vector representations are then fed into the Transformer encoder to learn a global feature representation. In our network, the features become very small after several downsample operations. We take the scaled features as input so the model parameters are few. We concatenate the features of different scales and feed them into the Transformer encoder blocks. Transformer's attention mechanism can measure the feature importance of different scales to obtain global information. The input $X \in R^{[b,n,d_x]}$ is firstly mapped to three quantities $Q \in R^{[b,n,d_q]}$, $K \in R^{[b,n,d_k]}$, $V \in R^{[b,n,d_v]}$, where b represents the batch size and n represents the number of patches. d_q, d_k and d_v represent the number of channels of Q, K and V matrices respectively. Q and the transpose of K is multiplied and divided by the square root of the scaling factor d_k, and multiplied by V after the softmax function to obtain attention. As shown in Eq. (3) and Eq. (4).

$$head_i = Attention = softmax(\frac{Q_i K_i^T}{\sqrt{d_k}})V_i \tag{3}$$

$$multihead = concat(head_i) \tag{4}$$

where $i \in \{1, 2, ..., N\}$, N is the number of heads. The concatenated multi-head attention features are added to the original input features. This residual connection can effectively prevent model degradation and enhance the representation ability of the model. In particular, in order to reduce model parameters, we replace the linear layer in the feed-forward neural network FFN with a 1×1 convolution operation. Following the method of CeiT [29], we add a depth-wise convolution layer which can make the model use more redundant information and improve its representation ability without introducing extra parameters. Finally, a residual connection and normalization are applied to the feature maps.

Generally, the multi-head attention mechanism computes intra- and inter-scale attention scores and uses them to enhance the feature map of each scale. Deeper features contain richer semantic information, more channels, and a larger weight in attention.

3.3   Multi-scale Fusion and Segmentation

UNet uses concatenation for upsampling feature fusion. However, it expands the number of feature channels and makes convolution operations more complicated. Therefore, we use the summation strategy for feature fusion. After the Transformer encoders, feature maps of different scales contain more global semantic information from different scales. We separate them along the channel dimension and upsample them by bilinear interpolation. The feature maps of different scales are reconstructed to the original feature size and added to the original features, to realize the fusion of shallow features and deep features. Then, the features of different scales are upsampled again through bilinear interpolation and reconstructed to the size of the original image. Finally, the reconstructed features of all scales are added together as input of the segmentation head to obtain the final segmentation result.

For the segmentation task, we use the cross-entropy loss function to get the final loss, as is shown in Eq. (5):

$$L = -\sum_{c=1}^{M} y_c log(p_c) \tag{5}$$

where M represents the number of class, y_c represents the label mask, and p_c represents the probability that the predicted sample belongs to class c.

4   Experiments

4.1   Datasets and Experimental Set up

The ISIC2016 [9] and ISIC2018 [4] datasets, released by the International Skin Imaging Collaboration (ISIC), are large-scale datasets of dermoscopic images. The ground truth data of the mask image is generated by several techniques and has been reviewed and curated by professional dermatologists. The datasets can be used in the research of automatic detection of skin diseases from medical images. We conduct experiments on ISIC2016, and ISIC2018 respectively (ISIC2019 and ISIC2020 are not available because they are currently in a Live Challenge), ISIC2016 includes 900 training images and 379 test images, and ISIC2018 includes 2594 training images, 100 validation images, and 1000 test images respectively.

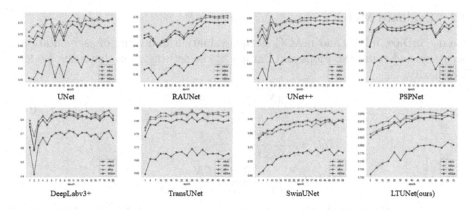

Fig. 2. The metrics trend of the 8 methods during training on ISIC2018.

We scale the images to a uniform size of 1024 * 1024 and feed them to the network, and then perform random cropping, random flipping, and optical distortion (PhotoMetricDistortion) for data augmentation. Our initial learning rate is 0.03, the Poly strategy is used for decay, and the Warm up [12] method is used to alleviate the early overfitting of the model to the mini-batch. Using the SGD optimizer, the momentum is set to 0.9. Batch Size is set to 8. The transformer encoders are set to 8 heads. The hardware platform is four Tesla P100 graphics cards with 16G memory. The experiment is based on the Pytorch framework and mmseg [5].

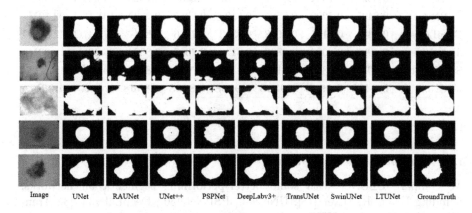

Fig. 3. Visualization of segmentation results of 8 networks on ISIC2016 and ISIC2018.

H. Guo et al.

4.2 Evaluate Metrics

We use mDice, mIoU, and mACC as the evaluation indicators of segmentation accuracy, and count the Flops and Params of the network to evaluate the model complexity.

$$Dice = \frac{2|X \cap Y|}{|X| + |Y|} \tag{6}$$

$$IoU = \frac{|X \cap Y|}{|X \cup Y|} \tag{7}$$

$$ACC = \frac{TP + TN}{TP + TN + FP + FN} \tag{8}$$

where X and Y represent the model prediction results and the real label results respectively. TP represents the true positive class, FN represents the false negative class, FP represents the false positive class, and TN represents the true negative class. $mDice$, $mIoU$ and $mACC$ represent the averages of foreground and background $Dice$, IoU and ACC respectively.

Table 1. Quantitative comparison of 8 methods on ISIC2016 and ISIC2018 datasets

| | ISIC2016 | | | ISIC2018 | | | | |
Method	mDice	mIoU	mACC	mDice	mIoU	mACC	Flops	Params
Unet	0.8766	0.7849	0.8680	0.8041	0.6820	0.7971	193.29G	28.99 M
RAUNet	0.081	0.6859	0.8113	0.7823	0.6411	0.7886	30.17G	21.89M
UNet++	0.8821	0.7634	0.8757	0.7837	0.6532	0.7926	139.63G	9.16M
PSPNet	0.8405	0.7299	0.8585	0.6837	0.5253	0.7333	101.46G	23.53M
DeepLabv3+	0.8335	0.7191	0.8618	0.8343	0.7216	0.8447	179.47G	42.52M
TransUNet	0.8968	0.8024	0.8889	0.8126	0.7002	0.8024	26.42G	19.03M
SwinUNet	0.9057	0.8300	0.9148	0.8237	0.7084	0.8158	10.28G	10.22M
LTUNet	**0.9432**	**0.8948**	**0.9348**	**0.9058**	**0.8138**	**0.8968**	**1.41G**	**4.96M**

4.3 Comparative Experiment

We compare LTUNet with some better-performing segmentation networks, including UNet, RAUNet [18], UNet++, PSPNet [30], DeepLabv3+ [3], TransUNet, SwinUnet, and LTUNet. Figure 2 shows the changes of three metrics and aACC during training on the ISIC2018 dataset. Compared with the baseline networks, our method converges faster and can also greatly improve accuracy. Compared with Transformer's methods, LTUNet can also effectively improve accuracy.

After the models converge, we visualize the results of the 8 segmentation networks, as shown in Fig. 3, the first three lines are images from ISIC2016, and the last two lines are images from ISIC2018. It can be seen that the outlines of the lesion areas in the images of row 1, row 4, and row 5 are relatively clear, and the segmentation results are not much different. Since there are some interference areas in the image of row 2, some methods will be affected by the interference areas, and the segmentation result is not ideal. However, our LTUNet can better segment the lesion area. The edge of the lesion area in the image of row 3 is blurred, and many methods cannot segment the contour, but LTUNet can still perform well.

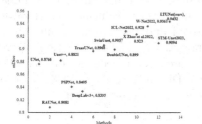
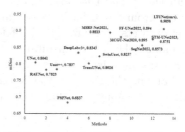

(a) The mDice of different methods on ISIC2016.

(b) The mDice of different methods on ISIC2018.

Fig. 4. Comparison of mDice with the baseline methods and the SOTA methods on ISIC2016 dataset (a) and ISIC2018 dataset (b).

The quantitative results on the ISIC2016 and ISIC2018 datasets are shown in Table 1. It can be seen that on the ISIC2016 dataset, the mDice, mIoU and mACC of our LTUNet are 0.9432, 0.8948 and 0.9438 respectively, which are the highest. On the ISIC2018 dataset, the mDice, mIoU, and mACC of our LTUNet are 0.9058, 0.8138, and 0.8968, respectively, all reaching the highest. What is more worth mentioning is that our LTUNet has fewer parameters and Flops. In addition, for a clearer comparison, we also quantitatively compare mDice with some SOTA methods on the ISIC2016 and ISIC2018 datasets in the past three years, as shown in Fig. 4(a) and Fig. 4(b). It can be clearly seen that LTUNet performs best.

4.4   Ablation Study

Our baseline model is a lightweight UNet (contains the inverted residual blocks) that uses a concatenate fusion strategy and Dice loss function. To verify the effectiveness of different parts, we add modules one by one to the baseline model. Firstly, we add Transformer encoder blocks in the middle of the network, with the input of the Transformer containing only the smallest scale features. Secondly, we introduce a multi-scale method and used features from four scales for fusion. Thirdly, we change the fusion strategy from concatenate to summation. Finally, we replace the Dice loss function with the CE loss function. The results on ISIC2016 are shown in Table 2.

After adding Transformer blocks to the Baseline model, mDice increases by 3.18%. It can be seen that the intra and inter-scale scores provided by the attention mechanism are effective. When we only use the smallest scale as the input of the Transformer, some objects with smaller shapes become smaller after downsampling, resulting in poor segmentation results. However, using the multi-scale fusion method effectively alleviates this problem, improving mDice by 3.7%. After changing the fusion method from concatenate to summation, the segmentation mDice does not change significantly, but the number of parameters are reduced, making the summation method more preferable. Lastly, we compare two loss functions. The Dice loss function is designed to deal with strongly unbalanced foreground and background samples in semantic segmentation. However, for our datasets, the proportion of foreground and background is almost the same, making Dice loss unsuitable. In contrast, the CE loss function regards each pixel as an independent sample and is more advantageous for our datasets. After changing the Dice loss to CE loss, the mDice increases by 1.06%. Ablation experiments demonstrate that each module of our method is effective.

Table 2. Ablation study on ISIC2016. The segmentation performance is measured by mDice.

Transformer	Multi-scale	Sum fusion	CE loss	mDice
				0.8634
✓				0.8952
✓	✓			0.9322
✓	✓	✓		0.9326
✓	✓	✓	✓	0.9432

5 Conclusion

Our proposed LTUNet is a lightweight model for skin lesion segmentation, which combines UNet and Transformer to develop an efficient multi-scale method. The multi-scale feature maps of images are extracted by the inverted residual blocks of UNet encoder. By concatenating the multi-scale features, we obtain intra- and inter-scale attention scores through Transformer blocks, which enhances the globality of the model. Deeper features have more semantic information and a higher number of channels, which makes them more important after attention. In addition, the downsampled multi-scale features are used as the input of ViT, which greatly reduces the number of parameters of the network.

Overall, LTUNet achieves excellent performance while significantly reducing the number of parameters, making it a promising model for skin lesion segmentation. Compared with 7 baseline methods and 5 state-of-the-art methods, our method achieves the best performance on both ISIC2016 and ISIC2018 datasets.

References

1. Cao, H., et al.: Swin-Unet: Unet-like pure transformer for medical image segmentation. In: Karlinsky, L., Michaeli, T., Nishino, K. (eds.) Computer Vision-ECCV 2022 Workshops: Tel Aviv, Israel, 23–27 October 2022, Proceedings, Part III, pp. 205–218. Springer, Cham (2023). https://doi.org/10.1007/978-3-031-25066-8_9
2. Chen, J., et al.: TransUNet: transformers make strong encoders for medical image segmentation. arXiv:abs/2102.04306 (2021)
3. Chen, L.-C., Zhu, Y., Papandreou, G., Schroff, F., Adam, H.: Encoder-decoder with atrous separable convolution for semantic image segmentation. In: Ferrari, V., Hebert, M., Sminchisescu, C., Weiss, Y. (eds.) ECCV 2018. LNCS, vol. 11211, pp. 833–851. Springer, Cham (2018). https://doi.org/10.1007/978-3-030-01234-2_49
4. Codella, N., et al.: Skin lesion analysis toward melanoma detection 2018: a challenge hosted by the international skin imaging collaboration (ISIC). arXiv preprint arXiv:1902.03368 (2019)
5. Contributors, M.: MMSegmentation: Openmmlab semantic segmentation toolbox and benchmark (2020). https://github.com/open-mmlab/mmsegmentation
6. Deng, J., Dong, W., Socher, R., Li, L.J., Li, K., Fei-Fei, L.: ImageNet: a large-scale hierarchical image database, pp. 248–255 (2009)
7. Diakogiannis, F.I., Waldner, F., Caccetta, P., Wu, C.: ResUNet-a: a deep learning framework for semantic segmentation of remotely sensed data. ISPRS J. Photogramm. Remote. Sens. **162**, 94–114 (2020)
8. Dosovitskiy, A., et al.: An image is worth 16×16 words: transformers for image recognition at scale. arXiv preprint arXiv:2010.11929 (2020)
9. Gutman, D., et al.: Skin lesion analysis toward melanoma detection: a challenge at the international symposium on biomedical imaging (ISBI) 2016, hosted by the international skin imaging collaboration (ISIC). arXiv preprint arXiv:1605.01397 (2016)
10. Han, K., Wang, Y., Tian, Q., Guo, J., Xu, C., Xu, C.: GhostNet: more features from cheap operations. In: Proceedings of the IEEE/CVF Conference on Computer Vision and Pattern Recognition, pp. 1580–1589 (2020)
11. Han, S., Pool, J., Tran, J., Dally, W.: Learning both weights and connections for efficient neural network. In: Advances in Neural Information Processing Systems, vol. 28 (2015)
12. He, K., Zhang, X., Ren, S., Sun, J.: Deep residual learning for image recognition. In: Proceedings of the IEEE Conference on Computer Vision and Pattern Recognition, pp. 770–778 (2016)
13. Hinton, G.E., Vinyals, O., Dean, J.: Distilling the knowledge in a neural network. arXiv:abs/1503.02531 (2015)
14. Howard, A.G., et al.: MobileNets: efficient convolutional neural networks for mobile vision applications. arXiv preprint arXiv:1704.04861 (2017)
15. Iandola, F.N., Han, S., Moskewicz, M.W., Ashraf, K., Dally, W.J., Keutzer, K.: SqueezeNet: AlexNet-level accuracy with 50x fewer parameters and <0.5 MB model size. arXiv preprint arXiv:1602.07360 (2016)
16. Lin, T.Y., Dollár, P., Girshick, R., He, K., Hariharan, B., Belongie, S.: Feature pyramid networks for object detection. In: 2017 IEEE Conference on Computer Vision and Pattern Recognition (CVPR), pp. 936–944 (2017). https://doi.org/10.1109/CVPR.2017.106
17. Marks, R., Staples, M., Giles, G.G.: Trends in non-melanocytic skin cancer treated in Australia: the second national survey. Int. J. Cancer J. Int. Du Cancer **53**(4), 585–590 (2010)

18. Ni, Z.L., et al.: RAUNet: residual attention U-Net for semantic segmentation of cataract surgical instruments. In: Gedeon, T., Wong, K., Lee, M. (eds.) Neural Information Processing: 26th International Conference, ICONIP 2019, Sydney, NSW, Australia, 12–15 December 2019, Proceedings, Part II, pp. 139–149. Springer, Cham (2019). https://doi.org/10.1007/978-3-030-36711-4_13

19. Oktay, O., et al.: Attention U-Net: learning where to look for the pancreas. arXiv preprint arXiv:1804.03999 (2018)

20. Punn, N.S., Agarwal, S.: Inception U-Net architecture for semantic segmentation to identify nuclei in microscopy cell images. ACM Trans. Multimedia Comput. Commun. Appl. (TOMM) **16**(1), 1–15 (2020)

21. Ronneberger, O., Fischer, P., Brox, T.: U-Net: convolutional networks for biomedical image segmentation. In: Navab, N., Hornegger, J., Wells, W., Frangi, A. (eds.) Medical Image Computing and Computer-Assisted Intervention-MICCAI 2015: 18th International Conference, Munich, Germany, 5–9 October 2015, Proceedings, Part III 18, pp. 234–241. Springer, Cham (2015). https://doi.org/10.1007/978-3-319-24574-4_28

22. Sandler, M., Howard, A., Zhu, M., Zhmoginov, A., Chen, L.C.: MobileNetV2: inverted residuals and linear bottlenecks. In: Proceedings of the IEEE Conference on Computer Vision and Pattern Recognition, pp. 4510–4520 (2018)

23. Sharma, N., Aggarwal, L.M.: Automated medical image segmentation techniques. J. Med. Phys./Assoc. Med. Physicists India **35**(1), 3 (2010)

24. Strudel, R., Garcia, R., Laptev, I., Schmid, C.: Segmenter: transformer for semantic segmentation. In: Proceedings of the IEEE/CVF International Conference on Computer Vision, pp. 7262–7272 (2021)

25. Szegedy, C., Vanhoucke, V., Ioffe, S., Shlens, J., Wojna, Z.: Rethinking the inception architecture for computer vision. In: Proceedings of the IEEE Conference on Computer Vision and Pattern Recognition, pp. 2818–2826 (2016)

26. Ullrich, K., Meeds, E., Welling, M.: Soft weight-sharing for neural network compression. arXiv preprint arXiv:1702.04008 (2017)

27. Vaswani, A., et al.: Attention is all you need. In: Advances in Neural Information Processing Systems, vol. 30 (2017)

28. Wang, R., Lei, T., Cui, R., Zhang, B., Meng, H., Nandi, A.K.: Medical image segmentation using deep learning: a survey. IET Image Process. **16**(5), 1243–1267 (2022). https://doi.org/10.1049/ipr2.12419

29. Yuan, K., Guo, S., Liu, Z., Zhou, A., Yu, F., Wu, W.: Incorporating convolution designs into visual transformers. In: Proceedings of the IEEE/CVF International Conference on Computer Vision, pp. 579–588 (2021)

30. Zhao, H., Shi, J., Qi, X., Wang, X., Jia, J.: Pyramid scene parsing network. In: Proceedings of the IEEE Conference on Computer Vision and Pattern Recognition, pp. 2881–2890 (2017)

31. Zhou, Z., Siddiquee, M.M.R., Tajbakhsh, N., Liang, J.: UNet++: redesigning skip connections to exploit multiscale features in image segmentation. IEEE Trans. Med. Imaging **39**(6), 1856–1867 (2019)

A Novel Online Multi-label Feature Selection Approach for Multi-dimensional Streaming Data

Zhanyun Zhang[1], Chuan Luo[1(✉)], Tianrui Li[2], Hongmei Chen[2], and Dun Liu[3]

[1] College of Computer Science, Sichuan University, Chengdu 610065, China
`cluo@scu.edu.cn`
[2] School of Information Science and Technology, Southwest Jiaotong University, Chengdu 610031, China
[3] School of Statistics, Southwest University of Finance and Economics, Chengdu 611130, China

Abstract. Online feature selection for streaming data has attracted much attention in the field of multi-label learning. Most of the existing online approaches can efficiently deal with the single-dimensional variation of a multi-label information system. However, multi-dimensional variations often occur in real-time streaming applications. Based on the improved Fisher score model for multi-label learning and feature redundancy analysis using symmetric uncertainty, we propose a novel online multi-label feature selection framework for both streaming feature and label spaces. For the situation of streaming features, the proposed framework calculates the Fisher score to obtain the importance of the feature, determines the redundancy of the newly arrived feature based on the symmetry uncertainty, and then obtains the position of the feature in the final feature rank list. For the newly arrived labels, we recalculates the weights of all current labels and updates the total Fisher score to update the current feature rank list. In the experiments, we compare the performance of our approach with four representative online feature selection algorithms for streaming features and labels, respectively. The extensive experimental results on nine multi-label benchmark datasets by using two evaluation metrics commonly used in multi-label classification demonstrate the effectiveness of the proposed framework.

Keywords: multi-label feature selection · streaming features · streaming labels · Fisher score · symmetrical uncertainty

1 Introduction

Nowadays, multi-label learning receives more and more attention, and make a big splash in various read-world applications (e.g., bioinformatics, text categorization, and music emotion recognition). High-dimensional features, often with thousands or even tens of thousands of features, have always been a curse for

L. Fang et al. (Eds.): CICAI 2023, LNAI 14474, pp. 159–171, 2024.
https://doi.org/10.1007/978-981-99-9119-8_15

multi-label learning [1,2]. Many excellent multi-label feature selection algorithms that can deal with high-dimensional feature problems have been proposed [3,4]. But in the past decade, with the explosive growth of data and time-sensitive processing, multi-label learning often requires feature selection can be adapted in the streaming feature scenario. The streaming features problem refers to the continuous growth of feature dimensions, features arrive one by one over time and need to be processed upon arrival. Several approaches to multi-label feature selection for streaming features have been proposed in the literature. Lin et al. introduced fuzzy mutual information for multi-label feature selection, and proposed the online feature selection algorithm consists of online relevance analysis and online redundancy analysis to effectively select the feature subset under the streaming feature problem [5]. Liu et al. proposed an online multi-label group feature selection algorithm, which considers the intrinsic relationship between feature groups in order to construct an online selecting criterion for feature groups, and then selects the best subset of features based on the interaction and redundancy of features [6]. Liu et al. proposed an online multi-label streaming feature selection framework based on multi-label neighborhood rough sets, which includes two components: online importance selection and online redundancy update [7]. Compared with the situation of streaming features, streaming labels are a relatively novel challenge for multi-label learning. Streaming labels problem refers to the whole label space is unknown, the label space increases over time, and the number of labels is variable, or even infinite. In the literature, a few of multi-label feature selection methods for streaming labels have been developed. Lin et al. [8] proposed an novel algorithm for multi-label feature selection in a dynamic label stream, which uses the mRMR strategy to generate a feature rank list for each arriving label, minimize the overall weight deviation between the lists and the final list, and finally output the optimized final feature list. Liu et al. [9] proposed a effective multi-label feature selection algorithm to create label-specific features based on between-class discrimination and within-class neighbor identification.

In this paper, based on the improved Fisher score for multi-label learning and feature redundancy processing using symmetric uncertainty, we propose a novel framework for multi-label feature selection based on the Fisher score model, which can perform feature selection effectively under the problem that both features and labels stream in dynamically. For the newly arrived features, the framework first calculates the total Fisher score between the feature and the current label space, and then calculates the symmetric uncertainty between the feature and the previously arrived feature set, so as to determine the importance of the feature. For the newly arrived label, the framework calculates the Fisher score between the current feature set and the newly arrived label to update the total Fisher score, and performs a new round of feature redundancy processing to obtain the current feature rank list. The experiments carried out on nine real-world datasets using two evaluation metrics for multi-label classification demonstrate the advantages of the proposed approach compared with existing widely used online multi-label feature selection algorithms.

2 Proposed Method

2.1 Label-Weight Based Fisher Score

Fisher score is supervised by class labels, which score features in the original feature space so as to find features with the greatest amount of discriminative strength [11–13]. The main idea is to identify the strong features with the distance within the class as small as possible and the distance between classes as large as possible. The Fisher score of the kth feature on the dataset can be defined as follows:

$$FS(k) = \frac{S_{outer}^{k}}{S_{inner}^{k}}, \tag{1}$$

where $S_{outer}^{k} = \sum_{i=1}^{C} \frac{|E_i|}{n} (\mu_i^k - \mu^k)^2$ is the sum of the between-class distances of the kth feature, $S_{inner}^{k} = \frac{1}{n} \sum_{i=1}^{C} \sum_{x \in E_i} (x^k - \mu_i^k)^2$ is the sum of the within-class distances of the kth feature, E_i is a set of samples of class i on the dataset, μ_i^k represents the mean of samples of class i on the kth feature, and μ^k represents the mean of samples of all classes on the kth feature, x^k represents the value of instance x on the kth feature.

In the following, we propose an enhanced Fisher score model based on label weights to fit well with the multi-label learning, which is a natural extension from single-label learning to multi-label learning. Suppose $MDS = <U, F \cup L>$ is a multi-label decision system, in which $U = \{x_1, x_2, \cdots, x_n\}$ are n instances, $F = \{f_1, f_2, \cdots, f_m\}$ are m features, and $L = \{l_1, l_2, \cdots, l_k\}$ are k labels, respectively. And labels associations usually are represented as a q-dimensional binary vector $y = (y_1, y_2, \cdots, y_k) = \{0, 1\}^k$.

Definition 1. Given $MDS = <U, F \cup L>$ with $f_k \in F$ and for any $l_i \in L$, the sum of the inter-label distances of the feature f_k with respect to the label l_i is defined as follows:

$$S_{outer}^{f_k, l_i} = \frac{|E_{l_i+}|}{n} (\mu_{l_i+}^{f_k} - \mu^{f_k})^2 + \frac{|E_{l_i-}|}{n} (\mu_{l_i-}^{f_k} - \mu^{f_k})^2, \tag{2}$$

where E_{l_i+}, E_{l_i-} denote the set of samples with and without l_i, $\mu_{l_i+}^{f_k}$, $\mu_{l_i-}^{f_k}$ denote the mean of E_{l_i+}, E_{l_i-} on the feature f_k, respectively. And μ^{f_k} denotes the mean of the feature f_k on the dataset.

Definition 2. Given $MDS = <U, F \cup L>$ with $f_k \in F$ and for any $l_i \in L$, the sum of the intra-label distances of the feature f_k with respect to the label l_i is defined as follows:

$$S_{inner}^{f_k, l_i} = \frac{1}{n} \left(\sum_{x \in E_{l_i+}} (x^{f_k} - \mu_{l_i+}^{f_k})^2 + \sum_{x \in E_{l_i-}} (x^{f_k} - \mu_{l_i-}^{f_k})^2 \right), \tag{3}$$

where x^{f_k} denotes the value of the sample x on the feature f_k.

Definition 3. Given $MDS = <U, F \cup L>$ with $f_k \in F$ and for any $l_i \in L$, the Fisher score of the feature f_k with respect to the label l_i is defined as follows:

$$FS(f_k, l_i) = \frac{S_{outer}^{f_k, l_i}}{S_{inner}^{f_k, l_i}}. \tag{4}$$

Definition 4. Given $MDS = <U, F \cup L>$. For $\forall l_i, l_j \in L$, the sum of the inter-label distances of the label l_i with respect to the label l_j is defined as follows:

$$S_{outer}^{l_i, l_j} = \frac{1}{n}\left(\frac{(n_{l_j+}^{l_i+})^2}{n_{l_j+}} - \frac{(n_{l_i+})^2}{n} + \frac{(n_{l_j-}^{l_i+})^2}{n_{l_j-}}\right), \tag{5}$$

where n_{l_j+}, n_{l_j-} denote the size of dataset with and without label l_j, n_{l_i+} denotes the size of dataset with label l_i, $n_{l_j+}^{l_i+}$ denote the size of dataset with label l_i and l_j, and $n_{l_j-}^{l_i+}$ denotes the size of dataset with label l_i and without label l_j.

Definition 5. Given $MDS = <U, F \cup L>$. For $\forall l_i, l_j \in L$, the sum of the intra-label distances of the label l_i with respect to the label l_j is defined as follows:

$$S_{inner}^{l_i, l_j} = \frac{1}{n}\left(n_{l_i+} - \frac{(n_{l_j+}^{l_i+})^2}{n_{l_j+}} - \frac{(n_{l_j-}^{l_i+})^2}{n_{l_j-}}\right). \tag{6}$$

Definition 6. Given $MDS = <U, F \cup L>$. For $\forall l_i, l_j \in L$, the Fisher score of l_i for l_j is defined as follows:

$$FS(l_i, l_j) = \frac{S_{outer}^{l_i, l_j}}{S_{inner}^{l_i, l_j}}. \tag{7}$$

Definition 7. Given $MDS = <U, F \cup L>$. For $\forall l_i \in L$, the label importance weight of l_i is defined as follows:

$$W(l_i) = \frac{\sum_{j=1}^{k} FS(l_i, l_j)}{\sum_{i=1}^{k} \sum_{j=1}^{k} FS(l_i, l_j)}, \tag{8}$$

where $\sum_{i=1}^{k} W(l_i) = 1$.

Definition 8. Given $MDS = <U, F \cup L>$ with $f_k \in F$, the total Fisher score of the feature f_k is defined as follows:

$$FS(f_k) = \sum_{i=1}^{|L|} W(l_i) * FS(f_k, l_i). \tag{9}$$

The importance weight of a label can reflect the significance of the label and its association with other labels. The higher the weight of the label, the more significant the label is in the label space, and the lower the weight of the label, the less significant the label is in the label space.

2.2   Symmetrical Uncertainty Based Redundancy Analysis

Fisher score model generates the score for each feature individually, the correlation between features is not considered. If the correlation between two features is strong, it means that they are redundant with each other, and if their Fisher scores are both large, it will affect the feature selection result and the classification performance. Therefore, we need to consider the mutual redundancy.

The uncertainty of a random variable $X = \{x_1, x_2, \cdots, x_n\}$ is measured by entropy, defined as follows:

$$H(X) = -\sum_{i=1}^{n} p(x_i) \log p(x_i), \tag{10}$$

where $p(x_i)$ is the probability of x_i.

The mutual dependency between random variables can be measured by mutual information. Given two variables $X = \{x_1, x_2, \cdots, x_n\}$ and $Y = \{y_1, y_2, \cdots, y_m\}$, the mutual information between X and Y can be defined as:

$$MI(X,Y) = -\sum_{i=1}^{n}\sum_{j=1}^{m} p(x_i, y_j) \log \frac{p(x_i|y_j)}{p(x_j)}, \tag{11}$$

where $p(x_i, y_j)$ is the joint probability of X and Y, $p(x_i|y_j)$ is the conditional probability.

Since mutual information tends to have features with more values, symmetrical uncertainty [14] was proposed to compensate this bias of mutual information, which is defined as:

$$SU(X,Y) = \frac{2MI(X,Y)}{H(X) + H(Y)}. \tag{12}$$

Consider treating features as variables, if the symmetrical uncertainty $SU(f_i, f_j)$ between any two random features is large, the two features are considered redundant, and if the symmetrical uncertainty between any two random features is small, the two features are considered independent.

Definition 9. The threshold for determining whether features are redundant with each other is calculated as follows:

$$\theta = \frac{1}{|F| \cdot |F|} \sum_{i=1}^{|F|}\sum_{j=1}^{|F|} SU(f_i, f_j), \tag{13}$$

where the threshold θ is continuously recalculated as the feature space increases.

Definition 10. Given $MDS = <U, F \cup L>$ and the currently newly arrived feature f_k, f_k is redundant if $\exists f_i \in F$ satisfies the following two conditions:

$$SU(f_k, f_i) \geq \theta \text{ and } FS(f_k) < FS(f_i). \tag{14}$$

For the newly arrived feature f_k, determine whether there is redundancy with all previous features in turn until there is another feature f_i such that $SU(f_k, f_i)$ is greater than the threshold θ. If $FS(f_k) < FS(f_i)$, then f_k is considered to be a redundant feature, otherwise f_i is considered redundant.

2.3   Algorithmic Framework

In order to solve the feature selection problem in both the streaming labels and the streaming labels scenario, the proposed algorithm framework, i.e., Algorithm 1, contains two components: online selection for streaming features as shown in Algorithm 2 and online selection for streaming features as shown in Algorithm 3. When a new feature arrives, the Fisher score obtained by the previously features in the current label space remains unchanged, and the previously obtained feature sort results can still be partially useful. We only need to calculate the Fisher score of the newly arrived feature for the current label space, and find the position of the newly arrived feature in the previously obtained feature sort list. The specific details are described on Algorithm 3. Steps 1–3 calculate the SU of the new feature f_k with the feature existed in the previously set, and steps 4–9 calculate the Fisher score of the arrival feature f_k with the current label space in the FSM, and adds the total Fisher score of the arrived feature to the SL. Step 10 finds the position in the feature rank R of the arrived feature f_k based on the total Fisher score. In steps 11–16, if the features f_j in 1 to $i-1$ are determined to be redundant with the arrived feature f_k, the arrived feature f_k will be directly inserted into the candidate feature list R' according to the total Fisher score. If the arrived feature f_k is not redundant with the features from 1 to $i-1$, the arrived feature f_k is inserted into the position i of R, but it is still necessary to determine whether the subsequent features of R are redundant with the arrived feature f_k. Steps 18–25 further determine whether the feature f_j after i is redundant with the arrived feature f_k, and if redundant, removes f_j from R and inserts it into the list of feature rank R' according to the total Fisher score.

Algorithm 1. Online multi-label feature selection for both streaming features and streaming labels (SFSL)

Require: f_i: Feature arrives at time t_i; l_i: Label arrives at time t_i;
Ensure: R: Features rank list.
 1: **if** f_i arrives at time t_i **then**
 2:    Update R by using OSSF(l_i);
 3: **else if** l_i arrives at time t_i **then**
 4:    Update R by using OSSL(f_i);
 5: **end if**
 6: **return** R.

If the new label arrives, the feature ranking result under the new label space may be inconsistent with the previous feature sorting result under the label space. We obtain a list of features sorted from the algorithm, which calculates the Fisher score separately based on all the labels in the current label space, and then combines the weights of all labels to obtain the total Fisher score, and finally sorts the features. Therefore, we choose to save the Fisher score of features for the previous label space, just calculate the Fisher score of features for the

---

Algorithm 2. Online selection for streaming features (OSSF)

Require: f_k: Feature arrives at time t_k;
Ensure: R: Features rank list.
1: **for** $i = 1 \rightarrow |F|$ **do**
2:     $SUM[f_k][f_i] \leftarrow$ Calculate the SU between f_k and f_i according to Eq. (12);
3: **end for**
4: $SL_{f_k} \leftarrow 0$
5: **for** $i = 1 \rightarrow |L|$ **do**
6:     $FSM[f_k][l_i] \leftarrow$ Calculate the Fisher score between f_k and l_i according to Eq. (4);
7:     $SL_{f_k} \leftarrow SL_{f_k} + W_{l_i} * FSM[f_k][l_i]$;
8: **end for**
9: $SL \leftarrow SL + SL_{f_k}$;
10: $i \leftarrow$ Find the position of f_k in R according to SL;
11: **for** $j = 1 \rightarrow i - 1$ **do**
12:     **if** $SUM[i][j] > \theta$ **then**
13:         Remove f_k from R and insert into $R^{'}$ accord to SL;
14:         Go to step 27;
15:     **end if**
16: **end for**
17: insert f_i into R;
18: $j = i + 1$
19: **while** $R[j]$ is not the last feature of the R **do**
20:     **if** $SUM[i][j] > \theta$ **then**
21:         Remove f_j from R and insert into $R^{'}$ accord to SL;
22:     **else**
23:         $j \leftarrow j + 1$;
24:     **end if**
25: **end while**
26: **return** $R + R^{'}$.

---

newly arrived label, and then update the total Fisher score. The specific details are described on Algorithm 2. Steps 1–4 update the weights of all labels. Steps 5–7 calculate all Fisher score of features for newly arrived label l_i, and steps 8–12 update the rank list SL for all features. Steps 13–25 check the redundancy of features. According to SL, the feature list R is ranked, and $R^{'}$ is used as the candidate feature list to save the features that are judged to be redundant. Traversing each feature f_i of R, if f_j satisfies $SUM[i][j] > \theta$ in 1 through $i - 1$, it means that f_i is redundant, and we remove the feature f_i from R and add it to $R^{'}$.

3   Experiments

In order to fully demonstrate the performance of SFSL, we designed experiments in the scenarios of streaming features, streaming labels, simultaneous streaming features and labels. For the streaming features, we chose the MSFS [5], OM-NRS [7] as the compared algorithms, and for the streaming labels, we chose the

---

Algorithm 3. Online selection for streaming labels (OSSL)

Require: l_i: Label arrives at time t_i;
Ensure: R: Features rank list.
 1: Calculate the entropy of l_i;
 2: **for** $i = 1 \rightarrow |L|$ **do**
 3:     $W_{l_i} \leftarrow$ Update the weight of l_i according to Eq. (8);
 4: **end for**
 5: **for** $k = 1 \rightarrow |F|$ **do**
 6:     $FSM[f_k][l_i] \leftarrow$ Calculate the Fisher score between f_k and l_i according to Eq. (4);
 7: **end for**
 8: **for** $k = 1 \rightarrow |F|$ **do**
 9:     **for** $i = 1 \rightarrow |L|$ **do**
10:         $SL[f_k] \leftarrow SL[f_k] + W_{l_i} * FSM[f_k][l_i]$;
11:     **end for**
12: **end for**
13: $R \leftarrow$ Sort SL and get the feature rank list from SL;
14: $i = 2, R' = \emptyset$;
15: **while** $R[i]$ is not the last feature of the R **do**
16:     **for** $j = 1 \rightarrow i - 1$ **do**
17:         **if** $SUM[i][j] > \theta$ **then**
18:             Remove f_i from R and add to R';
19:             $i \leftarrow i - 1$;
20:             break;
21:         **end if**
22:     **end for**
23:     $i \leftarrow i + 1$;
24: **end while**
25: **return** $R + R'$.

MLFSL [8] and FSSL [9] as the compared algorithms. Nine datasets from the Mulan Library were used to demonstrate the performance of algorithms. The specific description of these data sets is fully displayed in the Table 1. Preprocess the dataset by removing the rows containing the missing values before the experiment. The proposed algorithm, SFSL, selects 30% of the features sorted by the features obtained after each running. For MSFS, the parameter δ was set to be 0.08 and the significance levels δ in OM-NRS was set to be 0.1. For MLFSL, we set the labels arrived one by one. For FSSL, features with continuous value were discretized into 2 folds by using a monowidth strategy. In addition, ML-kNN was selected as the evaluation classifier, and the number of nearest neighbors was set to 10.

The average precision (AP) and hamming loss (HL) proposed in [10] were used to evaluate the performance of multi-label classification. Let $T = \{(x_i, Y_i) | 1 \leq i \leq N\}$ be the test set where $Y_i \in L$ is the label subset. AP evaluates the average fraction of labels ranked above a particular label $y \in Y_i$, which is defined as $AP = \frac{1}{N} \sum_{i=1}^{N} \frac{1}{|Y_i|} \sum_{y \in Y_i} \frac{|\{y' | rank_i(y') \leq rank_i(y), y' \in Y_i\}|}{rank_i(y)}$, where

Table 1. Characteristics of the experimental datasets.

Datasets	Characters				
	Instances	Nominal	Numeric	Labels	Cardinality
Birds	645	2	258	19	1.014
Cal500	502	0	68	174	26.044
Emotions	593	0	72	6	1.869
Enron	1702	1001	0	53	3.378
Flags	194	9	10	7	3.392
Genbase	662	1186	0	27	1.252
Medical	978	1449	0	45	1.245
Scene	2407	0	294	6	1.074
Yeast	2417	0	103	14	4.237

$rank_i(y)$ is the rank of label $y \in L$ predicted by the learner for x_i. HL evaluates the average time of misclassification in each instance-label pair. Which is defined as $HL = \frac{1}{N} \sum_{i=1}^{N} \frac{|Y_i \oplus h(x_i)|}{q}$, where operator $\oplus$ is XOR, and $h(x_i)$ denotes the predicted label set for instance x_i.

Firstly, we compared the proposed SFSL algorithm with the existing online multi-label feature selection algorithms for streaming features, i.e., MSFS and OM-NRS. The classification results of AP and HL by using ML-kNN on 9 datasets with the dynamic variation of features are shown in Table 2. From the experimental results, it can be find that the proposed algorithm achieved good classification results on 9 multi-label datasets. For indicators AP, SFSL achieved optimal performance on 5 datasets and suboptimal performance on 3 remaining datasets in the remaining datasets. For indicators HL, SFSL achieved optimal performance on 5 datasets.

Table 2. Comparison among online feature selection methods with streaming features.

Datasets	AP(↑)			HL(↓)		
	MSFS	*OM-NRS*	*SFSL*	*MSFS*	*OM-NRS*	*SFSL*
Birds	0.2484(3)	**0.2935(1)**	0.2486(2)	0.0539(3)	**0.0513(1)**	0.0535(2)
Cal500	0.3235(2)	0.3184(3)	**0.3265(1)**	0.1466(2)	0.1477(3)	**0.1463(1)**
Emotions	0.6653(2)	**0.6698(1)**	0.6498(3)	0.2633(2)	**0.2592(1)**	0.2761(3)
Enron	**0.3880(1)**	0.3016(3)	0.3774(2)	0.0545(2)	**0.0522(1)**	0.0565(3)
Flags	0.7671(2)	0.6916(3)	**0.7688(1)**	0.2950(3)	0.2868(2)	**0.2865(1)**
Genbase	0.4822(3)	**0.7285(1)**	0.6972(2)	0.0309(3)	**0.0163(1)**	0.0181(2)
Medical	0.3723(3)	0.4754(2)	**0.4825(1)**	0.0223(3)	0.0206(2)	**0.0202(1)**
Scene	0.7139(3)	0.7278(2)	**0.7485(1)**	0.1221(3)	0.1144(2)	**0.1096(1)**
Yeast	0.6622(3)	0.6628(2)	**0.6694(1)**	0.2236(3)	0.2234(2)	**0.2188(1)**
Mean Ranks	0.4820(2.5)	0.5253(1.9)	**0.5250(1.6)**	0.1147(2.5)	0.1113(1.9)	**0.1124(1.6)**

Secondly, we compare the SFSL with algorithms of MLFSL and FSSL in the scenario of streaming labels. The classification results of AP and HL are presented in Table 3. From the experimental results, it can be seen that our algorithm has achieved good classification results on 9 multi-label datasets. And, SFSL obtained the best performance on 6 datasets in terms of both AP and HL. Hence, we can conclude that SFSL is clearly superior to the existing online algorithms FSSL and MLFSL with streaming labels.

Table 3. Comparison among online feature selection methods with streaming labels.

Datasets	AP($\uparrow$)			HL($\downarrow$)		
	FSSL	*MLFSL*	*SFSL*	*FSSL*	*MLFSL*	*SFSL*
Birds	0.2522(2)	0.2428(3)	**0.2714(1)**	0.0538(2)	0.0543(3)	**0.0536(1)**
Cal500	0.3211(3)	0.3257(2)	**0.3320(1)**	0.1464(3)	0.1461(2)	**0.1453(1)**
Emotions	0.6701(2)	**0.6707(1)**	0.6497(3)	**0.2579(1)**	0.2640(2)	0.2786(3)
Enron	**0.4503(1)**	0.4257(2)	0.3661(3)	**0.0522(1)**	0.0538(2)	0.0575(3)
Flags	0.7587(3)	0.7629(2)	**0.7733(1)**	0.2978(3)	0.2897(2)	**0.2690(1)**
Genbase	0.7082(3)	0.8081(2)	**0.8508(1)**	0.0186(3)	0.0139(2)	**0.0096(1)**
Medical	0.4804(2)	0.4502(3)	**0.6323(1)**	0.0209(2)	0.0209(2)	**0.0156(1)**
Scene	0.7640(2)	**0.7666(1)**	0.7446(3)	0.1040(2)	**0.1028(1)**	0.1114(3)
Yeast	0.6698(2)	0.6696(3)	**0.6737(1)**	0.2188(3)	0.2169(2)	**0.2161(1)**
Mean Ranks	0.5395(2.1)	0.5449(2.1)	**0.5651(1.8)**	0.1091(2.1)	0.1091(2.0)	**0.1110(1.8)**

Considering the existing online algorithms are not suitable for dealing with the problem of streaming features and streaming labels at the same time, we re-run these algorithms at each time when a new feature or label is arrived for comparison. The classification results of SFSL and the comparison algorithms are shown in Figs. 1 and 2. From the results, it can be seen that the proposed SFSL has achieved good classification results. For example, on the dataset bird, SFSL always remain optimal on indicators AP and HL. On the datasets emotions, scene, and yeast, SFSL is the fastest one to reach the best level in terms of the indicators AP and HL. The performance of SFSL is on average level on the datasets cal500 and flag, but is the fastest algorithm for selecting excellent features on the dataset cal500 in terms of HL.

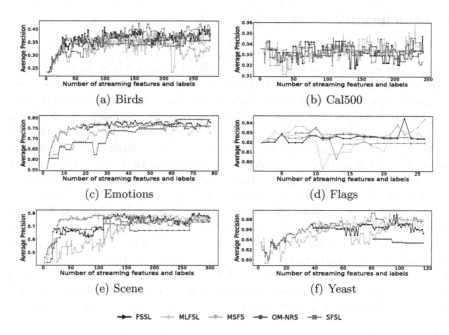

Fig. 1. Comparative results of indicator AP among five algorithms with streaming features and labels.

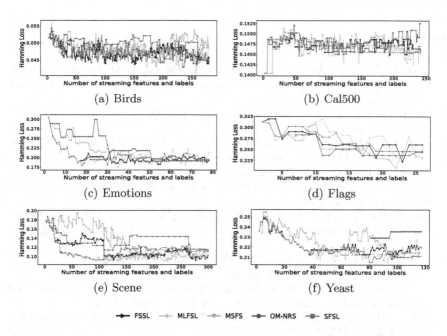

Fig. 2. Comparative results of indicator HL among five algorithms with streaming features and labels.

4 Conclusions

The conventional multi-label selection algorithm is not applicable to the challenge of streaming features and streaming labels concurrently. In view of this problem, based on the improved Fisher score model and symmetric uncertainty, this paper proposed a multi-label feature selection algorithm framework suitable for unknown feature space and label space, which consists of two parts: online streaming feature selection and streaming label feature selection. For newly arrived features, the online streaming feature selection part obtains the importance of the feature by calculating the Fisher score of the feature and the current label space, and calculates the symmetric uncertainty between the feature and all the features in the previous feature space to determine whether the feature is redundant, and finally obtains the position of the feature in the final feature rank list. For newly arrived label, the streaming labels feature selection part recalculates the weights of all current labels, calculates the Fisher scores of all current features and the label, and updates them to the total Fisher score, and re-obtains the current feature rank list after sorting and feature redundancy processing. Extensive experiments in comparison with existing online multi-label feature selection methods validated the performance of the proposed method.

Acknowledgments. This work was supported by the National Natural Science Foundation of China (Nos. 62076171, 62376230), and the Natural Science Foundation of Sichuan Province (2022NSFSC0898).

References

1. Chen, D., Yang, Y.: Attribute reduction for heterogeneous data based on the combination of classical and fuzzy rough set models. IEEE Trans. Fuzzy Syst. **22**, 1325–1334 (2014)
2. Javidi, M., Eskandari, S.: Streamwise feature selection: a rough set method. Int. J. Mach. Learn. Cybern. **1**, 1–10 (2016)
3. Spolaôr, N., Cherman, E., Monard, M., Lee, H.: ReliefF for multi-label feature selection. In: 2013 Brazilian Conference on Intelligent Systems, pp. 6–11 (2013)
4. Zhang, L., Hu, Q., Duan, J., Wang, X.: Multi-label feature selection with fuzzy rough sets. In: 2014 International Conference on Rough Sets and Knowledge Technology, pp. 121–128 (2014)
5. Lin, Y., Hu, Q., Liu, J., Li, J., Wu, X.: Streaming feature selection for multilabel learning based on fuzzy mutual information. IEEE Trans. Fuzzy Syst. **25**(6), 1491–1507 (2017)
6. Liu, J., Lin, Y., Wu, S., Wang, C.: Online multi-label group feature selection. Knowl.-Based Syst. **143**, 42–57 (2018)
7. Liu, J., Lin, Y., Li, Y., et al.: Online multi-label streaming feature selection based on neighborhood rough set. Pattern Recogn. **84**, 273–287 (2018)
8. Lin, Y., Hu, Q., Zhang, J., Wu, X.: Multi-label feature selection with streaming labels. Inf. Sci. **372**, 256–275 (2016)
9. Liu, J., Li, Y., Weng, W., Zhang, J., Wu, S.: Feature selection for multi-label learning with streaming label. Neurocomputing **387**, 268–278 (2020)

10. Schapire, R., Singer, Y.: BoosTexter: a boosting-based system for text categoriza-
    tion. Mach. Learn. **39**, 135–168 (2000)
11. Polat, K., Krmac, V.: Determining of gas type in counter flow vortex tube using
    pairwise Fisher score attribute reduction method. Int. J. Refrig **34**, 1372–1386
    (2011)
12. Gu, Q., Li, Z., Han, J.: Generalized Fisher score for feature selection. In: Conference
    on UAI, vol. 8, pp. 266–273. AUAI Press (2012)
13. Srividhya, S., Mallika, R.: Feature selection for high dimensional imbalanced
    datasets using game theory and Fisher score. J. Adv. Res. Dyn. Control Syst.
    9, 195–202 (2017)
14. Kannan, S., Ramaraj, N.: A novel hybrid feature selection via Symmetrical Uncer-
    tainty ranking based local memetic search algorithm. Knowl.-Based Syst. **23**, 580–
    585 (2010)

M²Sim: A Long-Term Interactive Driving Simulator

Zhengxiao Han[1,2], Zhijie Yan[2,3], Yang Li[2,4], Pengfei Li[2], Yifeng Shi[5], Nairui Luo[5], Xu Gao[5], Yongliang Shi[2], Pengfei Huang[2], Jiangtao Gong[2], Guyue Zhou[2], Yilun Chen[2], Hang Zhao[2], and Hao Zhao[2(✉)]

[1] Beijing University of Chemical Technology, Beijing, China
[2] Tsinghua University, Beijing, China
zhaohao@air.tsinghua.edu.cn
[3] Beihang University, Beijing, China
[4] University of California, San Diego, USA
[5] Baidu, Inc., Beijing, China

Abstract. Simulation now plays an important role in the development of autonomous driving algorithms as it can significantly reduce the economical cost and ethical risk of real-world testing. However, building a high-quality driving simulator is not trivial as it calls for realistic interactive behaviors of road agents. Recently, several simulators employ interactive trajectory prediction models learnt in a data-driven manner. While they are successful in generating short-term interactive scenarios, the simulator quickly breaks down when the time horizon gets longer. We identify the reason behind: existing interactive trajectory predictors suffer from the out-of-domain (OOD) problem when recursively feeding predictions as the input back to the model. To this end, we propose to introduce a tailored model predictive control (MPC) module as a rescue into the state-of-the art interactive trajectory prediction model M2I, forming a new simulator named M²Sim. Notably, M²Sim can effectively address the OOD problem of long-term simulation by enforcing a flexible regularization that admits the replayed data, while still enjoying the diversity of data-driven predictions. We demonstrate the superiority of M²Sim using both quantitative results and visualizations and release our data, code and models: https://github.com/0nhc/m2sim.

Keywords: Autonomous Driving · Interactive Simulator · MPC

1 Introduction

Collecting data for dangerous driving scenarios is challenging, making simulation the preferred choice for algorithm development. Thus, building a high-quality driving simulator with realistic interactive behaviors is crucial.

Figure 1-a shows a replayed data clip at an intersection, where the red line represents the future trajectory of a chosen ego car. Figure 1-b demonstrates the

Sponsored by Baidu Inc. through Apollo-AIR Joint Research Center.

drawback of a non-interactive simulator. The ego car, controlled by a to-be-tested algorithm, collides with other cars following the replayed data.

To address this issue, recent works integrate interactive trajectory predictors [1–4], using neural networks to capture the diverse distribution of real-world interactive trajectories. However, they encounter the out-of-distribution (OOD) [5] problem in long-term simulation due to their recursive nature [6]. As shown in Fig. 1-c, this leads to the unrealistic long-term simulation.

To alleviate the OOD issue, we propose using model predictive control (MPC) [7–12] as a rescue mechanism (Fig. 1-d). MPC introduces the replayed trajectory as a flexible regularization term, and strictly following the replayed data. This would reduce the simulator to a non-interactive one (Fig. 1-b). Thus, the outputs of the interactive trajectory predictor serve as another fundamental regularization term, offering diversity to the system.

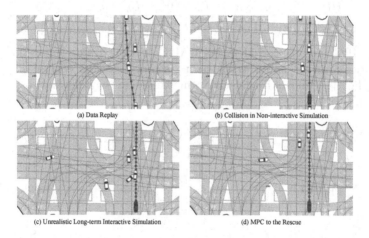

(a) Data Replay

(b) Collision in Non-interactive Simulation

(c) Unrealistic Long-term Interactive Simulation

(d) MPC to the Rescue

Fig. 1. (a) Data replay. (b) Collisions happen in a non-interactive simulator. (c) Other cars react according to a learned model. In the long term, their trajectories become unrealistic. (d) We leverage MPC to address the OOD issue in (c).

2 Method

As shown in Fig. 2-a, our simulator M²Sim addresses the OOD problem by integrating an MPC module with a data-driven interactive trajectory predictor. Here, we choose the state-of-the-art predictor M2I [2] originally trained on Waymo Open Motion Dataset (WOMD) [13]. In the MPC formulation, we add both M2I's output trajectories and ground truth trajectories (scenarios with collisions happen) to the optimization problem as constraints, which enforces a flexible regularization that admits the replayed ground truth while still enjoying the diversity of data-driven predictions.

As shown in Fig. 2-b, we design an MPC controller considering both M2I's output trajectories and ground truth trajectories as constraints for the optimization problem. In addition, MPC comes with a kinematic model, which also ensures the physical plausibility of its output trajectories.

We denote an autonomous vehicle(AV)'s state of [coordinate x, coordinate y, heading angle, speed] by $z(\tau) = [x(\tau), y(\tau), yaw(\tau), v(\tau)]$, and its control input of [acceleration, steering angle] by $u(\tau) = [a(\tau), \delta(\tau)]$. Then we denote the prediction horizon by $T_p \in \mathbb{N}$. Thus, the problem can be defined as solving the optimal control input $u(\tau) = [a(\tau), \delta(\tau)]$ at specific time τ. Our MPC formulations are as follows:

Optimization Problem Setup. At time τ, we denote control inputs of $U(\tau) \in U_b \subset \mathbb{R}^{2T_P}$ and corresponding states of $Z(\tau) \in Z_b \subset \mathbb{R}^{4T_P}$:

$$U(\tau) = [u(\tau), ..., u(\tau + T_p - 1)]^T, \quad U_b = [u_{min}, \ u_{max}]$$
$$Z(\tau) = [z(\tau + 1), ..., z(\tau + T_p)]^T, \quad Z_b = [z_{min}, \ z_{max}]$$

Our MPC module considers both M2I's output trajectories and ground truth trajectories as inputs, while taking smoothness and control efforts into account. Let $\| \cdot \|$ denotes Euclidean Norm, we design the following cost function $J(Z(\tau), \ U(\tau))$:

$$J = \sum_{\tau=1}^{T_p} \lambda_{z1} \|z(\tau) - z_{ref1}(\tau)\|^2 + \sum_{\tau=1}^{T_p} \lambda_{z2} \|z(\tau) - z_{ref2}(\tau)\|^2 \qquad \text{(State Error)}$$

$$+ \sum_{\tau=0}^{T_p-1} \lambda_u \|u(\tau)\|^2 \qquad \text{(Control Effort)}$$

$$+ \sum_{\tau=0}^{T_p-1} \lambda_{\Delta u} \|u(\tau) - u(\tau - 1)\|^2, \qquad \text{(Smoothness)}$$

where z_{ref1} and z_{ref2} denotes trajectories provided by M2I and ground truth respectively, and all λ denotes constant values corresponding to their weights in the cost function.

Thus, the optimization problem can be defined as solving:

$$U^* \triangleq \underset{U}{arg \ min} \ J(Z(\tau), U(\tau))$$

System Modeling. Motivated by [14], the car can be defined as a bicycle model with state $z(\tau)$ and control input $u(\tau)$ at specific time τ. Based on the bicycle model's kinematics, we can get the differential equation of z and u:

$$\dot{z} = \begin{bmatrix} \dot{x} \\ \dot{y} \\ ya\dot{w} \\ \dot{v} \end{bmatrix} = \begin{bmatrix} v \times cos(yaw) \\ v \times sin(yaw) \\ \frac{v \times tan(\delta)}{l} \\ a \end{bmatrix},$$

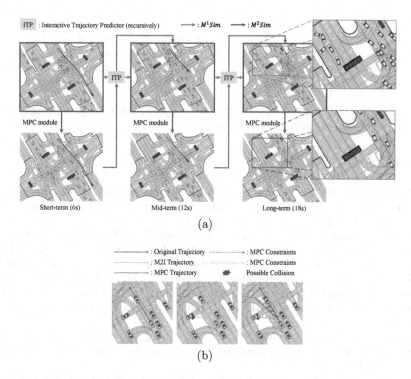

Fig. 2. (a) The structure of M²Sim. M¹Sim (M2I without the MPC module), denoted by the flow of pink arrows, can produce reasonable results in the short term, but its behaviors become unrealistic in the long term due to recursively feeding predictions into the model. However M²Sim (M2I with the MPC module), denoted by the flow of blue arrows, addresses the OOD problem by integrating the MPC module. Note that we conceptually use the same image to present recursively predicted results in red boxes, which are actually different for M¹Sim and M²Sim. (b) The MPC module considers two trajectories as constraints to the optimization problem. (Color figure online)

where l is the distance between the front and rear wheels of the vehicle. Then we linearize and discretize the kinematic model to obtain the equality constraint for the optimization problem. Suppose at time τ, we observed states $\hat{z}(\tau)$ and control inputs $\hat{u}(\tau)$ of the ego vehicle, then we can infer its future state:

$$\hat{z}(\tau + 1) = A \times \hat{z}(\tau) + B \times \hat{u}(\tau) + C,$$

where A, B and C are:

$$A = \begin{bmatrix} 1 & 0 & -v \times sin(yaw) \times dt & cos(yaw) \times dt \\ 0 & 1 & v \times cos(yaw) \times dt & sin(yaw) \times dt \\ 0 & 0 & 1 & \frac{tan(\delta) \times dt}{l} \\ 0 & 0 & 0 & 1 \end{bmatrix},$$

$$B = \begin{bmatrix} 0 & 0 \\ 0 & 0 \\ 0 & \frac{v}{l \times cos^2(\delta)} \\ 1 & 0 \end{bmatrix}, C = \begin{bmatrix} v \times sin(yaw) \times yaw \times dt \\ -v \times cos(yaw) \times yaw \times dt \\ -\frac{v \times \delta}{l \times cos^2(\delta)} \times dt \\ 0 \end{bmatrix}$$

At last the optimization problem can be solved with all these constraints. We use cvxpy to get optimal control inputs.

References

1. Ngiam, J., et al.: Scene transformer: a unified architecture for predicting future trajectories of multiple agents. In: International Conference on Learning Representations (2022)
2. Sun, Q., Huang, X., Gu, J., Williams, B.C., Zhao, H.: M2I: from factored marginal trajectory prediction to interactive prediction. In: Proceedings of the IEEE/CVF Conference on Computer Vision and Pattern Recognition, pp. 6543–6552 (2022)
3. Liu, X., Wang, Y., Jiang, K., Zhou, Z., Nam, K., Yin, C.: Interactive trajectory prediction using a driving risk map-integrated deep learning method for surrounding vehicles on highways. IEEE Trans. Intell. Transp. Syst. **23**(10), 19076–19087 (2022)
4. Zhang, K., Zhao, L., Dong, C., Wu, L., Zheng, L.: AI-TP: attention-based interaction-aware trajectory prediction for autonomous driving. IEEE Trans. Intell. Veh. **8**, 73–83 (2022)
5. Shen, Z., et al.: Towards out-of-distribution generalization: a survey. arXiv preprint arXiv:2108.13624 (2021)
6. Filos, A., Tigkas, P., McAllister, R., Rhinehart, N., Levine, S., Gal, Y.: Can autonomous vehicles identify, recover from, and adapt to distribution shifts? In: International Conference on Machine Learning, pp. 3145–3153. PMLR (2020)
7. Kerrigan, E.C.: Predictive control for linear and hybrid systems [bookshelf]. IEEE Control Syst. Mag. **38**(2), 94–96 (2018)
8. Falcone, P., et al.: Nonlinear model predictive control for autonomous vehicles (2007)
9. Carvalho, A., Gao, Y., Gray, A., Tseng, H.E., Borrelli, F.: Predictive control of an autonomous ground vehicle using an iterative linearization approach. In: 16th International IEEE Conference on Intelligent Transportation Systems (ITSC 2013), pp 2335–2340. IEEE (2013)
10. Gao, Y., Lin, T., Borrelli, F., Tseng, E., Hrovat, D.: Predictive control of autonomous ground vehicles with obstacle avoidance on slippery roads. In: Dynamic Systems and Control Conference, vol. 44175, pp. 265–272 (2010)
11. Beal, C.E., Gerdes, J.C.: Model predictive control for vehicle stabilization at the limits of handling. IEEE Trans. Control Syst. Technol. **21**(4), 1258–1269 (2012)
12. Levinson, J., et al.: Towards fully autonomous driving: systems and algorithms. In: 2011 IEEE Intelligent Vehicles Symposium (IV), pp. 163–168. IEEE (2011)
13. Ettinger, S., et al.: Large scale interactive motion forecasting for autonomous driving: the waymo open motion dataset. In: Proceedings of the IEEE/CVF International Conference on Computer Vision, pp. 9710–9719 (2021)
14. Kong, J., Pfeiffer, M., Schildbach, G., Borrelli, F.: Kinematic and dynamic vehicle models for autonomous driving control design. In: 2015 IEEE Intelligent Vehicles Symposium (IV), pp. 1094–1099. IEEE (2015)

Long-Term Interactive Driving Simulation: MPC to the Rescue

Zhengxiao Han[1,2], Zhijie Yan[2,3], Yang Li[2,4], Pengfei Li[2], Yifeng Shi[5],
Nairui Luo[5], Xu Gao[5], Yongliang Shi[2], Pengfei Huang[2], Jiangtao Gong[2],
Guyue Zhou[2], Yilun Chen[2], Hang Zhao[2], and Hao Zhao[2(✉)]

[1] Beijing University of Chemical Technology, Beijing, China
[2] Tsinghua University, Beijing, China
zhaohao@air.tsinghua.edu.cn
[3] Beihang University, Beijing, China
[4] University of California, San Diego, USA
[5] Baidu, Inc., Beijing, China

Abstract. Simulation now plays an important role in the development
of autonomous driving algorithms as it can significantly reduce the economical cost and ethical risk of real-world testing. However, building
a high-quality driving simulator is not trivial as it calls for realistic
interactive behaviors of road agents. Recently, several simulators employ
interactive trajectory prediction models learnt in a data-driven manner.
While they are successful in generating short-term interactive scenarios,
the simulator quickly breaks down when the time horizon gets longer.
We identify the reason behind: existing interactive trajectory predictors
suffer from the out-of-domain (OOD) problem when recursively feeding
predictions as the input back to the model. To this end, we propose to
introduce a tailored model predictive control (MPC) module as a rescue into the state-of-the art interactive trajectory prediction model M2I,
forming a new simulator named M^2Sim. Notably, M^2Sim can effectively
address the OOD problem of long-term simulation by enforcing a flexible regularization that admits the replayed data, while still enjoying the
diversity of data-driven predictions. We demonstrate the superiority of
M^2Sim using both quantitative results and visualizations and release our
data, code and models: https://github.com/0nhc/m2sim.

Keywords: Autonomous Driving · Interactive Simulator · MPC

1 Introduction

Autonomous driving [1–5] is one of the most important AI applications nowadays. Collecting data for dangerous driving scenarios is challenging, making simulation the preferred choice for algorithm development. Thus, building a high-
quality driving simulator with realistic interactive behaviors is crucial.

Figure 1-a shows a replayed data clip at an intersection, where the red line
represents the future trajectory of a chosen ego car. Figure 1-b demonstrates the

Sponsored by Baidu Inc. through Apollo-AIR Joint Research Center.

L. Fang et al. (Eds.): CICAI 2023, LNAI 14474, pp. 177–188, 2024.
https://doi.org/10.1007/978-981-99-9119-8_17

drawback of a non-interactive simulator. The ego car, controlled by a to-be-tested algorithm, collides with other cars following the replayed data.

To address this issue, recent works integrate interactive trajectory predictors [6–9], using neural networks to capture the diverse distribution of real-world interactive trajectories. However, they encounter the out-of-distribution (OOD) [10] problem in long-term simulation due to their recursive nature [11]. As shown in Fig. 1-c, this leads to the unrealistic long-term simulation.

To alleviate the OOD issue, we propose using model predictive control (MPC) [12–17] as a rescue mechanism (Fig. 1-d). This module introduces the replayed trajectory as a flexible regularization term. And the interactive trajectory predictor outputs serve as another fundamental regularization term, which offers diversity to the system. Other tailored add-ons like control effort and smoothness terms make the behavior of our agents more realistic.

In one word, the contribution of this study is a simulation system combining the advantages of data-driven interactive trajectory predictors and a tailored model predictive controller. To the best of our knowledge, M^2Sim is the first system that demonstrates realistic long-term interactive driving simulation results in the literature.

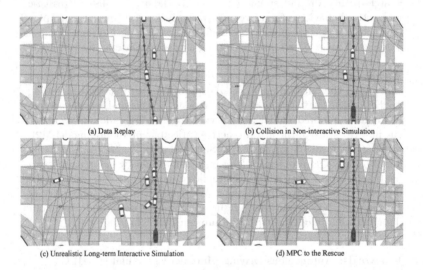

(a) Data Replay

(b) Collision in Non-interactive Simulation

(c) Unrealistic Long-term Interactive Simulation

(d) MPC to the Rescue

Fig. 1. (a) Data replay. (b) Collisions happen in a non-interactive simulator. (c) Other cars react according to a learned model. In the long term, their trajectories become unrealistic. (d) We leverage MPC to address the OOD issue in (c).

2   Related Work

2.1   Interactive Trajectory Prediction

For interactive trajectory prediction, some works utilize graph neural networks to learn the interaction relationship between different vehicles [18, 19], and other works use the attention in transformer to model the interaction relationship between vehicles [20,21]. M2I [7], which is the engine of choice in our simulator for its state-of-the-art performance, uses both rasterized map representation and polyline representation, and integrates graph neural network and attention mechanism for interactive prediction. Since it utilizes future trajectories of the ego vehicle, it has good prediction accuracy and high computational efficiency. However learning-based approaches lack interpretability and may break specific physical constraints. What's worse, their behaviors become unpredictable and unreliable when the time horizon gets long.

2.2 Traffic Simulation

As for simulating interactive behaviors, autonomous driving simulators such as SUMO [22], CityFlow [23], CommonRoad [24] focus on multi-agent traffic flow simulation, but due to the lack of realistic traffic data, they cannot simulate traffic flow with interactive behaviors. Recently some works such as TrafficSim [25], SimNet [26], etc. learn from data collected in real world to model interactive multi-agent behaviors. These approaches only learn from all the agents' past behaviors at once, without considering their future interactions. InterSim [27] simplifies the prediction of future interactions into a binary classification problem, which has better interpretability on relationship prediction and higher computational efficiency than counterparts using latent-variable [25] or cost functions [28] for to model future collisions. However when we test these learning-based approaches, we found them quickly breaking down when the time horizon of traffic scenarios gets longer, which is caused by the out-of-domain(OOD) problem [10].

3 Method

3.1 Overview

As shown in Fig. 2, our simulator M^2Sim addresses the OOD problem by integrating an MPC module with a data-driven interactive trajectory predictor. Here, we choose the state-of-the-art predictor M2I [7]. In the MPC formulation, we add both M2I's output trajectories and ground truth trajectories (explained later) to the optimization problem as constraints, which enforces a flexible regularization that admits the replayed ground truth while still enjoying the diversity of data-driven predictions.

3.2 M2I

As shown in Fig. 3, the trajectory predictor has three modules: relation predictor, marginal trajectory predictor, and conditional trajectory predictor. The relation predictor predicts whether each road agent will be an influencer (and yield), a reactor (and be yielded), or neither. Then the marginal trajectory predictor of it predicts the future 8 s of trajectories of all road agents without considering their

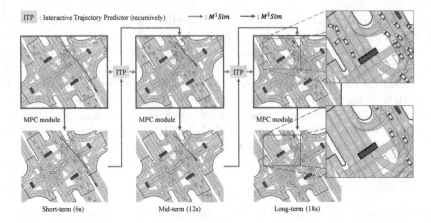

Fig. 2. The structure of M^2Sim. M^1Sim (M2I without the MPC module), denoted by the flow of pink arrows, can produce reasonable results in the short term, but its behaviors become unrealistic in the long term due to recursively feeding predictions into the model. As shown in the enlarged pink box, agents form a cluster in the long term. However M^2Sim (M2I with the MPC module), denoted by the flow of blue arrows, addresses the OOD problem by integrating the MPC module. Note that we conceptually use the same image to present recursively predicted results in red boxes, which are actually different for M^1Sim and M^2Sim. (Color figure online)

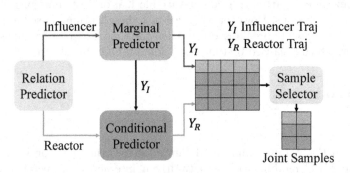

Fig. 3. A brief recap of the structure of M2I, which is our data-driven interactive trajectory prediction engine. It outputs several trajectories with confidence values, so that we can randomly sample from them to generate diverse trajectories.

potential interactions. Finally, with the predicted relations and marginal trajectories, its conditional trajectory predictor modifies the trajectories of reactors to account for their interactions with influencers.

In addition to the three modules for predicting relations and trajectories, it also has a module for selecting trajectories. Specifically, for N single vehicle trajectories, we have N^2 pairs of interactions. The sample selector outputs the top K out of the N^2 pairs based on the ranking of confidence scores. We can generate diverse trajectories with interactive trajectory predictor by selecting trajectories randomly according to the confidence values.

The interactive trajectory predictor takes 1.1 s of past trajectories and predicts 8 s of future trajectories (represented as a set of coordinates x, y). By differential operation, we can get yaw and v from discrete coordinates x, y of future trajectories. The final output trajectory can be described as $z_{ref} = [x, y, yaw, v]$ denoting [coordinate x, coordinate y, heading angle, speed].

3.3 MPC

As shown in Fig. 4, we design an MPC controller considering both M2I's output trajectories and ground truth trajectories as constraints for the optimization problem. In addition, MPC comes with a kinematic model, which also functions as a constraint and ensures the physical plausibility of its output trajectories.

We denote an autonomous vehicle (AV)'s state of [coordinate x, coordinate y, heading angle, speed] by $z(\tau) = [x(\tau), y(\tau), yaw(\tau), v(\tau)]$, and its control input of [acceleration, steering angle] by $u(\tau) = [a(\tau), \delta(\tau)]$. Then we denote the prediction horizon by $T_p \in \mathbb{N}$. Thus, the problem can be defined as solving the optimal control input $u(\tau) = [a(\tau), \delta(\tau)]$ at specific time τ. Our MPC formulations are as follows:

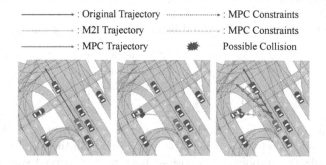

Fig. 4. The MPC module considers both the original replay trajectory and the predicted interactive trajectory as constraints to the optimization problem.

Optimization Problem Setup. At time τ, we denote control inputs of $U(\tau) \in U_b \subset \mathbb{R}^{2T_P}$ and corresponding states of $Z(\tau) \in Z_b \subset \mathbb{R}^{4T_P}$:

$$U(\tau) = [u(\tau), ..., u(\tau + T_p - 1)]^T, \quad U_b = [u_{min}, \ u_{max}]$$
$$Z(\tau) = [z(\tau + 1), ..., z(\tau + T_p)]^T, \quad Z_b = [z_{min}, \ z_{max}]$$

Our MPC module considers both M2I's output trajectories and ground truth trajectories as inputs, while taking smoothness and control efforts into account. Let $\| \cdot \|$ denotes Euclidean Norm, we design the following cost function $J(Z(\tau),\ U(\tau))$:

$$
\begin{aligned}
J &= \sum_{t=\tau+1}^{\tau+T_p} \lambda_{z1}\|z(t) - z_{ref1}(t)\|^2 + \sum_{t=\tau+1}^{\tau+T_p} \lambda_{z2}\|z(t) - z_{ref2}(t)\|^2 \quad \text{(State Error)} \\
&+ \sum_{t=\tau}^{\tau+T_p-1} \lambda_u\|u(t)\|^2 \quad\quad\quad\quad\quad\quad\quad\quad\quad\quad\quad\quad\quad\quad\text{(Control Effort)} \\
&+ \sum_{t=\tau}^{\tau+T_p-1} \lambda_{\Delta u}\|u(t) - u(t-1)\|^2, \quad\quad\quad\quad\quad\quad\quad\quad\quad\text{(Smoothness)}
\end{aligned}
$$

where z_{ref1} and z_{ref2} denotes trajectories provided by M2I and ground truth trajectories respectively, and all λ denotes constant values corresponding to their weights in the cost function.

Thus, the optimization problem can be defined as solving:

$$
U^* \triangleq \underset{U}{arg\ min}\ J(Z(\tau), U(\tau))
$$

System Modeling. Motivated by [29], the car can be defined as a bicycle model with state $z(\tau)$ and control input $u(\tau)$ at specific time τ. Based on the bicycle model's kinematics, we can get the differential equation of z and u:

$$
\dot{z} = \begin{bmatrix} \dot{x} \\ \dot{y} \\ y\dot{a}w \\ \dot{v} \end{bmatrix} = \begin{bmatrix} v \times cos(yaw) \\ v \times sin(yaw) \\ \frac{v \times tan(\delta)}{l} \\ a \end{bmatrix},
$$

where l is the distance between the front and rear wheels of the vehicle. Then we linearize and discretize the kinematic model to obtain the equality constraint for the optimization problem. Suppose at time τ, we observed states $\hat{z}(\tau)$ and control inputs $\hat{u}(\tau)$ of the ego vehicle, then we can infer its future state:

$$
\hat{z}(\tau+1) = A \times \hat{z}(\tau) + B \times \hat{u}(\tau) + C,
$$

where A, B and C are:

$$
A = \begin{bmatrix} 1 & 0 & -v \times sin(yaw) \times dt & cos(yaw) \times dt \\ 0 & 1 & v \times cos(yaw) \times dt & sin(yaw) \times dt \\ 0 & 0 & 1 & \frac{tan(\delta) \times dt}{l} \\ 0 & 0 & 0 & 1 \end{bmatrix},
$$

$$
B = \begin{bmatrix} 0 & 0 \\ 0 & 0 \\ 0 & \frac{v}{l \times cos^2(\delta)} \\ 1 & 0 \end{bmatrix}, C = \begin{bmatrix} v \times sin(yaw) \times yaw \times dt \\ -v \times cos(yaw) \times yaw \times dt \\ -\frac{v \times \delta}{l \times cos^2(\delta)} \times dt \\ 0 \end{bmatrix}
$$

At last the optimization problem can be solved with all these constraints. We use a off-the-shelf toolbox cvxpy to get optimal control inputs.

3.4 Collision Avoidance

To reduce collisions, integrating inequality constraints into the optimization problem is a natural choice. Inequality constraints such as the Euclidean distance between the ego vehicle and its surrounding vehicles may be useful for collision avoidance [30]. However, too many inequality constraints can cause high computational cost in optimization.

Thus, we designed a collision avoidance mechanism outside the optimization process by only changing the acceleration of control inputs. Suppose we have the ego vehicle of state $z(\tau) = [x(\tau), y(\tau), yaw(\tau), v(\tau)]$ and surrounding vehicles of states $z_i(\tau) = [x_i(\tau), y_i(\tau), yaw_i(\tau), v_i(\tau)], i \in N$, corresponding to [coordinate xy, heading angle, speed] at time τ, then we can get the final control inputs $u(\tau) = [a(\tau), \delta(\tau)]$ corresponding to [acceleration, steering angle] through Algorithm 1.

Algorithm 1: Framework of Collision Avoidance.

Input: Ego Vehicle of State: $z(\tau)$; Number of Surrounding Vehicles N;
Surrounding Vehicles of States: $z_i(\tau)$ where $i \in N$; Safety Distance D;
Coordinate Vector $\vec{s}$; Constant Value λ_f
Output: Control Inputs $u(\tau)$ of the Ego Vehicle

Get $z(\tau)$, $z_i(\tau), i \in N$ at time τ; $i = 0$; $\vec{s} = [0, 0]$; **foreach** i in N **do**
> Calculate the Euclidean distance d_i between the ego vehicle and the surrounding vehicle:
> $d_i = \sqrt{[x(\tau) - x_i(\tau)]^2 + [y(\tau) - y_i(\tau)]^2}$
> **if** $d_i < D$ **then**
>> Sum up force vector $\vec{s}$.
>> $\vec{s_i} = [x_i(\tau), y_i(\tau)] - [x(\tau), y(\tau)]$
>> $\vec{s} = \vec{s} + \vec{s_i}$

The unit vector $\vec{e}$ of the ego vehicle:
$\vec{e} = v(\tau) \cdot [cos(yaw(\tau)), sin(yaw(\tau))]$
The projection p of $\vec{s}$ in the direction of $\vec{e}$: $p = \frac{\lambda_f}{\vec{s} \cdot \vec{e}}$
if $p < 0$ **then**
> Only consider the case of deceleration.
> $a(\tau) = a(\tau) + p$
> Limit $a(\tau)$ within its boundary.
> $a(\tau) = max(a(\tau), a_{min})$

Return $u(\tau) = [a(\tau), \delta(\tau)]$

---

4 Experiment

In this section, we introduce model details, and provide quantitative results and qualitative examples of different simulators to demonstrate the performance of M^2Sim.

4.1 Model Details

We used the original M2I model trained on Waymo Open Motion Dataset (WOMD) from our baseline InterSim for evaluation.

As for the MPC and collision avoidance module, we initialize default constant values of the optimization problem $U^* \triangleq arg\ min_U\ J(Z(\tau), U(\tau))$ where $\lambda_{z1} = \lambda_{z2} = 0.5$, $\lambda_u = \lambda_{\Delta u} = 1.0$. And in the collision avoidance algorithm, we initialize safety distance with $D = 8$ and $\lambda_f = 1.0$. We tuned these hyperparameters on a small validation set.

4.2 Simulation Task

The simulation task is to test the performance of agents in M^2Sim and our baseline InterSim [27] on driving scenarios generated by editing the data replay clips to make collisions happen (as ground truth trajectories). The editing process involves selecting a random car as the ego vehicle, generate a random trajectory for it using the marginal predictor in Fig. 3.

The reason why we choose InterSim is, InterSim uses the state-of-the-art interactive prediction model M2I, and the difference between InterSim and M^2Sim is, M^2Sim has an MPC and collision avoidance module in addition to the M2I model. Because InterSim only uses an M2I model for predicting trajectories, while M^2Sim leverages both an M2I model and an MPC module, we call our re-implemented InterSim as M^1Sim.

4.3 Quantitative Results

Table 1. Performance of interactive prediction.

Method	minADE ↓	minFDE ↓	missRate ↓	mAP ↑
M^1Sim (re-implemented InterSim)	9.349	25.917	0.906	0.004
M^2Sim	**1.089**	**2.161**	**0.154**	**0.175**

Motivated by [7,31], we use these metrics to evaluate the performance of simulators, including:

- **minADE** (Minimum Average Displacement Error). The minADE metric computes the mean of the L2 norm between the ground truth future trajectory and the closest predicted output trajectory from M2I out of $K = 6$ (number of M2I's outputs) samples.
- **minFDE** (Minimum Final Displacement Error). The minFDE is the same as minADE to compute the displacement error, but the minFDE only computes the displacement of the final positions of ground truth trajectory and M2I's predicted trajectory.
- **missRate** (Miss Rate). A **miss** is defined as the state when none of the individual K predictions for an object are within a given lateral and longitudinal threshold of the ground truth trajectory. The missRate is calculated as the total number of misses divided by $K = 6$ (number of M2I's outputs) for M2I.
- **mAP** (Mean Average Precision). The mAP computes the area under the precision-recall curve of the prediction samples by applying confidence score thresholds.

Table 2. Different types of collision rate for M^1Sim and M^2Sim.

Method	Agent-agent ↓	Agent-environment ↓
M^1Sim (re-implemented InterSim)	0.343	0.447
M^2Sim	**0.075**	**0.326**

The results are summarized in Table 1 representing how the simulators' behaviors adhere to the ground truth data replay. We observed that M^2Sim achieves the lowest errors, the lowest miss rate and the highest precision in predicting trajectories. InterSim, also utilizing an M2I model, suffers from the out-of-domain (OOD) problem in our long-term experiments. As the simulation time gets longer, environmental vehicles in InterSim will begin to deviate from ground truth trajectories, collide with each other, and even drive out of the lane. But we add an MPC controller along with a collision avoidance module to process M2I's output trajectories, so that vehicles' trajectories are constrained to be as close as possible to ground truth trajectories.

4.4 Qualitative Examples

In Fig. 5, we present two representative scenarios to show our method's performance of long-term simulation. We present M^2Sim without MPC module as M^1Sim. In both scenarios, M^1Sim suffers from the out-of-domain (OOD) problem when recursively feeding predictions as the input back to the model, failing to keep environmental vehicles from tracking ground truth trajectories and avoiding collisions over time. But M^2Sim achieves to keep track of environmental vehicles' original trajectories and avoid collisions by adding an MPC and collision avoidance module with original trajectories as optimization constraints.

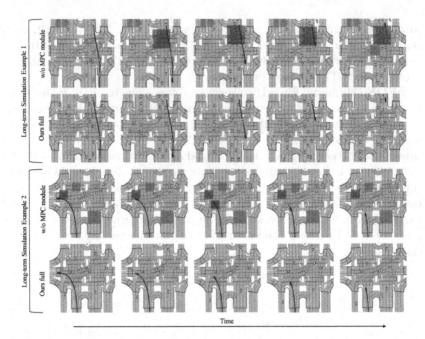

Fig. 5. Examples of M^2Sim successfully keeping environmental vehicles from tracking their original trajectories and avoiding collisions. Failures of keeping original trajectories with weird heading angles (the OOD problem) are highlighted in yellow boxes. Collisions are highlighted in red boxes. (Color figure online)

4.5   Ablation Study

We present ablation study on the MPC controller along with the collision avoidance module by comparing the collision rate of M^2Sim and M^1Sim for the same simulation task. The results are summarized in Table 2. The agent-agent collision rate computes the number of colliding agent pairs divided by the number of simulated agents. Similarly, the agent-environment collision rate computes the number of agents colliding with lanes or pedestrians divided by the number of simulated agents. It is not surprising to see M^2Sim achieves better performance due to the constraints of following ground truth trajectories and collision avoidance mechanism.

5   Conclusion

In conclusion, we present an interactive traffic simulator with an advanced interactive trajectory predictor (M2I) and address the OOD problem in long-term simulation. M^2Sim improved its performance by adding an MPC controller along with a collision avoidance module. In the experiments, we test M^2Sim and InterSim on the same simulation task to demonstrate the superiority of our M^2Sim. In the ablation study, we show the effectiveness of our proposed MPC module.

Limitations. However, to avoid too much computational consumption, our proposed MPC module does not consider obstacle avoidance in the cost function. Further research could go deeper in optimization.

References

1. Tian, B., Liu, M., Gao, H.-A., Li, P., Zhao, H., Zhou, G.: Unsupervised road anomaly detection with language anchors. In: 2023 IEEE International Conference on Robotics and Automation (ICRA), pp. 7778–7785. IEEE (2023)
2. Li, P., et al.: LODE: locally conditioned Eikonal implicit scene completion from sparse LiDAR. arXiv preprint arXiv:2302.14052 (2023)
3. Jin, B., et al.: ADAPT: action-aware driving caption transformer. arXiv preprint arXiv:2302.00673 (2023)
4. Zheng, Y., et al.: Steps: joint self-supervised nighttime image enhancement and depth estimation. arXiv preprint arXiv:2302.01334 (2023)
5. Hu, Y., et al.: Planning-oriented autonomous driving. In: Proceedings of the IEEE/CVF Conference on Computer Vision and Pattern Recognition, pp. 17 853–17 862 (2023)
6. Ngiam, J., et al.: Scene transformer: a unified architecture for predicting future trajectories of multiple agents. In: International Conference on Learning Representations (2022)
7. Sun, Q., Huang, X., Gu, J., Williams, B.C., Zhao, H.: M2I: from factored marginal trajectory prediction to interactive prediction. In: Proceedings of the IEEE/CVF Conference on Computer Vision and Pattern Recognition, pp. 6543–6552 (2022)
8. Liu, X., Wang, Y., Jiang, K., Zhou, Z., Nam, K., Yin, C.: Interactive trajectory prediction using a driving risk map-integrated deep learning method for surrounding vehicles on highways. IEEE Trans. Intell. Transp. Syst. **23**(10), 19 076–19 087 (2022)
9. Zhang, K., Zhao, L., Dong, C., Wu, L., Zheng, L.: AI-TP: attention-based interaction-aware trajectory prediction for autonomous driving. IEEE Trans. Intell. Veh. (2022)
10. Shen, Z.: Towards out-of-distribution generalization: a survey. arXiv preprint arXiv:2108.13624 (2021)
11. Filos, A., Tigkas, P., McAllister, R., Rhinehart, N., Levine, S., Gal, Y.: Can autonomous vehicles identify, recover from, and adapt to distribution shifts? In: International Conference on Machine Learning, pp. 3145–3153. PMLR (2020)
12. Kerrigan, E.C.: Predictive control for linear and hybrid systems [bookshelf]. IEEE Control Syst. Mag. **38**(2), 94–96 (2018)
13. Falcone, P., et al.: Nonlinear model predictive control for autonomous vehicles (2007)
14. Carvalho, A., Gao, Y., Gray, A., Tseng, H.E., Borrelli, F.: Predictive control of an autonomous ground vehicle using an iterative linearization approach. In: 16th International IEEE Conference on Intelligent Transportation Systems (ITSC 2013), pp. 2335–2340. IEEE (2013)
15. Gao, Y., Lin, T., Borrelli, F., Tseng, E., Hrovat, D.: Predictive control of autonomous ground vehicles with obstacle avoidance on slippery roads. In: Dynamic Systems and Control Conference, vol. 44175, pp. 265–272 (2010)
16. Beal, C.E., Gerdes, J.C.: Model predictive control for vehicle stabilization at the limits of handling. IEEE Trans. Control Syst. Technol. **21**(4), 1258–1269 (2012)

17. Levinson, J., et al.: Towards fully autonomous driving: systems and algorithms. In: 2011 IEEE Intelligent Vehicles Symposium (IV), pp. 163–168. IEEE (2011)
18. Mohamed, A., Qian, K., Elhoseiny, M., Claudel, C.: Social-STGCNN: a social spatio-temporal graph convolutional neural network for human trajectory prediction. In: Proceedings of the IEEE/CVF Conference on Computer Vision and Pattern Recognition, pp. 14 424–14 432 (2020)
19. Casas, S., Gulino, C., Suo, S., Luo, K., Liao, R., Urtasun, R.: Implicit latent variable model for scene-consistent motion forecasting. In: Vedaldi, A., Bischof, H., Brox, T., Frahm, J.-M. (eds.) ECCV 2020. LNCS, vol. 12368, pp. 624–641. Springer, Cham (2020). https://doi.org/10.1007/978-3-030-58592-1_37
20. Kamra, N., Zhu, H., Trivedi, D.K., Zhang, M., Liu, Y.: Multi-agent trajectory prediction with fuzzy query attention. In: Advances in Neural Information Processing Systems, vol. 33, pp. 22 530–22 541 (2020)
21. Nayakanti, N., Al-Rfou, R., Zhou, A., Goel, K., Refaat, K.S., Sapp, B.: Wayformer: motion forecasting via simple & efficient attention networks. arXiv preprint arXiv:2207.05844 (2022)
22. Lopez, P.A.: Microscopic traffic simulation using sumo. In: 2018 21st International Conference on Intelligent Transportation Systems (ITSC), pp. 2575–2582. IEEE (2018)
23. Zhang, H.: CityFlow: a multi-agent reinforcement learning environment for large scale city traffic scenario. In: The World Wide Web Conference, pp. 3620–3624 (2019)
24. Althoff, M., Koschi, M., Manzinger, S.: CommonRoad: composable benchmarks for motion planning on roads. In: 2017 IEEE Intelligent Vehicles Symposium (IV), pp. 719–726. IEEE (2017)
25. Suo, S., Regalado, S., Casas, S., Urtasun, R.: TrafficSim: learning to simulate realistic multi-agent behaviors. In: Proceedings of the IEEE/CVF Conference on Computer Vision and Pattern Recognition, pp. 10 400–10 409 (2021)
26. Bergamini, L., et al.: SimNet: learning reactive self-driving simulations from real-world observations. In: 2021 IEEE International Conference on Robotics and Automation (ICRA), pp. 5119–5125. IEEE (2021)
27. Sun, Q., Huang, X., Williams, B.C., Zhao, H.: InterSim: interactive traffic simulation via explicit relation modeling. In: 2022 IEEE/RSJ International Conference on Intelligent Robots and Systems (IROS), pp. 11 416–11 423. IEEE (2022)
28. Zhou, J.: Exploring imitation learning for autonomous driving with feedback synthesizer and differentiable rasterization. In: 2021 IEEE/RSJ International Conference on Intelligent Robots and Systems (IROS), pp. 1450–1457. IEEE (2021)
29. Kong, J., Pfeiffer, M., Schildbach, G., Borrelli, F.: Kinematic and dynamic vehicle models for autonomous driving control design. In: 2015 IEEE Intelligent Vehicles Symposium (IV), pp. 1094–1099. IEEE (2015)
30. Cho, M., Lee, Y., Kim, K.-S.: Model predictive control of autonomous vehicles with integrated barriers using occupancy grid maps. IEEE Rob. Autom. Lett. (2023)
31. Kühner, T., Kümmerle, J.: Large-scale volumetric scene reconstruction using LiDAR. In: 2020 IEEE International Conference on Robotics and Automation (ICRA), pp. 6261–6267. IEEE (2020)

Text-Oriented Modality Reinforcement Network for Multimodal Sentiment Analysis from Unaligned Multimodal Sequences

Yuxuan Lei[1], Dingkang Yang[1], Mingcheng Li[1], Shunli Wang[1], Jiawei Chen[1], and Lihua Zhang[1,2,3(✉)]

[1] Academy for Engineering and Technology, Fudan University, Shanghai, China
{yxlei22,mingchengli21,jwchen22}@m.fudan.edu.cn,
{dkyang20,slwang19,lihuazhang}@fudan.edu.cn
[2] Jilin Provincial Key Laboratory of Intelligence Science and Engineering, Changchun, China
[3] Engineering Research Center of AI and Robotics, Ministry of Education, Shanghai, China

Abstract. Multimodal Sentiment Analysis (MSA) aims to mine sentiment information from text, visual, and acoustic modalities. Previous works have focused on representation learning and feature fusion strategies. However, most of these efforts ignored the disparity in the semantic richness of different modalities and treated each modality in the same manner. That may lead to strong modalities being neglected and weak modalities being overvalued. Motivated by these observations, we propose a Text-oriented Modality Reinforcement Network (TMRN), which focuses on the dominance of the text modality in MSA. More specifically, we design a Text-Centered Cross-modal Attention (TCCA) module to make full interaction for text/acoustic and text/visual pairs, and a Text-Gated Self-Attention (TGSA) module to guide the self-reinforcement of the other two modalities. Furthermore, we present an adaptive fusion mechanism to decide the proportion of different modalities involved in the fusion process. Finally, we combine the feature matrices into vectors to get the final representation for the downstream tasks. Experimental results show that our TMRN outperforms the state-of-the-art methods on two MSA benchmarks.

Keywords: Multimodal sentiment analysis · Attention mechanism · Representation learning · Multimodal fusion · Modality reinforcement

1 Introduction

Recognizing the research value of sentiments, numerous studies [3,20,22,23,25] in recent years have focused on identifying and analyzing human sentiments. Compared with traditional unimodal sentiment analysis, Multimodal Sentiment

L. Fang et al. (Eds.): CICAI 2023, LNAI 14474, pp. 189–200, 2024.
https://doi.org/10.1007/978-981-99-9119-8_18

Analysis (MSA) attempts to mine sentiment information from multiple data sources to more comprehensively and accurately understand and predict a wide range of complex human emotions.

While data from multiple modalities can be complementary, the asynchrony between different modality sequences caused the distress of fusion. To address this problem, most prior works have manually aligned visual and acoustic sequences at the resolution of text words [16,18], but this has also resulted in high labor costs and ignored long-term dependencies between different modal elements. Recent efforts like [9,15] have tended to deal with unaligned multimodal sequences by cross-modal attention. They often digest inter-modality correlations through sufficient interactions between each pair of modalities. However, this results in a surge in the number of parameters and redundant information in the modalities. They treat all modalities with the same weight without regard to the fact that the semantic richness of distinct modalities is different, which may lead to strong modalities being neglected and weak modalities being overvalued. Observing previous works [1,19], we found that text modality dominates the MSA task. On the one hand, the text modality is naturally highly structured and semantically condensed; on the other hand, due to the maturity of natural language processing techniques, modeling techniques for text data are relatively mature. In this situation, it is crucial to balance the contributions of different modalities. Moreover, the vanilla Transformer [17] also has some drawbacks. The self-attention mechanism incorporates redundancy and noise while focusing on the information within the modality, especially for the visual and acoustic modalities. Unlike spoken words that can be encoded directly, acoustic and visual modalities are pre-processed before being fed into the network, and noise is inevitably introduced during the pre-processing process [1]. Secondly, the redundancy in time series between visual and acoustic sequences is very high.

Inspired by the above observations, we propose a Text-oriented Modality Reinforcement Network (TMRN) to refine multimodal representations effectively. The core strategy of the TMRN is to interact between modalities with the text modality at the center and to guide the reinforcement process of the other two modalities by text modality. For the inter-modal intersection, we propose a text-centered cross-modal attention module to make full interaction for text/acoustic and text/visual pairs. We also present an adaptive fusion mechanism to measure the weights of the different modalities during fusion. For the intra-modal reinforcement, we design a text-gated self-attention module to introduce prior knowledge of textual semantics in the process of feature reinforcement of visual/acoustic modalities. This aims to mine the semantic information on time series better and to ignore the noise of visual/acoustic modalities. Overall, we make the following three contributions:

- We propose TMRN, a method that focuses on the dominance of the text modality in MSA task. The TMRN interacts and reinforces the other two modalities with the text modality as the main thread to obtain a low redundancy and denoised feature representation.

- We present a Text-Centred Cross-modal Attention (TCCA) module and a Text-Gated Self-Attention (TGSA) module to mine inter-modal and intra-modal contextual relationships.
- We perform a comprehensive set of experiments on two human multimodal language benchmarks MOSI [29] and MOSEI [30]. Our experiments show that our method achieves state-of-the-art methods on these two datasets.

2   Related Work

Human multimodal sentiment analysis is to infer human emotional attitudes from the various modality information in video clips. Compared to multimodal fusion from static modalities like images [10], the key technique for this task is how to fuse time-series sequences from different modalities such as natural language, video frames, and acoustic signals [16], especially when these sequences are temporally unaligned. Some recent works [16,18] have focused on manually aligning the visual and acoustic sequences in the resolution of textual words before training. However, manual word alignment is costly, and there is inevitably some loss of information in the multimodal fusion after alignment.

Furthermore, some researchers have worked on unaligned multimodal data. These works can be classified into two categories: discarding the time series dimension and retaining the time series dimension in the subsequent modal interactions. For the former, they usually take one row of the two-dimensional features as a feature vector for subsequent interaction and fusion [4,24,27]. [4,24] learned modality-invariant and modality-specific representations to give a comprehensive and disentangled view of the multimodal data. [27] jointed training the multimodal and unimodal tasks to learn the consistency and difference, respectively. For the latter, they tend to use the attention mechanism to implement interactions between non-aligned sequences [9,15].

A great deal of attention to attention mechanism has been triggered by the Transformer [17]. Transformer networks have been successfully applied to many tasks like semantic role labeling [13] and word sense disambiguation [14]. And now, Transformer is also widely used in the multimodal field. [15] presented Multimodal Transformer (MulT), which uses cross-modal attention to capture the bimodal interactions without manually aligning the three modalities. [9] proposed PMR, which is a further improvement of the interaction between the three modalities based on MulT. Following the latter approaches, our work is also based on the attention mechanism.

3   Problem Statement and Model Structure

3.1   Problem Statement

In this work, the multimodal sentiment analysis task focuses on using the same video clip from the text (t), visual (v), and acoustic (a) modalities as

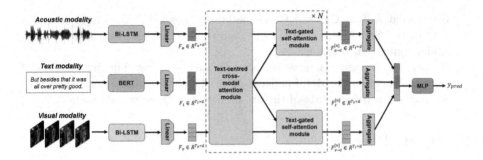

Fig. 1. The overall architecture of the proposed model TMRN.

inputs to the model, which is represented as $X_m \in R^{T_m \times d_m}$ for each modality $m \in \{t, v, a\}$. For the rest of the paper, T_m and d_m are used to represent sequence length and feature dimension of modality m, respectively. The goal of our model is to fully explore and fuse sentiment-related information from these input unaligned multimodal sequences to obtain a text-driven multimodal representation and thus predict the final sentiment analysis results.

3.2   Overall Architecture

The overall architecture of our TMRN is shown in Fig. 1, which consists of three main components: 1) *Unimodal feature extraction module*: we utilize pre-trained BERT [2] to generate the extravagant representation of input words and process visual and acoustic features with Bi-LSTM [5]; 2) *Modality reinforcement*: this part is composed of cross-stacking TCCA and TGSA modules to interact and reinforce the features. We divide the features into visual-text and acoustic-text pairs for cross-attention with the text modality as the query, while self-attention is performed on the text modality. Then, we fuse the pairs with an adaptive fusion mechanism. After that, we use the text modality as a gate to adding prior knowledge to the process of self-reinforcement of visual/acoustic modalities; 3) *Fusion and output module*: we aggregate the final two-dimension features into one-dimension vectors and concatenate them for the downstream tasks. Our aim is to further guide and interact with acoustic and visual modalities through the text modality to obtain a text-dominated implicitly aligned fusion feature.

Unimodal Feature Extraction. To obtain a stronger feature representation of the text, we use a pre-trained BERT [2] model to extract the feature of the sentences:

$$F_t = BERT\left(X_t; \theta_t^{BERT}\right) \in R^{T_t \times d_t}. \tag{1}$$

In acoustic and visual modalities, following [26,28], we use pre-trained ToolKits to extract the initial features X_m from raw data. Then, we use Bi-directional Long Short-Term Memory (BiLSTM) [5] to capture the timing characteristics:

$$F_a = BiLSTM\left(X_a; \theta_a^{LSTM}\right) \in R^{T_a \times d_a}, \tag{2}$$

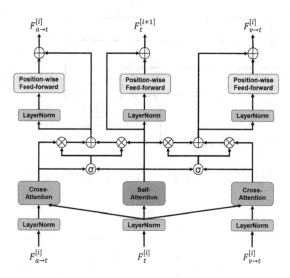

Fig. 2. The architecture of the Text-Centred Cross-modal Attention (TCCA) module.

$$F_v = BiLSTM\left(X_v; \theta_v^{LSTM}\right) \in R^{T_v \times d_v}. \tag{3}$$

For subsequent calculations, we use one fully connected layer to project the features into a fixed dimension as $F_m \in R^{T_m \times d}$, where $m \in \{t, a, v\}$.

Modality Reinforcement. This part includes two key modules: a Text-Centred Cross-modal Attention (TCCA) module and a Text-Gated Self-Attention (TGSA) module. The architecture of TCCA is shown in Fig. 2. Unlike [9], the visual and acoustic modalities share the same text self-attention block to reduce the amount of computation in our TCCA module. This unit is composed of two cross-attention blocks and one self-attention block. The cross-attention block takes $F_t^{[i]}$ and $F_{m \to t}^{[i]}$ as its inputs, where $m \in \{a, v\}$, and the superscript $[i]$ indicates the i-th modality reinforcement processes. First, we perform a layer normalization (LN) on the features like $F_{m \to t}^{[i]} = LN\left(F_{m \to t}^{[i]}\right)$ and $F_t^{[i]} = LN\left(F_t^{[i]}\right)$, and then we put them into a Cross-Attention (CA) block:

$$
\begin{aligned}
F_{m \to t}^{[i]} &= CA_{m \to t}^{[i]}\left(F_{m \to t}^{[i]}, F_t^{[i]}\right), \\
&= softmax\left(\frac{F_t^{[i]} W_{Q_t} W_{K_m}^T F_{m \to t}^{[i]}{}^T}{\sqrt{d}}\right) F_{m \to t}^{[i]} W_{V_m},
\end{aligned} \tag{4}
$$

where $F_{m \to t}^{[0]} = F_m \in R^{T_m \times d}$ and $F_t^{[0]} = F_t \in R^{T_t \times d}$. Note that the sequence length of $F_{m \to t}^{[i]}$ is updated to T_t after the first CA block. And the Self-Attention (SA) block takes $F_t^{[i]}$ as input to obtain $F_t^{[i+1]} \in R^{T_t \times d}$:

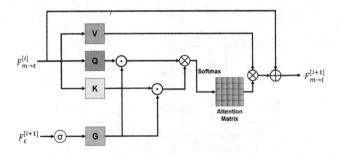

Fig. 3. The architecture of the Text-Gated Self-Attention (TGSA) module.

$$F_t^{[i+1]} = SA_t^{[i]}\left(F_t^{[i]}\right),$$

$$= softmax\left(\frac{F_t^{[i]}W_{Q_t}W_{K_t}^T F_t^{[i]T}}{\sqrt{d}}\right)F_t^{[i]}W_{V_t}. \tag{5}$$

Then the reinforced features $F_t^{[i+1]}$ and $F_{m\rightarrow t}^{[i]}$ are processed via the following adaptive fusion mechanism:

$$G^{[i]} = \sigma\left(F_t^{[i+1]} * W_t^{[i]} + F_{m\rightarrow t}^{[i]} * W_{m\rightarrow t}^{[i]} + b^{[i]}\right), \tag{6}$$

$$F_{m\rightarrow t}^{[i]} = G^{[i]} \odot F_t^{[i+1]} + \left(1 - G^{[i]}\right) \odot F_{m\rightarrow t}^{[i]}, \tag{7}$$

where σ denotes the sigmoid non-linearity function, $\odot$ denotes element-wise multiplication. We can determine the passed proportions of $F_t^{[i+1]}$ and $F_{m\rightarrow t}^{[i]}$ via the learnable parameters $W_t^{[i]}$, $W_{m\rightarrow t}^{[i]}$, and $b^{[i]}$. This operation can filter the incorrect information produced by the cross-modal interactions, and measure the fusion ratio of two modalities. After that, we process $F_t^{[i+1]}$ and $F_{m\rightarrow t}^{[i]}$ by a Position-wise Feed-Forward layer (PFF) with skip connection, as in the Transformer [17]:

$$F_{m\rightarrow t}^{[i]} = PFF\left(LN\left(F_{m\rightarrow t}^{[i]}\right)\right) + F_{m\rightarrow t}^{[i]}, \tag{8}$$

$$F_t^{[i+1]} = PFF\left(LN\left(F_t^{[i+1]}\right)\right) + F_t^{[i+1]}. \tag{9}$$

After the TCCA module, we obtain unified dimensional features of three modalities. We think that the relationships within each modality are complementary to the cross-modal relations, so we do self-attention for $F_{v\rightarrow t}^{[i]}$ and $F_{a\rightarrow t}^{[i]}$, while using the $F_t^{[i+1]}$ as a gate to activate or deactivate the corresponding key and value channels:

$$g^{[i]} = \sigma\left(Linear\left(F_t^{[i+1]}; \theta_g\right)\right), \tag{10}$$

$$gF_{m\rightarrow t}^{[i]} = \left(1 + g^{[i]}\right) \odot F_{m\rightarrow t}^{[i]}. \tag{11}$$

The query and key from visual/acoustic modalities are then modulated by the gate from the text modality:

$$F_{m \to t}^{[i+1]} = TGSA_m^{[i]} \left(F_{m \to t}^{[i]}, \, gF_{m \to t}^{[i]} \right),$$

$$= softmax \left(\frac{gF_{m \to t}^{[i]} W_{Q_{m \to t}} W_{K_{m \to t}}^T gF_{m \to t}^{[i]}}{\sqrt{d}} \right)^T F_{m \to t}^{[i]} W_{V_{m \to t}} + F_{m \to t}^{[i]}.$$

$$(12)$$

The architecture of the TGSA is shown in Fig. 3.

Fusion and Output Module. Here, we utilize a simple attention approach to aggregate the reinforced features of the three modalities. Specifically, given the feature $F_m^{[n]} \in R^{T_m \times d}$ for modality m output by the last TGSA module, we get the attention weight matrix:

$$a_m = softmax \left(\frac{F_m^{[n]} W_m}{\sqrt{d}} \right)^T \in R^{1 \times T_m}, \tag{13}$$

where $W_m \in R^d$ denotes the linear projection parameter, and a_m denotes the attention weight matrix for the feature $F_m^{[n]}$. Then we aggregate the features with the attention weights:

$$f_m = a_m F_m^{[n]} \in R^{1 \times d}. \tag{14}$$

Eventually, we concatenate all of the three modalities' features as $f = [f_t; f_a; f_v] \in R^{1 \times 3d}$ as the fused feature passing through a Multi-Layer Perceptron (MLP) to make the final prediction y_{pred}:

$$y_{pred} = \Phi \left(f; \theta_\Phi \right), \tag{15}$$

where the $\Phi(\cdot)$ is a MLP parameterized by θ_Φ.

4 Experiments

In this section, we empirically evaluate our model on two datasets that are frequently used to benchmark the MSA task in prior works, and we introduce the datasets, implementation details, and the results of our experiments.

4.1 Datasets and Evaluation Metrics

MOSI [29] dataset is a widely used benchmark dataset for the MSA task. It comprises 2,199 short monologue video clips taken from 93 Youtube movie review videos. Its predetermined data partition has 1,284 samples in the training set, 229 in the validation set, and 686 in the testing set. MOSEI [30] dataset is an

Table 1. Comparison results on the MOSI. For Acc_2 and $F1$, we have two sets of non-negative/negative (left) and positive/negative (right) evaluation results.

Method	$MAE \downarrow$	$Corr \uparrow$	$Acc_7 \uparrow$	$Acc_2 \uparrow$	$F1 \uparrow$
TFN	0.901	0.698	34.9	-/80.8	-/80.7
LMF	0.917	0.695	33.2	-/82.5	-/82.4
MulT	0.861	0.711	-	81.5/84.1	80.6/83.9
MISA	0.783	0.761	42.3	81.8/83.4	81.7/83.6
MAG-BERT	0.731	0.789	-	82.5/84.3	82.6/84.3
Self-MM	0.718	**0.796**	46.04	82.62/84.45	82.55/84.44
TMRN(ours)	**0.704**	0.784	**48.68**	**83.67/85.67**	**83.45/85.52**

Table 2. Comparison results on the MOSEI.

Method	$MAE \downarrow$	$Corr \uparrow$	$Acc_7 \uparrow$	$Acc_2 \uparrow$	$F1 \uparrow$
TFN	0.593	0.700	50.2	-/82.5	-/82.1
LMF	0.623	0.677	48.0	-/82.0	-/82.1
MulT	0.580	0.703	-	82.5	-/82.9
MISA	0.568	0.724	-	82.59/84.23	82.67/83.97
MAG-BERT	0.539	0.753	-	83.8/85.2	83.7/85.1
Self-MM	0.536	**0.763**	**54.5**	82.59/84.95	82.9/84.85
TMRN(ours)	**0.535**	0.762	53.65	**83.39/86.19**	**83.67/86.08**

improvement over MOSI. It contains 22,856 annotated video segments (utterances) from 5,000 videos, 1,000 distinct speakers, and 250 different topics. Its predetermined data partition has 16,326 samples in the training set, 1,871 in the validation set, and 4,659 in the testing set. Each sample in both MOSI and MOSEI is manually annotated with a sentiment score between $[-3, 3]$, which indicates the polarity and relative strength of expressed sentiment. The polarity is indicated by positive/negative, and strength is indicated by absolute value. As in the previous works [7,9], we evaluate the model performances by the 7-class accuracy (Acc_7), the binary accuracy (Acc_2), mean absolute error (MAE), the correlation of the model's prediction with human ($Corr$), and the $F1$ score.

4.2 Implementation Details

All models are built on the Pytorch toolbox [11] with two Quadro RTX 8000 GPUs. The Adam optimizer [6] is adopted for network optimization. For the MOSI and MOSEI datasets, the training setting follows: the batch sizes are $\{128, 64\}$, the epochs are $\{100, 40\}$, the learning rates are $\{1e^{-3}, 2e^{-3}\}$, and the hidden dimension d is 128. The number of TCCA and TGSA is $N = 3$.

Table 3. Ablation results of our TMRN on the MOSI.

Model	MAE ↓	$Corr$ ↑	Acc_7 ↑	Acc_2 ↑	$F1$ ↑
Full method	**0.7041**	**0.7844**	**48.68**	**83.67/85.67**	**83.45/85.52**
w/o A	0.8114	0.7426	45.48	81.05/81.86	81.09/81.96
w/o V	0.8452	0.7382	41.69	80.61/81.71	80.67/81.82
Acoustic-oriented	0.7508	0.7658	43.00	81.92/83.38	81.85/83.36
Visual-oriented	0.7956	0.7309	41.69	82.07/83.23	82.06/83.27
w/o TCCA	0.7498	0.7817	44.75	83.09/84.76	83.03/84.75
w/o TGSA	0.7467	0.7824	45.33	80.45/81.71	80.50/81.79

4.3 Comparison with State-of-the-Art Methods

The proposed approach is compared to the existing state-of-the-art (SOTA) baselines, including TFN [28], LMF [8], Mult [15], MISA [4], MAG-BERT [12], and Self-MM [27]. Table 1 and 2 show the comparison results on the MOSI and MOSEI, respectively. The result of Self-MM [27] is reproduced from open-source code with hyper-parameters provided in the original paper.

The proposed TMRN significantly outperforms most previous methods [4, 8, 12, 15, 28] by considerable margins on all metrics in both datasets, demonstrating the superiority of our method. In addition, our model is superior to the current SOTA Self-MM [27] in most metrics (*i.e.*, $MAE, Acc_7, Acc_2, F1$ scores on the MOSI, and $MAE, Acc_2, F1$ scores on the MOSEI.) with better or competitive performance, suggesting the effectiveness of our text-oriented design philosophy.

4.4 Ablation Study

The overall performance has proven the superiority of TMRN. To understand the necessity of the different components and the dominance of the text modality, we conduct systematic ablation experiments on the MOSI, as shown in Table 3.

Importance of Modality. We remove a modality separately to explore the performance of our model. Both declining results indicate the importance of visual and acoustic modalities when removing the visual or acoustic sequences. Furthermore, the performance degradation is more severe when the visual modality is removed. This is in line with the previous work [21]. This result suggests that the information in nonverbal modalities complements the text modality.

Importance of Center Modality. To demonstrate the dominance of the text modality, we replace the other two modalities as the dominant modality for the experiments. The acoustic- and visual-oriented models invariably suffer from significant performance degradation. These observations demonstrate that the text modality is richer in semantics and less noisy, which leads to better feature reinforcement of the other two modalities.

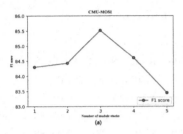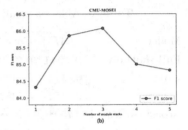

Fig. 4. Performance of TMRN with different parameter N on MOSI and MOSEI.

Importance of Module. Finally, we explore the importance of the proposed components by removing the TCCA and TGSA modules separately. For the TCCA module, we remove the cross-attention block and only do self-attention for text modality. We can see that the gain degrades when removing one of the modules. These observations suggest that adequate guidance of the text modality is necessary and indispensable.

4.5 Sensitivity of Parameter

In order to explore the effect of parameter N on the model performance, we conducted experiments on MOSI and MOSEI datasets with different parameters N. The results are summarized in Fig. 4. With the increase of N, we find that the $F1$ scores show a trend of increasing and then decreasing, and the network performs best when $N = 3$. In our conjecture, the larger N can result in better modality reinforcement. However, experiments show us that too many layers may bottleneck the ability of the text modality to guide the other two modalities. We should choose the appropriate network for different datasets, which is exactly what our proposed TMRN can flexibly do. If migrating our model to a more complex dataset, we can properly increase the number of TCCA and TGSA modules to achieve the best performance.

5 Conclusion

This paper presents a text-oriented multimodal sequence reinforcement network to achieve interaction and fusion over unaligned sequences of three modalities in the context of multimodal human sentiment analysis. The work is based on inter- and intra-modal attention mechanisms, and the attention of the other two modalities is guided throughout by sequences from the text modality, enabling the alternate transfer of information within and across modalities. We experimentally observe that our approach can achieve better performance than the baselines in MSA benchmarks.

Acknowledgments. This work was supported in part by the National Key R&D Program of China (2021ZD0113503), and in part by the Shanghai Municipal Science and Technology Major Project (2021SHZDZX0103).

References

1. Chen, M., Wang, S., Liang, P.P., Baltrušaitis, T., Zadeh, A., Morency, L.P.: Multimodal sentiment analysis with word-level fusion and reinforcement learning. In: Proceedings of the 19th ACM International Conference on Multimodal Interaction, pp. 163–171 (2017)
2. Devlin, J., Chang, M.W., Lee, K., Toutanova, K.: BERT: pre-training of deep bidirectional transformers for language understanding. arXiv preprint arXiv:1810.04805 (2018)
3. Du, Y., Yang, D., Zhai, P., Li, M., Zhang, L.: Learning associative representation for facial expression recognition. In: IEEE International Conference on Image Processing, pp. 889–893 (2021)
4. Hazarika, D., Zimmermann, R., Poria, S.: MISA: modality-invariant and-specific representations for multimodal sentiment analysis. In: Proceedings of the 28th ACM International Conference on Multimedia, pp. 1122–1131 (2020)
5. Hochreiter, S., Schmidhuber, J.: Long short-term memory. Neural Comput. **9**(8), 1735–1780 (1997)
6. Kingma, D.P., Ba, J.: Adam: a method for stochastic optimization. arXiv preprint arXiv:1412.6980 (2014)
7. Liang, T., Lin, G., Feng, L., Zhang, Y., Lv, F.: Attention is not enough: mitigating the distribution discrepancy in asynchronous multimodal sequence fusion. In: Proceedings of the IEEE/CVF International Conference on Computer Vision, pp. 8148–8156 (2021)
8. Liu, Z., Shen, Y., Lakshminarasimhan, V.B., Liang, P.P., Zadeh, A., Morency, L.P.: Efficient low-rank multimodal fusion with modality-specific factors. arXiv preprint arXiv:1806.00064 (2018)
9. Lv, F., Chen, X., Huang, Y., Duan, L., Lin, G.: Progressive modality reinforcement for human multimodal emotion recognition from unaligned multimodal sequences. In: Proceedings of the IEEE/CVF Conference on Computer Vision and Pattern Recognition, pp. 2554–2562 (2021)
10. Mittal, T., Guhan, P., Bhattacharya, U., Chandra, R., Bera, A., Manocha, D.: EmotiCon: context-aware multimodal emotion recognition using Frege's principle. In: Proceedings of the IEEE/CVF Conference on Computer Vision and Pattern Recognition, pp. 14234–14243 (2020)
11. Paszke, A., et al.: PyTorch: an imperative style, high-performance deep learning library. In: Advances in Neural Information Processing Systems, vol. 32 (2019)
12. Rahman, W., et al.: Integrating multimodal information in large pretrained transformers. In: Proceedings of the Conference Association for Computational Linguistics. Meeting, vol. 2020, p. 2359. NIH Public Access (2020)
13. Strubell, E., Verga, P., Andor, D., Weiss, D., McCallum, A.: Linguistically-informed self-attention for semantic role labeling. arXiv preprint arXiv:1804.08199 (2018)
14. Tang, G., Müller, M., Rios, A., Sennrich, R.: Why self-attention? A targeted evaluation of neural machine translation architectures. arXiv preprint arXiv:1808.08946 (2018)
15. Tsai, Y.H.H., Bai, S., Liang, P.P., Kolter, J.Z., Morency, L.P., Salakhutdinov, R.: Multimodal transformer for unaligned multimodal language sequences. In: Proceedings of the Conference Association for Computational Linguistics. Meeting, vol. 2019, p. 6558. NIH Public Access (2019)

16. Tsai, Y.H.H., Liang, P.P., Zadeh, A., Morency, L.P., Salakhutdinov, R.: Learning factorized multimodal representations. arXiv preprint arXiv:1806.06176 (2018)
17. Vaswani, A., et al.: Attention is all you need. In: Advances in Neural Information Processing Systems, vol. 30 (2017)
18. Wang, Y., Shen, Y., Liu, Z., Liang, P.P., Zadeh, A., Morency, L.P.: Words can shift: dynamically adjusting word representations using nonverbal behaviors. In: Proceedings of the AAAI Conference on Artificial Intelligence, pp. 7216–7223 (2019)
19. Wu, Y., Lin, Z., Zhao, Y., Qin, B., Zhu, L.N.: A text-centered shared-private framework via cross-modal prediction for multimodal sentiment analysis. In: Findings of the Association for Computational Linguistics: ACL-IJCNLP 2021, pp. 4730–4738 (2021)
20. Yang, D., et al.: Context de-confounded emotion recognition. In: Proceedings of the IEEE/CVF Conference on Computer Vision and Pattern Recognition, pp. 19005–19015, June 2023
21. Yang, D., Huang, S., Kuang, H., Du, Y., Zhang, L.: Disentangled representation learning for multimodal emotion recognition. In: Proceedings of the 30th ACM International Conference on Multimedia, pp. 1642–1651 (2022)
22. Yang, D., Huang, S., Liu, Y., Zhang, L.: Contextual and cross-modal interaction for multi-modal speech emotion recognition. IEEE Signal Process. Lett. **29**, 2093–2097 (2022)
23. Yang, D., et al.: Emotion recognition for multiple context awareness. In: Avidan, S., Brostow, G., Cissé, M., Farinella, G.M., Hassner, T. (eds.) Proceedings of the European Conference on Computer Vision, vol. 13697, pp. 144–162. Springer, Cham (2022). https://doi.org/10.1007/978-3-031-19836-6_9
24. Yang, D., Kuang, H., Huang, S., Zhang, L.: Learning modality-specific and -agnostic representations for asynchronous multimodal language sequences. In: Proceedings of the 30th ACM International Conference on Multimedia, pp. 1708–1717 (2022)
25. Yang, D., et al.: Target and source modality co-reinforcement for emotion understanding from asynchronous multimodal sequences. Knowl.-Based Syst. **265**, 110370 (2023)
26. Yu, W., et al.: CH-SIMS: a Chinese multimodal sentiment analysis dataset with fine-grained annotation of modality. In: Proceedings of the 58th Annual Meeting of the Association for Computational Linguistics, pp. 3718–3727 (2020)
27. Yu, W., Xu, H., Yuan, Z., Wu, J.: Learning modality-specific representations with self-supervised multi-task learning for multimodal sentiment analysis. In: Proceedings of the AAAI Conference on Artificial Intelligence, pp. 10790–10797 (2021)
28. Zadeh, A., Chen, M., Poria, S., Cambria, E., Morency, L.P.: Tensor fusion network for multimodal sentiment analysis. arXiv preprint arXiv:1707.07250 (2017)
29. Zadeh, A., Zellers, R., Pincus, E., Morency, L.P.: MOSI: multimodal corpus of sentiment intensity and subjectivity analysis in online opinion videos. arXiv preprint arXiv:1606.06259 (2016)
30. Zadeh, A.B., Liang, P.P., Poria, S., Cambria, E., Morency, L.P.: Multimodal language analysis in the wild: CMU-MOSEI dataset and interpretable dynamic fusion graph. In: Proceedings of the 56th Annual Meeting of the Association for Computational Linguistics, pp. 2236–2246 (2018)

PSDD-Net: A Dual-Domain Framework for Pancreatic Cancer Image Segmentation with Multi-scale Local-Dense Net

Dongying Yang[1], Cong Xia[2], Ge Tian[3], Daoqiang Zhang[1], and Rongjun Ge[4(✉)]

[1] Key Laboratory of Brain-Machine Intelligence Technology, Ministry of Education, Nanjing University of Aeronautics and Astronautics, Nanjing 211106, China
[2] Jiangsu Key Laboratory of Molecular and Functional Imaging, Department of Radiology, Zhongda Hospital, Medical School of Southeast University, Nanjing, China
[3] Beijing Institute of Tracking and Telecommunications Technology, Beijing, China
[4] School of Instrument Science and Engineering, Southeast University, Nanjing, China
rongjun_ge@126.com

Abstract. Pancreatic ductal adenocarcinoma (PDAC) is one of the deadliest cancers in the word. However, the diverse microenvironment, unclear boundaries, integrity destruction inter the slices, and enormous individual differences of tumors pose tremendous challenges to the segmentation process. To address these challenges, we proposed a physical-spiral dual-domain network (PSDD-Net) that combines the advantages of the spiral domain and the physical domain. First of all, the physical domain promotes integral representations of the tumor features, and the spiral domain protrudes the tumor region under CT multi-directions. As a result, the dual-domain framework makes the dual-domain feature simultaneously sent to the network to promote greater attention to the pancreatic region and reduce the interference of redundant background information. Secondly, we also present a multi-scale local-dense net (MSLD-Net) in the physical domain which contains local-channel dense block (LCDB) and multi-scale semantic feature extraction (MSSFE) module. The MSLD-Net grasps more multi-scale geometric information of the tumors and facilitates feature map fusion. Thirdly, a cross-domain aggregation (CDA) module is designed to interact bridging the two domains to interleave and integrate dual-domain complementary visual information. The extensive experiments on the clinical dataset show that our method obtained the DSC of 76.00% in abdominal CT, which outperformed the other state-of-the-art on pancreatic cancer segmentation results and demonstrated strong potential for clinical applications.

D. Yang and C. Xia—Contribute equally to this work.

1 Introduction

Accurate segmentation of pancreatic lesions in computed tomography (CT) is a crucial and challenging task in the identification and treatment of pancreatic ductal adenocarcinoma. According to the World Cancer Report [1], pancreatic ductal cancer, once detected, deteriorates rapidly and is recognized as one of the cancers with a high mortality rate. In addition, PDAC can also easily cause infiltrate with other organs and tissues of the abdomen leading to lesions, resulting in poor prognosis (the 5-year survival rate is lower than 8%). CT scanning is one of the commonly used diagnostic methods for PDAC. Therefore, accurate tumor segmentation on CT is crucial to monitor abnormal volume changes and growth of tumors, and plays an important role in diagnosis, prognosis, and intraoperative guidance. However, the inherent characteristics of pancreatic tumors and the variability of their surrounding environment pose significant challenges for accurate segmentation, the challenges are listed as follows:

Challenge 1: As shown in Fig. 1(a), as the tumor rapidly deteriorates and its volume continues to increase, it is prone to infiltration with peripheral blood vessels and duodenum. Such diversity of the tumor microenvironment encounters difficulties in the accurate separation of the tumor from the surrounding tissues.

Challenge 2: As shown in Fig. 1(b), the fibrillar connective tissues and organs surrounding with tumors share similar intensity and texture distribution. That because they have similar imaging characteristics in CT. Thus, it leads to very low contrast or even no visible boundary.

Challenge 3: As shown in Fig. 1(c), the shape of tumors has specificity. The integrity of tumor anatomical structure is easily disrupted inter the slices. The topology errors of tumor can lead to some clinical issues being overlooked, making subsequent diagnosis difficult.

Challenge 4: As shown in Fig. 1(d), different from abdominal organs, the tumors possess the appearance properties of diverse shapes, various orientations, and different aspect ratios. Therefore, it is difficult to accurately describe and segment tumors.

As far as we know, with the rapid development of automated segmentation algorithms, the existing segmentation methods of pancreatic cancer can be divided into three categories, based on the single-slices process, physical-volume process, and spatial-transformation process, respectively. 1) Single-slice process segment tumor from each slice to obtain the segmentation results of the physical entity of the tumor [2,3]. V-mesh FCN [4] though a graph-based visual saliency (GBVS) algorithm to enhance contrast between tumors and surrounding tissues in slices which alleviate the problem of unclear tumor boundaries in a large extent. UDA framework [5] used GCN and meta-learning strategy to pay more attention to tumor context information to alleviate a series of problems caused by individual differences in tumors. 2) Physical-volume process can obtain volumetric information between slices, which includes complete size, shape, and texture information of the tumors [6,7]. M^3Net [8] utilizes CT images of different phases to obtain more comprehensive tumor characteristics from various perspectives. A transformer guided progressive fusion network (TGPFN) [9] utilizes transformer

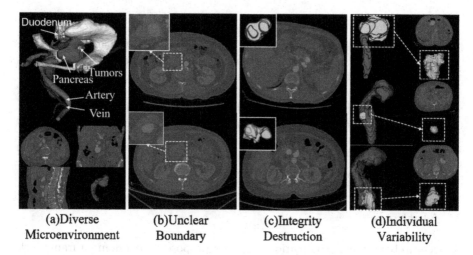

| (a)Diverse | (b)Unclear | (c)Integrity | (d)Individual |
| Microenvironment | Boundary | Destruction | Variability |

Fig. 1. The challenges of pancreatic cancer segmentation(Green is the tumor, red is the pancreas). (a) The tumors embed into surrounding tissues, blood vessels, and duodenum; (b) Low contrast between tumors and its neighboring structures; (c) The integrity of tumor anatomical structure is easily disrupted inter the slices; (d) Large individual differences in tumors. (Color figure online)

to capture global attention and improve long-range dependencies under different receptive fields. 3) Spatial-transformation process converts the physical-volume process into a transformation spatial process. In [10], spiral transformation uses a spherical coordinate system to alleviate a series of problems caused by small sample sizes in medical datasets.

However, currently available methods for segmentation tumors may have certain limited representation capabilities due to the inherent characteristics of the tumor. As a result, we proposed a physical-spiral dual-domain network (PSDD-Net) that combines the advantages of the spiral domain and the physical domain to overcome the aforementioned challenges. First of all, we design a dual-domain framework in order to alleviate the above challenges. The spiral domain give prominence to the tumor region under multi-directions (axial, sagittal, and coronal) information in CT scans. The physical domain preserve the integrity representations of the tumor. Secondly, a multi-scale local-dense net (MSLD-Net) is designed in the physical domain which contains local-channel dense block (LCDB) and multi-scale semantic feature extraction (MSSFE) module. The overall schemes in MSLD-Net obtain tumor information under different receptive fields and preserving the detailed features of the tumor to better ensure the accuracy of tumor boundary segmentation. Thirdly, in order to achieve information fusion between the two domains and better leverage their characteristics, we design a cross-domain aggregation (CDA) module to ensure feature consistency.

Our main contributions are highlighted as follows:

- We design a dual-domain framework called PSDD-Net which combines the advantages of the spiral domain and the physical domain to overcome the challenges caused by the inherent characteristics of tumors and promote greater attention to the pancreatic region.
- We conduct a MSLD-Net in the physical domain which contains LCDB and MSSFE module to capture the multi-scale geometric information of the tumors and facilitate feature fusion.
- We present a CDA module to interact bridging the two domains to interleave and integrate dual-domain complementary visual information.

2   Methods

To accurately segment pancreatic cancer, we propose a dual-domain framework called PSDD-Net. The whole pipeline is illustrated in Fig. 2. Specifically, it is implemented with three special designs: 1) We design a dual-domain framework (detailed in Sect. 2.1), which combines the advantages of the spiral domain and the physical domain to obtain comprehensive tumor features. 2) We conduct a MSLD-Net (detailed in Sect. 2.2) in the physical domain, which consists of LCDB and MSSFE module to grasp multi-scale context information. 3) We present a CDA module (detailed in Sect. 2.3), which mix cross-feature to maximize the characteristics of the spiral domain and physical domain.

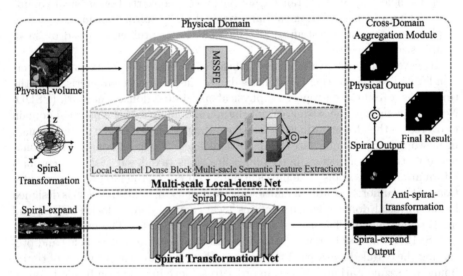

Fig. 2. The pipeline of the physical-spiral dual-domain net (PSDD-Net). 1) A dual-domain framework is applied on the spiral domain and physical domain. 2) A multi-scale local-dense Net (MSLD-Net) which includes local-channel dense block (LDCB) and multi-scale semantic feature extraction (MSSFE) module in the physical domain. 3) A cross-domain aggregation (CDA) module is designed to mix cross-feature of the dual-domain.

2.1   Physical-Spiral Domain Framework

Physical-spiral domain framework give prominence to the greater attention of the pancreatic tumor region and reducing the invading of redundant surrounding features of the tumor to enhance segmentation accuracy. The physical domain contains the 3D CT region of the tumor, which includes tumors, surrounding tissues, organs, and microenvironment of tumors. The physical domain preserve the integrity of the tumor. Simultaneously, the spiral domain protrudes the tumor region under multi-directions (axial, sagittal, and coronal) in CT scans. The spiral domain preserve the specificity of tumors. Compared to the single-slice process and physical-volume process methods, the dual-domain method can combine the advantages of physical-spiral domain features to pay more attention to the tumor region.

Since the pancreatic tumor region is less than 5% of the abdominal CT area [11], many methods may encounter issues such as imbalanced categories in segmentation. However, we use the coarse-fine stage is used to achieve rough localization and polishes segmentation. What is more, after the coarse stage, we can get the center of the tumor to crop the pancreatic tumor in the raw CT images and put it and the spiral-expend image as dual input in the network.

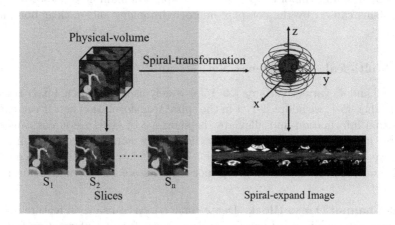

Fig. 3. Through spiral transformation, the tumor area (surrounded by the red line) is stretched into a strip, and the tumor area is protruded. Compared to slices, spiral transformation can preserve the specificity of tumors. (Color figure online)

The spiral transformation creatively converts the tumor and adjacent structural information into a spiral domain. Through these, the tissue around the tumor is mapped into different radii. Thus, spiral transformation can protrude tumor regions under multi-directions (axial, sagittal, and coronal) information in CT scans. Specifically, the special-designed spatial spherical coordinate system is embedded in the CT image. Then the tumor and adjacent tissue around are mapped into the rectangular coordinate system. The detailed procedure of spiral transformation can be seen in Fig. 3. Firstly, select any point inside the tumor which depend on the coarse stage. The point is setting as the origin of the spherical coordinate system. The direction of the x, y, and z, as well as the initial values of azimuth angle and polar angle would be select to initialization. Therefore, based on the conversion between the systems, the coordinates of any point in the CT image can be expressed as:

$$\begin{cases} x = r\sin\alpha\cos\beta \\ y = r\sin\alpha\sin\beta \\ z = r\cos\alpha \end{cases} \quad where \begin{cases} 0 \le \alpha \le \pi \\ 0 \le \beta \le 2\pi \\ -R \le r \le R \end{cases} \quad (1)$$

x, y, and z represent the coordinate in the CT image, r represents the radius, R is the selectable radius range, α is the polar angle, and β is the azimuth angle. After that, we can simultaneously increase the polar and azimuth angles in the spherical coordinate to spiral sample the tumor region. Then, the sampled spiral lines map into a rectangular coordinate system sequentially to obtain the spiral-expand image.

The physical domain crop tumor region is based on the coarse stage in the raw CT images and put into the physical domain net. The dual-domain models play their respective roles. The physical domain helps to solve a series of problems caused by irregular tumor shapes, while the spiral domain helps to solve a series of problems caused by the complex microenvironment and unclear boundaries of tumors.

2.2 Multi-scale Local-Dense Net

U-Net [2] and its various variants [12–15] is widely used in medical image segmentation. Therefore, we adopt U-Net in the spiral transformation net. It can extract contextual information and alleviate the shortage of GPU resources caused by 3D large images. Moreover, inspired by the densely connected convolutional network, we designed a MSLD-Net in the physical domain. It mainly combines LCDB and MSSFE module to grasp more multi-scale geometric information.

Local-Channel Dense Block. Dense connections transmit all feature maps in each layer to every forward layer to preserve more feature information from the original image. However, due to the repeated links cause excessive GPU resource usage and interference with the model due to redundant information, we propose a LCDB. The detailed structure configuration of the LCDB module is shown in Fig. 2. It not only simplifying the network and saving memory, but also keep the advantages of dense connection. The LCDB adopts dense offset connections as skip connections, and local channel feature maps are transmitted from each layer to each forward layer. It can be expressed as follows:

$$F_i = N(C(F_{i-1}, F_{i-2}[0:k_{i-2}], \cdots, F_0[0:k_0])) \quad (2)$$

where N denotes a composite function of operations. We can use group normalization (GN), rectified linear units (ReLU), or convolution (Conv) to replace it. C denotes the concatenation operation. Thus, the different resolutions and different receptive fields map are fused through LCDB. The features of different semantic levels expression of the model has been improved. By utilizing these, the details of tumor features are also well preserved, which helps with subsequent clinical treatment and diagnosis.

Multi-scale Semantic Feature Extraction. Tumors are different from abdominal organs, the diverse distribution of tumors in CT images make enhancing the feature representation capability is crucial. Hence, we propose a MSSFE module embedded at the bottom of MSLD-Net, which can efficiently integrate multi-scale features. The detailed structure configuration of the MSSFE module is shown in Fig. 2. We set up four different dilation rates in the dilated convolutions and the different dilated convolutions get a more abundant feature map with various receptive fields. The MSSFE module can be expressed as follows:

$$f_i' = S(DilConv(f_i)) \bigotimes DilConv(f_i) \tag{3}$$

$$F = C(f_1', f_2', f_3', f_5') \tag{4}$$

where i $(i = 1, 2, 3, 5)$ is dilated rate of MSSFE module, f_i represent input feature under dilated rate i, $DliConv$ express Dilated Convolution, S denotes sigmoid function, C denotes the concatenation operation, F is final output in the MSSFE module.

2.3 Cross-Domain Aggregation Module

Cross-domain aggregation (CDA) module is proposed to capture the dependencies between dual-domain features. Two domain preserving the characteristics of tumors in different senses. Therefore, the fusion of information between the two domains is crucial. The detailed structure configuration of the CDA module is shown in Fig. 2.

According to the principle of spiral transformation, we utilize anti-spiral-transformation to project the rectangular coordinate system into the spherical coordinate system. It can be indicated by the formula as follows:

$$\begin{cases} x_{spiral} = r + R \\ y_{spiral} = (\beta(\alpha) - \beta_0(\alpha_0)) \times 180 \div \pi \end{cases} \tag{5}$$

where x_{spiral}, y_{spiral} denote the coordinate in spiral-expand image, β, β_0 can be substituted for α, α_0, $\beta, \beta_0, \alpha, \alpha_0$, r, R represent the polar angle, initialization value of polar angle, azimuth angle, initialization value of azimuth angle, radius, selectable radius range individually. After anti-spiral-transformation, we coupled two domain outputs to capture Integrity and specificity of tumors which can obtain the final result.

2.4 Multi-domain Mixed Loss

The overall cost function of PSDD-Net can be expressed as:

$$Loss = \mathcal{L}_{spiral} + \lambda \mathcal{L}_{physical} \tag{6}$$

where λ are balance factors in the model. The cost function of the $\mathcal{L}_{spiral}$ in spiral domain and the $\mathcal{L}_{physical}$ in physical domain consists of two parts:

$$\mathcal{L}_{spiral} = \mathcal{L}_{CE-spiral} + \lambda_1 \mathcal{L}_{DSC-spiral} \tag{7}$$

$$\mathcal{L}_{physical} = \mathcal{L}_{CE-physical} + \lambda_2 \mathcal{L}_{DSC-physical} \tag{8}$$

where λ_1 and λ_2 are hyper-parameters for balancing the two parts in the domain. The $\mathcal{L}_{CE-physical}$ and $\mathcal{L}_{CE-spiral}$ mean the cost function of cross entropy (CE) in the physical domain and spiral domain. The $\mathcal{L}_{DSC-physical}$ and $\mathcal{L}_{DSC-spiral}$ mean the cost function of dice similarity coefficient (DSC) in the physical domain and spiral domain.

Cross entropy (CE) is used to minimize the differential between the predicted segmentation P^i and the ground truth G^i, which is defined as follows:

$$CE(P^i, G^i) = -\frac{1}{N} \sum_{i=1}^{n} -[g_i \cdot \log(p_i) + (1 - g_i) \cdot \log(1 - p_i)] \tag{9}$$

where $p_i, g_i \in 0, 1$ indicates whether pixel belongs to the P^i and G^i region respectively. The dice coefficient is a similarity measurement function. Its value is between 0 and 1. 1 represents a coincidence of up to 100% between the predicted results and the ground truth, while 0 represents the opposite. According to it, the dice similarity coefficient (DSC) can be used to alleviate sample imbalance caused by tumors. All of these are defined as follows:

$$DSC = 1 - \frac{2|P^i \cap G^i|}{|P^i| + |G^i|} \tag{10}$$

3 Experiments

3.1 Experimental Settings

Datasets and Evaluation Metrics. We conducted experiments on clinical private dataset. The private abdominal dataset contains 401 cases of pancreatic ductal cancer. We randomly divide the dataset into training set, validation set, and testing set. The training set contains 240 patients CT, validation set contains 82 patients CT, testing set contains 79 patients CT. The private abdominal dataset are composed of (34–1073) slices of (512 × 512) images. Each CT have voxel spatial resolution of ([0.501–0.976] × [0.501–0.976] × [0.300–5.000]) mm^3. All CT scans are contrast-enhanced images obtained. Three representative evaluation indicators are used to evaluate the effectiveness of our method. Firstly, Dice ($Dice = \frac{2|P \cap G|}{|P| + |G|}$) illustrates the degree of overlap between the prediction and the ground truth. Secondly, hausdorff distance

$(HD = max\{max_{p \in S_p}min_{g \in S_g}\|p - g\|, max_{g \in S_g}min_{p \in S_p}\|g - p\|\})$ is used to evaluate the boundary similarity between two images, where $\| \cdot \|$ denotes the Euclidean distance, S_p and S_g denote the voxel sets within the predicted segmentation and ground truth boundary, respectively. Thirdly, ASD $(ASD = \frac{1}{|S_g|}\sum_{g \in S_g}min_{p \in S_p}\|g - p\|)$ is the most representative and valuable metric for evaluating the segmentation's performance. The $|\cdot|$ denotes the cardinality.

Implementation Details. Due to the fact that tumor region only account for a small portion of abdominal CT, it is very easy to encounter imbalanced sample categories during the segmentation process. Thus, we adopt prior knowledge to crop images on the CT slices and use U-Net to achieve coarse segmentation. During the spiral transformation, we set the center point of the tumor which depends on the coarse segmentation output as the origin of the spherical coordinate system. We also set the center point of the tumor as the center of the clipping region in the physical domain for dual-domain feature fusion. Moreover, the PSDD-Net training with SGD optimizer and setting the batch size was 1, the initial learning rate is set as 0.001 and use reduceLronplateau strategy to change the learning rate timely. The network code is written based on the Pytorch.

3.2 Results and Analysis

Table 1. The quantitative analysis of the ablation experiment of the PSDD-Net demonstrates the effectiveness of each component.

Method	Dice(%)	HD(mm)	ASD(mm)
Physical domain(U-Net)	58.57 ± 30.12	7.57 ± 7.68	3.10 ± 7.80
Spiral domain(U-Net)	60.92 ± 22.08	25.17 ± 16.47	3.10 ± 12.77
Physical domain(U-Net+LCDB)	69.25 ± 21.85	6.22 ± 6.02	1.96 ± 5.92
Physical domain(MSLD-Net)	71.01 ± 18.85	6.62 ± 10.40	2.04 ± 8.56
Physical-Spiral domain(U-Net, U-Net)	66.39 ± 22.37	6.04 ± 7.23	2.16 ± 3.16
Our Method	$\mathbf{76.00 \pm 16.78}$	$\mathbf{5.120 \pm 5.23}$	$\mathbf{1.62 \pm 3.51}$

Ablation Study. As shown in Table 1, we conduct ablation experiments on our method to analyze the effectiveness of each component in the PSDD-Net. We also demonstrated the effectiveness of each component in the network through six visualized cases in Fig. 4. According to the first, second, fifth, and last rows of Table 1, a dual-domain framework achieves complementary features in the physical domain and spiral domain, which has better performance than a single domain. According to the second, third, and fourth rows of Table 1, we can find that the LCDB and MSSFE modules in the MSLD-Net play an important role in the multi-scale feature extraction of tumors, which makes MSLD-Net outperform the U-Net. Thus, through these architectures, we will better address the challenges brought by the characteristics of tumors.

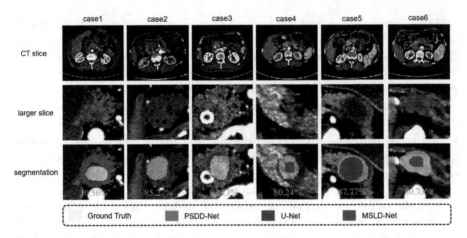

Fig. 4. The final visual segmentation result of our method. The ground truth is the yellow areas, the PSDD-Net indicates the green area, the MSLD-Net indicates the blue area (only in the physical domain), and the U-Net indicates the red area (only in the physical domain). The numerical value represents the dice of the corresponding case under our method. (Color figure online)

Comparison with Existing Methods. We selected five networks with high robustness and good performance to compare with our method. The five state-of-the-art segmentation models including the dual-domain model (M^3Net), multi-scale model (such as CAN, CAS-Net, VGGU-Net), and classic abdominal organ segmentation method (nnUNet). Note that for a fair comparison, all segmentation models introduced a rough stage that provides extra tumor location information, and inputs of models were cropped to a size of $96 \times 96 \times 64$ respectively with the tumor as the center. Table 2 shows that PSDD-Net achieves the highest DSC, and the lowest HD and ASD, compared with the other five models, illustrating the exceptional and stable segmentation performance of the proposed PSDD-Net.

Table 2. The quantitative analysis of representative experimental methods compared with our proposed method.

Method	Dice(%)	HD(mm)	ASD(mm)
M^3Net	62.57 ± 27.68	8.41 ± 9.81	4.06 ± 6.34
VGGU-Net	67.28 ± 34.71	11.12 ± 14.17	4.61 ± 4.10
CAS-Net	64.68 ± 31.69	13.21 ± 9.50	5.70 ± 4.25
CAN	57.53 ± 34.60	18.35 ± 14.75	6.89 ± 4.67
nnUNet	71.44 ± 20.04	10.715 ± 14.21	6.25 ± 6.63
Our Method	$\mathbf{76.00 \pm 16.78}$	$\mathbf{5.120 \pm 5.23}$	$\mathbf{1.62 \pm 3.51}$

4 Conclusion

The segmentation of pancreatic cancer plays an important role in determining the treatment plan for PDAC in clinical diagnosis, we propose a physical-spiral dual-domain net (PSDD-Net) to address the challenge brought by the characteristics of pancreatic tumors in CT images. First of all, a dual-domain architecture is presented based on the physical domain and spiral domain to highlight the tumor area and preserve its structural integrity. Thus achieving efficient application of tumor features. Besides, a novel multi-scale local-dense net (MSLD-Net) is proposed in the physical domain to effectively utilize multi-scale features of the PSDD-Net. In the end, a cross-domain aggregation (CDA) module is designed to mix cross domain feature fusion. Extensive ablation experiments and comparative experiments on the private dataset, and compare corresponding indicators and visualization results. All of these both reveal our method has great clinical potential in PDAC segmentation.

Acknowledgements. This study was supported by the Natural Science Foundation of Jiangsu Province (No. BK20210291), the National Natural Science Foundation (No. 62101249 and No. 62136004), and the China Postdoctoral Science Foundation (No. 2021TQ0149 and No. 2022M721611).

References

1. Guire, S.: World cancer report 2014. Geneva, Switzerland: world health organization, international agency for research on cancer, WHO press, 2015. Adv. Nutr. **7**(2), 418–419 (2016)
2. Ronneberger, O., Fischer, P., Brox, T.: U-Net: convolutional networks for biomedical image segmentation. In: Navab, N., Hornegger, J., Wells, W.M., Frangi, A.F. (eds.) MICCAI 2015. LNCS, vol. 9351, pp. 234–241. Springer, Cham (2015). https://doi.org/10.1007/978-3-319-24574-4_28
3. Zaiwang, G., et al.: CE-Net: context encoder network for 2D medical image segmentation. IEEE Trans. Med. Imaging **38**(10), 2281–2292 (2019)
4. Wang, Y., et al.: Pancreas segmentation using a dual-input v-mesh network. Med. Image Anal. **69**, 101958 (2021)
5. Li, J., Feng, C., Lin, X., Qian, X.: Utilizing GCN and meta-learning strategy in unsupervised domain adaptation for pancreatic cancer segmentation. IEEE J. Biomed. Health Inform. **26**(1), 79–89 (2021)
6. Havaei, M., et al.: Brain tumor segmentation with deep neural networks. Med. Image Anal. **35**, 18–31 (2017)
7. Kim, H.: Abdominal multi-organ auto-segmentation using 3D-patch-based deep convolutional neural network. Sci. Rep. **10**(1), 6204 (2020)
8. Taiping, Q., et al.: M3Net: a multi-scale multi-view framework for multi-phase pancreas segmentation based on cross-phase non-local attention. Med. Image Anal. **75**, 102232 (2022)
9. Taiping, Q., et al.: Transformer guided progressive fusion network for 3D pancreas and pancreatic mass segmentation. Med. Image Anal. **86**, 102801 (2023)
10. Chen, X., Chen, Z., Li, J., Zhang, Y.-D., Lin, X., Qian, X.: Model-driven deep learning method for pancreatic cancer segmentation based on spiral-transformation. IEEE Trans. Med. Imaging **41**(1), 75–87 (2021)

11. Zhou, Y., Xie, L., Shen, W., Wang, Y., Fishman, E.K., Yuille, A.L.: A fixed-point model for pancreas segmentation in abdominal CT scans. In: Descoteaux, M., Maier-Hein, L., Franz, A., Jannin, P., Collins, D.L., Duchesne, S. (eds.) MICCAI 2017. LNCS, vol. 10433, pp. 693–701. Springer, Cham (2017). https://doi.org/10.1007/978-3-319-66182-7_79

12. He, K., Zhang, X., Ren, S., Sun, J.: Deep residual learning for image recognition. In: Proceedings of the IEEE Conference on Computer Vision and Pattern Recognition, pp. 770–778 (2016)

13. Zongwei Zhou, Md., Siddiquee, M.R., Tajbakhsh, N., Liang, J.: UNet++: redesigning skip connections to exploit multiscale features in image segmentation. IEEE Trans. Med. Imaging **39**(6), 1856–1867 (2019)

14. Huang, G., Liu, Z., Van Der Maaten, L., Weinberger, K.Q.: Densely connected convolutional networks. In: Proceedings of the IEEE Conference on Computer Vision and Pattern Recognition, pp. 4700–4708 (2017)

15. He, Y., et al.: Dense biased networks with deep priori anatomy and hard region adaptation: semi-supervised learning for fine renal artery segmentation. Med. Image Anal. **63**, 101722 (2020)

Airport Boarding Bridge Pedestrian Detection Based on Spatial Attention and Joint Crowd Density Estimation

Xu Han[1], Hao Wan[1,2], Wenxiao Tang[1], and Wenxiong Kang[1,3,4]($\boxtimes$)

[1] School of Automation Science and Engineering, South China University
of Technology, Guangzhou, China
`auwxkang@scut.edu.cn`
[2] Guangdong Airport Baiyun Information Technology Co., Ltd. Postdoctoral
Innovation Practice Base, Guangzhou, China
[3] Pazhou Lab, Guangzhou, China
[4] Guangdong Enterprise Key Laboratory of Intelligent Finance, Guangzhou, China

Abstract. Pedestrian detection serves as the cornerstone of pedestrian tracking and re-identification, playing a pivotal role in the realm of intelligent transportation. Accurate identification of pedestrians with diverse identities, such as passengers, crew members, and cleaning staff, is of utmost importance in high-security-demand scenarios like airport boarding bridges. The varied poses of pedestrians, occlusions, and small appearance differences pose significant challenges for accurately detecting individuals with different identities in boarding bridge scenarios. Existing object detectors exhibit limited prowess in extracting discriminative features tailored specifically for pedestrians, hampering their ability to fulfill the requirements of precise localization and classification. In this paper, we propose a method based on spatial attention and joint crowd density estimation. By incorporating spatial attention, our network selectively focuses on salient regions corresponding to different pedestrian categories, thereby enhancing classification accuracy. Moreover, through introducing an auxiliary task of crowd density estimation, the supervision of pedestrian head position information is added to the network. This significantly alleviates the missed detection problems caused by perspective distortion and occlusion, leading to significant improvements in detection accuracy. In our study, we use YOLO as the baseline model. The improved model shows a 5.81% increase in mAP and significantly outperforms several common object detectors.

Keywords: Pedestrian Detection · Attention Mechanism · Crowd Density Estimation

1 Introduction

Pedestrian detection is an important branch in the field of object detection. In many situations, it is not only necessary to accurately locate pedestrians but also

L. Fang et al. (Eds.): CICAI 2023, LNAI 14474, pp. 213–228, 2024.
https://doi.org/10.1007/978-981-99-9119-8_20

to precisely classify their identities. At the airport boarding bridge, it is essential to accurately detect pedestrians with different identities, such as passengers, crew members and cleaning staff. It is the basis for many downstream tasks, such as passenger trajectory analysis and anomaly recognition. The appearance of different pedestrians have small distinctions, making it difficult for detectors to accurately classify them. Meanwhile, due to the varied poses of pedestrians and the occlusion between pedestrians, localization is also challenging.

Existing pedestrian detection methods [5,6] mainly focus on the occlusion and shape changes of pedestrians. However, these methods treat pedestrians as a single category, and the pedestrian detection task is degraded into a single-category object detection task. For multi-category pedestrian detection tasks, in addition to accurately locating pedestrians, it's also necessary to correctly distinguish pedestrian categories. Therefore, the enhancement of pedestrian position information and appearance information are equally important.

Common object detection frameworks typically utilize a backbone network to extract features for a detection head, which then predicts the location and category of the objects. For multi-category pedestrian detection tasks, there are small appearance differences between different categories. Common backbone networks (such as ResNet [7]) struggle to extract discriminative features, and detection heads find it challenging to accurately detect pedestrians with varying shapes. Therefore, we propose a multi-category pedestrian detection algorithm based on spatial attention mechanism and joint crowd density estimation. Crowd density estimation methods based on heatmaps [37] use pedestrian head positions as supervisory signals and output position estimates for pedestrians' heads. Pedestrian head areas are generally not affected by varying perspectives and pedestrian postures, making them a robust supervisory signal for pedestrian locations [31]. This paper employs a multi-task learning approach, adding an extra branch to the detection model to perform crowd density estimation tasks, providing additional information on pedestrian locations and enhancing localization accuracy. Meanwhile, to address the small appearance differences between different categories of pedestrians, we utilizes a attention module to make the network focus on key areas of different category pedestrians, thus increasing inter-category differences. The improved model proposed in this paper achieves excellent results in multi-category pedestrian detection tasks under airport boarding bridge scenarios. Compared to the baseline model, the mAP is improved by 5.81%, and it significantly outperforms common object detectors.

2   Related Work

2.1   Object Detection

Object detection based on deep learning [2,17,19,23,25,32,39] have achieved great success in recent years, resulting in various applications in fields such as smart security [10], remote sensing [36], and autonomous driving [12]. CNN-based object detection algorithms can be mainly divided into one-stage object detection algorithms [17,19,23,32,39] and two-stage object detection algorithms

[16,25]. Faster RCNN [25] represents the two-stage object detection algorithm, which first generates proposals through the Region Proposal Network (RPN) and then sends the proposals to the detection head for classification and localization to obtain the final detection boxes. Compared to two-stage methods, one-stage detectors, exemplified by the YOLO series [1,23,24], eliminate the RPN by using the redundant prediction approach and generate final prediction boxes by setting Anchors. Due to the simplicity of the one-stage networks, they often resulting in faster speeds than two-stage detection networks. Simultaneously, with targeted network structure design and optimization strategies [17,29,30], one-stage networks have achieved performance nearly comparable to two-stage networks. Although Transformer-based object detection methods [2,20,40] have been emerging in recent years, their deployment complexity is higher than that of CNN-based architectures due to the Transformer's structure, and their practical applications are not yet widespread.

Since its release in 2014, the YOLO series of object detection networks has gone through many iterations, and the paradigm has expanded to areas such as 3D object detection [27] and human pose estimation [21]. It has also been used as a basis for improvement in some specific detection tasks [18,26]. YOLO uses deep convolutional networks for feature extraction, adopts FPN networks [16] for feature fusion, and generates detection boxes based on Anchors on multi-layer feature maps. The sophisticated and robust network structure is the key to YOLO's success in various tasks.

2.2 Attention Mechanism in Convolutional Neural Networks

According to the feature map processing strategy of the attention module, the attention mechanisms in convolutional neural networks can be divided into channel attention [9,13], spatial attention [33], temporal attention [14], and hybrid attention [22,34]. SENet [9], as an early attention method, used the Squeeze-and-Excitation structure to extract weight parameters for each channel of the feature map, thereby obtaining a channel-enhanced feature map. Hu et al. [8], inspired by SENet, performed aggregation and activation operations in space to generate spatial feature maps. Considering the similarities between channel attention and spatial attention, Park et al. [22] proposed a BAM model that simultaneously calculates channel and spatial attention using dilated convolution to expand the receptive field of the spatial attention module, eventually obtaining the final channel-spatial attention. Woo et al. [34] proposed a CBAM module that sequentially calculates channel and spatial attention in series and combines average pooling and max pooling information. Attention modules can significantly enhance a model's feature extraction capability in classification, detection, and segmentation tasks, with a small computational overhead.

2.3 Crowd Density Estimation

Early crowd density estimation methods [35] generally built upon detection, counting the number of detection boxes as the number of pedestrians. However, this approach struggles to handle densely populated scenes and often has limitations on detection count. Chan et al. [3] proposed a regression-based method that abandons the detection paradigm and directly regresses the number of people in the image. However, the learned features of the network are not strongly correlated with pedestrians, resulting in poor robustness and a lack of interpretability. Current mainstream crowd density estimation methods are generally based on heatmap [15,28,38], which estimate the positions of pedestrians' heads in images through heatmap and obtain the number of people in the crowd by summing the heatmap values.

3   Method

To address the challenges of varying shapes and small appearance differences among pedestrians in boarding bridge scenes, we enhances the backbone network's feature extraction and representation capabilities by incorporating attention mechanisms on YOLOv5 [11]. Additionally, crowd density is introduced as auxiliary supervision, using pedestrian head information to improve the network's ability to extract pedestrian location information and thus enhance the localization accuracy of detection. We call the improved model density-yolo, and its overall structure is shown in Fig. 1.

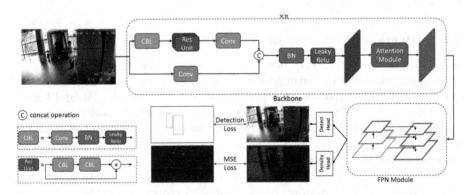

Fig. 1. Illustration of the density-yolo model.

3.1   Attention Module

The attention mechanism has been proven to help models focus on important information and improve their representation capabilities [34]. Spatial attention mechanisms enable networks to focus on key regions of objects in the spatial dimension, which are generally the most distinctive areas of the object. This is crucial for achieving accurate classification. The data distribution between different categories of pedestrians in boarding bridge scenes is relatively similar, and the discriminative regions for specific categories are rather small. Conventional backbone networks lack the ability to extract robust features; therefore, this paper uses CBAM (Convolutional Block Attention Module) to enhance the backbone network's feature extraction and representation capabilities.

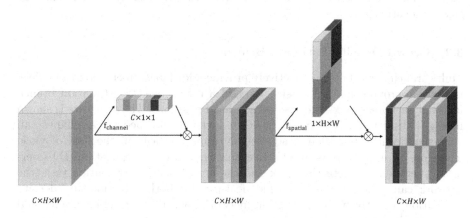

Fig. 2. CBAM architecture.

CBAM is a lightweight attention module, consisting of channel attention and spatial attention components (show in Fig. 2). Both channel and spatial dimensions use max pooling and average pooling for dimension reduction, followed by feature extraction and fusion. The excitation at the corresponding positions is obtained through a Sigmoid function. The feature map first passes through the channel attention to obtain a channel-enhanced feature map and then goes through the spatial attention to obtain a spatially-enhanced feature map. The calculation of attention can be expressed as:

$$X_{att} = (X \odot f_{channel}(X)) \odot f_{spatial}(X \odot f_{channel}(X)) \tag{1}$$

In this equation, X represents the original feature map, $f_{channel}$ and $f_{spatial}$ denote the channel attention module and the spatial attention module, respectively, and $\odot$ indicates the element-wise multiplication operation. For channel attention, the feature extraction and fusion operations are implemented using fully connected layers. For spatial attention, the feature extraction and fusion operations are implemented using channel concatenation and 1×1 convolution.

As a plug-and-play attention module, CBAM can be inserted at any position in the network; however, the impact on performance varies depending on the insertion location. For the feature fusion and detection head parts of the network, the feature maps have been fused with multi-scale information. It is difficult for spatial attention to enhance key information and fade redundant information at this stage, which may even lead to a decline in accuracy. For the feature extraction backbone network of the detection model, the shallow layers mainly extract information such as color, texture, and edges, while the deeper layers primarily extract more abstract semantic information. Therefore, enhancing the expression of semantic information is more helpful for improving classification accuracy. This paper chooses to place the CBAM module in the deeper position of the backbone network. The experimental results show that inserting CBAM into the deep location of the backbone network significantly outperforms the shallow location.

3.2 Crowd Density Estimate Branch

Multi-branch structure can effectively provide additional supervision to models, thereby improving multiple tasks. We add a crowd density estimation branch followed backbone, predicting pedestrian head positions. The structure is shown in Fig. 3. Under different perspectives and viewpoints, pedestrian shapes can vary significantly, but their heads usually appear clearly in the field of view. Therefore, it can serve as an effective auxiliary way to localization. By interacting between the detection task and the crowd density estimation task, the network can simultaneously consider features for both detection and density estimation, effectively enhancing the network's feature extraction ability and improving localization accuracy.

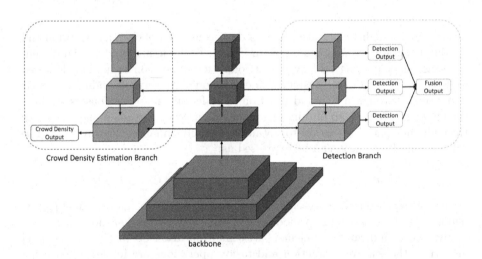

Fig. 3. Crowd density estimation branch and detection branch in density-yolo.

The density estimation branch shares a similar structure with the detection branch, with both sharing the same backbone network. To fuse pedestrian features at different scales, the density estimation branch uses an FPN structure for multi-scale feature fusion. To ensure sufficient resolution, the density estimation branch task head does not use a multi-scale structure but predicts using the highest resolution layer in the FPN. The crowd density estimation task head adopts a fully convolutional structure, using 1×1 convolutions to reduce the dimensionality of the input feature map, generating a density estimation map with one channel.

During the network training process, the density estimation branch is used as auxiliary supervision and is trained together with the detection branch. In the inference stage, the density estimation branch can be detached, and only the detection branch is used for inference, which does not affect the network's inference speed.

3.3   Loss Function

The loss function of the network consists of two parts: detection task loss and density estimation task loss. The detection task loss follows YOLO's settings, which consist of localization loss, classification loss, and confidence loss. The crowd density estimation loss uses the mean squared error loss (MSE) between the density estimation map and the Ground Truth.

$$L = L_{det} + \alpha L_{density} \tag{2}$$

$$L_{density} = \sum_{i=0}^{w} \sum_{j=0}^{h} (f_{i,j} - GT_{i,j})^2 \tag{3}$$

In this equation, w and h represent the width and height of the output density estimation map, $f_{i,j}$ denotes the density estimation value at position (i, j) on the output density map, and $GT_{i,j}$ represents the true density value at position (i, j) on the actual label map.

Following the crowd density estimation method [37], to construct GT, we smooth each annotation point using Gaussian functions, modeling each annotation point as a Gaussian distribution, allowing the network to learn density information more easily. GT construction is as follows:

$$GT(x) = H(x) \odot G_\sigma(x) \tag{4}$$

where:

$$H(x) = \sum_{i=1}^{n} \delta(x - x_i) \tag{5}$$

$$G_\sigma(x) = e^{-\frac{(x-x_i)^2}{2\sigma^2}} \tag{6}$$

In this equation, δ represents the Dirac delta function, x_i refers to the annotation points of pedestrian head positions, and σ is the Gaussian kernel. For low crowd density scenarios in boarding bridges, we fix σ at 0.5.

4   Experiments

4.1   Dataset

We have constructed a pedestrian dataset from real boarding bridge scenes, which includes pedestrians with a variety of perspectives and different postures. The dataset contains 35,361 images collected from 17 real scenarios, with a total of 102,370 pedestrian instances, divided into three categories: crew members, cleaning staff, and other pedestrians. The number ratio of three types of pedestrians is about 2:1:2. During the training process, we use data from 13 scenarios in the dataset, which accounts for about 80% of the images as the training set, and the remaining 4 scenarios with about 20% of the images serve as the test set. All the following experiments are completed on this dataset.

4.2   Implementation Details

We use YOLOv5-s v6.2 [11] as our baseline model, the structure of the crowd density estimation branch adopts the original FPN structure combined with the fully convolutional density estimation head. To facilitate network training, the FPN in the density estimation branch uses the same convolution parameters as the detection branch, and the density estimation head structure uses a 1×1 convolution to obtain a single-channel density estimation map. The density estimation branch performs three upsampling operations, eventually obtaining a density estimation map with 1/8 of the original scale.

In terms of training strategy, we use stochastic gradient descent (SGD) with momentum to optimize the model, setting the learning rate to 0.01 and the momentum parameter to 0.937. The training strategy employs Warmup, with a warmup epoch of 5. The batch size is set to 32 and input image size is set to 640×640. Random HSV jitter, random flipping, and mosaic augmentation strategies are enabled. The balance parameter α in the loss function is set to 0.1. All experiments are performed using an Nvidia Tesla T4 device.

Pedestrian detection datasets often lack additional annotations for pedestrian head positions, and adding head position annotations to datasets is costly and difficult to extend conveniently to different scenarios. Compared to pedestrian head annotations, using the midpoint of the upper boundary of the annotation box as an alternative supervisory signal achieves similar performance [31]. Therefore, we use the midpoint of the upper boundary of the annotation box as the supervisory signal for crowd density estimation.

As the purpose of enabling the crowd density estimation branch is to improve detection performance, we focus on the enhancement of detection performance and use mean Average Precision (mAP) for model evaluation.

4.3   Results

Compare with Common Object Detectors. To demonstrate the effectiveness of our proposed method, we compared the performance of our density-yolo with common object detectors on the airport boarding bridge pedestrian dataset. We compared several common object detectors of different paradigms, including two-stage Faster RCNN [25], one-stage SSD [19], anchor-free FCOS [32] and CenterNet [39], and Transformer-based DETR [40]. To ensure a fair comparison, we used the same input size and data augmentation strategies, and the implementation of the compared models was based on the MMDetection [4] framework.

Table 1 shows that the performance of common object detection frameworks with different paradigms is broadly consistent. Our density-yolo significantly outperforms other models and achieves better results with fewer parameters than other models.

Table 1. Comparisons of our method with other common object detectors. All models use the same data preprocessing and augmentation techniques.

Model	mAP@0.5/%	mAP/%	Paras(M)
Faster RCNN [25]	79.86	42.47	15.45
SSD [19]	60.02	35.34	24.01
FCOS [32]	88.80	61.93	31.84
CenterNet [39]	74.07	47.83	14.21
RetinaNet [17]	89.35	57.27	36.86
Deformable-DETR [40]	86.61	61.67	39.82
yolo [11](baseline)	85.34	68.06	7.2
density-yolo	91.15	70.79	7.2(+0.4)

To validate the effectiveness of our method and eliminate the impact of different model structure designs on the results, we added attention mechanisms and the crowd density estimation branch to the models mentioned above for comparison. Table 2 shows that after adding attention and density estimation branches, the performance of different detection models on the airport boarding bridge multi-category pedestrian detection dataset also significantly improves. Based on these results, we can conclude that our method is significantly effective in multi-category pedestrian detection tasks and is not influenced by model structure, detection paradigm, or model size.

Ablation Study. To further analyze the roles of different parts of the model, we performed a ablation study to separately validate the effects of the attention mechanism and the crowd density estimation branch (show in Table 3). To demonstrate the effect of the attention mechanism, we visualize the feature maps (Fig. 4). After adding the attention module, the activation value distribution of the feature maps is more concentrated, effectively focusing on key positions for distinguishing pedestrian categories, thereby enhancing the model's classification capability. For the density estimation branch, adding it separately results in a more significant improvement in the mAP metric compared to adding the

Table 2. The effectiveness of our proposed method on other models. (* indicates improved model using our method)

Model	mAP@0.5/%	mAP/%
FasterRCNN [25]	79.86	42.47
FasterRCNN*	85.36	53.90
FCOS [32]	88.80	61.93
FCOS*	89.09	63.80

attention module alone, proving that the density estimation branch can significantly enhance the network's acquisition of pedestrian location information and improve localization accuracy.

Table 3. Ablation study on the effects of different components.

Model	mAP@0.5/%	mAP/%
yolo	85.34	68.06
yolo+CBAM	90.36	68.26
yolo+density branch	88.82	69.47
yolo+CBAM+density branch	91.15	70.79

(a) Original feature map   (b) Feature map with attention module

Fig. 4. Visualization of the attention module

Crowd Density Estimation. In addition to improving the performance of the detection task, we also tested the effectiveness of the model's crowd density estimation branch and visualized the output of the network density map as shown in Fig. 5. Our model achieve MSE = 0.554 and MAE = 0.576 in crowd density estimation task. And according to the density map, the network can accurately estimate the pedestrian head positions, thereby assisting in the enhancement of detection task performance. As for the density estimation task itself, it can achieve relatively accurate estimation in boarding bridge scenarios as well.

Fig. 5. Visualization of crowd density estimation

Hyperparameters. As described in Sect. 4.1, the attention module has different effects on network performance at different positions. We tried inserting the CBAM module at multiple locations and compared its impact on network performance. As shown in Table 4, placing the attention module in a deeper position within the network can better enhance the model's accuracy. In terms of network structure design, we compared the results of density estimation branches with different downsampling factors. Higher-resolution density estimation maps can bring more significant improvements in detection task accuracy (show in Table 5), with only slight increases in model training duration. For the Gaussian kernel parameter, different values have little impact on network performance (show in Table 6), and the network's robustness is not greatly affected by hyperparameters.

Table 4. Effect of different insertion positions of attention modules on model performance. P2, P3, P4, P5 are the different depths of yolo's backbone.

CBAM insertion position				mAP@0.5/%	mAP/%
P2	P3	P4	P5		
✓				88.47	68.90
	✓			88.15	69.87
		✓		88.75	69.72
			✓	89.45	70.16
✓	✓			89.60	70.05
		✓	✓	91.15	70.79
✓	✓	✓	✓	88.29	69.04

Table 5. Effect of density map resolution on model performance. (Where 1/8 represents that the resolution of the density map is 1/8 of the resolution of the input image.)

Resolution	mAP@0.5/%	mAP/%
Big(1/8)	91.15	70.79
Middle(1/16)	90.38	71.04
Small(1/32)	88.23	68.65

Table 6. Effect of Gaussian kernel parameters on model performance.

σ	mAP@0.5/%	mAP/%
0.4	90.92	71.82
0.5	91.15	70.79
0.6	91.59	71.12

Visualization. From the visualization results (Fig. 6), the improved model can not only distinguish different pedestrian categories more accurately, but also has a significant improvement in positioning accuracy. In scenarios with large changes in pedestrian posture and the presence of occlusions, the model can significantly reduce the occurrence of missed detections.

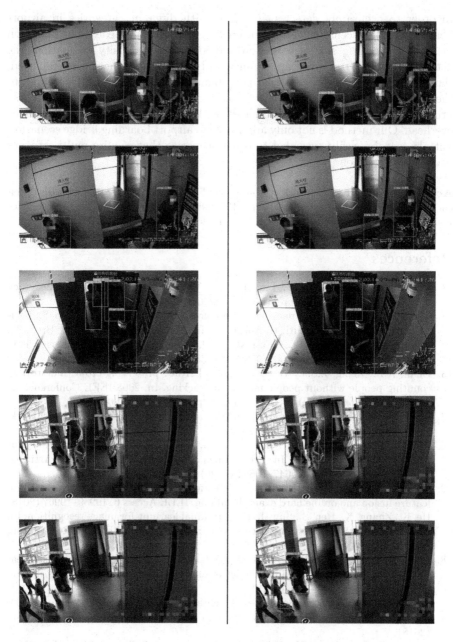

Fig. 6. Comparison of the baseline (yolo [11]) and our improved model (density-yolo), where the left image is the baseline model and the right one is our improved model.

5 Conclusions

In order to accurately detect pedestrians with different categories in airport boarding bridge scenarios, we proposes a multi-category pedestrian detection method. By adding spatial attention and joint crowd density estimation, the detector's ability to acquire appearance and location information is enhanced, thus achieving accurate pedestrian positioning and classification. Compared to the baseline, our method has achieved 5.81% improvement on mAP, and it also brings considerable gains to other models, with almost no additional performance overhead. Our method is not only applicable to airport boarding bridge scenarios, but also can be easily extended to other scenarios that require multi-category pedestrian detection.

Acknowledgement. This work was supported in part by the National Natural Science Foundation of China under Grant 61976095 and in part by the Natural Science Foundation of Guangdong Province, China, under Grant 2022A1515010114.

References

1. Bochkovskiy, A., Wang, C.Y., Liao, H.Y.M.: YOLOv4: optimal speed and accuracy of object detection. arXiv preprint arXiv:2004.10934 (2020)
2. Carion, N., Massa, F., Synnaeve, G., Usunier, N., Kirillov, A., Zagoruyko, S.: End-to-end object detection with transformers. In: Vedaldi, A., Bischof, H., Brox, T., Frahm, J.-M. (eds.) ECCV 2020. LNCS, vol. 12346, pp. 213–229. Springer, Cham (2020). https://doi.org/10.1007/978-3-030-58452-8_13
3. Chan, A.B., Liang, Z.S.J., Vasconcelos, N.: Privacy preserving crowd monitoring: counting people without people models or tracking. In: 2008 IEEE Conference on Computer Vision and Pattern Recognition, pp. 1–7. IEEE (2008)
4. Chen, K., et al.: MMDetection: open MMLab detection toolbox and benchmark. arXiv preprint arXiv:1906.07155 (2019)
5. Chi, C., Zhang, S., Xing, J., Lei, Z., Li, S.Z., Zou, X.: Relational learning for joint head and human detection. In: Proceedings of the AAAI Conference on Artificial Intelligence, vol. 34, pp. 10647–10654 (2020)
6. Chu, J., Guo, Z., Leng, L.: Object detection based on multi-layer convolution feature fusion and online hard example mining. IEEE Access **6**, 19959–19967 (2018)
7. He, K., Zhang, X., Ren, S., Sun, J.: Deep residual learning for image recognition. In: Proceedings of the IEEE Conference on Computer Vision and Pattern Recognition, pp. 770–778 (2016)
8. Hu, J., Shen, L., Albanie, S., Sun, G., Vedaldi, A.: Gather-excite: exploiting feature context in convolutional neural networks. In: Advances in Neural Information Processing Systems, vol. 31 (2018)
9. Hu, J., Shen, L., Sun, G.: Squeeze-and-excitation networks. In: Proceedings of the IEEE Conference on Computer Vision and Pattern Recognition, pp. 7132–7141 (2018)
10. Jha, S., Seo, C., Yang, E., Joshi, G.P.: Real time object detection and tracking system for video surveillance system. Multimedia Tools Appl. **80**, 3981–3996 (2021)
11. Jocher, G., et al.: ultralytics/YOLOv5: V6.2 - YOLOv5 Classification Models, Apple M1, Reproducibility, ClearML and Deci.ai integrations, August 2022. https://doi.org/10.5281/zenodo.7002879

12. Ku, J., Mozifian, M., Lee, J., Harakeh, A., Waslander, S.L.: Joint 3D proposal generation and object detection from view aggregation. In: 2018 IEEE/RSJ International Conference on Intelligent Robots and Systems (IROS), pp. 1–8. IEEE (2018)
13. Lee, H., Kim, H.E., Nam, H.: SRM: a style-based recalibration module for convolutional neural networks. In: Proceedings of the IEEE/CVF International Conference on Computer Vision, pp. 1854–1862 (2019)
14. Li, J., Wang, J., Tian, Q., Gao, W., Zhang, S.: Global-local temporal representations for video person re-identification. In: Proceedings of the IEEE/CVF International Conference on Computer Vision, pp. 3958–3967 (2019)
15. Li, Y., Zhang, X., Chen, D.: CSRNet: dilated convolutional neural networks for understanding the highly congested scenes. In: Proceedings of the IEEE Conference on Computer Vision and Pattern Recognition, pp. 1091–1100 (2018)
16. Lin, T.Y., Dollár, P., Girshick, R., He, K., Hariharan, B., Belongie, S.: Feature pyramid networks for object detection. In: Proceedings of the IEEE Conference on Computer Vision and Pattern Recognition, pp. 2117–2125 (2017)
17. Lin, T.Y., Goyal, P., Girshick, R., He, K., Dollár, P.: Focal loss for dense object detection. In: Proceedings of the IEEE International Conference on Computer Vision, pp. 2980–2988 (2017)
18. Liu, G., Nouaze, J.C., Touko Mbouembe, P.L., Kim, J.H.: YOLO-Tomato: a robust algorithm for tomato detection based on YOLOv3. Sensors 20(7), 2145 (2020)
19. Liu, W., et al.: SSD: single shot MultiBox detector. In: Leibe, B., Matas, J., Sebe, N., Welling, M. (eds.) ECCV 2016. LNCS, vol. 9905, pp. 21–37. Springer, Cham (2016). https://doi.org/10.1007/978-3-319-46448-0_2
20. Liu, Z., et al.: Swin transformer: hierarchical vision transformer using shifted windows. In: Proceedings of the IEEE/CVF International Conference on Computer Vision, pp. 10012–10022 (2021)
21. Maji, D., Nagori, S., Mathew, M., Poddar, D.: YOLO-Pose: enhancing YOLO for multi person pose estimation using object keypoint similarity loss. In: Proceedings of the IEEE/CVF Conference on Computer Vision and Pattern Recognition, pp. 2637–2646 (2022)
22. Park, J., Woo, S., Lee, J.Y., Kweon, I.S.: BAM: bottleneck attention module. arXiv preprint arXiv:1807.06514 (2018)
23. Redmon, J., Divvala, S., Girshick, R., Farhadi, A.: You only look once: unified, real-time object detection. In: Proceedings of the IEEE Conference on Computer Vision and Pattern Recognition, pp. 779–788 (2016)
24. Redmon, J., Farhadi, A.: YOLOv3: an incremental improvement. arXiv preprint arXiv:1804.02767 (2018)
25. Ren, S., He, K., Girshick, R., Sun, J.: Faster R-CNN: towards real-time object detection with region proposal networks. In: Advances in Neural Information Processing Systems, vol. 28 (2015)
26. Sasagawa, Y., Nagahara, H.: YOLO in the dark - domain adaptation method for merging multiple models. In: Vedaldi, A., Bischof, H., Brox, T., Frahm, J.-M. (eds.) ECCV 2020. LNCS, vol. 12366, pp. 345–359. Springer, Cham (2020). https://doi.org/10.1007/978-3-030-58589-1_21
27. Simon, M., Milz, S., Amende, K., Gross, H.-M.: Complex-YOLO: an Euler-Region-Proposal for real-time 3D object detection on point clouds. In: Leal-Taixé, L., Roth, S. (eds.) ECCV 2018. LNCS, vol. 11129, pp. 197–209. Springer, Cham (2019). https://doi.org/10.1007/978-3-030-11009-3_11

28. Song, Q., et al.: Rethinking counting and localization in crowds: a purely point-based framework. In: Proceedings of the IEEE/CVF International Conference on Computer Vision, pp. 3365–3374 (2021)
29. Tan, M., Pang, R., Le, Q.V.: EfficientDet: scalable and efficient object detection. In: Proceedings of the IEEE/CVF Conference on Computer Vision and Pattern Recognition, pp. 10781–10790 (2020)
30. Tan, Z., Wang, J., Sun, X., Lin, M., Li, H., et al.: GiraffeDet: a heavy-neck paradigm for object detection. In: International Conference on Learning Representations (2021)
31. Tang, W., Liu, K., Shakeel, M.S., Wang, H., Kang, W.: DDAD: detachable crowd density estimation assisted pedestrian detection. IEEE Trans. Intell. Transp. Syst. (2022)
32. Tian, Z., Shen, C., Chen, H., He, T.: FCOS: fully convolutional one-stage object detection. In: Proceedings of the IEEE/CVF International Conference on Computer Vision, pp. 9627–9636 (2019)
33. Wang, X., Girshick, R., Gupta, A., He, K.: Non-local neural networks. In: Proceedings of the IEEE Conference on Computer Vision and Pattern Recognition, pp. 7794–7803 (2018)
34. Woo, S., Park, J., Lee, J.-Y., Kweon, I.S.: CBAM: convolutional block attention module. In: Ferrari, V., Hebert, M., Sminchisescu, C., Weiss, Y. (eds.) ECCV 2018. LNCS, vol. 11211, pp. 3–19. Springer, Cham (2018). https://doi.org/10.1007/978-3-030-01234-2_1
35. Wu, X., Liang, G., Lee, K.K., Xu, Y.: Crowd density estimation using texture analysis and learning. In: 2006 IEEE International Conference on Robotics and Biomimetics, pp. 214–219. IEEE (2006)
36. Xiao, C., et al.: DSFNet: dynamic and static fusion network for moving object detection in satellite videos. IEEE Geosci. Remote Sens. Lett. 19, 1–5 (2021)
37. Zhang, L., Shi, M., Chen, Q.: Crowd counting via scale-adaptive convolutional neural network. In: 2018 IEEE Winter Conference on Applications of Computer Vision (WACV), pp. 1113–1121. IEEE (2018)
38. Zhang, Y., Zhou, D., Chen, S., Gao, S., Ma, Y.: Single-image crowd counting via multi-column convolutional neural network. In: Proceedings of the IEEE Conference on Computer Vision and Pattern Recognition, pp. 589–597 (2016)
39. Zhou, X., Wang, D., Krähenbühl, P.: Objects as points. arXiv preprint arXiv:1904.07850 (2019)
40. Zhu, X., Su, W., Lu, L., Li, B., Wang, X., Dai, J.: Deformable DETR: deformable transformers for end-to-end object detection. arXiv preprint arXiv:2010.04159 (2020)

A Novel Neighborhood-Augmented Graph Attention Network for Sequential Recommendation

Shuxiang Xu[1,2], Qibu Xiang[1,2], Yushun Fan[1,2(✉)], Ruyu Yan[1,2],
and Jia Zhang[3]

[1] Beijing National Research Center for Information Science and Technology,
Beijing, China
[2] Department of Automation, Tsinghua University, Beijing, China
{xsx22,xqb22,yanry18}@mails.tsinghua.edu.cn, fanyus@tsinghua.edu.cn
[3] Department of Computer Science, Southern Methodist University, Dallas, TX, USA
jiazhang@smu.edu

Abstract. In recent years, sequential recommender systems have been widely applied for alleviating information overload. Some solutions employ graph attention networks (GAT) to aggregate rich neighborhood information for the representation learning of items. However, how to sufficiently exploit graph structure deserves careful examination due to two challenges. Firstly, highly related items may not appear in the same interaction sequence due to the data sparsity issue. Secondly, the connection weights among items are randomly initialized, which brings significant uncertainty for information propagation. To tackle these challenges, we propose a novel Neighborhood-Augmented Graph ATtention network (NA-GAT). For the former challenge, we globally screen a fixed number of potential neighbors for each item node based on the attention mechanism. For the latter challenge, we devise a two-stage learning strategy to make full use of the transition frequency and the attention score, to achieve sufficient utilization of graph structure. Extensive experimental results have demonstrated the necessity of neighborhood augmentation and the effectiveness of the proposed NA-GAT framework.

Keywords: Sequential Recommendation · Graph Neural Network · Graph Augmentation · Attention Mechanism

1 Introduction

As a bridge connecting content creators and users on Internet platforms, sequential recommender systems (SRSs) have been widely applied to alleviating information overload, which can capture users' preferences from their sequential behaviors and then predict their next interactions [11]. To leverage complex intrinsic associations that exist among items, some studies adopt graph neural networks (GNN) to perform information propagation and aggregation for

L. Fang et al. (Eds.): CICAI 2023, LNAI 14474, pp. 229–241, 2024.
https://doi.org/10.1007/978-981-99-9119-8_21

representation learning of user and item nodes [22,25,26]. The graph attention network (GAT) [21] has been widely used because of its good generalization ability and flexibility [1,2,9], which leverages the attention mechanism to distinguish the contributions of neighbors.

Despite the progress of these solutions, how to sufficiently exploit the graph structure to aggregate neighborhood information based on GAT has not been fully studied. Two major challenges exist. The first challenge is the data sparsity issue. Highly related items may not always appear in the same interaction sequence, so it is hard to fully utilize the social information among items based on the original graph. In other words, the original graph structure shall be enriched. We refer to the original neighbors of a node as **structural neighbors** and the mined potential neighbors as **augmented neighbors**. Most studies treat them equally, while structural neighbors tend to contain more realistic and reliable information than augmented neighbors, so it is necessary to distinguish their importance. The second challenge is how to calculate the connection weights between nodes. Knyazev et al. [7] find that for GAT-based models, the effect of attention on performance may be negative under typical conditions. Most current approaches to graph construction neglect to initialize connection weights. As a result, at the start of the training process, GAT-based models with randomly initialized parameters assign almost random attention scores as connection weights, which brings a large uncertainty for information propagation and aggregation and thus makes the training process unstable.

To tackle the aforementioned two challenges, in this paper, we propose a novel Neighborhood-Augmented Graph ATtention network (NA-GAT).

For the first challenge, except for structural neighbors, we screen a fixed number of potential neighbors among all items for each item node. Specifically, based on GAT, we obtain highly correlated potential neighbors by ranking attention scores. With this method, we can perform global-wise graph structure augmentation and discover new augmented neighbors at each training epoch. For the second challenge, we propose a two-stage learning strategy to stabilize the learning process. In the first stage, we combine the transition frequency and attention score as the final connection weight to provide a warm start for graph learning. In the second stage, we only utilize the attention scores to aggregate information from structural neighbors and augmented neighbors, which can take full advantage of the attention mechanism and make the information propagation and aggregation process more precise.

Our contributions are summarized as follows:

- We augment the original item graph by discovering potential neighbors with high attention scores for each item node.
- We devise a two-stage learning strategy to stabilize the training process.
- Experimental results on two publicly available datasets demonstrate the superiority of our model and the effectiveness of the components.

2   Related Work

2.1   Sequential Recommendation

Existing solutions for sequential recommendation can be divided into three categories: conventional models, RNN-based models, and attention-based models.

With an assumption that the next behavior depends only on the most recent interaction, Markov Chain (MC) is a typical conventional approach to predict the next item of interest to a user [17]. Recently, RNN-based models [5,8] have attracted a wide range of attention in sequential recommendation due to their superiority in modeling temporal patterns. To identify the importance of historical interactions, attention-based models employ the attention mechanism to assign weights to different interactions and then generate collective preferences [3,6].

2.2   Graph Neural Networks

In sequential recommendation, users' interaction data have sequential nature and the user/item graph can be constructed by connecting adjacent items. Therefore, many solutions have applied graph neural networks (GNN) to obtain high-order features of users and items [18,20,27]. The main purpose of GNN-based models is to update node representations with neighborhood information [10,21].

Due to the incapacity of GNN to effectively capture long-term dependencies, some GNN-based models also adopt attention mechanisms to integrate all the item representations in the sequence. SRGNN [23] introduces a gated GNN layer to obtain node representations and generates the session embedding via the attention mechanism. GC-SAN [24] utilizes a graph neural network to extract local dependencies and the self-attention mechanism to obtain global dependencies.

To enrich neighborhood information, some solutions treat more than one adjacent item as neighbors [14] or aggregate information from cross-hop neighbors [13]. However, these local augmentation operations struggle to integrate global-wise information. Therefore, we utilize the attention mechanism to globally screen highly correlated potential neighbors for each node, which can enrich the social relationships between items and alleviate the data sparsity issue.

3   Methodology

3.1   Problem Statement

Let $\mathcal{U} = \{u_1, u_2, \ldots, u_{|\mathcal{U}|}\}$ and $\mathcal{I} = \{i_1, i_2, \ldots, i_{|\mathcal{I}|}\}$ denote the set of users and the set of items, where $|\mathcal{U}|$ and $|\mathcal{I}|$ are the number of users and items, respectively. Each user u has sequential interactions $\mathcal{Q}^u = \{i_1^u, \ldots, i_t^u, \ldots, i_{|\mathcal{Q}^u|}^u\}$, where $i_t^u \in \mathcal{I}$ denotes the item interacted by user u at time step t. Given the interaction sequence $\mathcal{Q}^u$, the sequential recommendation aims to predict an item from $\mathcal{I}$ with which user u may interact at the next time step, i.e., $S_{|\mathcal{Q}^u|+1}^u$.

3.2   Framework Overview

Fig. 1 illustrates the overview of our proposed method. Based on all the item sequences, we first construct a directed item graph, taking the transition frequency matrix between items as the adjacency matrix. During the training process, for a particular item shown on the lower half, we obtain its structural neighbors based on the transition matrix and screen highly correlated items via the attention mechanism as its augmented neighbors, and then we integrate the information from both types of neighbors into the neighborhood vector. Next, as shown on the upper half, we utilize the gating mechanism to aggregate the neighborhood information and finally employ a position-aware attention mechanism to generate the user's interest preferences to score the candidate items.

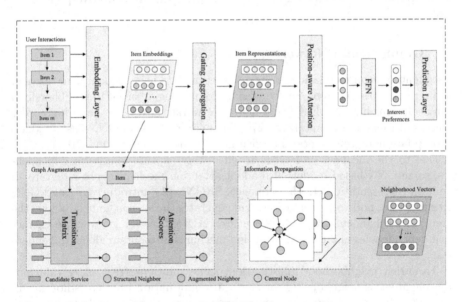

Fig. 1. The structure of the proposed NA-GAT. For a particular item, we first get its structural neighbors based on the transition matrix and adopt the attention mechanism to globally screen highly related items as its augmented neighbors. Next, we utilize the gating mechanism to aggregate neighborhood information to obtain the updated item representation. Finally, we employ a position-aware attention mechanism to capture the sequential features and generate the user's preferences to predict the next item.

3.3   Graph Construction and Augmentation

Graph Construction. A common approach to graph construction is to connect two adjacent items in user-item interactions. Given an interaction sequence $\mathcal{Q}^u = \{i_1^u, i_2^u, \ldots, i_{|\mathcal{Q}^u|}^u\}$, we denote $\langle i_k^u, i_{k+1}^u \rangle$ as an edge from item i_k^u to item i_{k+1}^u.

For a pair of nodes $i,\ j \in \mathcal{I}$, we traverse all the interaction sequences to count the occurrence times of item i and the consecutive co-occurrence times of item i and j, and then we can obtain the transition frequency from i to j:

$$f(i,\ j) = \frac{\sum_{u \in \mathcal{U}} \sum_{k=1}^{|\mathcal{Q}^u|-1} [\![i_k^u = i, i_{k+1}^u = j]\!]}{\sum_{u \in \mathcal{U}} \sum_{k=1}^{|\mathcal{Q}^u|-1} [\![i_k^u = i]\!]}, \tag{1}$$

where $[\![\cdot]\!]$ is the indicator function. The transition frequency can be used to approximate the transition probability which determines the process of information propagation, so we apply the transition frequency to form the adjacency matrix of the item graph: $\mathbf{F} \in \mathbb{R}^{|\mathcal{I}| \times |\mathcal{I}|}$, where $\mathbf{F}_{i,j} = f(i,\ j)$. For a particular item $i \in \mathcal{I}$, its structural neighbors can be formulated as $\mathcal{N}_i^s = \{j \in \mathcal{I} | \mathbf{F}_{j,i} > 0, j \neq i\}$.

Graph Augmentation. Considering that highly related items may not appear in the same interaction sequence, we utilize the attention mechanism to obtain the similarity between nodes and augment the original item graph.

For a particular item node $i \in \mathcal{I}$, we denote its embedding as $\mathbf{e}_i \in \mathbb{R}^d$, where d is the hidden size. Following [21], we adopt a feedforward layer to obtain the attention score between node i and j:

$$\alpha_j^i = \text{LeakyReLU}(\mathbf{q}^T [\mathbf{W} \mathbf{e}_i || \mathbf{W} \mathbf{e}_j]), \tag{2}$$

where $\text{LeakyReLU}(\cdot)$ is the activation function, $||$ indicates the concatenation operation, and $\mathbf{W} \in \mathbb{R}^{d \times d}$, $\mathbf{q} \in \mathbb{R}^{2d}$ are learnable parameters. Next, we rank α_*^i to filter potential neighbors:

$$\mathcal{N}_i^a = \text{argtop}_k \left(\alpha_j^i | j \in \mathcal{I} \backslash \mathcal{N}_i^s \right), \tag{3}$$

where $\mathcal{N}_i^a$ denotes the augmented neighbors of node i, and k is the number of augmented neighbors. By this method, we can discover nodes that are highly related to node i from a global scope.

3.4   Information Propagation and Aggregation

The key to information propagation is to assign different weights to neighbors so as to control the influence of neighbors. To this end, we propose a two-stage learning strategy to utilize the information from both types of neighbors.

Specifically, we first normalize the attention scores of structural/augmented neighbor j:

$$\alpha_j^i = \frac{\exp(\alpha_j^i)}{\sum_{k \in \mathcal{N}_i^s \cup \mathcal{N}_i^a} \exp(\alpha_k^i)}. \tag{4}$$

At the initial stage of the training process, the transition frequency can model the influence of neighbors more accurately than the random attention score calculated by the randomly initialized parameters. However, using only the fixed transition frequency lacks flexibility, while the attention score can accurately represent the influence after sufficient training. Therefore, we weight the transition frequency and the attention score to obtain the final connection strength:

$$w_j^i = \lambda * \alpha_j^i + (1 - \lambda) * \mathbf{F}_{j,i}, \tag{5}$$

where $\lambda \in [0, 1]$ is the importance coefficient of attention scores and $j \in \mathcal{N}_i^s \cup \mathcal{N}_i^a$ is a neighbor of node i. In addition, to take full advantage of the attention mechanism, we gradually increase λ during the training process. Specifically, we set the epoch threshold as L and make λ increase linearly until the second stage:

$$\lambda = \min(\text{epoch}/L, 1). \tag{6}$$

In this way, the attention mechanism gradually plays a dominant role in modeling the connection strength. In addition, for augmented neighbors, the transition frequency is 0 and thus they just propagate a small amount of information in the initial training process, which ensures that the model can adequately learn the original graph structure in the first stage. Compared with augmented neighbors, structural neighbors have real connections with the central node and thus contain more reliable information. Therefore, by setting up the first training stage, not only can the transition frequency be utilized to guide the graph learning process, but also the influence of structural neighbors can be emphasized to prevent the original graph structure from being damaged by augmented neighbors.

In the second stage, that is, after L epochs of training, we only take attention scores calculated by sufficiently trained parameters as connection weights due to their flexibility to control the influence of neighbors. At this stage, structural neighbors and augmented neighbors have the same status, which allows each node to obtain rich global social information.

For item i, the final representation of neighborhood information can be formulated as:

$$\mathbf{e}_i^{\mathcal{N}} = \sum_{j \in \mathcal{N}_i^s \cup \mathcal{N}_i^a} w_j^i \mathbf{e}_j. \tag{7}$$

Inspired by the gate unit in GRU [5], we adopt a gating mechanism to aggregate neighborhood information to obtain the final representation of node i:

$$\mathbf{g}_i = \sigma(\mathbf{W}_1 \mathbf{e}_i + \mathbf{W}_2 \mathbf{e}_i^{\mathcal{N}}), \tag{8}$$

$$\widetilde{\mathbf{e}}_i = \mathbf{g}_i \odot \mathbf{e}_i + (1 - \mathbf{g}_i) \odot \mathbf{e}_i^{\mathcal{N}}, \tag{9}$$

where $\sigma(\cdot)$ denotes the sigmoid function, $\odot$ denotes the element-wise product, and $\mathbf{W}_1, \mathbf{W}_2 \in \mathbb{R}^{d \times d}$ are trainable parameters.

3.5   Preference Extraction

Let m denote the maximum sequence length that the model can handle, and we truncate (or pad) the user-item interactions into $\{i_1^u, i_2^u, \ldots, i_m^u\}$. After the embedding layer and the graph learning layer, we can obtain the item sequence representations $\widetilde{\mathbf{E}}^u := \{\widetilde{\mathbf{e}}_1^u, \widetilde{\mathbf{e}}_2^u, \ldots, \widetilde{\mathbf{e}}_m^u\} \in \mathbb{R}^{m \times d}$. Next, we utilize the attention mechanism to obtain the collective preferences of a user:

$$\mathbf{z}^u = \sum_{k=1}^{m} \beta_k \widetilde{\mathbf{e}}_k^u, \tag{10}$$

where β_k indicates the importance of item i_k^u. Considering that recent interactions tend to influence user behaviors more than earlier interactions, we introduce the learnable position embeddings $\mathbf{P} := \{p_1, p_2, \ldots, p_m\} \in \mathbb{R}^{m \times d}$ to the item representations:

$$\mathbf{H}^u = \widetilde{\mathbf{E}}^u + \mathbf{P}, \tag{11}$$

where $\mathbf{H}^u = \{\mathbf{h}_1^u, \mathbf{h}_2^u, \ldots, \mathbf{h}_m^u\} \in \mathbb{R}^{m \times d}$. The last item representation contains the information of previous items through the graph learning process, so we calculate the attention score between each item and the last interaction:

$$\beta_k = \mathbf{v}^T \sigma(\mathbf{W}_3 \mathbf{h}_k^u + \mathbf{W}_4 \mathbf{h}_m^u]), \tag{12}$$

where $\sigma(\cdot)$ denotes the sigmoid function and $\mathbf{v} \in \mathbb{R}^d$, $\mathbf{W}_3$, $\mathbf{W}_4 \in \mathbb{R}^{d \times d}$ are trainable parameters.

To enhance the learning ability of our model, we use a two-layer fully-connected network with residual connections to obtain the final user preferences:

$$\mathbf{o}^u = \mathbf{z}^u + \mathbf{W}_6 \text{ReLU}\left(\mathbf{W}_5 \mathbf{z}^u + \mathbf{b}_1\right) + \mathbf{b}_2, \tag{13}$$

where $\mathbf{W}_5$, $\mathbf{W}_6 \in \mathbb{R}^{d \times d}$, $\mathbf{b}_1$, $\mathbf{b}_2 \in \mathbb{R}^d$ are learnable parameters of the model.

3.6 Prediction and Optimization

In the phase of prediction, we utilize the user's interest preferences $\mathbf{o}^u$ and to score candidate item c:

$$y_c^u = \mathbf{o}^u \mathbf{e}_c^T. \tag{14}$$

We apply the Binary Cross-Entropy loss function to train our model:

$$\mathcal{L} = -\sum_{u \in \mathcal{U}} [\log(\sigma(y_{i^u}^u)) + \sum_{j \notin \mathcal{Q}^u} \log(1 - \sigma(y_j^u))], \tag{15}$$

where $\sigma(\cdot)$ denotes the sigmoid function, i^u is the ground truth item, and j represents the negative sampling items.

4 Experiments

4.1 Experimental Setup

To evaluate the performance of models, we selected two real-world datasets, Yelp[1] and Amazon[2]. The detailed descriptions of both datasets after preprocessing are shown in Table 1. We set the maximum sequence length that the model can handle as 50. For each sequence, we used the last item for testing, the penultimate item for validation, and the rest items for training.

Table 1. Statistics of the datasets

Statistics	Yelp	Amazon
#user	2,233	2,070
#item	2,428	2,698
#interaction	94,730	47,850
avg.length	42.42	23.12

We compared our proposed NA-GAT with the following 10 baseline models to evaluate performance: BPR-MF [16], GRU4Rec [5], NARM [8], Caser [19], STAMP [12], NISER [4], SR-GNN [23], GC-SAN [24], LESSR [2], SGNN-HN [15]. To evaluate our proposed NA-GAT and baseline models, we randomly sampled 100 negative items for each positive item. For the evaluation, we adopted two types of metrics: HR@K and NDCG@K.

For a fair comparison, we implemented all the models by PyTorch and set the hidden size as 128 and the learning rate as 0.001 over both datasets. We set the number of augmented neighbors k as 30 for Yelp and 40 for Amazon, and set the epoch threshold L as 40 and 60 for Yelp and Amazon, respectively.

4.2 Overall Performance

The experimental results of all models on both datasets are shown in Table 2, where the models above the dashed line only adopt classical methods such as RNNs and attention mechanisms to extract user preferences, and the models below employ GNN to enrich a node's representation. From the results, we have the following observations:

1. Models utilizing GNN tend to perform better than models that capture sequential features directly. The GNN-based models can capture the transition pattern among items and thus precisely extract users' interest preferences.
2. On both datasets, our proposed NA-GAT outperforms the baseline models in terms of all metrics. Specifically, compared with the best baseline, NA-GAT has a significant improvement of 5.12%-9.84% and 7.44%-8.59% on Yelp and Amazon, respectively. The reasons may be three-fold for the performance improvement. Firstly, we globally mine highly correlated neighbors of each node, which effectively alleviates the data sparsity issue and makes the information flow process more efficient. Secondly, we initialize the connection weight with the transition frequency, which provides a warm start for model

---

[1] https://www.kaggle.com/datasets/z5025122/yelp-csv.

[2] https://cseweb.ucsd.edu/~jmcauley/datasets/amazon_v2/.

Table 2. Performance comparison between different models. The best performance of all models is boldfaced and the best performance of baselines is underlined.

Model	Yelp				Amazon			
	HR@5	NDCG@5	HR@10	NDCG@10	HR@5	NDCG@5	HR@10	NDCG@10
BPR-MF	0.3457	0.2345	0.5186	0.2902	0.2671	0.1860	0.3696	0.2191
GRU4Rec	0.3945	0.2674	0.5580	0.3202	0.3198	0.2411	0.4275	0.2755
NARM	0.3972	0.2697	0.5647	0.3237	0.3248	0.2442	0.4301	0.2782
Caser	0.3565	0.2467	0.5179	0.2989	0.3029	0.2226	0.4111	0.2573
STAMP	0.3865	0.2423	0.5419	0.2920	0.2860	0.2004	0.3638	0.2256
NISER	0.3996	0.2739	0.5613	0.3259	0.3087	0.2274	0.4126	0.2609
SR-GNN	0.4138	0.2751	0.5759	0.3269	0.3242	0.2427	0.4271	0.2757
GC-SAN	<u>0.4324</u>	<u>0.2947</u>	<u>0.5815</u>	<u>0.3428</u>	<u>0.3377</u>	<u>0.2506</u>	0.4363	<u>0.2834</u>
LESSR	0.4090	0.2792	0.5720	0.3320	0.3261	0.2348	0.4208	0.2648
SGNN-HN	0.4118	0.2846	0.5739	0.3369	0.3305	0.2435	<u>0.4406</u>	0.2792
Ours	**0.4595**	**0.3237**	**0.6113**	**0.3728**	**0.3667**	**0.2720**	**0.4734**	**0.3066**
Improv.	6.27%	9.84%	5.12%	8.75%	8.59%	8.54%	7.44%	8.19%

training, and the introduction of transition frequency also emphasizes the importance of structural neighbors, allowing the model to adequately learn the original graph structure. Thirdly, after training smoothly, we utilize only the attention mechanism to calculate the connection weights, fully exploiting the role of augmented neighbors and aggregating rich global information.

4.3 The Effect of Neighborhood Augmentation

Recall that we utilized the attention mechanism to mine k potential neighbors for each node. To explore the role of neighborhood augmentation, we tested the effect of different k values on the model performance.

From the results shown in Fig. 2, if $k = 0$, which means that we aggregate neighborhood information only through structural neighbors, the performance is much worse than with augmented neighbors, which indicates that appropriately extending the neighborhood has a positive effect on node representation learning. As k increases, the metrics first rise and then fall, which indicates that the significant augmentation tends to cover nodes with low relevance and bring redundant information, resulting in inaccurate neighborhood information.

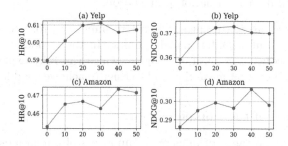

Fig. 2. Performance with different numbers of augmented neighbors (k) on two datasets

4.4   The Effect of the Two-Stage Learning Strategy

Recall that we devised a two-stage learning strategy to stabilize the training process. To verify the effectiveness of this strategy, we designed several variants:

- **TF** employs only structural neighbors and takes the <u>T</u>ransition <u>F</u>requency as the fixed connection weight.
- **AS** employs only structural neighbors and takes <u>A</u>ttention <u>S</u>cores as connection weights.
- **AS-NA** performs <u>N</u>eighborhood <u>A</u>ugmentation and takes only attention scores as connection weights.
- **TF-AS-NA** employs both types of neighbors and consistently averages the transition frequencies and attention scores to obtain the connection weights.

Table 3. Performance with different learning strategies

Model	Yelp		Amazon	
	HR@10	NDCG@10	HR@10	NDCG@10
TF	0.5553	0.3235	0.4340	0.2737
AS	0.5796	0.3440	0.4501	0.2796
AS-NA	0.5981	0.3622	0.4594	0.2894
TF-AS-NA	0.6002	0.3637	0.4491	0.2828
NA-GAT	**0.6113**	**0.3728**	**0.4734**	**0.3066**

From the results shown in Table 3, we can draw the following conclusions:

1. **TF** achieves the worst performance, indicating that fixed connection weights cannot precisely model the relationship between items.
2. Extending neighbors can achieve better performance than using only structural neighbors, e.g., **AS-NA** outperforms **AS**.
3. **TF-AS-NA** introduces the transition frequency, yet the results on Amazon are worse than **AS-NA**, which suggests that consistently taking the transition frequency as part of connection weights cannot fully exploit the role of the attention mechanism in dynamically modeling connection weights and thus may be harmful for information propagation.

Overall, the transition frequency plays an important role in the first stage by providing a warm start for model training. In the second stage, we utilize only attention scores as connection weights, which can fully exploit the role of augmented neighbors and make the model easier to achieve good local optimum.

4.5   The Effect of the Components

To explore the importance of different components of our proposed NA-GAT, we conducted ablation experiments by designing the following variants:

- **w/o GNN** only employs the position-aware attention mechanism to obtain the final representation.
- **w/o Position** utilizes GNN to aggregate information and adopts the attention mechanism to obtain user preferences without position embeddings.
- **w/o Attn** aggregates neighborhood information via GNN and only uses the representation of the last item to score the candidate items.

Table 4. Performance of different variants

Model	Yelp		Amazon	
	HR@10	NDCG@10	HR@10	NDCG@10
w/o GNN	0.5746	0.3305	0.4466	0.2728
w/o Position	0.5864	0.3461	0.4467	0.2872
w/o Attn	0.4384	0.2433	0.3382	0.2224
NA-GAT	**0.6113**	**0.3728**	**0.4734**	**0.3066**

From the experimental results shown in Table 4, **w/o Attn** achieves the worst performance, which indicates that the attention mechanism is useful for capturing long-term dependencies. In addition, both GNN which aggregates neighborhood information, and position embeddings which differentiate the impact of sequential positions can effectively improve the model performance.

5 Conclusions

To further enhance the neighborhood information of item nodes, in this work, we propose a novel Neighborhood-Augmented Graph ATtention network (NA-GAT). Specifically, we introduce the transition frequency to initialize the connection weight among items, providing a warm start for model training. Meanwhile, we adopt the attention mechanism to globally identify highly correlated nodes as augmented neighbors for each item node. To achieve full utilization of structural neighbors and augmented neighbors, we devise a two-stage learning strategy, that is, we first emphasize the importance of structural neighbors to sufficiently learn the original structure and then only take attention scores as connection weights to effectively aggregate global-wise information. Experimental results validate the effectiveness of the proposed model.

Acknowledgements. This research has been partially supported by the National Natural Science Foundation of China (No. 62173199).

References

1. Chang, J., et al.: Sequential recommendation with graph neural networks. In: Proceedings of The 44th International ACM SIGIR Conference on Research and Development in Information Retrieval, pp. 378–387 (2021)
2. Chen, T., Wong, R.C.W.: Handling information loss of graph neural networks for session-based recommendation. In: Proceedings of The 26th ACM SIGKDD International Conference on Knowledge Discovery & Data Mining, pp. 1172–1180 (2020)
3. Fan, X., Liu, Z., Lian, J., Zhao, W.X., Xie, X., Wen, J.R.: Lighter and better: low-rank decomposed self-attention networks for next-item recommendation. In: Proceedings of The 44th International ACM SIGIR Conference on Research and Development in Information Retrieval, pp. 1733–1737 (2021)
4. Gupta, P., Garg, D., Malhotra, P., Vig, L., Shroff, G.: Niser: normalized item and session representations to handle popularity bias. arXiv preprint arXiv:1909.04276 (2019)
5. Hidasi, B., Karatzoglou, A., Baltrunas, L., Tikk, D.: Session-based recommendations with recurrent neural networks. arXiv preprint arXiv:1511.06939 (2015)
6. Kang, W.C., McAuley, J.: Self-attentive sequential recommendation. In: Proceedings of IEEE International Conference on Data Mining (ICDM), pp. 197–206. IEEE (2018)
7. Knyazev, B., Taylor, G.W., Amer, M.: Understanding attention and generalization in graph neural networks. In: Advances in Neural Information Processing Systems, vol. 32 (2019)
8. Li, J., Ren, P., Chen, Z., Ren, Z., Lian, T., Ma, J.: Neural attentive session-based recommendation. In: Proceedings of the 2017 ACM on Conference on Information and Knowledge Management, pp. 1419–1428 (2017)
9. Li, Y., Chen, T., Luo, Y., Yin, H., Huang, Z.: Discovering collaborative signals for next poi recommendation with iterative seq2graph augmentation. arXiv preprint arXiv:2106.15814 (2021)
10. Li, Y., Tarlow, D., Brockschmidt, M., Zemel, R.: Gated graph sequence neural networks. arXiv preprint arXiv:1511.05493 (2015)
11. Liang, G., Liao, J., Zhou, W., Wen, J.: Integrating the pre-trained item representations with reformed self-attention network for sequential recommendation. In: 2022 IEEE International Conference on Web Services (ICWS), pp. 27–36. IEEE (2022)
12. Liu, Q., Zeng, Y., Mokhosi, R., Zhang, H.: Stamp: short-term attention/memory priority model for session-based recommendation. In: Proceedings of The 24th ACM SIGKDD International Conference on Knowledge Discovery & Data Mining, pp. 1831–1839 (2018)
13. Liu, Z., Meng, L., Jiang, F., Zhang, J., Yu, P.S.: Deoscillated graph collaborative filtering. arXiv preprint arXiv:2011.02100 (2020)
14. Ma, C., Ma, L., Zhang, Y., Sun, J., Liu, X., Coates, M.: Memory augmented graph neural networks for sequential recommendation. In: Proceedings of the AAAI conference on Artificial Intelligence, vol. 34, pp. 5045–5052 (2020)
15. Pan, Z., Cai, F., Chen, W., Chen, H., De Rijke, M.: Star graph neural networks for session-based recommendation. In: Proceedings of The 29th ACM International Conference on Information & Knowledge Management, pp. 1195–1204 (2020)
16. Rendle, S., Freudenthaler, C., Gantner, Z., Schmidt-Thieme, L.: Bpr: bayesian personalized ranking from implicit feedback. arXiv preprint arXiv:1205.2618 (2012)

17. Rendle, S., Freudenthaler, C., Schmidt-Thieme, L.: Factorizing personalized markov chains for next-basket recommendation. In: Proceedings of The 19th International Conference on World Wide Web, pp. 811–820 (2010)
18. Sun, J., et al.: Multi-graph convolution collaborative filtering. In: Proceedings of IEEE International Conference on Data Mining (ICDM), pp. 1306–1311. IEEE (2019)
19. Tang, J., Wang, K.: Personalized top-n sequential recommendation via convolutional sequence embedding. In: Proceedings of The 11st ACM International Conference on Web Search and Data Mining, pp. 565–573 (2018)
20. Tian, Y., Chang, J., Niu, Y., Song, Y., Li, C.: When multi-level meets multi-interest: a multi-grained neural model for sequential recommendation. In: Proceedings of The 45th International ACM SIGIR Conference on Research and Development in Information Retrieval, pp. 1632–1641 (2022)
21. Veličković, P., Cucurull, G., Casanova, A., Romero, A., Lio, P., Bengio, Y.: Graph attention networks. arXiv preprint arXiv:1710.10903 (2017)
22. Wei, C., Fan, Y., Zhang, J., Lin, H.: A-hsg: neural attentive service recommendation based on high-order social graph. In: 2020 IEEE International Conference on Web Services (ICWS), pp. 338–346. IEEE (2020)
23. Wu, S., Tang, Y., Zhu, Y., Wang, L., Xie, X., Tan, T.: Session-based recommendation with graph neural networks. In: Proceedings of the AAAI Conference on Artificial Intelligence, vol. 33, pp. 346–353 (2019)
24. Xu, C., et al.: Graph contextualized self-attention network for session-based recommendation. In: IJCAI, vol. 19, pp. 3940–3946 (2019)
25. Yan, R., Fan, Y., Zhang, J., Zhang, J., Lin, H.: Service recommendation for composition cration based on collaborative attention convolutional network. In: 2021 IEEE International Conference on Web Services (ICWS), pp. 397–405. IEEE (2021)
26. Zhang, M., Liu, J., Zhang, W., Deng, K., Dong, H., Liu, Y.: CSSR: a context-aware sequential software service recommendation model. In: Hacid, H., Kao, O., Mecella, M., Moha, N., Paik, H. (eds.) ICSOC 2021. LNCS, vol. 13121, pp. 691–699. Springer, Cham (2021). https://doi.org/10.1007/978-3-030-91431-8_45
27. Zhu, T., Sun, L., Chen, G.: Graph-based embedding smoothing for sequential recommendation. IEEE Trans. Knowl. Data Eng. **35**(1), 496–508 (2021)

Reinforcement Learning-Based Algorithm for Real-Time Automated Parking Decision Making

Xiaoyi Wei[1], Taixian Hou[1], Xiao Zhao[1], Jiaxin Tu[1], Haiyang Guan[1], Peng Zhai[1,2,3(✉)], and Lihua Zhang[1,4,5(✉)]

[1] Academy for Engineering and Technology, Fudan University, Shanghai 200433, China
{pzhai,lihuazhang}@fudan.edu.cn
[2] Ji Hua Laboratory, Foshan 528251, China
[3] Engineering Research Center of AI and Robotics, Ministry of Education, Shanghai 200433, China
[4] Institute of Meta-Medical, Fudan University, Shanghai 200433, China
[5] Jilin Provincial Key Laboratory of Intelligence Science and Engineering, Changchun 130013, China

Abstract. Recently, automated parking has gained attention for its ability to enhance parking accuracy and provide a comfortable experience for car owners. However, with the increasing number of vehicles in the parking lot, the traditional automatic parking algorithm face the dual challenges brought by narrow parking spaces and random vehicle obstacles. To address these issues, this paper proposes Curriculum Learning RL for automatic parking decision making in unregulated parking lots. Our approach involves SAC, a reinforcement learning (RL) algorithm, for curriculum learning, where the vehicle learns to park and avoid the obstacles separately through two courses. We incorporate a reward function that considers both location and safety, facilitating continuous learning of optimal actions. In addition, we develop a simulation platform for unregulated parking lots, and we train the algorithm on this platform. Comparing our algorithm with one that learns both actions simultaneously, we observe superior results in shorter timesteps. Furthermore, experiments conducted under various parking conditions demonstrate the algorithm's strong generalization capabilities.

Keywords: Reinforcement Learning (RL) · Automatic Parking Decision Making · SAC · Curriculum Learning

1 Introduction

In recent years, the rise in car ownership has resulted in issues like traffic congestion and road safety [1]. The emergence of autonomous driving technology holds increasing importance as it enhances traffic efficiency and reduces accidents [2]. One significant application of this technology is automatic parking,

L. Fang et al. (Eds.): CICAI 2023, LNAI 14474, pp. 242–252, 2024.
https://doi.org/10.1007/978-981-99-9119-8_22

which greatly improves the accuracy and safety without manual operation by the driver. However, current autonomous parking technology lack adaptability due to their reliance on pre-written programs, which may lead to crashes in new scenarios [9]. Moreover, the growing number of cars in cities has caused a shortage of parking spaces and increased randomness in parking, making control more challenging. Therefore, this paper focuses on planning of parking movement in parking lots with narrow parking spaces and random parking (called **unregulated parking lot**), which is crucial for the development of urban transportation.

Generally, an automated parking system contains the following key technologies: **environment sensing, automatic decision making, control execution** and **monitoring**. Automatic decision making plays a vital role, utilizing environment models and driving knowledge to generate simulated paths for parking through real-time path planning and obstacle avoidance algorithms [2].

1.1   Related Works

Current research approaches for automatic decision making involve curve fitting-based, control theory-based, and RL-based methods.

Curve fitting-based method uses sensor data to perform spline curve fitting for parking operations. For instance, Macek et al. [7] employs RRT and B splines to generate collision-free trajectories for parking between moving obstacles. However, this method requires pre-planning and extensive calculations, limiting its applicability to simple parking scenarios.

Control theory-based method achieves automatic parking by designing controllers and utilizing feedback control based on control theory principles like PID. For instance, Ballina et al. [3] develops an algorithm for parking with spatial constraints, employing a fuzzy PD+I controller to drive the error between the current position and the desired target position towards zero. However, this approach needs precise theoretical derivation and parameter design, and is less suitable for parking scenarios that are sensitive to environmental changes.

RL-based method utilize intelligent interaction with the environment to learn optimal parking strategies. Li et al. [5] employed a kinematic model, using vehicle position, velocity, and facing angle as observations, and using the vehicle acceleration and the front wheel steering angle as actions, achieving vertical parking with the DDPG algorithm. Some improvements have also been proposed, like enhancing robustness by employing a DQN-based curriculum learning approach [15], using MCTS for data generation, evaluation, and network training [13], and designing more appropriate reward functions [11].

RL-based method autonomously learns and optimizes the parking strategy, offering higher robustness and reliability. It applies to complex and diverse parking scenarios, adapting to unforeseen situations not encountered during training. This paper adopts RL method as the algorithm to address the automatic parking decision problem, reflecting its significance in advancing future automatic parking technology.

1.2   Objectives and Contributions

Current research on RL-based method primarily focuses on parking lot environments with easy traffic access, spacious parking spaces, and few obstacles. However, in unregulated parking lots, the effectiveness of automatic parking decisions diminishes.

This paper aims to address this issue by developing a decision algorithm based on deep reinforcement learning. Using an environment sensing system and LIDAR distributed around the vehicle, partial information about the surrounding environment in an unregulated parking lot is obtained. The algorithm plans real-time vehicle movements to avoid randomly generated obstacles until the vehicle is safely parked in the target space. The main contributions are summarized as follows:

(1) We propose a novel RL method combining curriculum learning with SAC. By setting two training courses, our method trains the vehicle separately for parking and obstacle avoidance actions, which generates better parking actions in shorter training timesteps.
(2) We build a simulation platform for automatic decision making, which satisfies the setting of unregulated parking lots, and we train and test the algorithm on this platformfacilitating simulation experiments and meeting environmental requirements.

2   Establish Parking Kinematics Model and Environment

The parking kinematic model captures vital vehicle information like position, speed, acceleration, and steering angle [12]. With an accurate model, the parking trajectory can be predicted precisely. The parking environment directly influences the planning complexity. This paper utilizes the Gym-based Highway-env environment [6] as a simulation platform, suitably modified to align with unregulated parking lots.

2.1   Parking Kinematics Model

Table 1. Simplified vehicle parameters.

Physical Quantity	Symbol	Parameter Value	Unit
Width	W	2	m
Wheelbase	L	3	m
Front Overhang	L_f	1	m
Rear Overhang	L_r	1	m
Body Steering Angle	ϕ	$[-30, 30]$	deg
Front Wheel Steering Angle	δ	$[-45, 45]$	deg

The vehicle is a complex system, so both the parking environment and the vehicle's internal transmission influence motion planning outcomes in the complex vehicle system. To address this, the vehicle's external profile is simplified as a rectangular rigid body. This preserves the vehicle's kinematic characteristics while preventing collisions with obstacles. Table 1 presents the simplified vehicle parameters, and Fig. 1(a) illustrates the simplified vehicle model.

Ackermann Steering Theory [14] is extensively employed in the design and manufacturing of modern automobiles. Due to the small disparity in turning angles between the inner and outer wheels on the front side of the car, the four-wheel car model can be simplified to a two-wheel model for analyzing parking kinematics, which is called Single-Vehicle Model [10]. This kinematics is depicted in Fig. 1(b).

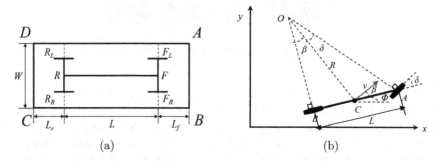

(a)   (b)

Fig. 1. (a)Simplified vehicle model. A, B, C, D represent the car body vertices. The projections of the wheel centers on the ground are denoted as F_L, F_R, R_L, R_R, and the ground projections of the two axes are represented by F and R. (b)Kinematic diagram of Single-Vehicle Model. A and B represent the front and rear points of the single vehicle model, C represents the vehicle's center of mass, and O represents the instantaneous center of rotation.

As the vehicle is typically propelled by the rear wheels and steered by the front wheels, we denote the steering angle of the rear wheels $\delta_f = 0$ and that of the front wheels $\delta_r = \delta$. R represents the instantaneous radius of rotation and β denotes lateral sway angle. x and y represent the two components of vehicle displacement in the ground coordinate system, while v and a denote the vehicle's velocity and acceleration respectively. With this, the kinematic equations of Single-Vehicle Model can be derived as

$$
\begin{cases}
\frac{dx}{dt} = v \cos(\beta + \phi) \\
\frac{dy}{dt} = v \sin(\beta + \phi) \\
\frac{dv}{dt} = a \\
\frac{d\phi}{dt} = \frac{2v \sin \beta}{L} \\
\beta = \arctan\left(\frac{1}{2}\delta\right)
\end{cases}
\tag{1}
$$

Single-Vehicle Model completes the control of the vehicle by the vehicle acceleration a and the front wheel steering angle δ. Therefore, accurately controlling these two parameters can effectively represent the vehicle's kinematics.

2.2 Setting of Unregulated Parking Lot

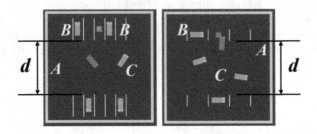

Fig. 2. Simulation environment of unregulated parking lot.

Figure 2 illustrates the simulation environment for an unregulated parking lot, consisting of green vehicle, yellow vehicles and squares, white lines, and blue dots. The **green vehicle** represents the controlled vehicle, aiming to park in target space. The **yellow vehicle and square** represent potential obstacles in the parking lot. The **white lines** indicate the boundary of the parking space, while the **blue dot** signifies an empty parking space, which serves as the target position for the green vehicle.

In this simulation environment, the 8m-long and 4m-wide parallel and vertical parking spaces are designed. The parking distance ($d = 20$ m) is calculated based on the minimum parking radius. Additionally, three types of parking lot obstacles are designed: A represents the parking lot boundary, B represents obstacles in parking spaces, and C represents obstacles in parking lot.

3 Algorithm Framework

To handle the continuous action space in parking motion, this paper adopts the Soft Actor Critic algorithm (SAC) [4] as the basic algorithm, known for its stability in training. The framework of SAC comprises 5 neural networks: 1 Actor network and 4 Critic networks. In addition, we use a curriculum learning approach to speed up and better the training process. Figure 3 illustrates our approach.

3.1 Curriculum Learning

Curriculum learning is an effective approach for humans to progress from simple to complex problem-solving. By systematically organizing lessons of increasing difficulty, complex knowledge can be effectively acquired. Jeffery Elman proposes using curriculum learning to train neural networks, which enhances model convergence and performance [8]. Thus, designing a set of incrementally challenging and easily learnable courses becomes crucial.

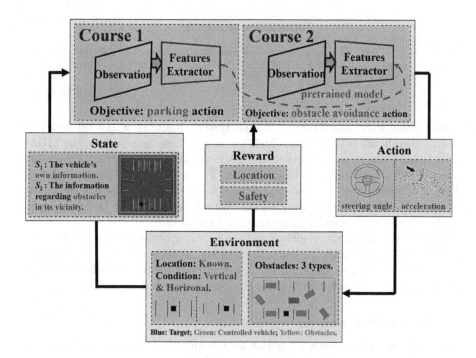

Fig. 3. Algorithm framework of our approach.

Therefore, we adopt a curriculum learning approach to accelerate and better the training process. The training process includes two courses. The objective of Course 1 is to make the vehicle learn to park in a parking lot without obstacles. Using the training results of Course 1 as the pretrained model, the objective of Course 2 is to make the vehicle learn to avoid obstacles by adding them to the parking lot. Each course is trained using the SAC algorithm. Further details on course settings are provided in Table 2.

3.2   State and Action Spaces

According to Eq.(1), the vehicle's control can be achieved by manipulating the vehicle acceleration a and the front wheel steering angle δ, resulting in an action space A represented as

$$A = \{a, \delta\}. \tag{2}$$

The state space S encompasses the vehicle's own information S_1 and the information regarding obstacles in its vicinity S_2. For the vehicle's information

S_1, we consider the position x, y, speed v_x, v_y, and heading angle ϕ as three sets of relevant data, forming the state space

$$S_1 = \{x, y, v_x, v_y, \cos\phi, \sin\phi\}. \tag{3}$$

Table 2. Course settings.

Course	Initial Pose		Obstruct Vehicles	Number of Parking Spaces		Obstacles in Parking Spaces		Obstacles in Parking Lot	
				Vertical	Horizontal	Vertical	Horizontal	Vertical	Horizontal
1	Fixed position	Arbitrary posture	No	10	6	0	0	0	0
2	Fixed position	Arbitrary posture	Yes	10	6	6	4	3	2

Regarding the obstacles S_2, a LIDAR system mounted on the vehicle monitors the surroundings. With the vehicle's center of mass as the origin, the LIDAR emits 16 laser beams uniformly within a 20-meter range. The corresponding state space for this component is a 16×2 matrix. The first column represents the distance of a nearest obstacle from the vehicle, as detected by a specific beam, and the second column indicates whether the obstacle is detected by that beam, with 0 indicating detection and 1 indicating no detection.

Full state space of the vehicle $S = S_1 \cup S_2$. After obtaining the observations, they need to pass through the features extractor, as shown in Fig. 3. The features extractor flattens extractsfeature vectors from high-dimensional observations to one 50-dimensional vector to facilitate learning through SAC network.

3.3   Reward Function

When formulating the reward function for the automatic parking planning task, we consider both **location**, which ensures the controlled vehicle consistently enters the parking space with the correct orientation, and **safety**, which prevents collisions with obstacles. The reward function is represented as

$$r = r_{\text{location}} + r_{\text{safety}}. \tag{4}$$

Location. Location r_{location} is crucial that enables the controlled vehicle to progressively approach the target parking space through training. r_{location} is a numerical value equal to or less than 0. When $r_{\text{location}} = 0$, it indicates that the vehicle has successfully parked in the predefined position.

$$r_{\text{location}} = -\|S_1 - S_g\|_{W,p}^{p}. \tag{5}$$

where S_1 is the vehicle's own information, S_g is the goal's information, $S_g = \{x_g, y_g, 0, 0, \cos\phi_g, \sin\phi_g\}$. Define the operator

$$\|x\|_{W,p} = \left(\sum_{i=1}^{k} |W_i x_i|^p\right)^{\frac{1}{p}}, \tag{6}$$

where $W = [1, 1, 0, 0, 0.02, 0.02]$.

Safety. During automatic parking, the vehicle detects obstacles based on S_2. In the event of a collision, a significantly low reward value is assigned, prompting the controlled vehicle to learn collision avoidance behavior.

$$r_{\text{safety}} = r_{\text{safe}} = \begin{cases} 0, \text{no collision} \\ -5, \text{ if collision} \end{cases} \tag{7}$$

The corresponding training procedure is shown in Algorithm 1.

Algorithm 1. SAC with Curriculum Learning

1: **INPUT**: initial policy parameters θ for each course, Q-function parameters ϕ_1, ϕ_2
    for each course, empty replay buffer $\mathcal{D}$, total number of course n.
2: **for** $n = 1, 2, \dots$ **do**
3:    Load the model trained in the previous course, and skip if $n = 1$.
4:    Use the SAC algorithm to update θ and ϕ_1, ϕ_2 in this course.
5: **end for**
6: **OUTPUT**: θ and ϕ_1, ϕ_2 from the last course.

---

4 Experiments

We perform experiments by using our method (called **Curriculum Learning RL**) in the unregulated parking lot depicted in Fig. 2 using a PC equipped with a 3080 graphics card. We utilize the parameters from Table 3 for the algorithm. To assess our approach's performance, we compare it with a **Baseline Algorithm**, which directly learns parking and obstacle avoidance actions in the unregulated parking lot using the SAC algorithm. Furthermore, after completing training, we design parking experiments under various parking conditions to verify the algorithm's generalization.

Table 3. Algorithm parameters.

Parameters	Values	Parameters	Values
policy	MultiInputPolicy	learning_rate	3×10^{-4}
buffer_size	10^6	batch_size	256
gamma	0.99	tau	0.005
network structure of SAC	2 layers, with 256 neurons each	activation function of network	ReLU

We conduct separate training for vertical and horizontal parking spaces using Curriculum Learning RL. The total duration of this algorithm is 3,150,000 timesteps, with Course 1 lasting 1,900,000 timesteps and Course 2 lasting 1,250,000 timesteps. As a contrast, Baseline Algorithm has a total duration of

5,000,000 timesteps. The evaluation metrics for training are **average reward** and **success rate**. Figure 4 shows the training progress for both algorithms.

Using Curriculum Learning RL, as depicted in Fig. 4, the vehicle learns parking action in Course 1 and obstacle avoidance action in Course 2 for both vertical and horizontal parking. In 3,150,000 timesteps, Curriculum Learning RL achieves higher average reward and success rate for parking actions compared to the Baseline Algorithm, demonstrating its effectiveness.

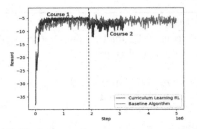 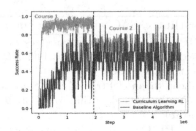

(a) Variation of average reward under vertical parking.

(b) Variation of success rate under vertical parking.

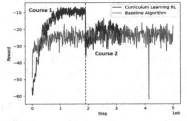 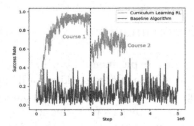

(c) Variation of average reward under horizontal parking.

(d) Variation of success rate under horizontal parking.

Fig. 4. Comparison between Curriculum Learning RL and Baseline Algorithm.

To validate the effectiveness of our Curriculum Learning RL algorithm, we test 1000 diverse parking scenarios for **vertical parking** on this platform. Each scenario features unique initial angle, target parking space, and obstacles. Comparisons are made against the Baseline Algorithm and **Curve Interpolation**. In Curve Interpolation, paths are fitted using fifth-degree polynomials. Evaluation metrics includes success rate, collision rate, and parking time. Table 4 demonstrates that Curriculum Learning RL algorithm achieves higher success rate, lower collision rate, and shorter parking time, affirming its effectiveness.

In addition, to assess the model's generalization, we conduct experiments in various parking conditions. Using Table 2 as reference, we modify the parking starting point to a random coordinate (x-axis: $[-6, 6]$ m, y-axis: $[-6, 6]$ m) in the parking coordinate system, with an arbitrary pose. Additionally, we introduce 4

Table 4. Evaluation metrics on different algorithms.

	Curriculum Learning RL	Baseline Algorithm	Curve Interpolation
success rate	79.8%	46.9%	39.1%
collision rate	20.2%	49.9%	50.2%
parking time	13.28 s	23.91 s	31.32 s

cars in the parking space and 6 cars in the parking lot, placed at random positions and poses. Figure 5 presents some test results, demonstrating the successful application of Curriculum Learning RL in diverse parking conditions with strong generalization.

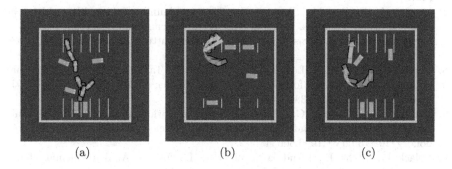

      (a)   (b)   (c)

Fig. 5. Results of the generalizability experiments.

5 Conclusion

In this paper, we propose a RL algorithm based on curriculum learning for automatic parking in unregulated parking lots. By decomposing the complex problem into multi-stage tasks with shorter-term rewards, and utilizing two courses for parking and obstacle avoidance, our method outperforms the Baseline Algorithm, which directly learns under the unregulated parking lots, in terms of higher reward functions and success rates within a shorter time. The experiments conducted under different parking conditions demonstrate the algorithm's generalization. Future work involves extensive simulations of vehicle kinematics and dynamics using a larger engine, as well as validation in real-world scenarios.

References

1. Al-Nussairi A K J, R.A.T.: Autonomous car driving using neural networks. In: International Conference on Applied Artificial Intelligence and Computing (ICAAIC), pp. 213–217 (2022)

2. Badue, C., et al.: Self-driving cars: a survey. Expert Syst. Appl. **165**, 113816 (2020). https://doi.org/10.1016/j.eswa.2020.113816

3. Ballinas, E., Montiel, O., Castillo, O., Rubio, Y., Aguilar, L.T.: Automatic parallel parking algorithm for a car-like robot using fuzzy PD+I control. Eng. Lett. **26**, 447–454 (2018)

4. Haarnoja, T., Zhou, A., Abbeel, P., Levine, S.: Soft actor-critic: off-policy maximum entropy deep reinforcement learning with a stochastic actor. In: Dy, J., Krause, A. (eds.) Proceedings of the 35th International Conference on Machine Learning. Proceedings of Machine Learning Research, vol. 80, pp. 1861–1870. PMLR (2018)

5. Junzuo, L., Qiang, L.: An automatic parking model based on deep reinforcement learning. In: Journal of Physics: Conference Series, vol. 1883, p. 012111. IOP Publishing (2021)

6. Leurent, E.: An Environment for Autonomous Driving Decision-making. GitHub (2018)

7. Macek, K., Becker, M., Siegwart, R.: Motion planning for car-like vehicles in dynamic urban scenarios. In: 2006 IEEE/RSJ International Conference on Intelligent Robots and Systems, pp. 4375–4380 (2006). https://doi.org/10.1109/IROS.2006.282013

8. Narvekar, S., Peng, B., Leonetti, M., Sinapov, J., Taylor, M.E., Stone, P.: Curriculum learning for reinforcement learning domains: a framework and survey. J. Mach. Learn. Res. **21**, 7382–7431 (2020)

9. Paromtchik, I., Laugier, C.: Autonomous parallel parking of a nonholonomic vehicle. In: Proceedings of Conference on Intelligent Vehicles, pp. 13–18 (1996). https://doi.org/10.1109/IVS.1996.566343

10. Polack, P., Altché, F., d'Andréa Novel, B., de La Fortelle, A.: The kinematic bicycle model: a consistent model for planning feasible trajectories for autonomous vehicles? In: 2017 IEEE Intelligent Vehicles Symposium (IV), pp. 812–818 (2017). https://doi.org/10.1109/IVS.2017.7995816

11. Thunyapoo, B., Ratchadakorntham, C., Siricharoen, P., Susutti, W.: Self-parking car simulation using reinforcement learning approach for moderate complexity parking scenario. In: 2020 17th International Conference on Electrical Engineering/Electronics, Computer, Telecommunications and Information Technology (ECTI-CON), pp. 576–579. IEEE (2020)

12. Wang, W., Song, Y., Zhang, J., Deng, H.: Automatic parking of vehicles: a review of literatures. Int. J. Automot. Technol. **15**, 967–978 (2014). https://doi.org/10.1007/s12239-014-0102-y

13. Zhang, J., Chen, H., Song, S., Hu, F.: Reinforcement learning-based motion planning for automatic parking system. IEEE Access **8**, 154485–154501 (2020)

14. Zheng, H., Yang, S.: Research on race car steering geometry considering tire side slip angle. Proc. Inst. Mech. Eng. P J. Sports Eng. Technol. **234**, 72–87 (2020)

15. Zhuang, Y., Gu, Q., Wang, B., Luo, J., Zhang, H., Liu, W.: Robust auto-parking: Reinforcement learning based real-time planning approach with domain template. In: 32nd Conference on Neural Information Processing Systems (NIPS 2018) (2018)

FAI: A Fraudulent Account Identification System

Yixin Tian, Yufei Zhang, Fangshu Chen$^{(\boxtimes)}$, Bingkun Wang, Jiahui Wang,
and Xiankai Meng

Shanghai Polytechnic University, Shanghai, China
`fschen@sspu.edu.cn`

Abstract. Fraudulent account detection is essential for businesses and online Internet enterprises, which can help to avoid financial loss and improve user experience. However, conventional solutions suffer from two main challenges which remain unresolved; first, it's hard to monitor and detect fraud behaviors in real-time, and second, the features of the cheaters keep changing dynamically, which makes it hard to capture the most relevant features for the detection models. In this demonstration, we present a fraudulent account identification system called FAI, which can help to address the above challenges by exploring a multi-granularity sliding window strategy to construct the dynamic features, and both dynamic and static features are embedded together as the input of pre-training models. FAI also provides an interface that allows users to select sets of features in the spatio-temporal dimension flexibly, visualize the feature aggregation results, and assess the quality of fraud detection results. Demo video click here.

Keywords: Fraud detection · Dynamic feature extraction ·
Anomalous data flow monitoring · Spark SQL

1 Introduction

Online Internet enterprises are now confronted with potential multi-party organized attacks, such as massive tests, fake registrations, and invitations. Fraud behaviors demonstrate characteristics such as large-scale, spatial, and temporal aggregation, which have seriously affected the security of enterprises' property and the experience of normal users [1,12]. Thus, detecting latent fraudsters has become a vital security task [2].

The rule-based decision-making approach and machine-learning inference approach are two mainstream fraudulent account detection methods. However, they both rely on domain knowledge and delicate feature engineering. Moreover, the rule-based decision-making approach must update promptly to adapt to the dynamically changing activities of fraudsters, and rules are easy to be found and avoid by massive tests. On the other hand, the changeable fraudulent

L. Fang et al. (Eds.): CICAI 2023, LNAI 14474, pp. 253–257, 2024.
https://doi.org/10.1007/978-981-99-9119-8_23

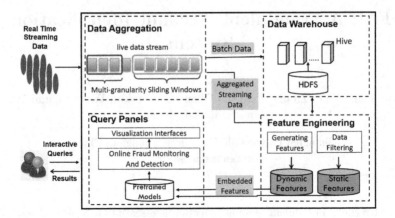

Fig. 1. FAI Architecture

activities could also cause degradation of machine-learning models [3,4]. Therefore, we design a **F**raudulent **A**ccount **I**dentifying (**FAI**) system to address the aforementioned challenges.

Users of FAI and conference audiences, both there could interact with the system in real-time to query and visualize various data analyses on fraudulent accounts, including the distribution of fraudulent accounts countrywide, a real-time view of the number of fraudulent and normal accounts, and the user can view the results of the classification of the given data through the model prediction interface. They can observe the aggregation of fraudulent accounts through a visual display. The key contributions in this paper are listed as follows:

- FAI encapsulates multiple types of feature construction methods, including dynamic feature and static feature construction strategies, which help practitioners discover complex anomalous aggregation patterns more quickly and comprehensively.
- FAI explores a multi-granularity sliding window strategy to construct the dynamic features in a variety of granularities (ranging from a second to a week) which can better adapt to both online fraud monitoring and offline prediction model training.
- FAI explores self-adaption strategies, such as incremental online learning Random Forest to deal with the degradation of models.
- FAI allows practitioners to casually combine multiple suspicious data fields to analyze the weird correlation of account information at the spatial and temporal levels.

2 Fai Architecture

Figure 1 depicts the architecture of FAI. It is composed of the following four main modules:

Data Aggregation. This module is responsible for aggregating the streaming data from the server, and a multi-granularity sliding window strategy is used to aggregate data in different periods for dynamic feature construction. Various pre-processing tasks are performed on the data, including filtering irrelevant data, filling in missing values, etc. [5,10].

Data Warehouse. This module partitions large amounts of historical data (derived from a large granularity sliding window) into smaller data cubes which can be processed efficiently in Spark RDDs in memory. We perform a partitioning operation that reorders the dates in the data and restores them in a partition, where each day corresponds to one partition [7]. The partitioned data will reduce the storage and indexing response time, allowing users to interact live with FAI more efficiently.

Feature Engineering. This module is designed to process the real-time streaming data derived from user customization or a sliding window in default granularity. See Sect. 3 for more details [9].

Query Panels. This module is user-oriented and supplies users with the choice of the functions to be performed.

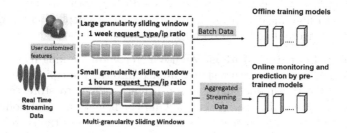

Fig. 2. Generation Features

3   Feature Engineering

Once receiving a command from the user, the module first reads the required data from memory, then aggregates the feature information selected by the user and constructs the relevant dynamic features. And then, these dynamic features will be transformed into Spark RDDs and processed by subsequent prediction models under a parallel computing framework. The module consists of two parts:

Generating Features. This module is designed to generate the corresponding dynamic features according to different granularity and also provides an interface for users to customize the window size. For example, suppose the user observes a more pronounced aggregation of fraudulent accounts in the *uid* and *phone* features in the *ip address* over an hour. In that case, the user can combine the two features and count the times the same field appears over the hour. The combined dynamic features are then stored in memory and used for later prediction models [11]. All feature analysis methods, including those integrated with temporal level (e.g., *request_time* combined with *user_id*), spatial level (e.g., *ip_province* combined with *user_id*), spatial-temporal level(e.g., *request_time* combined with *ip_province*) and other aggregations with casually multiple suspicious features, can better facilitate the observation of fraudulent account characteristics and dedicate the construction of useful features (Fig. 2).

Data Filtering. The main task of this part is to filter the static features and display them according to the user's selection. The static features allow the user to understand better a range of behavioral information about the fraudulent accounts, such as the fact that fraudulent accounts prefer to log in at night.

4   Prediction Models

FAI explored the following forecasting models, FAI integrates pre-trained models with online incremental learning models.

Pre-trained Models. We choose the decision tree-based models LightGBM and Random Forest, which work better when dealing with our data as decision trees can learn non-linear relationships and are easy to deploy on Spark. These models are trained offline based on large-scale historical data sets.

Incremental Learning Models. We use incremental learning methods where both static and dynamic feature splices are fed into the pre-trained model. Each leaf node in the tree model maintains a list of samples to store the samples categorized to that leaf node. When the Gini coefficient on a leaf node exceeds a set threshold, the current leaf node needs to be split and the stored samples are used to construct a new sub-incremental tree, thus extending the whole incremental tree with new features that work well for classification [6,8].

5   Demo Scenes

As shown in Fig. 3, the scenario of the system is divided into several parts, such as data visualization, feature generation, and real-time fraud monitoring and prediction. A detailed demonstration can be viewed in the video. Click here to watch.

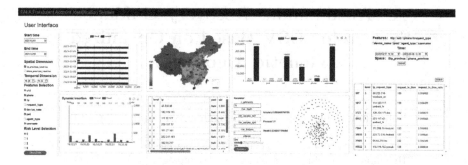

Fig. 3. User Interface of FAI

Acknowledgement. This work is supported by the National Natural Science Foundation of China (Grant No. 62002216), the Shanghai Sailing Program (Grant No. 20YF1414400), the Collaborative Innovation Platform of Electronic Information Master of Shanghai Polytechnic University (Grant NO. A10GY21F015), the Research Projects of Shanghai Polytechnic University (Grant No. EGD23DS05).

References

1. Abdallah, A., Maarof, M.A., Zainal, A.: Fraud detection system: a survey. J. Netw. Comput. Appl. **68**, 90–113 (2016)
2. Baesens, B., Höppner, S., Verdonck, T.: Data engineering for fraud detection. Decis. Support Syst. **150**, 113492 (2021)
3. Bera, D., Ogbanufe, O., Kim, D.J.: Towards a thematic dimensional framework of online fraud: an exploration of fraudulent email attack tactics and intentions. Decis. Support Syst. 113977 (2023)
4. Bierstaker, J.L., Brody, R.G., Pacini, C.: Accountants' perceptions regarding fraud detection and prevention methods. Manag. Audit. J. (2006)
5. Chang, V., Di Stefano, A., Sun, Z., Fortino, G., et al.: Digital payment fraud detection methods in digital ages and industry 4.0. Comput. Electr. Eng. **100**, 107734 (2022)
6. Gepperth, A., Hammer, B.: Incremental learning algorithms and applications. In: European Symposium on Artificial Neural Networks (ESANN) (2016)
7. Laender, A.H., Ribeiro-Neto, B.A., Da Silva, A.S., Teixeira, J.S.: A brief survey of web data extraction tools. ACM SIGMOD Rec. **31**(2), 84–93 (2002)
8. Luo, Y., Yin, L., Bai, W., Mao, K.: An appraisal of incremental learning methods. Entropy **22**(11), 1190 (2020)
9. Naeem, M., et al.: Trends and future perspective challenges in big data. In: Pan, J.S., Balas, V.E., Chen, C.M. (eds.) Advances in Intelligent Data Analysis and Applications. Smart Innovation, Systems and Technologies, vol. 253, pp. 309–325. Springer, Singapore (2022). https://doi.org/10.1007/978-981-16-5036-9_30
10. Salekshahrezaee, Z., Leevy, J.L., Khoshgoftaar, T.M.: The effect of feature extraction and data sampling on credit card fraud detection. J. Big Data **10**(1), 1–17 (2023)
11. Salloum, S., Dautov, R., Chen, X., Peng, P.X., Huang, J.Z.: Big data analytics on apache spark. Int. J. Data Sci. Anal. **1**, 145–164 (2016)
12. Zhang, Z., et al.: Temporal burstiness and collaborative camouflage aware fraud detection. Inf. Process. Manag. **60**(2), 103170 (2023)

An Autonomous Recovery Guidance System for USV Based on Optimized Genetic Algorithm

Lulu Zhou, Xiaoming Ye$^{(\boxtimes)}$, Pengzhan Xie, and Xiang Liu

School of Energy and Power Engineering,
Huazhong University of Science and Technology, Wuhan, China
xmye@hust.edu.cn

Abstract. As a kind of flexible and efficient device that can autonomously complete tasks without human intervention, Unmanned Surface Vehicle (USV) has gained increasing attention in the research field recently. Path planning is an essential hotspot in the study of the USV. Unlike traditional robotic path planning, the path planning of the USV needs to consider the dynamic impact of the water environment as well as the constraints of its own vessel's kinematics. For the sake of enhancing the practical operability of the navigation, an optimized Genetic Algorithm (GA) based on three-dimensional environment modeling is proposed. By simplifying the 3D coordinate, the algorithm can efficiently deal with the avoidance of dynamic and static obstacles. Population initialization is improved to reduce the calculating load, and the Elitism Strategy is combined to ensure convergence. An innovative Sacrifice Strategy and intraspecific hybrid methods are proposed to further increase the genetic diversity and convergence rate. We also propose a new penalty fitness function. Through simulation results and experiments in a water surface environment, the effectiveness and rationality of this method were verified, providing new ideas for path planning research of the USV.

Keywords: Unmanned surface vehicle · Genetic algorithm · Path planning · Evolution strategy

1 Introduction

Since the launch of the first lunar probe in 1959, an increasing number of intelligent astronomical devices have been on a mission to explore the vast and magnificent universe, while the journey of human beings continues to advance toward the deep and boundless oceans. With the upgrading and iteration of artificial intelligence, research in various fields such as meteorology and oceanography is gradually expanding in breadth as well as depth [6]. But the demand for environmental monitoring and its safety precautions is also increasing at the same time. As a new type of surface vessel, the unmanned surface vehicle (USV)

L. Fang et al. (Eds.): CICAI 2023, LNAI 14474, pp. 258–270, 2024.
https://doi.org/10.1007/978-981-99-9119-8_24

which offers the benefits of higher efficiency and lower casualty risk has raised worldwide interest [15].

However, the performance of the USV largely depends on its core systems, which are the navigation system, the control system, and the guidance system. Making use of the navigation system and the control system, real-time navigational data can be obtained and accurate motion control can be achieved [8]. Nonetheless, most of USVs are semi-automatic so they can be flexibly relied upon various types of missions. By adjusting the heading and speed, the guidance system assists in mission planning and execution for the USV and guides the USV to navigate along waypoints accurately. Based on the tasks' requirements, it provides effective and available path planning, allowing the USV to take proactive measures to prevent collisions. It should be noted that the path planning algorithm is the fundamental part of the guidance system, which was consisted of global path planning and local path planning [11].

Unlike maritime navigation, the guidance of the USV during the recovery situation doesn't consider the International Regulations for Preventing Collisions at Sea (COLREGS) for its immediate environment where close to the stern ramp of its large target ship without the interference of other vessels. In spite of this, how to operate collision avoidance remains a challenge for path planning because of the reefs and floats. After a comprehensive investigation of various algorithms, we choose the genetic algorithm (GA) as the global path planning and improvements are going to be made to ensure real-time collision avoidance and operation effectiveness. Inspired by the previous works [16,18], we propose a recovery guidance system based on optimized genetic algorithm, an upper computer software is created as well as the correspondence tracking program.

Our main contributions are summarized as followed:

(1) We propose the movement rules for the USV based on 3-D map information in order to optimize the dynamic collision avoidance and relief the computational load.
(2) We optimize the initialization process of the genetic algorithm and propose a new type of crypto theory that enhances the algorithm's iteration efficiency resulting in high-quality outcomes.
(3) We propose a novel genetic algorithm execution process and experiments of simulation and water surface to demonstrate the effectiveness of our method based on the smoothing and time costs aspects.

2   Related Work

In this section, we will discuss some other algorithms which have been applied in the guidance system of the USV and introduce the different methods to optimize the traditional genetic algorithm proposed by other researchers.

2.1   Path Planning Algorithms

The quality, efficiency, and convergence of path planning algorithms are crucial. Choosing suitable algorithms for different scenarios and leveraging their advantages is a major concern of researchers.

Dijkstra is a classical graph algorithm which is popular for its good robustness and easy implementation. But it has high spatio-temporal complexity and low efficiency. As another classical graph algorithm, A* algorithm has also become popular in recent years due to its excellent scalability [14,15].

Virtual Force Field (VFF) is a time-efficient approach suitable for dynamic collision avoidance, particularly in local path planning. But it often yields local optimal solutions. By modifying the physical model or combining it with other algorithms, VFF can overcome this limitation [13,17].

Random sampling methods have different performances in terms of convergence rate, making them widely used in Multi-Agent Path Finding (MAPF) due to their purposefulness. To enhance dynamic collision avoidance, Ouyang et al. [12] optimize the RRT algorithm, while Hou et al. [4] propose a penalty-based method considering multiple constraints and average sailing time.

For comparison with the genetic algorithm more intuitively, the presentative intelligence algorithm will be introduced in the next part.

2.2   Optimization Methods of Traditional Genetic Algorithm

Represented by Ant Colony Optimization (ACO), some intelligent algorithms tend to get trapped in local optimal solutions. Path planning has always been a multi-objective problem, not merely linear [5,7]. In complex conditions, the genetic algorithm outperforms others due to its unique selection mechanism, which enables it to quickly search for multiple target results.

Despite this, the traditional genetic algorithm still has a low convergence rate for path planning of USVs. Wang et al. [16] combine the genetic algorithm with fuzzy APF to develop a hierarchical path planning method that generates optimally sparse waypoints and smooth paths for unpredictable environments. However, it remains challenging when dealing with time-varying dynamic obstacles and unforeseen marine scenarios. To overcome inherent shortcomings like premature population and slow convergence, Xin et al. [18] propose a strategy to increase the number of advantageous offspring using multi-domain inversion and conducting a second fitness evaluation. However, this method adds complexity to the algorithm and its encoding process.

Based on a reference from task relation networks [20,21], we propose a novel 3D modified genetic algorithm giving a new encoding strategy and upgraded genetic manipulations to improve the convergence rate and quality of planning.

3   Method

3.1   Overview

The guidance system we propose is able to find an optimal unobstructed path based on the start and end points entered by the upper computer, in combination with GPS and other navigation devices. As shown in Fig. 1, with the support of the data transceiver, effective communication can be established between the navigation system and the guidance system for smooth route planning and execution of the USV.

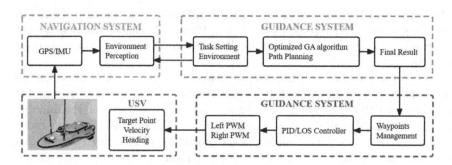

Fig. 1. Communication architecture of the main control system of our USV.

The optimized genetic algorithm, which incorporates the time dimension into geographic coordinates, enhances real-time communication and promotes the feedback mechanism between systems. Computation time is reduced by optimizing coding mechanisms and improving genetic manipulations. The multidimensional evaluation of the fitness function provides high-quality path results and reduces the tracking burden on the control system.

3.2   Map and Avoidance

In genetic algorithms, the distribution of waypoint locations is random and it takes time to iterate. On the other hand, modeling with the ordinal grid method requires a large amount of memory and has to consider irregular edges of the experimental environment during sequencing [1]. To overcome these issues, we propose using a three-dimensional Cartesian Coordinate System to generate waypoints, which reduces the code size by more than half and increases the convergence rate.

We represent all obstacles as circular bounding boxes. Taking into account the irregularity of the obstacles and the fact that the USV itself has a volume, which was considered as a prime point during planning, we compensate the radius of the boxes as shown in Fig. 2(a). The radius compensation could be calculated by:

$$R_l = R_{obs} + D_{USV} + R_c \tag{1}$$

where R_l denotes the minimum radius of collision avoidance distance from the USV to the center of the circular bounding box, R_{obs} represents the initial radius of the box, and the whole compensation radius is the length of the hull D_{USV} plus the possible displacement R_c caused by environmental interference which is usually taken as 1 m.

Introducing the time dimension to environmental modeling, we implement collision avoidance strategies for both static and dynamic obstacles [19]. The expression of the obstacles can be given by:

$$(X - X_{obs} + V_x t)^2 + (Y - Y_{obs} + V_y t)^2 = R_{obs}^2 \tag{2}$$

where X_{obs} and Y_{obs} denote the abscissa and ordinate of the center of the circular bounding boxes, R_{obs} represents its radius, V_x and V_y denote the velocity in x-direction and velocity in y-direction. When it comes to the static obstacles, $V_x = 0\,\text{m/s}$, $V_y = 0\,\text{m/s}$.

For static obstacles, the distance is calculated from the center of the circular bounding box to the segment of the navigation line formed by the current and the previous waypoint. This computed distance is then compared with the minimum collision radius. If the distance is greater than the minimum collision radius, then the collision will be avoided. This procedure is illustrated in Fig. 2(b). For dynamic obstacles, the planned recovery path is considered as a uniform acceleration process, and the group of equations is solved to determine whether there are positive real solutions with respect to T to tell if a collision will occur. The equation group is expressed as follows:

$$\begin{cases} X = X_{USV} + \overline{V} \cdot \cos \theta \cdot T \\ Y = Y_{USV} + \overline{V} \cdot \cos \theta \cdot T \\ \overline{V} = \frac{V_i + V_{i+1}}{2} \\ (X - X_{obs} + V_{obs,x} \cdot T)^2 + (Y - Y_{obs} + V_{obs,y} \cdot T)^2 = R_{obs}^2 \end{cases} \tag{3}$$

where θ denotes the heading of the USV, and $\overline{V}$ presents the average velocity of the current segment, of which the index is i.

During the experiments, we preset four static obstacles and four dynamic obstacles, and how they are represented in the three-dimensional Cartesian coordinate system are shown in Fig. 2(c).

3.3   Encoding and Population Initialization

We define the recovery environment as a three-dimensional map with a unit distance of 1m. During design, individuals' gene loci are designated as Point class, which stores the three-dimensional waypoint coordinates (X, Y, T) at real-time speed. We formulate the X-axis direction as the direction of the step and determine the range of intervals generated for each waypoint by the length of step size D_{SL}, which is different according to various scenarios and the accuracy requirements. The number of gene loci of each individual N can be calculated by:

$$N = \frac{X_{PS} - X_{PE}}{D_{SL}} + 1 \tag{4}$$

where X_{PS} and X_{PE} denote the abscissa of the start point and the end point in USV's mission. Then we take the reminder of the horizontal distance between start and end points, which is the compensation interval D_C, as shown in Eq. 5:

$$D_C = (X_{PE} - X_{PS}) \backslash D_{SL} \tag{5}$$

and based on the performance of USV during the experiments, we determine D_{SL} to be 10 m.

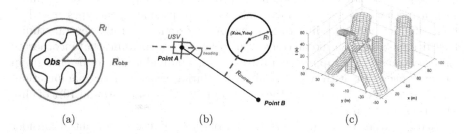

(a)   (b)   (c)

Fig. 2. Detail parts of the map modeling. (a) Example of the circular bounding box for obstacle; (b) Avoidance strategy for static obstacles; (c) The default obstacles for our simulation and field trial in three-dimensional Cartesian coordinate system.

To reduce inconsistencies in navigation such as 'unnecessary' jags and large span, we omit the start and end points of each interval when generating way-points. To address real number overflow resulting from genetic mutations during binary code decompilation, we design a function that maps the horizontal coordinates of the waypoints to a range between 0 and 7. The ordinate and velocity are also processed in a similar manner. The specific mapping process is shown in Eq. 6:

$$\begin{cases} X_{tr} = X_i - D_{SL} \cdot (i-1) - 1 \\ Y_{tr} = Y_i + 64 \\ V_{tr} = V_i \cdot 5 \end{cases} \tag{6}$$

where X_{tr} and X_i denote the abscissa of binary system and decimal system. Ordinate and velocity are presented in a similar way.

After each round of variation, the corresponding time for the related waypoints is updated. Each navigation segment is considered as a uniformly accelerated process, and T_i can be calculated is by:

$$T_i = \sum_{n=1}^{i-1} \frac{\sqrt{(X_{n+1} - X_n)^2 + (Y_{n+1} - Y_n)^2} \cdot 2}{V_n + V_{n+1}} \tag{7}$$

where i is the index of currently updated waypoint.

During population initialization, the generation of waypoints is not performed randomly. Instead, an obstacle collision avoidance check is executed within the loop for preparing individuals, to ensure that the initial paths exhibit desirable quality. It is worth noting that, due to the potential for unsolvable calculations, the threshold for a single individual loop is set alternately. And the coordinates are randomly assigned if the loop exceeds 1000 times in case of a program crash.

3.4 Improved Genetic Manipulation

The traditional genetic algorithm is not completely globally convergent. During genetic variation, individuals will undergo crossover and mutation, resulting in gene sequence changes and modification of their phenotypes. However, under such a mechanism, genetic individuals with exceptional phenotypes may lose their superior genes, leading to a reduction in search speed [9]. Several studies have demonstrated the global convergence of genetic algorithms that incorporate Elitism Strategy. This approach involves copying individuals from the elite group to next generation with a probability of 1 during every genetic manipulation, thereby continuously updating the elite group [2]. Generally, around 10% of elite sequences are retained.

In order to ensure optimal performance of the USV, it is important to consider various factors such as heading angle, rotational speed, and water environment conditions. To address these concerns, we propose a multi-penalty mechanism that takes into account factors such as time, distance, and orientation [10]. Furthermore, we suggest implementing a hard kinetic evaluation index to determine whether a particular gene should be eliminated [3]. Quantitative indicators and fitness can be calculated by:

$$
\begin{cases}
Fitness\,(x) = G_{acc}G_{vel}G_{col}\dfrac{A}{a\cdot Fit_d + b\cdot Fit_T + c\cdot Fit_a + d\cdot Fit_m} \\[2mm]
Fit_d = \dfrac{\sum D_{segment,i}}{\sqrt{(x_{end}-x_{start})^2+(y_{end}-y_{start})^2}} \\[2mm]
Fit_T = \dfrac{\sum \Delta T_{segment,i}}{N-1} \\[2mm]
Fit_m = \dfrac{\sum\limits_{i=1}^{N} |y_i|}{N} \\[2mm]
Fit_a = \dfrac{\sum\limits_{i=1}^{N} (\pi-\theta)}{\pi}
\end{cases}
\tag{8}
$$

where Fit_d, Fit_T, Fit_m, and Fit_a denote the evaluation functions for distance, time cost, malposition, and turning angle, respectively. And G_{acc}, G_{vel}, and G_{col} present the hard kinetic evaluation index of acceleration, velocity, and collision avoidance. a,b,c,d are parameters set by researcher.

In the process of evolution, unqualified individuals can take up computational memory, resulting in longer execution times and reduced benefits for real-time planning. To solve this issue, we suggest a new approach called the sacrifice strategy. This method imitates natural processes such as birth, aging, illness, and death of individuals to enhance the algorithm's convergence. After that, the memory of the eliminated individuals is released for the generation of new individuals. The new population will undergo intraspecific hybridization in the next generation. Moreover, this strategy ensures that the generated recovery planning paths are practical and functional. By improving convergence speed while maintaining practicality, our proposed strategy offers an efficient solution to this problem.

4   Experiments

In this section, we will introduce the simulation and field experiments using our method. We will first introduce the settings and implement details. Then the comparison with the traditional genetic algorithm will be presented. At the last, the result of ablation study will be shown to compare the contribution of various components of our algorithm.

4.1   Settings and Implement Details

The experimental simulation is conducted in a Windows 10 Professional 64-bit operating system using MATLAB R2018a software and Visual Studio 2022 C++ programming language. The virtual environment for the USV's operation is represented by a 100×128 raster map, which is symmetrical to the stern ramp. However, a final generated path map of 100×100 is sufficient. The genetic algorithm is implemented with the following operating constants: a population size of 100, a maximum number of iterations of 300, a tournament competition group of 2 genes, a crossover probability of 0.6, and a single point mutation probability of 0.01. The fitness function parameters are set as follows: $A = 150$, $a = 1.75$, $b = 0.05$, $c = 4.50$, $d = 1.50$.

The details of circular bounding boxes are as shown in Table 1.

Table 1. Details of the dynamic and static obstacles.

Type of Obstacles	X(m)	Y(m)	Radius(m)	V_X(m/s)	V_Y(m/s)
Dynamic	12	25	7	-0.2	0.5
	55	25	10	-0.2	0.5
	85	35	6	-1.0	-0.2
	10	-36	10	0.5	0.5
Static	25	10	6	0	0
	40	-18	8	0	0
	70	-30	10	0	0
	90	-15	8	0	0

The end point of the axis of the stern ramp, which we have identified as the coordinate origin, serves as the endpoint for our planning. The starting point can be entered directly into our self-designed software through the upper computer. We have designed four cases with different characteristics to analyze the impact of the optimized genetic algorithm on the length of the journey and the difference in longitudinal distance between the starting point and the end point. The four coordinates are as follows: (85, 10), (55, −15), (43, 6), and (92, 20).

After generating the paths, for testing the integrity and consistency of the paths, the compiled code is then transferred to the T30 upper computer. The performance of the field experiment is shown in Fig. 3(a) and components of the hardware are illustrated in Fig. 3(b).

(a)   (b)

Fig. 3. (a) USV during the field test on the lake; (b) Hardware components of our USV.

4.2   Comparisons

To control the variables, we implemented the Simple Genetic Algorithm (SGA) code on a 3D map model and conducted a comparative analysis using proposed quantitative metrics. The resulting path diagram is displayed in Fig. 4.

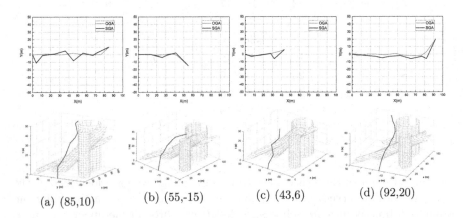

(a) (85,10)   (b) (55,-15)   (c) (43,6)   (d) (92,20)

Fig. 4. The comparative results of path planning using SGA and OGA and the planning path in 3D coordinate using OGA where t indicates time remaining.

The longitudinal distribution of waypoints in SGA appears to be more scattered as a result of the higher randomness present in the initialization process.

Specifically, when aligning with the stern ramp, the path generated by SGA exhibits a lack of gradual approach towards the central axis of the stern ramp. This characteristic hinders the execution of recovery maneuvers. The comparative results of the quantitative indicators are presented in the Table 2 where the red values indicate worse results and the comparison results of time cost are shown in Table 3.

Table 2. Comparative results of quantitative indicators (O-OGA, S-SGA).

Case	Fit_{a_O}	Fit_{a_S}	Fit_{m_O}	Fit_{m_S}	Fit_{d_O}	Fit_{d_S}	Fit_{T_O}	Fit_{T_S}
1	0.48	2.63	1.30	4.10	1.05	1.32	5.18	5.75
2	0.42	1.26	2.57	3.71	1.08	1.15	5.75	5.71
3	0.37	1.22	2.17	3.00	1.03	1.23	4.96	5.08
4	1.44	1.55	4.55	4.82	1.19	1.25	5.09	4.92

Table 3. Comparison of the time consumption for iterations between OGA and SGA.

Case	$T_{SGA}(s)$	$T_{OGA}(s)$	Improvement (%)
1	2.067	10.846	80.9423
2	2.683	8.561	68.6602
3	1.556	7.218	78.4428
4	3.792	13.074	70.9959

In terms of malposition, OGA performs better in cases 1 and 4 due to the complex planning in length and vertical difference. Although the indicators of time show OGA does better, they do not differ much in overall performance. And in each case, the convergence time of SGA is significantly longer than that of OGA, and this time difference between the two algorithms increases as the planning distance and the vertical difference between the start and end points increase.

After the field test, the actual path is further fitted with the β-sample curve. A comparison between the actual path and the planning results is made to evaluate the smoothness and consistency of the planning process under the influence of the water environment and the constraints imposed by the dynamic conditions of the USV. The specific experimental results are shown in Fig. 5.

It can be seen that the diversion disadvantage of the actual path is magnified. And it also takes some time to return to the original planned route after the track on one side of the corner which is shifted, so there is a greater diversion disadvantage compared to the planned path. Additionally, the planned path

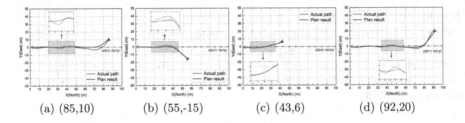

(a) (85,10)   (b) (55,-15)   (c) (43,6)   (d) (92,20)

Fig. 5. The comparative results of simulation and field tests.

shows a significant increase in length due to unexpected corners, failing to converge within the desired range as intended. That is because the water surface has interference and the USV's automatic dynamic tuning needs the contribution of time and distance. Generally, OGA demonstrates commendable performance in terms of the smoothness and consistency of the actual path.

4.3   Ablation Study

Since population initialization and fitness evaluation systems impact the algorithm framework, for the ablation study, we exclude elitism and sacrifice strategies in genetic manipulation to evaluate their impact on algorithm performance. Based on the results in Table 4, the complete algorithm structure demonstrates advantages in time consumption, distance, malposition, and path smoothing for path planning. Particularly, path length and smoothness are significantly optimized. This is due to the improvement in genetic manipulation which facilitates a better distribution of waypoints, thereby impacting path length and smoothness to a greater extent.

Table 4. Comparative results of the average values of quantitative indicators.

Models	Baseline	Without SS	Without ES	Ours
Fit_a	2.42	0.90	1.36	0.68
Fit_m	3.91	2.97	2.69	2.65
Fit_d	1.24	1.12	1.12	0.68
Fit_T	5.37	5.42	5.36	5.25

5   Conclusion

In this paper, we propose an optimized genetic algorithm based on 3D environment and build a navigation system for the USV. To address the problems of traditional genetic algorithms with high memory consumption and slow convergence, we determine a new coding mechanism, introduce an Elitism Strategy, and propose a sacrifice strategy to improve the genetic operation. In setting the fitness function, a multidimensional evaluation index is used to more closely match the recovery environment for the stern ramp. After simulation and field tests, the effectiveness of the method is demonstrated. It has the reference value under the influence of a realistic water surface environment.

References

1. Akshya, J., Priyadarsini, P.L.K.: Graph-based path planning for intelligent UAVs in area coverage applications. J. Intell. Fuzzy Syst. **39**(6), 8191–8203 (2020)
2. De Jong, K.A.: An analysis of the behavior of a class of genetic adaptive systems. University of Michigan (1975)
3. Deb, K.: An efficient constraint handling method for genetic algorithms. Comput. Methods Appl. Mech. Eng. **186**(2–4), 311–338 (2000)
4. Hou, Y.Q., Tao, H., Gong, J.B., Liang, X., Zhang, N.: Cooperative trajectory planning for unmanned boat and UAV clusters under multiple constraints. Chin. Ship Res. **16**(1), 74–82 (2021)
5. Lazarowska, A.: Multi-criteria ACO-based algorithm for ship's trajectory planning. TransNav: Int. J. Marine Navigat. Saf. Sea Transp. **11**(1) (2017)
6. Lazarowska, A.: Research on algorithms for autonomous navigation of ships. WMU J. Marit. Aff. **18**(2), 341–358 (2019)
7. Li, S., Luo, T., Wang, L., Xing, L., Ren, T.: Tourism route optimization based on improved knowledge ant colony algorithm. Compl. Intell. Syst. **8**(5), 3973–3988 (2022)
8. Liu, W., Liu, Y., Bucknall, R.: A robust localization method for unmanned surface vehicle (USV) navigation using fuzzy adaptive Kalman filtering. IEEE Access **7**, 46071–46083 (2019)
9. Medeiros, D.R.d.S., Fernandes, M.A.C.: Distributed genetic algorithms for low-power, low-cost and small-sized memory devices. Electronics **9**(11), 1891 (2020)
10. Ni, S., Liu, Z., Cai, Y., Wang, X.: Modelling of ship's trajectory planning in collision situations by hybrid genetic algorithm. Polish Maritime Res. **25**(3 (99)), 14–25 (2018)
11. Nunia, V., Poonia, R.C.: A review and comparative study on surface vehicle path planning algorithm. In: Proceedings of the International Conference on Data Science, Machine Learning and Artificial Intelligence, pp. 106–109 (2021)
12. Ouyang, Y., Wang, Z., Huang, X., Yang, L.: Unmanned boat formation path planning technique based on improved RRT algorithm. Chin. Ship Res. **15**(3), 18–24 (2020)
13. Sang, H., You, Y., Sun, X., Zhou, Y., Liu, F.: The hybrid path planning algorithm based on improved a* and artificial potential field for unmanned surface vehicle formations. Ocean Eng. **223**, 108709 (2021)
14. Singh, Y., Sharma, S., Sutton, R., Hatton, D., Khan, A.: A constrained a* approach towards optimal path planning for an unmanned surface vehicle in a maritime environment containing dynamic obstacles and ocean currents. Ocean Eng. **169**, 187–201 (2018)
15. Song, R., Liu, Y., Bucknall, R.: Smoothed a* algorithm for practical unmanned surface vehicle path planning. Appl. Ocean Res. **83**, 9–20 (2019)

16. Wang, N., Hongwei, X., Li, C., Yin, J.: Hierarchical path planning of unmanned surface vehicles: a fuzzy artificial potential field approach. Int. J. Fuzzy Syst. **23**, 1797–1808 (2021)

17. Zhenyu, W., Guang, H., Feng, L., Jiping, W., Liu, S.: Collision avoidance for mobile robots based on artificial potential field and obstacle envelope modelling. Assem. Autom. **36**(3), 318–332 (2016)

18. Xin, J., Zhong, J., Yang, F., Cui, Y., Sheng, J.: An improved genetic algorithm for path-planning of unmanned surface vehicle. Sensors **19**(11), 2640 (2019)

19. Xu, X., Cai, P., Ahmed, Z., Yellapu, V.S., Zhang, W.: Path planning and dynamic collision avoidance algorithm under colregs via deep reinforcement learning. Neurocomputing **468**, 181–197 (2022)

20. Zhao, F., Zhao, J., Yan, S., Feng, J.: Dynamic conditional networks for few-shot learning. In: Ferrari, V., Hebert, M., Sminchisescu, C., Weiss, Y. (eds.) ECCV 2018. LNCS, vol. 11219, pp. 20–36. Springer, Cham (2018). https://doi.org/10.1007/978-3-030-01267-0_2

21. Zhao, J., Li, J., Zhao, F., Yan, S., Feng, J.: Marginalized CNN: learning deep invariant representations (2017)

UAV Path Planning Based on Enhanced PSO-GA

Hongbo Xiang[1,2], Xiaobo Liu[1,2(✉)], Xinsheng Song[1,2], and Wen Zhou[1,2]

[1] Key Laboratory of Geological Survey and Evaluation of Ministry of Education, China University of Geosciences, Wuhan 430074, China
{hbxiang,xbliu,xinsheng_song,1202211346}@cug.edu.cn
[2] School of Automation, China University of Geosciences, Wuhan 430074, China

Abstract. Path planning for unmanned aerial vehicles (UAV) is a key technology for UAV intelligent system in the aspect of model construction. In order to improve the rapidity and optimality of UAV path planning, we propose a hybrid approach for UAV path planning in 2D environment. First, an enhanced particle swarm optimization algorithm (EPSO) combine with genetic algorithm (GA) which named as EPSO-GA is utilized to obtain the initial paths of UAV. In EPSO-GA, a hybrid initialization of Q-learning and random initial solutions is adopted to find the better initial paths for the UAV, which improves the quality of initial paths and accelerates the convergence of the EPSO-GA. The acceleration coefficients of EPSO-GA are designed as adaptive ones by the fitness value to make full use of all particles and strengthen the global search ability of the algorithm. Finally, the effectiveness of the proposed algorithm is proved by the experiments of UAV path planning.

Keywords: UAV path planning · PSO-GA · Hybrid initialization · Q-learning

1 Introduction

Unmanned aerial vehicle (UAV) path planning refers to find a path between the starting position and the destination under the conditions of terrain, radar, and other factors [1]. In last decades, many methods have been proposed to solve the issue of UAV path planning in 2D environment, which includes Graph-based algorithm [2,3], Rapidly-exploring Random Trees algorithm (RRT) [4], artificial potential field algorithm (APF) [5,6], and reinforcement learning [7]. Population-based algorithms are the most commonly methods used to plan path for the UAV, such as genetic algorithm (GA) [8,9], particle swarm optimization (PSO) [10], and ant colony algorithm (ACO) [11,12]. This kind of algorithms

This work was supported by the National Natural Science Foundation of China (61973285, 62076226), Hubei Provincial Natural Science Foundation of China (2022CFB438), the Opening Fund of Key Laboratory of Geological Survey and Evaluation of Ministry of Education (Grant No. GLAB 2023ZR08).

L. Fang et al. (Eds.): CICAI 2023, LNAI 14474, pp. 271–282, 2024.
https://doi.org/10.1007/978-981-99-9119-8_25

has the advantages of fast convergence speed, good parallelism and facilitate collaboration among multiple populations.

Compared to other algorithms, PSO is often utilized in the path planning of UAV. However, conventional PSO is confronted with several challenges. Firstly, the initialization process of the particles is complex, leading to suboptimal initial paths. Secondly, these algorithms exhibit slow convergence rates and are prone to getting trapped in local optima. In order to solve the above problems, we propose an EPSO-GA, and furthermore adopted it on the UAV path planning. The main contribution of this paper can be summarized as follows.

(1) A hybrid approach of Q-learning and random initial solutions is applied to find the initial paths for the UAV, which improves the quality of initial paths and accelerates the convergence of the EPSO-GA.

(2) The acceleration coefficients of EPSO-GA are designed as adaptive ones by the fitness value to make full use of all particles and strengthen the global search ability of the algorithm. In addition, a heuristics factor is recommended into mutation to speed up the convergence of the algorithm.

2   The Proposed Method

In this section, we construct the objective function for UAV path planning. To enhance the quality of the initial population in the particle swarm algorithm, we introduce Q-learning for hybrid initialization. Furthermore, we accelerate the convergence speed and improve the global search capability of the algorithm by incorporating mutation and crossover mechanisms.

2.1   Objective Function Construction

In this article, path length is defined as follows:

$$f_L = \sum_{i=1}^{n} \sqrt{(x_{i+1} - x_i)^2 + (y_{i+1} - y_i)^2} \tag{1}$$

where x_i and y_i are the coordinates of path point i, and n is the total number of path points.

To simplify the calculation, the path from the starting point to the destination is divided into three segments. For each segment, the 1/4, 2/4 and 3/4 points of the outbound leg are used as reference points. The formula for calculating the threat intensity of the jth radar against the drone is as follows:

$$T_j = \sum_{m=1}^{3} K \left(\frac{1}{d_{1/4,m,j}^4} + \frac{1}{d_{2/4,m,j}^4} + \frac{1}{d_{3/4,m,j}^4} \right) \tag{2}$$

where m represents the path segment, K is the threat intensity coefficient, $\frac{1}{d_{1/4,m,j}}$ represents the distance between the jth radar and a quarter of the m path segment, and T_j represents the threat intensity of the jth radar against the drone. Therefore, the threat intensity of a single drone is described as follows:

$$f_T = \sum_{j=1}^{N} T_j \tag{3}$$

where N is the number of radars.

In order to reduce computational complexity, the path planning problem for UAV is transformed into a constrained multi-objective function. By assigning weights to each objective, transforming the multi-objective problem into a single-objective problem. The objective function can be represented as follows:

$$f_{obj} = w_1 * f_L + w_2 * f_T \tag{4}$$

where w_1 and w_2 are the weights for path length and threat intensity respectively.

2.2 Initialization Method of PSO Particle Swarm Based on Q-Learning

The hybrid initialization method of Q-learning and random initial population is used in this paper to obtain the initial swarm of particles. In order to ensure that the paths generated by Q-learning have lower threat intensity and shorter path length compared to those generated randomly, the reward function of Q-learning is shown in Table 1.

Table 1. Design of reward function.

State of UAV	Reward
Starting point	0
End position	20
Obstacles and radar	-1
Feasible point	r

In Q-learning, all feasible points, obstacles and radars are distinguished in the form of positive and negative values. Each cell represents a different value, where 0 represents the starting point, r represents free space, -1 represents an obstacle or radar occupied cell, and 20 represents the destination. The reward values of all feasible points are adaptively adjusted based on their positions, where $r > 0$. The reward value of feasible points increases as they approach the destination and decreases as they approach a threat. The reward values of all feasible points are shown below:

$$r = r_0 + r_{\max} * \exp\left[\frac{-\sqrt{(x_u - x_g)^2 + (y_u - y_g)^2}}{k_g}\right]$$

$$-r_t * \exp\left[\frac{-\sum_{i=1}^{m}\sqrt{(x_u - x_{it})^2 + (y_u - y_{it})^2}}{k_t}\right] \tag{5}$$

where r_0 represents the basic reward value, $r_{\max}$ represents the maximum reward value without considering threats, r_t is the threat coefficient, (x_u, y_u) and (x_g, y_g) are the coordinates of the UAV and the target point respectively, (x_{it}, y_{it}) is the coordinates of the ith radar, k_g and k_t are the control coefficients of the reward function and m is the number of threats.

The balance between exploration and exploitation is another key part in Q-learning. This paper adopts Boltzmann distribution [13], and its expression is as follows:

$$p(a \mid s) = \frac{e^{Q(s,a)/T}}{\sum_{a_i \in A} e^{Q(s,a_i)/T}} \tag{6}$$

$$T = \lambda^k T_0 \tag{7}$$

where $p(a \mid s)$ represents the probability that action a is selected in state s. λ is a constant satisfying $0 < \lambda < 1$, k is the current iteration number and T is a control parameter. At the beginning of the training process, T has a large value to ensure strong exploration capability, which decreases as the number of iterations increases to ensure that the algorithm focuses on exploitation. The update formula for Q values is as follows:

$$Q(s,a) \leftarrow Q(s,a) + \alpha\left[r + \gamma \max Q\left(s', a'\right) - Q(s,a)\right] \tag{8}$$

where α is the learning rate, γ is the discount factor, r is the immediate reward value of the current action and $Q(s, a)$ is the estimated value of taking the current action.

Using the Q-learning method based on the fitness values of all solutions, m paths are generated, and n paths are obtained by random initialization. The initial population size $S = m + n$.

2.3   PSO Mutation Crossover Strategy

Inspired by the way of multiple UAV task assignment in [14], the updating strategy of EPSO-GA can be defined as follows:

$$x_{i,j}(t+1) = c_2 \bullet f_3\left(c_1 \bullet f_2\left(w \bullet f_1\left(x_{i,j}(t), p_{i,j,\text{ best}}(t)\right), g_{i,j,\text{ best}}(t)\right)\right) \tag{9}$$

where $x_{i,j}(t)$, $p_{i,j,best}(t)$ represent the position and personal optimal value of particle in t iterations respectively, while $g_{i,j,\text{ best}}(t)$ represents the global best value of the particle in the t-th generation. w is the learning factor of the particle with respect to itself, c_1 and c_2 are the learning factors of the current particle with respect to its individual best value and global best particle, respectively. f_1 is the operation of the particle with respect to itself, while f_2 and f_3 are the operations of the particle based on its individual best value and global best value, respectively.

f_1 is defined as a mutation where the mutation probability of the particle i is w, f_1 is formulated as follows:

$$f_1 = w \bullet f_1\left(x_{i,j}(t)\right) \tag{10}$$

where w is the inertia weight which has a great influence on performance of the algorithm. In order to accelerate the convergence of the algorithm, we adopt the linear time-varying inertia weight updating strategy in [15], which is expressed in the following formula:

$$w(t) = \frac{T-t}{T}\left(w_{\max} - w_{\min}\right) + w_{\min} \tag{11}$$

where the value of w is linearly decreased from $w_{\max}$ to the final value $w_{\min}$. t is the current iteration of the algorithm and maxiter is the maximum iterations of the algorithm. At the beginning of the iteration, the particle exhibits a strong global search ability, while towards the end of the iteration, it acquires a local search capability.

As shown in the Fig. 1(a), the mutation operation begins by randomly selecting a point in the path (excluding the start and end points). Assuming the previous coordinate is N, there are 8 possible choices for the mutated coordinate. To improve the efficiency of search, the distance D_i between each candidate i and the destination is calculated, and the probability P_i of the i being selected is calculated as follows:

$$P_i = k_i * \frac{1/D_i}{\sum_{i=1}^{8} 1/D_i} \tag{12}$$

where k_i is the adjusted adaptively according to the angle between coordinate i and the destination. In particular, the probability is 0 when i represents an obstacle.

To address the coordinate transformation issue, this paper introduces the crossover mechanism from genetic algorithms while retaining the individual and global learning strategies of the particle swarm.

f_2 is defined as the intersection between particle $x_{i,j}$ and the individual best particle with a probability of c_1. The formula for f_2 is as follows:

$$f_2 = c_1 \bullet f_2\left(f_1, p_{i,j,\text{ best}}(t)\right) \tag{13}$$

f_3 is defined as the intersection between particle $x_{i,j}$ and the global best particle with a probability of c_2. The formula for f_3 is as follows:

$$f_3 = c_2 \bullet f_3\left(f_2, g_{i,j,best}(t)\right) \tag{14}$$

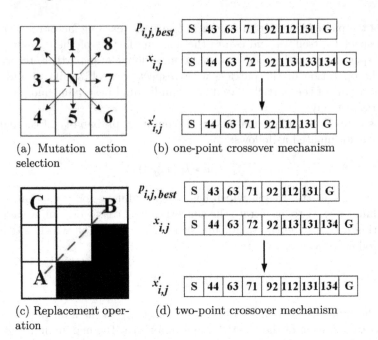

(a) Mutation action selection

(b) one-point crossover mechanism

(c) Replacement operation

(d) two-point crossover mechanism

Fig. 1. Crossover and mutation mechanisms.

If there is an overlap between the path points of the current particle and either the individual best particle or the global best particle (excluding the start and end points), a crossover operation is performed. In Fig. 1(b), particle x shares two crossover points (63, 92) with the individual best particle. This operation involves swapping the path point 71 between the two crossover points and swapping the subsequent crossover point with the path points (112, 131) between the individual best particle and particle x's target point. In Fig. 1(d), particle x shares more than two crossover points (63, 92, 131) with the individual best particle. In this case, only the path points 71 and 112 between the single best particle and particle x's crossover point are swapped.

In EPSO-GA, the values of c_1 and c_2 are updated based on their fitness values to fully utilize all particles and enhance the algorithm's ability to escape local optima. In each iteration, all particles are sorted based on their fitness values, with the top half being stored in set A and the bottom half in set B. In the next iteration, particles in set A have a higher probability of crossover. The formulas for updating the learning factors c_1 and c_2 are as follows:

$$\eta = \frac{min\_\text{fitness}}{max\_\text{fitness}} - \frac{min\_\text{fitness}}{\text{fitness }(x_{ij})} \tag{15}$$

$$c_1 = a + (1-a)e^{\eta} \tag{16}$$

$$c_2 = b + (1-b)e^{\eta} \tag{17}$$

In the formulas above, $max\_$fitness and $min\_$fitness represent the maximum and minimum fitness values of all particles in the population, respectively. fitness (x_{ij}) represents the fitness value of particle j after i iterations, and the values of a and b are experimentally determined constants. For particles in set A, higher fitness values result in higher crossover probabilities. The particle with the highest fitness value will always undergo crossover, while the probabilities of the other particles decrease exponentially with their fitness values, but still satisfy $c_1 > a$, $c_2 > b$. Through these operations, the top half of the particles with higher fitness values are given an increased probability of crossover, allowing them to focus more on searching, improving the algorithm's global search capability, and helping to escape local optima.

For the bottom half of the particles with lower fitness values, the values of c_1 and c_2 are designed as follows:

$$c_1 = c_{\max} - \frac{c_{\max} - c_{\min}}{max\_iter} * t \tag{18}$$

$$c_2 = c_{\min} + \frac{c_{\max} - c_{\min}}{max\_iter} * t \tag{19}$$

Here, both $c_{\max}$ and $c_{\min}$ are set to $(c_{\max} + c_{\min})/2 = c_1^c = c_2^c$, where c is a constant satisfying $1 > c_{\max} > c_{\min}$. max_iter represents the maximum number of iterations, and t represents the current iteration number. As the number of iterations increases, the value of c_1 gradually increases while the value of c_2 gradually decreases, causing the algorithm to shift from global search to local search.

Additionally, this article simplifies the paths by performing simplification operations when there are no obstacles or threat points blocking a straight line between two non-contiguous positions. As shown in Fig. 1(c), if the UAV can fly directly from the current point to another waypoint in a straight line, the path A-C-B will be replaced by the simpler path of A-B.

3   Simulation

To demonstrate the performance of proposed method, simulations and comparisons are carried out in 2D static environment. They are implemented in the MATLAB environment and compared performance with CIPSO [16] and CIGA [17], the simulation is running on a platform with a 3.2 GHz CPU and 8.0 GB of RAM.

3.1   Parameters Setting

The weight of objectives are set as $w_1 = 0.8$, $w_2 = 0.2$, the size of population is set as $S = 50$ and $max\_iter = 100$. The parameters of the EPSO-GA algorithm are set as $w_{max} = 0.8$, $w_{min} = 0.2$, $c_{max} = a = b = 0.8$, $c_{min} = 0.6$. The parameters of the CIPSO algorithm are set as $w_{max} = 0.9$, $w_{min} = 0.4$, $c_{max} = 3.5$, $c_{min} = 0.5$, $V_1 = 0.5$, $V_2 = 0.1$, $a = 2$, $\mu = 4$. The parameters of the CIGA algorithm are set as $c_r = 0.8$, $pc = 0.15$. Coordinates of starting position, destination and radars in the case are listed in Table 2.

Table 2. The starting position and the destination of UAV and the coordinates of the radars.

Environment type	Start position	End position	Radar1	Radar2	Radar3
Z	381	20	55	356	×
Complex	1	400	131	217	309

3.2   Result of UAV Path Planning

To compare the effect of the proportion of paths obtained by the Q-learning algorithm on the optimization of the algorithm, the number of paths m obtained by the Q-learning and the number of paths n obtained by random initialization were adjusted respectively, and five groups of experiments were conducted in the complex environment, with each group of experiments running independently for 20 times. The experimental results are shown in Table 3. Where $m = 0$ and $n = 50$ means purely random initialization, the initial best fitness is reduce with the proportion of m increase, and the convergence result of the algorithm is the best when $m = 5$ and $n = 45$. However, the average iteration number increases as m increases when $m > 5$. Therefore, compared with the random initialization, the path obtained by introducing Q-learning reduces the initial best fitness value of the population, which can accelerate the convergence speed of the algorithm. However, if the proportion of m is too large, the algorithm may fall into local optimal and the number of iterations of the algorithm will be increased.

Table 3. Results of EPSO-GA under different m and n values.

Indicator	$m = 0, n = 50$	$m = 3, n = 47$	$m = 5, n = 45$	$m = 10, n = 40$	$m = 15, n = 35$
Initial best fitness	33.7546	32.2565	30.7846	30.0512	**28.3654**
Average running time	16.21	15.84	**15.36**	15.38	16.02
Average iterations	21	18	**17**	20	26

The comparative simulations among different algorithms are carried out in two different environments, including Z-type and a complex environment, each method is repeated 50 times independently and the best results are chosen. The generated paths and corresponding convergence curves under different environments are presented in Fig. 2 and Fig. 3. The statistical results of simulation are shown in Table 4. As observed in Fig. 2(a) and Fig. 3(a), CIGA, CIPSO and our proposed method are complete the mission from the staring position to the destination without any collision with obstacles and radars. In the Z-type environment, the path generated by CIGA has a longest path and closest to the radars which the value of path length and threat intensity are 42.1637 and 17.0397 respectively, while CIPSO has a shortest path with 39.2854 and a second highest threat intensity with 16.6979. Our method is slightly longer than CIPSO where the path length is 39.7274, but maintain a relative longer distance

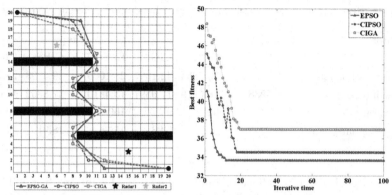

(a) Generated path under Z-type environment.

(b) Best fitness curves of different algorithms.

Fig. 2. Simulation results under Z-type environment.

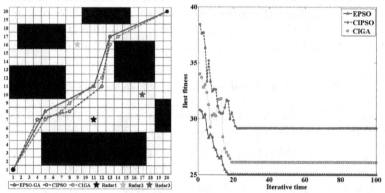

(a) Generated path under complex environment.

(b) Best fitness curves of different algorithms.

Fig. 3. Simulation results under complex environment.

from radars which threat intensity is only 9.5938. In the complex environment, our method complete the mission with the shortest path length and the lowest threat intensity.

In the Fig. 2(b) and Fig. 3(b), the best fitness of EPSO-GA are obviously smaller than that of the other algorithms at the initial stage, which are the results of hybrid initialization. Besides, EPSO-GA holds the fastest convergence compare with the other methods, while reach its optimal and remain stable in 12th and 17th iteration under two different environments respectively.

Table 4. Results comparison between the algorithms under different environments.

Indicator	Z-type			Complex type		
	EPSO-GA	CIPSO	CIGA	EPSO-GA	CIPSO	CIGA
Shortest path length	39.7274	**39.2854**	42.1637	**28.1069**	29.5298	28.7108
Threat intensity	**9.5938**	16.6979	17.0397	**13.4810**	27.8362	16.0080
Success rate (%)	**98**	96	98	**100**	98	100
Average running time (s)	6.24	6.87	**5.83**	15.36	18.91	**14.46**

A successful search is defined as finding the optimal solution after 200 iterations. As shown in Table 4, the EPSO-GA algorithm we proposed ranks first in the success rate. In terms of threat intensity, EPSO-GA are 43.7%, 42.5% better than CIGA and CIPSO in the Z-type environment, 15.8%, 51.6% in the complex environment respectively. Although our proposed method consumes slightly more time than the CIGA which due to the adaptive adjustment of parameters and hybrid initialization, the optimality of the solutions we have obtained. Our proposed EPSO-GA can effectively generate optimal path and is more practical in off-line path planing.

Table 5. Results comparison based on different weights under complex environment.

Weights	Algorithms	Best fitness	Average fitnes	Average iterations
$w_1 = 0.9$, $w_2 = 0.4$	EPSO-GA	**26.3028**	**27.0015**	**17**
	CIPSO	30.1475	30.7854	20
	CIGA	29.1311	30.1227	18
$w_1 = 0.8$, $w_2 = 0.2$	EPSO-GA	**25.1817**	**25.4910**	**17**
	CIPSO	29.1911	32.1429	21
	CIGA	26.1702	27.8914	19
$w_1 = 0.6$, $w_2 = 0.4$	EPSO-GA	27.4030	**28.0321**	**19**
	CIPSO	28.9213	29.1231	23
	CIGA	**26.5410**	28.2356	19
$w_1 = 0.5$, $w_2 = 0.5$	EPSO-GA	**25.3211**	**25.8940**	21
	CIPSO	29.1010	29.7414	21
	CIGA	26.3007	26.5641	**19**

In order to compare the influence of the weight of path length and threat intensity on the optimization process, four groups of experiments are conducted in complex environment by adjusting the weights of two sub-objectives, and each group of experiments is run independently for 20 times. The experimental results are shown in Table 5. It can be seen that the average iteration number of each algorithm is the least, the best fitness and the average fitness of ESPO-GA algorithm are the minimum, and the obtained path quality is the best when w_1 = 0.8, w_2 = 0.2.

4   Conclusion

In this paper, a hybridization of EPSO-GA algorithm is proposed for path planning of UAV in 2D static environment. Q-learning is utilized to initialize paths of UAV which improves the quality of initial paths and accelerates the convergence of the EPSO-GA. The acceleration coefficients of EPSO-GA are designed as adaptive ones by the fitness value to make full use of all particles and strengthen the global search ability of the algorithm, and a heuristics factor is introduced into mutation to speed up the convergence of algorithm. Through experiments in two different environments, it has been shown that our proposed algorithm has the advantages of path security and fast convergence in the path planning of single UAV. In the future work, we will concentrate on reducing time consumption of path planning for UAV and 3D environment will be introduced with dynamic obstacles.

References

1. Gao, X., Zhu, X., Zhai, L.: Minimization of aerial cost and mission completion time in multi-UAV-enabled iot networks. IEEE Trans. Commun. (2023)
2. Liu, Y., Chen, B., Zhang, X., Li, R.: Research on the dynamic path planning of manipulators based on a grid-local probability road map method. IEEE Access 9, 101186–101196 (2021)
3. Li, Z., You, K., Sun, J., Song, S.: Fast trajectory planning for dubins vehicles under cumulative probability of radar detection. Signal Process. 210, 109085 (2023)
4. Wang, J., Meng, M.Q.H., Khatib, O.: Eb-rrt: optimal motion planning for mobile robots. IEEE Trans. Autom. Sci. Eng. 17(4), 2063–2073 (2020)
5. Chen, Y., Bai, G., Zhan, Y., Hu, X., Liu, J.: Path planning and obstacle avoiding of the USV based on improved ACO-APF hybrid algorithm with adaptive early-warning. IEEE Access 9, 40728–40742 (2021)
6. Sun, J., Tang, J., Lao, S.: Collision avoidance for cooperative UAVs with optimized artificial potential field algorithm. IEEE Access 5, 18382–18390 (2017)
7. Guo, X., Peng, G., Meng, Y.: A modified Q-learning algorithm for robot path planning in a digital twin assembly system. Int. J. Adv. Manuf. Technol. 119, 3951–3961 (2022)
8. Yuan, J., Liu, Z., Lian, Y., Chen, L., An, Q., Wang, L., Ma, B.: Global optimization of UAV area coverage path planning based on good point set and genetic algorithm. Aerospace 9(2), 86 (2022)
9. Lee, J., Kim, D.W.: An effective initialization method for genetic algorithm-based robot path planning using a directed acyclic graph. Inf. Sci. 332, 1–18 (2016)
10. Mesquita, R., Gaspar, P.D.: A novel path planning optimization algorithm based on particle swarm optimization for uavs for bird monitoring and repelling. Processes 10(1), 62 (2021)
11. Zeng, M.R., Xi, L., Xiao, A.M.: The free step length ant colony algorithm in mobile robot path planning. Adv. Robot. 30(23), 1509–1514 (2016)

12. Stodola, P.: Hybrid ant colony optimization algorithm applied to the multi-depot vehicle routing problem. Nat. Comput. **19**(2), 463–475 (2020)
13. Das, P., Jena, P.K.: Multi-robot path planning using improved particle swarm optimization algorithm through novel evolutionary operators. Appl. Soft Comput. **92**, 106312 (2020)
14. Yin, G., Zhou, S., Mo, J., Cao, M., Kang, Y.: Multiple task assignment for cooperating unmanned aerial vehicles using multi-objective particle swarm optimization. Comput. Modernization **8**, 7–11 (2016)
15. Tian, D., Shi, Z.: Mpso: Modified particle swarm optimization and its applications. Swarm Evol. Comput. **41**, 49–68 (2018)
16. Shao, S., Peng, Y., He, C., Du, Y.: Efficient path planning for UAV formation via comprehensively improved particle swarm optimization. ISA Trans. **97**, 415–430 (2020)
17. Qu, H., Xing, K., Alexander, T.: An improved genetic algorithm with co-evolutionary strategy for global path planning of multiple mobile robots. Neurocomputing **120**, 509–517 (2013)

Machine Learning

How to Select the Appropriate One from the Trained Models for Model-Based OPE

Chongchong Li[1], Yue Wang[1], Zhi-Ming Ma[2], and Yuting Liu[1(✉)]

[1] Beijing Jiaotong University, Beijing, China
{18118002,ytliu}@bjtu.edu.cn
[2] Academy of Mathematics and Systems Science, Chinese Academy of Sciences,
Beijing, China

Abstract. Offline Policy Evaluation (OPE) is a method for evaluating and selecting complex policies in reinforcement learning for decision-making using large, offline datasets. Recently, Model-Based Offline Policy Evaluation (MBOPE) methods have become popular because they are easy to implement and perform well. The model-based approach provides a mechanism for approximating the value of a given policy directly using estimated transition and reward functions of the environment. However, a challenge remains in selecting an appropriate model from those trained for further use. We begin by analyzing the upper bound of the difference between the true value and the approximated value calculated using the model. Theoretical results show that this difference is related to the trajectories generated by the given policy on the learned model and the prediction error of the transition and reward functions at these generated data points. We then propose a novel criterion inspired by the theoretical results to determine which trained model is better suited for evaluating the given policy. Finally, we demonstrate the effectiveness of the proposed method on both simulated and benchmark offline datasets.

Keywords: Reinforcement Learning · Model Based · Offline Policy Evaluation

1 Introduction

Reinforcement learning (RL) is a powerful approach for solving the sequential decision making problems [28]. It has demonstrated excellent performance on various complex tasks such as Go [27], Atari games [21], and control tasks [9]. However, RL has a limitation as it requires online interactions with the environment, making it impractical for real-world problems where data collection is costly [6]. Offline RL aims to overcome this limitation by using only offline datasets, prohibiting the agent from interacting with the actual environment [1,7,14,16]. Offline Policy Evaluation (OPE) is a challenge in Offline RL that evaluates the expected performance of policies solely based on offline data

L. Fang et al. (Eds.): CICAI 2023, LNAI 14474, pp. 285–297, 2024.
https://doi.org/10.1007/978-981-99-9119-8_26

[6, 24, 37]. This evaluation technique is essential for many practical fields, including medicine [22], recommendation systems [18], and education [20]. Model-based OPE (MBOPE) [6, 37] is an important class of methods for OPE.

MBOPE offers a direct approach to approximate the value of a specific policy using Monte Carlo methods and the trained model. Generally, feed-forward neural networks are utilized to model complex environments and perform well in standard RL tasks [4, 34]. Nevertheless, previous MBOPE techniques encountered difficulties in choosing a suitable model from the set of trained models. Specifically, the training process produces a collection of models with different hyperparameters, and only one model or an ensemble of models is selected for OPE. At present, the common method is to compare the validation errors of the models and select the model with smaller validation errors. However, the local errors of the model and the degree of fit with the policy to be evaluated are ignored. To address this problem, we explore the criterion for selecting models from trained models in this study. Specifically, our contributions can be summarized as follows: (1) We theoretically analyze the error upper bound between the true value and the estimated value by using MBOPE; (2) We design a new method for picking models based on the theoretical result; (3) We verify the effectiveness of the proposed method on simulated and real offline data.

Next, we introduce the work of each part in detail.

Firstly, we analyze the upper bound of the difference between the ground truth value and the approximated value by MBOPE. This upper bound is dependent on the learned model and describes the quality of the model when applied to MBOPE. A tighter bound indicates that the model is more suitable for evaluating the given policy. The theoretical results of this study show that the difference is related to the trajectories generated by the given policy on the learned model. Additionally, it is related to the prediction error of the transition and reward functions at these generated data points.

Then, we propose a new criterion, inspired by theoretical results, to determine which trained model is better for evaluating a given policy. Designing the criterion using theoretical results is challenging because the error of the learned model on the data point generated by unrolling the given policy on the learned model is unavailable. To overcome this challenge, we first generate fake data using the model and policy, and then find the nearest data point in the actual offline dataset to the generated fake data. We then use the prediction error of the trained model on the found actual data to replace the prediction error on the generated data. After solving this problem, we can calculate the criterion using only the model, policy, and offline data.

At last, in this study, we empirically demonstrate the performance of the proposed method on simulation data and benchmark offline datasets. To demonstrate the validity, we first design a simulation dataset to show the validation loss is not enough to choose a suitable model for MBOPE. However, our method can consider the local error of the model and pick out the most suitable model for MBOPE for a given policy. We then apply our method on a real offline dataset, namely the RL Unplugged dataset [8] which contains a suite of benchmarks for

DeepMind Control Suite [29], a continuous action reinforcement learning benchmark. We learn a series of models with different hyperparameters and then use them to evaluate given policies. We compare the performance of the MBOPE using the model chosen by validation loss with the performance of the MBOPE using the model selected by the criterion we designed. In this study, we use the policies given from an offline policy evaluation benchmark [6]. The results show that using the model chosen by the proposed criterion can achieve better OPE performance than using the validation loss.

2 Background

Reinforcement Learning: In this paper we consider a discrete-time Markov decision process (MDP) [28] which is defined by the tuple $(\mathcal{S}, \mathcal{A}, M, r, \gamma, p_0)$, where $\mathcal{S}$ and $\mathcal{A}$ are the state and action spaces, respectively, M defines the transition probability $P(s_{t+1}|s_t, a_t)$ and $s_{t+1} \sim M(s_t, a_t)$, $r : r_t = r(s_t, a_t)$ defines the reward function, p_0 is the initial state distribution and γ represents the discount factor. Return $(R = \sum_{t=0}^{H} \gamma^t r(s_t, a_t))$ is defined as the expected sum of discounted rewards along a trajectory $\tau = (s_0, a_0, \ldots, s_H, a_H)$ of horizon length H. A policy which is a conditional distribution over actions conditioned on states is defined as $\pi(a|s)$. Reinforcement learning (RL) is dedicated to finding the policy π_ϕ that maximizes the expected sum of discounted rewards, i.e., $\max_\phi \mathbb{E}_{\tau \sim p_\pi(\tau)} \left[\sum_{t=0}^{H} \gamma^t r(s_t, a_t) \right]$, where ϕ are parameters of the policy, $p_\pi(\tau)$ is the trajectory distribution and $p_\pi(\tau) = p_0(s_0) \prod_{t=0}^{H-1} \pi(s_t|s_t)P(s_{t+1}|s_t, a_t)$. Note that the objective is usually defined as the value function of the policy [28]: $V(\pi) = \mathbb{E}_{\tau \sim p_\pi(\tau)} \left[\sum_{t=0}^{H} \gamma^t r(s_t, a_t) \right]$. **Model-Based RL (MBRL):** MBRL is characterized by learning the transition model using samples collected [34]. And we use a parametric function $\hat{M}_\theta$ to denote the learned transition model. We define the state predicted by the learned model as $\hat{s}_{t+1}$ and $\hat{s}_{t+1} \sim \hat{M}_\theta(s_t, a_t)$. Similarly, we use $\hat{r}_\theta$ to denote the learned reward function. Usually, the model predicts the distribution of the next state and reward as a multivariate Gaussian with a diagonal covariance structure [12]. And the model is usually trained by supervised learning [4,12,34], e.g., via maximum likelihood: $J_{\hat{M}}(\theta) = \mathbb{E} \left[\log \hat{P}(s_{t+1}, r_t|s_t, a_t) \right]$, where (s_t, a_t, r_t, s_{t+1}) are samples collected by interacting with the true model, $\hat{P}$ is the learned transition probability on $\hat{M}$. After model learning, the policy π can be optimized with data generated by the learned model and samples from the actual environment. **Offline Policy Evaluation (OPE):** For offline RL, the agent is not allowed to interact with the real environment, however, can access an offline dataset of transitions $\mathcal{D} = \{(s_t^i, a_t^i, r_t^i, s_{t+1}^i)_{i=1}^{N}\}$. The goal of OPE is to estimate the value $V(\pi)$ by only using the given dataset of transitions $\mathcal{D}$. A more exhaustive definition can be found in [16,17]. **Model-based OPE (MBOPE):** In MBOPE, we learn the model using the offline data. Then the value function of a given policy is directly estimated by Monte-Carlo methods using the learned model. Specifically, we use the given policy to interact with the learned model to get some fake trajectories. Then we can calculate the expected sum of the discounted rewards. We show the pseudo-code of MBOPE in Appendix (https://github.com/CCreal/C4MBOPE-materials).

3   Related Works

In this study, we follow the research of OPE whose goal is to prevent the risks and costs related to the online evaluation. Many methods have been proven effective on OPE, including methods which are based on importance weighting [19,25] and Lagrangian duality [23,32,36]. In this study, we mainly focus on the model-based approach that belongs to the direct method [5,13,23,33], a category of algorithms which approximate the value of the given policy by using the learned transition and reward functions of the environment directly. For simple cases with finite state and action spaces, we don't need function approximation [11,30]. By contrast, for challenging continuous domains, previous works [10,37] have provided an extensive demonstration of the model-based OPE approach. Usually, a feed-forward neural network is used to portray the transition and reward function [10]. Auto-regressive models are used to avoid assumptions about the independence of different state dimensions and improve the effectiveness of MBOPE [37]. In practice, we usually train a set of models and then choose one of them for further use when applying MBOPE methods. An important problem is how to choose an appropriate model from the trained models. Now existing works [10,37] usually use the validation loss of the model as a criterion. Our work is different from these works by designing a totally new method for model choosing. The idea of choosing the suitable model is similar to setting the suitable hyperparameter for offline RL [24] which is vital in practical problems. Note that choosing an appropriate model is also considered in online Model-Based Reinforcement Learning [15], discussing in which cases we should use a probabilistic model rather than a deterministic one. Our work is different from [15] since we consider how to select the model from a set of learned models rather than how to decide which type of model we should use.

4   Methods

4.1   Theoretical Analysis

In this study, we aim to address the problem of selecting a suitable model from a set of trained models to evaluate a given policy. Suppose we have a set of models $\{\hat{M}_k\}_{k=1}^{K}$ trained using an offline data set $\mathcal{D}$. Given a policy π_j from a set $\{\pi_j\}_{j=1}^{J}$, we need to evaluate π_j using the model chosen from the set of trained models, by using Monte-Carlo methods. It is common to use validation loss, which is the prediction error on the validation set (such as MSE loss), to select the model for further use. However, selecting the model by the validation loss is not related to the given policy. Therefore, this method does not consider the local accuracy of the region corresponding to the data produced by the interaction of the given policy with the actual environment. In this study, we desire to select the corresponding model for each policy to make the evaluation more accurate when using the chosen model.

To solve the above problem, we first analyze the upper bound of the discrepancy between the actual value and the approximated value calculated using the learned model. This discrepancy serves as an indicator of the model's quality when applied to MBOPE. The upper bound is dependent on both the trained model and the given policy and the Theorem 1 shows the result. We place the proof in the appendix.

Theorem 1. *Let η_π be the true value of the policy π, $\hat{\eta}_\pi$ be the estimated value using the learned model, then we have*

$$|\eta_\pi - \hat{\eta}_\pi| \leq C \cdot \mathbb{E}_{t \sim Gemo(\gamma)} \mathbb{E}_{s' \sim P^t_{mix}, a' \sim \pi(s')} \mathcal{L}(s', a') \tag{1}$$

where $P^t_{mix} = \beta P^t_\pi + (1-\beta)\hat{P}^t_\pi$, $C = \frac{2\gamma r_{max}}{(1-\gamma)^2}$. $\mathcal{L}(s', a')$, the error of the model at (s', a'), is $D_{TV}\left(P(s|s', a')\|\hat{P}(s|s', a')\right)$ and $Gemo(\gamma)$ is a geometric distribution with parameter γ.

In this paper, P^t_{mix} represents the mixed distribution of P^t_π and $\hat{P}^t_\pi$, where P^t_π is the state distribution of the Markov process at time step t given policy π and the environment, while $\hat{P}^t_\pi$ is the state distribution given the learned model. C is a constant. r_{max} is the maximum of the reward. D_{TV} is the total variation distance and here portrays the error of the model. This theoretical results indicate that the error of the estimated value of a given policy can be upper bounded by the expected error of the learned model over the distribution of trajectories produced by that policy. Specifically, the error depends on how the agent generates trajectories on the learned model and actual environment. The error of the learned model at these generated data points also plays a role. The geometric distribution shows that the error of the estimated value of the given policy is more relevant to the front part of the resulting trajectory. This is reasonable because the value of a policy is defined with a discount factor. In the study, β is set to zero to provide a more convenient bound since it is not possible to gather trajectories using the given policy on offline policy evaluation tasks. However, $\hat{P}^t_\pi$ can be obtained by unrolling the policy on the learned model. After proving Theorem 1, we will show how to design and calculate a criterion inspired by this result in the next part.

4.2   Proposed Criterion for Choosing a Trained Model

In this study, we first obtain the theoretical bound of the estimated value and the actual value of a given policy. Then we introduce how to design and calculate the criterion based on the theoretical bound. We will also discuss how to use the proposed criterion to judge which model is better for MBOPE from a series of trained models. According to Theorem 1, the bound is related to the distribution

P^t_{mix} of the state and action at time t. However, in our study, we set β to zero because the distribution induced by the real environment is not available. Although we can achieve samples from $\hat{P}^t$ by unrolling the given policy on the learned model, there are still two challenges for calculating the bound. Firstly, it is difficult to obtain the model error on the sampled data point from $\hat{P}^t$ since we only know the model error on data points from the given offline data set. To address this problem, we replace the model error on the data point sampled from the generated trajectories with the model error on the data point that was picked from the validation set of the offline data. Specifically, we choose the data point that is closest to the data point of which we want to know the model error. Secondly, the TV divergence needed in the theoretical result is not available. To solve this problem, we use the mean squared error instead to calculate the criterion.

Algorithm 1. A Criterion for Choosing the Trained Model

Require: the learned models $\{\hat{M}_k\}^K_{k=1}$, policy π_j, discount factor γ, horizon H, set of initial states S_0, batch size N, r_{max} which is the maximum of the reward in offline dataset.

1: **for** k=1,...,K **do**
2:     $B_k \leftarrow 0$
3:     **for** $i = 1, 2, \ldots, N$ **do**
4:         $B^i_k \leftarrow 0$
5:         Sample initial state s_0 from S_0.
6:         $\hat{s}^i_0 = s_0$
7:         **for** $t = 0, 1, \ldots, H - 1$ **do**
8:             Sample action using policy π, $a^i_t \sim \pi(\hat{s}^i_t)$
9:             Sample next state and reward using $\hat{M}_k$: $\hat{s}^i_{t+1}, \hat{r}^i_{t+1} \sim \hat{M}_k(\hat{s}_t, a^i_t)$
10:            Find the nearest data of $(\hat{s}^i_t, a^i_t)$ in validation: $(\hat{s}^{i*}_t, a^{i*}_t)$
11:            Calculated the distance d of $(\hat{s}^i_t, a^i_t)$ and $(\hat{s}^{i*}_t, a^{i*}_t)$.
12:            Find the model prediction error l on $(\hat{s}^{i*}_t, a^{i*}_t)$ in the validation set
13:            $B^i_k \leftarrow B^i_k + (1 - \gamma)\gamma^t (d + l)$
14:        **end for**
15:        $B^i_k \leftarrow B^i_k \cdot \frac{2\gamma r_{max}}{(1-\gamma)^2}$
16:    **end for**
17:    $B_k = \frac{1}{N} \sum^N_{i=1} B^i_k$
18: **end for**
19: $k^* = \arg \min_k B_k$
20: **return** the index k^* of the best model for π_j

After addressing the challenges mentioned above, we propose a practical method to calculate the criterion (we call it geometric loss criterion). Suppose we have a set of learned models and know the prediction error (MSE) of each model on the validation dataset. For a given policy, we first interact with the learned model using the given policy and generate N trajectories with horizon H. Then for each pair $(\hat{s}_t, a_t)$, we find the nearest pair (s^*_t, a^*_t) from the validation set of the offline dataset. We can then obtain the distance of (s^*_t, a^*_t) and $(\hat{s}_t, a_t)$, and the prediction error of the model on (s_t^*, a_t^*) is known. Combined with the probability of the geometric distribution, we obtain the criterion by:

$$\text{Geometric Loss Criterion} := C \frac{1}{N} \sum_{i=1}^{N} \sum_{t=0}^{H} g_t \left(d \left((\hat{s}_t^i, a_t^i), (s_t^{i*}, a_t^{i*}) \right) + l(s_t^{i*}, a_t^{i*}) \right), \quad (2)$$

where C is $\frac{2\gamma r_{max}}{(1-\gamma)^2}$. The variable g_t represents the probability of the geometric distribution for sampling t, i.e., $(1-\gamma)\gamma^t$. We calculate the distance of the generated data and the corresponding nearest data point in validation $d\left((s_t^i, a_t^i), (s_t^{i*}, a_t^{i*})\right)$ by taking the Mean Squared Error (MSE) of these two data points. The prediction loss of the learned model on (s_t^{i*}, a_t^{i*}), which we also use the MSE loss in this paper, is denoted by $l(s_t^{i*}, a_t^{i*})$. To choose the best model, we need to calculate the criterion of each model for the given policy and then use these criteria. In this paper, we present the overall algorithm pseudo-code in Algorithm 1.

5 Experiments

In this study, we designed experiments to answer two questions: (1) Is the validation loss sufficient to select the appropriate model for MBOPE? (2) Can the method proposed in this paper find the model for a given policy, which yields a more precise value estimation? We evaluated our approach on synthetic data and realistic model-based continuous control benchmarks to answer these questions.

5.1 Experiments on Synthetic Data

In this section, we demonstrate the effectiveness of the proposed criterion by designing a simulation experiment to evaluate its performance in a one-dimensional movement problem. **Experimental Setup** The agent starts at the origin and has the ability to move either left or right, with the objective of arriving at either -5 or 5. The true model is symmetric with respect to the origin, and we need to evaluate two policies. Policy π_1 directs the agent to move towards the negative half-axis and spend the majority of its time there, while policy π_2 instructs the agent to spend most of its time on the positive axis. We intend to use fake models to evaluate the aforementioned policies. We configure the fake model to have distinct errors on the positive and negative axes while maintaining a consistent total error. This setting implies that the fake model with greater errors on the positive half-axis will have reduced errors on the negative half-axis. By modifying the error on the negative half-axis, we develop a range of fake models that we then employ for OPE. Further information about the experimental settings can be found in the Appendix. **Results** It is evident that a model exhibiting a lower error on the negative half-axis is better suited for evaluating policy π_1, while conversely, a model with a lower error on the positive half-axis is more appropriate for evaluating policy π_2. Notably, the validation loss of the fake model remains constant, rendering it unsuitable for selecting an appropriate model. Our objective is to demonstrate that our proposed method can identify appropriate models. Based on Fig. 1, it can be observed that the estimated value approaches the actual value when the computed geometric loss criterion is lower. The outcomes reveal the efficacy of the proposed approach. Given a specific policy, the suggested method outlined in this study can be employed to select the appropriate model for the evaluation process.

5.2 Experiments on Benchmark Data

In this part, we present our experiments on the DeepMind control suite [29], a suit of continuous control tasks implemented in MuJoCo [31]. In this study, we consider the agents using the states as input like in previous works [6,12,34,37].

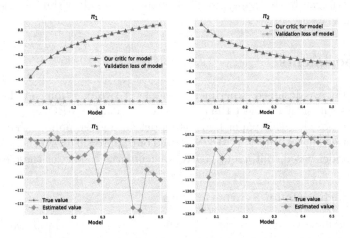

Fig. 1. The figure displays the results of experiments conducted on synthetic data. The left column shows the calculated geometric loss criterion, the validation loss for each model, and the estimated value for π_1 using each model. The models are identified by their prediction error on the negative half-axis. The right column displays the results for π_2. The results indicate that when the proposed criterion is small, the evaluation error is also small. However, the validation loss of the model remains constant.

Experimental Setup. In this paper, we use datasets from RL Unplugged [8], an offline RL benchmark suite based on Deepmind control-suite [29], to train models. For each environment, we train a series of models with different hyperparameters such as the epoch number, learning rate, and number of hidden layers. Further details on the experimental settings for training the models can be found in the Appendix. Subsequently, we evaluate given policies using the learned models. As for the policies to be evaluated, we use the benchmark from a previous work [6]. The benchmark comprises a total of 96 policies generated by four different algorithms: behavioral cloning [2], D4PG [3], Critic Regularized Regression [35], and RABM [26]. The actual value of each policy can be obtained by interacting with the realistic environment. We estimate the value of each policy by selecting a model from the learned models and then approximating it using the Monte Carlo method. We investigate whether the proposed criterion can better indicate the quality of a learned model for OPE compared to the validation loss when evaluating a given policy. We also compare the performance of our proposed method against a variation of our method (Ours w\o d) that only use $l(s_t^{i*}, a_t^{i*})$ in Eq. 2.

Table 1. Correlation coefficient results.

Environment	Validation loss	Ours w/o d	**Ours**
cartpole_swingup	-0.01 ± 0.29	0.28 ± 0.15	**0.45 ± 0.20**
cheetah_run	0.21 ± 0.14	0.18 ± 0.15	**0.33 ± 0.13**
finger_turn_hard	-0.30 ± 0.16	**-0.17 ± 0.09**	-0.19 ± 0.09
fish_swim	0.15 ± 0.12	0.12 ± 0.10	**0.42 ± 0.26**
walker_stand	-0.20 ± 0.07	-0.20 ± 0.07	**0.18 ± 0.21**
walker_walk	0.15 ± 0.24	**0.16 ± 0.28**	**0.16 ± 0.29**
humanoid_run	-0.06 ± 0.04	-0.06 ± 0.04	**0.01 ± 0.03**

Results. First, we see whether the geometric loss criterion, rather than the validation loss, can better point out the quality of a learned model when used for OPE. To demonstrate the result, we calculate the correlation coefficient of different assessment criteria and the value error. By utilizing a set of models, we can estimate the value of a given policy, with the validation loss known. We can calculate the criterion using the proposed method in this paper. The correlation coefficient $\rho_{C,E}$ shows the correlation between value error and the proposed criterion, and $\rho_{L,E}$ represents the correlation between value error and validation loss. The average correlation coefficients of all policies are calculated. The results in Table 1 indicate that our proposed criterion has better average performance than using validation loss in most environments. Specifically, for the cartpole_swimngup environment, we draw the criterion and validation loss for each model when used to evaluate a given policy. We present the results of a portion of the policies generated by D4PG [3] and RABM [26] with different training stages. From Fig. 2, we observe that the validation loss cannot determine which model is better when used for OPE. However, the proposed criterion can provide useful insights.

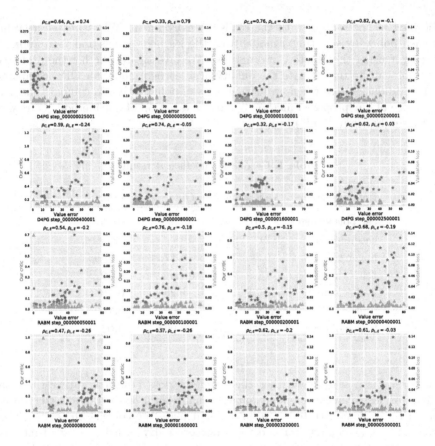

Fig. 2. Each subplot corresponds to a particular policy. In each subplot, blue stars represent a learned model, where the x-coordinate corresponds to the estimated value error using that model, and the y-coordinate represents the calculated geometric loss criterion for the model. Orange triangles also correspond to a learned model, while the y-coordinate represents the validation loss of the model. The correlation coefficient between the value error and the proposed criterion is denoted by $\rho_{C,E}$, while the correlation between the value error and the validation loss is denoted by $\rho_{L,E}$ (Color figure online).

We aim to determine whether the proposed method's selected model can achieve better OPE performance than using the validation loss. For each policy, we initially select the best model using the validation loss and use it to estimate the value. Simultaneously, we also use the model with the smallest criterion to estimate the value of each policy. To demonstrate whether the proposed method provides more precise value estimates than using the validation loss, we compare the absolute error between the ground truth and the estimated value. The absolute error of all policies is averaged, and we can observe from Table 2 that the proposed method has a smaller error compared to using the validation loss

Table 2. Average Absolute Error results.

Environment	Validation loss	Ours w/o d	**Ours**
cartpole_swingup	26.9±15.1	24.2±19.5	**17.5**±24.2
cheetah_run	13.4±8.37	13.2±8.37	**8.52**±8.49
finger_turn_hard	**31.0**±19.4	35.5±24.6	34.2±29.0
fish_swim	27.5±14.2	27.7±14.3	**20.1**±15.5
walker_stand	66.5±27.5	**55.5**±27.6	58.5±27.3
walker_walk	66.7±28.3	**57.1**±30.5	59.6±30.0
humanoid_run	34.3±22.4	**32.2**±24.8	35.4±27.1

in most environments. In addition, we also compare the performance of different methods by calculating Spearman's rank correlation between the ground truth values and the estimated values. The results are presented in the appendix, which further exhibits the performance of the proposed method.

6 Conclusion and Future Work

This study aims to discuss the criterion for selecting the best-trained model for model-based Off-Policy Evaluation (OPE). Initially, we analyze the upper bound of the difference between the ground truth value and the approximated value obtained using the model. Subsequently, we propose a new criterion inspired by the theoretical results to determine which trained model is better suited for evaluating a given policy. Finally, we demonstrate the proposed method empirically using simulation datasets and benchmark offline datasets. In the future, it would be interesting to combine the proposed method with model-based offline policy optimization problems, instead of only focusing on policy evaluation problems.

Acknowledgments. This work was supported in part by the Beijing Natural Science Foundation (L222051) and in part by the Fundamental Research Funds for the Central Universities (2022JBMC049).

References

1. Agarwal, R., Schuurmans, D., Norouzi, M.: An optimistic perspective on offline reinforcement learning. In: ICML, November 2020
2. Bain, M., Sammut, C.: A framework for behavioural cloning. In: Machine Intelligence 15, Intelligent Agents, pp. 103–129. Oxford University, GBR (1999)
3. Barth-Maron, G., et al.: Distributional policy gradients. In: ICLR (2018)
4. Chua, K., Calandra, R., McAllister, R., Levine, S.: Deep reinforcement learning in a handful of trials using probabilistic dynamics models. In: NeurIPS (2018)
5. Dudík, M., Langford, J., Li, L.: Doubly robust policy evaluation and learning. In: ICML, pp. 1097–1104. Madison, WI, USA, June 2011

6. Fu, J., et al.: Benchmarks for deep off-policy evaluation. In: ICLR (2021)
7. Fujimoto, S., Meger, D., Precup, D.: Off-policy deep reinforcement learning without exploration. In: ICML, pp. 2052–2062, May 2019
8. Gulcehre, C., et al.: RL Unplugged: a suite of benchmarks for offline reinforcement learning. In: NeurIPS, vol. 33, pp. 7248–7259 (2020)
9. Haarnoja, T., Zhou, A., Abbeel, P., Levine, S.: Soft actor-critic: off-policy maximum entropy deep reinforcement learning with a stochastic actor. In: ICML, pp. 1861–1870, July 2018
10. Hallak, A., Schnitzler, F., Mann, T., Mannor, S.: Off-policy model-based learning under unknown factored dynamics. In: ICML, pp. 711–719, June 2015
11. Hanna, J.P., Stone, P., Niekum, S.: Bootstrapping with models: confidence intervals for off-policy evaluation. In: AAAI, February 2017
12. Janner, M., Fu, J., Zhang, M., Levine, S.: When to trust your model: model-based policy optimization. In: NeurIPS, vol. 32 (2019)
13. Kostrikov, I., Nachum, O.: Statistical Bootstrapping for Uncertainty Estimation in Off-Policy Evaluation. arXiv:2007.13609 [cs, stat], July 2020. arXiv: 2007.13609
14. Kumar, A., Fu, J., Soh, M., Tucker, G., Levine, S.: Stabilizing off-policy Q-learning via bootstrapping error reduction. In: NeurIPS, vol. 32 (2019)
15. Kégl, B., Hurtado, G., Thomas, A.: Model-based micro-data reinforcement learning: what are the crucial model properties and which model to choose? In: ICLR, September 2020
16. Lange, S., Gabel, T., Riedmiller, M.: Batch reinforcement learning. In: Reinforcement Learning: State-of-the-Art, pp. 45–73. Adaptation, Learning, and Optimization. Springer, Heidelberg (2012)
17. Levine, S., Kumar, A., Tucker, G., Fu, J.: Offline Reinforcement Learning: Tutorial, Review, and Perspectives on Open Problems. arXiv:2005.01643 (Nov 2020)
18. Li, L., Chu, W., Langford, J., Wang, X.: Unbiased offline evaluation of contextual-bandit-based news article recommendation algorithms. In: WSDM, February 2011
19. Li, L., Munos, R., Szepesvari, C.: On Minimax Optimal Offline Policy Evaluation. arXiv:1409.3653 [cs], September 2014. arXiv: 1409.3653
20. Mandel, T., Liu, Y.E., Levine, S., Brunskill, E., Popovic, Z.: Offline policy evaluation across representations with applications to educational games. In: AAMAS, May 2014
21. Mnih, V., et al.: Asynchronous Methods for Deep Reinforcement Learning. In: ICML, pp. 1928–1937, June 2016
22. Murphy, S.A., van der Laan, M.J., Robins, J.M.: Marginal mean models for dynamic regimes. J. Am. Stat. Assoc. **96**(456), 1410–1423 (2001)
23. Nachum, O., Chow, Y., Dai, B., Li, L.: DualDICE: behavior-agnostic estimation of discounted stationary distribution corrections. In: NeurIPS, vol. 32 (2019)
24. Paine, T.L., et al.: Hyperparameter Selection for Offline Reinforcement Learning. arXiv:2007.09055 [cs, stat], July 2020
25. Precup, D., Sutton, R.S., Singh, S.P.: Eligibility traces for off-policy policy evaluation. In: ICML, pp. 759–766. San Francisco, CA, USA, June 2000
26. Siegel, N., et al.: Keep doing what worked: Behavior modelling priors for offline reinforcement learning. In: ICLR (2020)
27. Silver, D., et al.: Mastering the game of Go with deep neural networks and tree search. Nature **529**(7587), 484–489 (2016)
28. Sutton, R.S., Barto, A.G.: Reinforcement Learning: An Introduction. MIT Press, Cambridge (1998)
29. Tassa, Y., et al.: DeepMind Control Suite. arXiv:1801.00690 [cs], January 2018

30. Thomas, P., Brunskill, E.: Data-efficient off-policy policy evaluation for reinforcement learning. In: ICML, pp. 2139–2148, June 2016, iSSN: 1938–7228
31. Todorov, E., Erez, T., Tassa, Y.: MuJoCo: a physics engine for model-based control. In: IROS, pp. 5026–5033, October 2012
32. Uehara, M., Huang, J., Jiang, N.: Minimax weight and Q-function learning for off-policy evaluation. In: ICML, November 2020
33. Voloshin, C., Le, H.M., Jiang, N., Yue, Y.: Empirical Study of Off-Policy Policy Evaluation for Reinforcement Learning. arXiv:1911.06854 [cs, stat], November 2021
34. Wang, T., et al.: Benchmarking Model-Based Reinforcement Learning. arXiv:1907.02057 [cs, stat], July 2019. arXiv: 1907.02057
35. Wang, Z., et al.: Critic regularized regression. In: NeurIPS, vol. 33, pp. 7768–7778 (2020)
36. Yang, M., Nachum, O., Dai, B., Li, L., Schuurmans, D.: Off-policy evaluation via the regularized Lagrangian. In: NeurIPS, vol. 33, pp. 6551–6561 (2020)
37. Zhang, M.R., Paine, T., Nachum, O., Paduraru, C., Tucker, G., ziyu wang, Norouzi, M.: Autoregressive dynamics models for offline policy evaluation and optimization. In: ICLR (2021)

A Flexible Simplicity Enhancement Model for Knowledge Graph Completion Task

Yashen Wang[1,2], Xuecheng Zhang[3(✉)], Tianzhu Chen[2], and Yi Zhang[1,4]

[1] National Engineering Laboratory for Risk Perception and Prevention (RPP), China Academy of Electronics and Information Technology, Beijing 100041, China
[2] Key Laboratory of Cognition and Intelligence Technology (CIT), Artificial Intelligence Institute of CETC, Beijing 100144, China
chentianzhu@iie.ac.cn
[3] Khoury College of Computer Science, Northeastern University, Boston, MA 02115, USA
zhang.xueche@northeastern.edu
[4] CETC Academy of Electronics and Information Technology Group Co., Ltd., Beijing 100041, China

Abstract. Knowledge graph (KG) has gradually become the cornerstone of many Artificial Intelligence (AI) tasks, as one of the most effective ways to represent world knowledge, while these KGs still hardly cover the massive emerging knowledge in the real world. Knowledge Graph Completion (KGC) tries to reason over known facts and infer the missing links, to improve KG's coverage. The conventional KGC algorithms usually map each entity to a unique embedding vectors, which incurs a linear growth in memory consumption for saving embedded matrices, and results in high computational costs when modeling real-world KG. Hence, this paper aims to investigate how to strengthen the *simplicity* (i.e., reduce complexity) of KGC model and strike a reasonable *balance* between accuracy and complexity. Especially, this paper proposes a novel concept-enhanced anchor-based entity representation method to learn a fixed-size vocabulary in condition of the collapsed KG, which is built by the selected anchor entities, concept semantics respect to these anchors and relation types. This work can be viewed as a flexible plug-in unit to serve many current KGC models. Experiments show that our model performs competitively in KGC task while retaining less than 10% of explicit entities in a given KG and less than 10% of parameters.

Keywords: Knowledge Graph · Knowledge Graph Completion · Representation Learning · Simplicity Enhancement · Complexity Reduction

1 Introduction

Knowledge Graphs (KGs) are viewed as collections of real-world fact represented in form of triple (h, r, t), consisting of head entity, relation type and tail entity, respectively. Recently, KGs have significantly boosted the developments of various vertical fields, such as question answering, machine reading, information retrieval and dialogue systems. Since there exists a remarkable increasing-demand for the remarkable of

L. Fang et al. (Eds.): CICAI 2023, LNAI 14474, pp. 298–309, 2024.
https://doi.org/10.1007/978-981-99-9119-8_27

high-quality real-world knowledge, the *reliability* of KG has become increasingly critical [30]. Therefore, Knowledge Graph Completion (KGC) task and the corresponding methods, which aims at identifying what the missing relation in an incomplete triple is or whether the triple in KG is valid or not, has been widely researched.

Recently, various KGC models have been developed and discussed for KG reasoning, and have achieved satisfactory results in many KG-oriented tasks [3,8], demonstrating their ability to understand and explore diverse and complex relationships ability [12,33]. These methods generally represent entities and relations in triples, as real-valued vectors and assess triples' plausibility with these vectors. Take translation-based KGC model as examples, conventional TransE [1] utilizes the structural signals of known triples ($\mathbf{h} + \mathbf{r} \approx \mathbf{t}$) to project KGs into a continuous semantic vector space, which triumphantly opens up a new waterway for representing knowledge based on translation-based methodology. Inspired by this, many other translation-based methods are introduced, including TransH [34], TransD [11], TransR [13], TrasG [38], RotatE [22] and QuatDE [7].

The conventional KGC algorithms usually map each entity to a unique embedding vectors. This kind of shallow search strategy, unfortunately leads to a linear increase in memory consumption for storing embedded matrices, and incurs high computational costs when modeling real-world KG [18] [33]. The Natural language processing (NLP) field has also faced the problem of this kinds of *vocabulary* being too large to embed and model all words, and WordPiece technique has been used to solve it. It no longer sees a word as a whole, but divides it into several sub-words [9]. E.g., the three words "loved", "loving" and "loves" actually have the same semantic meaning. However, if we use word as an unit, they are considered as different words. WordPiece divides the abovementioned three words into parts such as "love", "ed", "ing" and 'es', which can distinguish the meaning and tense of the words themselves and availably reduce the size of vocabulary. By comparing with the commonly used sub-word tokenization strategy in NLP field, [6] explores a more parameter efficient node embedding strategy to overcome KGC task, which is regarded as a kind of compositional method for depicting nodes (i.e., entities) in multi-relational KG with a fixed-size vocabulary. Similar to sub-word units, it try to tokenize each node as a combination of selected proxy entities and relations. Although the number of parameters decreased significantly in its experiment, its performance also decreased significantly.

To overcome this problem, this paper aims at improving performance while maintaining an equivalent number of parameter magnitude with [6]. Especially, an concept-enhanced anchor-based method is utilized here to learn a fixed-size vocabulary for the collapsed KG, which is built by the anchor entity nodes, concept semantics respect to these anchors and relation types. Given this kind of fixed-size vocabulary, the proposed model can guide the encoding and embedding of any entities. This work successfully slashes the scale of trainable parameters, and simultaneously pullups the model's performance for KGC task. Experiments have shown that the proposed work performs well in entity prediction task and relation prediction task, while merely retaining *less* than $\sim$10% of explicit entities as anchors in the given original KG, and typically has $\sim$10 times *fewer* parameters.

2 Preliminary

2.1 Knowledge Graph

Knowledge Graph (KG, denoted as $\mathcal{G} = \{E, R, \mathcal{E}, \mathcal{R}, \phi_{\mathcal{E}}, \phi_{\mathcal{R}}\}$), usually consists of a set of triples (h, r, t), wherein $h \in E$ represents the head entity, $r \in R$ represents the relation type, and $t \in E$ represents the tail entity. E and R indicate the entity set and relation set, respectively. It is also associated with an entity type mapping function $\phi_{\mathcal{E}} : E \rightarrow \mathcal{E}$ and a relation type mapping function $\phi_{\mathcal{R}} : R \rightarrow \mathcal{R}$, wherein notations $\mathcal{E}$ and $\mathcal{R}$ separately describe the set of entity types and the set of relation types.

2.2 Knowledge Graph Completion

Knowledge Graph Completion (KGC) is one of the main challenges in the KG field, since most KGs are *incomplete*. It mainly aims to predict the missing element in the given separately triple, which usually consists of several main sub-tasks: entity prediction (EP), relation prediction (RP), etc.,. Wherein, entity prediction task attempts at predicting the missing entity when given an entity and a relation, i.e., we identify tail entity t when given $(h, r, ?)$, or similarly identify head entity h when given $(?, r, t)$. Relation prediction task tries to discriminate the missing relation when given two entities, i.e., we identify which r is true when given $(h, ?, t)$.

2.3 Concept

Following [10,24], this work depicts a "concept" as a set (or a class, a category, etc.,) of "entities" or "things" within a domain, such that words belonging to same (as well as similar) classes get similar semantic representations [23] [36] [5]. E.g., "microsoft" and "amazon" could be represented by concept COMPANY. Probase [37] is selected in our study as lexical KG to generate concept for entity, to enhance the performance of simplified KGC model [30]. Recently, Probase has been widely used as valuable extra-resource in research about text understanding [19,20,37], text representation [10,28], information retrieval [27,31], because entity's concepts are simpler, highly-structural and more specific when compared with other kinds of semantic information (such as entity type and entity description, etc.,) [15,16,39]. Uniformly, this work utilizes upper-case notation C to represent the set of concepts, and utilizes lowercase notation $c \in C$ to describe the concept pre-defined in the lexical KG Probase [21,37].

2.4 Conceptualization

Given a word-formed entity e, *instance conceptualization* algorithm readily contributes to distilling the optimal open-domain concepts $C_e = \{< c_j, p_j >\}$ from the lexical KG Probase which own the optimal ability for discriminatively representing the given entity e [29]. The aforementioned probability p_j describes the confidence level of the concept c_j for entity e. E.g., given entity "London" in KG, we could generate the concepts for word(s)-formed entity mention "London" as $C_{\text{London}} = \{<$"City",0.503$>$, $<$"Place",0.075$>$, $<$"Area",0.072$>$, $<$"Location",0.057$>$, $\cdots\}$ from Probase. This paper succinctly adopts the state-of-the-art instance conceptualization algorithm proposed in [10], because it is not central to this study.

3   Methodology

This work can be viewed as a flexible plug-in unit to be loaded into the many kinds of current KGC models. Hence, we faithfully follows the native representation learning procedure and scoring function of current KGC model, and mainly focus on how to simplify parameter scale for entity representation. To calculate the embedding of a target entity e, we firstly select top-n anchor entities ($\{a_1, a_2, a_3, a_4\}$ in green circles as examples) closest to it (left side of Fig. 1, discussed in Sect. 3.1), and then add their embeddings and their concept embeddings, their respective distances to the entity as well as the relation types directly connected to the target entity e (right side of Fig. 1, discussed in Sect. 3.2), to the encoder and finally output the results.

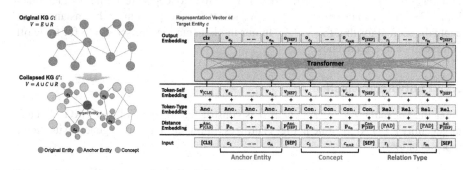

Fig. 1. The architecture of the proposed model. left: Collapsed KG Construction; right: Entity Representation (take Transformer encoder as an example).

3.1   Collapsed KG Construction

We first construct a Collapsed KG G' based on the given original KG G (left side of Fig. 1). Especially, in this Collapsed KG G', we utilize anchor entities (noted as A with size of $|A|$) and their corresponding concepts (noted as C with size of $|C|$, discussed in Sect. 2.3), to represent the overall original entities (E with size of $|E|$) in native KG G, with condition of $|A| + |C| \ll |E|$.

From another perspective: (i) in original KG G, the vocabulary V consists of all the native entities E and the native relation types R (i.e., $V = E \cup R$); while (ii) in our Collapsed KG G', the vocabulary V consists of the anchor entities A, the fix-sized concepts C and the native relation types R (i.e., $V = A \cup C \cup R$), wherein the parameter scale apparently becomes much smaller.

Anchor Entity Set Construction: For constructing anchor entity set A (green circles in Fig. 1), we follow the intuitive strategy proposed in [6], and randomly select a set of entities with size of $|A|$ from native entities E. Hence, $A \subset E$ and $|A| \ll |E|$. Previous pre-training models determine sub-word units according to the frequency of lexical co-occurrence. In KG' situation, the corresponding metric is *centrality*. [6] has intriguingly shows that, the effectiveness of selecting anchor entity nodes based on centrality is consistent with that of random selection. Besides, other kinds of metrics could be discussed which will be future work.

Concept Set Construction: For compensating performance of previous KGC simplicity enhancement, this work seeks helps from extra structural high-level semantics, i.e., concept. For constructing concept set C (grey circles in Fig. 1), there exist different choices to construct the concept set C. Choice #1: We can straightly use all the concepts defined in Probase to construct C, which potentially results in high computational-complexity because this strategy has cover large amount native concepts and multi-granularity semantics. Choice #2: For balance, we can adopt a clustering algorithm to project all the concepts in Probase into 5,000 disjoint concept clusters [26,35], wherein each concept cluster indicates "concept of concepts" which merely represents one sense or a general topic ($|C| = 5,000$), since many native concepts are similar to each other [32] . For each anchor entity $a_i \in A$, we select k concept from C, by instance conceptualization algorithm (Sect. 2.4). In our implementation, we set $k = 5$ for trade off complexity and effectiveness.

3.2   Entity Representation

After the vocabulary $V = A \cup C \cup R$ is constructed for the Collapsed KG G', each target entity $e \in E$ (red circle in Fig. 1) utilizes following elements to generate its vector representation:

(i) The top-n nearest *anchor* entities respect to target entity e, with BFS or the other methods;
(ii) The top-k *concepts* respect to each selected anchor entity;
(iii) The distances between the top-n closest anchor entities and the target entity e;
(iv) The top-m *relation types* between target entity e and top-n nearest anchor.

Note that, for selecting top-m relation types, an entity can utilize up to m relation types that are connected to it: (i) If there exist more than m relation types, we randomly sample m relation types; and (ii) if there exist less than m relation types, we use [PAD] instead. Given a target entity e and an anchor entity a , this work defines "anchor distance" $p_a \in [0; diameter(G)]$ as an integer indicating the shortest path distance between anchor entity a and target entity e in the original KG G. We then project each integer-formed anchor distance p_a to a d-dimensional vector with a relative distance encoding scheme [6]: $\mathbf{p}_a = f_{dist.}(p_a)$. Moreover, for each concept c, it shares distance vector with the anchor entity which c belongs to.

For universality, Transformer-based encoder[1] is utilized here to generate entity embeddings $\mathbf{e}$ for the given target entity e (shown in right side of Fig. 1):

$$\mathbf{e} = \text{Transformer}[\{\mathbf{a}_1, \mathbf{a}_2, \cdots, \mathbf{a}_n\}; \{\mathbf{p}_1, \mathbf{p}_2, \cdots, \mathbf{p}_n\}; \{\mathbf{c}_1, \mathbf{c}_2, \cdots, \mathbf{c}_{n \times k}\}; \{\mathbf{r}_1, \mathbf{r}_2, \cdots, \mathbf{r}_n\}] \quad (1)$$

Finally, the output of [CLS] token in this Transformer-based encoder (yellow rectangle in Fig. 1), is viewed as the vector representation of the target entity e. Since this

---

[1] Note that, other kinds of encoder such as MLP, can be used here, which is not the focus of this work.

work can be viewed as a flexible entity representation plug-in unit to be loaded into current KGC models, we faithfully follows the native optimization procedure and scoring function of current KGC model (TransE [1], TransH [34], TransR [13], DKRL [39], RotatE [22] emphasized in this paper as discussed in details in Sect. 4), and we'll ignore that details for this article.

4 Experiments

4.1 Datasets

For the evaluations of entity prediction task and link prediction task, this paper conducts experiments on the WN18RR (subset of WordNet) and FB15k-237 (subset of Freebase), which are widely used. Wherein, WN18RR is a lexical database of English [4], and FB15k-237 includes general human knowledge [25]. Each dataset consists of abundant relational patterns (such as *symmetry*, *inversion*, *composition*) and complex 1-N, N-1 and N-N of relations. Table 1 describes the summary of the these two datasets. Besides, as mentioned before, concept signals of entities and relations respect to these datasets, are generated by instance conceptualization algorithm proposed by [10] based on Probase [17,37].

Table 1. Summary of WN18RR (WordNet) and FB15k-237 (Freebase) used for entity prediction task and link prediction task evaluations.

Dataset	$|E|$	$|\mathcal{R}|$	#train	#valid	#test
WN18RR	40,943	11	86,835	3,034	3,134
FB15k-237	14,541	237	272,115	17,535	20,466

4.2 Baselines

The baselines include conventional translation-based KGC models, including TransE [1], TransH [34], TransR [13], DKRL [39], RotatE [22]. For each competitive model mentioned above, we compare the native KGC model with its variant with help of this work as a plug-in unit (denoted with suffix of "+(Ours)"). Note that, for each variant, we faithfully follows the optimization procedure and scoring function of the corresponding native KGC model.

4.3 Experimental Settings

This section directly reuses the empirical results of several baselines from the previous literature [13,34,40], since the datasets (as well as the corresponding splitting manners) and the "unif." sampling strategy are the same. KG G has performed *inverse* operations on all relation type to ensure that each relation type in the given KG G is bidirectional.

Several hyper-parameter settings have been attempted on the validation dataset, for helping us to investigate the best configuration. Under the "unif." sampling strategy

[34], the optimal configurations are concluded as follows: the mini-batch size is set as 32, Adam's learning rate $\epsilon = 2e - 5$, dropout rate is 0.1, vector dimension $d = 500$, size of anchor entity set $|A| = 2,000$, size of contextual anchor entity for each target entity $n = 30$, size of contextual relation type $m = 5$, numbers of concepts respect to each anchor entity $k = 5$ for trade-off, on WN18RR dataset; the mini-batch size is set as 32, $\epsilon = 2.5e - 5$, dropout rate is 0.1, $d = 550$, $|A| = 1,000$, $n = 20$, $m = 15$, $k = 5$, on FB15K-237 dataset. For all the comparative models, we train the model until convergence. For avoiding dataset-specific tuning, most hyper-parameters in addition to learning rate and training epochs etc., are shared across all datasets. Generally, we cautiously run all the comparative models 20 times and observe the deviations are less than 0.10, and meanwhile the improvements are significant.

For evaluation metrics, we mainly use these widely-used automatic evaluation metrics: (i) Mean Reciprocal Rank (MRR), indicating the average reciprocal rank of all tested triples; (ii) HITS@n ($n \in \{1, 3, 10\}$, calculating the ratio of ground-truth entities ranked among the top-n.

4.4   Evaluations on Entity Prediction Task

Table 2 and Table 3 summary the overall experimental results respect to entity prediction task, which is intended to predict the missing entity when given an entity and a relation, i.e., we identify tail entity t when given $(h, r, ?)$, or similarly identify head entity h when given $(?, r, t)$. Wherein, $|V|$ denotes vocabulary size (i.e., $V = A \cup C \cup R$), as discussed in Sect. 3.1. #*par.* is a total parameter count (millions), which is concluded based on [18]. MRR% denotes the MRR ratio based on the strongest model.

Table 2. Evaluation results of entity prediction (EP) task on WN18RR.

Method	WN18RR						
	$|V|$	#*par.*	MRR↑	MRR%	H@1↑	H@3↑	H@10↑
TransE [1]	40k+22+0	20.0	24.3	100.00	4.3	44.1	53.2
TransE [1]+(Ours)	2k+22+5k	2.2	21.4	88.41	4.2	43.4	52.3
TransH [34]	40k+22+0	20.0	23.1	100.00	4.1	41.9	50.5
TransH [34]+(Ours)	2k+22+5k	2.2	20.3	88.43	3.8	38.5	46.4
TransR [13]	40k+22+0	25.0	23.2	100.00	4.1	42.0	50.7
TransR [13]+(Ours)	2k+22+5k	2.8	20.4	88.43	3.8	38.6	46.6
DKRL [39]	40k+22+0	23.5	26.0	100.00	30.4	47.1	56.9
DKRL [39]+(Ours)	2k+22+5k	2.6	22.9	86.46	28.0	43.3	52.3
RotatE [22]	40k+22+0	41.0	47.6	100.00	42.8	49.2	57.1
RotatE [22]+(Ours)	2k+22+5k	4.6	41.9	88.21	39.4	45.2	52.5

Table 3. Evaluation results of entity prediction (EP) task on FB15k-237.

Method	$|V|$	#par.	MRR↑	MRR%	H@1↑	H@3↑	H@10↑
	FB15k-237						
TransE [1]	14.5k+0.5k+0	7.5	27.9	100.00	19.8	37.6	44.1
TransE [1]+(Ours)	1k+0.5k+5k	0.9	22.9	85.58	16.3	30.9	36.2
TransH [34]	14.5k+0.5k+0	7.8	30.7	100.00	21.8	41.4	48.6
TransH [34]+(Ours)	1k+0.5k+5k	0.9	25.2	85.25	17.9	34.0	39.9
TransR [13]	14.5k+0.5k+0	132.3	30.7	100.00	21.8	41.4	48.6
TransR [13]+(Ours)	1k+0.5k+5k	15.5	25.2	85.25	17.9	34.0	39.9
DKRL [39]	14.5k+0.5k+0	11.0	29.8	100.00	21.2	40.2	47.1
DKRL [39]+(Ours)	1k+0.5k+5k	1.3	24.5	75.29	17.4	33.0	38.7
RotatE [22]	14.5k+0.5k+0	29.0	33.8	100.00	24.1	37.5	53.3
RotatE [22]+(Ours)	1k+0.5k+5k	3.4	27.7	84.96	19.8	30.8	43.8

From the result, we observe that: Our work with a fixed-size vocabulary $|V|$ of <10% of entities sustains more than 84% of MRR metric compared to ~10x larger competitive baseline models. E.g., our RotatE variant (i.e., RotatE+(Ours)) achieves 84.95% MRR performance with only maintaining 6.89% entities and 11.82% parameter scale, when comparing the native RotatE baseline. Little performance loss is expected according to the compositional and compressive nature of KG's collapse (i.e., $E \rightarrow A$ and $|A| \ll |E|$). Similar to the inspections from [14,41], the advantages of our work over baselines on WN18RR (WordNet) is larger than FB15k-237 (Freebase). Since dataset FB15k-237 embeds more diverse relation types and has a more complex structure than WN18RR, translation based models methods are reported likely to hold an advantage for this condition. However, the proposed method achieves stable performance on FB15k-237 with help of extra concept semantics. Especially, we leverage high-quality and structural concept semantics from lexical KG Probase, and these encouraging results shows the advantage of discrete and structural entity's concept towards continuous and unstructural entity's description or other kinds of extra semantics.

Beside, we find that many errors generated by our method, are possibly caused by the so called closed-world assumption (CWA) [14], assuming that any knowledge *unseen* in given KG is *incorrect* and most existing translation-based KGC models are usually built and evaluated under this setting. Although there have existed works investigating open-world assumption (OWA) [2], it requires a lot of labour to manually annotate unseen triples. Evaluations under OWA, will be our future direction.

4.5 Evaluations on Relation Prediction Task

Table 4 and Table 5 summary the overall experimental results respect to relation prediction task, which tries to discriminate the missing relation when given two entities, i.e., we identify which r is true when given $(h, ?, t)$. From the preliminary result, we could observe conclusions similar to previous entity prediction tasks, i.e., our work with a fixed-size vocabulary V of <10% of entities sustains more than 86% of MRR metric compared to ~10x larger competitive baseline models. On bigger KGs, parameter efficiency is still pronounced, i.e., on FB15k-237, our work sustains more than 82% of MRR metric compared to ~10x larger competitive baseline models. While on WN18RR, the value of MRR% is more than 88%.

Table 4. Evaluation results of relation prediction (RP) task on WN18RR.

Method	WN18RR						
	$\|V\|$	#par.	MRR↑	MRR%	H@1↑	H@3↑	H@10↑
TransE [1]	40k+22+0	20.0	46.56	100.00	18.31	22.36	24.95
TransE [1]+(Ours)	2k+22+5k	2.2	41.4	89.21	38.4	46.9	52.3
TransH [34]	40k+22+0	20.0	44.26	100.00	17.46	21.32	23.79
TransH [34]+(Ours)	2k+22+5k	2.2	39.4	89.23	15.5	19.0	21.2
TransR [13]	40k+22+0	25.0	43.87	100.00	17.23	21.04	23.48
TransR [13]+(Ours)	2k+22+5k	2.8	39.0	89.23	15.3	18.7	20.9
DKRL [39]	40k+22+0	23.5	50.11	100.00	52.08	63.62	70.98
DKRL [39]+(Ours)	2k+22+5k	2.6	44.6	88.20	46.4	56.6	63.2
RotatE [22]	40k+22+0	41.0	91.20	100.00	72.89	89.03	96.30
RotatE [22]+(Ours)	2k+22+5k	4.6	81.2	89.11	64.9	79.2	85.7

Table 5. Evaluation results of relation prediction (RP) task on FB15k-237.

Method	FB15k-237						
	$\|V\|$	#par.	MRR↑	MRR%	H@1↑	H@3↑	H@10↑
TransE [1]	14.5k+0.5k+0	7.5	63.79	100.00	69.30	69.99	85.60
TransE [1]+(Ours)	1k+0.5k+5k	0.9	54.2	86.57	58.3	58.9	72.0
TransH [34]	14.5k+0.5k+0	7.8	75.71	100.00	82.82	83.64	83.73
TransH [34]+(Ours)	1k+0.5k+5k	0.9	64.4	86.32	69.6	70.3	70.4
TransR [13]	14.5k+0.5k+0	132.3	74.00	100.00	81.60	82.42	82.50
TransR [13]+(Ours)	1k+0.5k+5k	15.5	62.9	86.35	68.6	69.3	69.4
DKRL [39]	14.5k+0.5k+0	11.0	72.63	100.00	80.8	81.6	91.7
DKRL [39]+(Ours)	1k+0.5k+5k	1.3	61.7	82.25	68.0	68.6	77.1
RotatE [22]	14.5k+0.5k+0	29.0	88.91	100.00	82.60	83.43	83.51
RotatE [22]+(Ours)	1k+0.5k+5k	3.4	75.6	86.12	69.5	70.2	70.2

5   Conclusion

The emergence of large-scale KGs makes it very difficult to learn the vector representations of all entity nodes. For enhancing the simplicity of entity representation in KGC task, this work proposes a novel semantic-enhanced anchor-based entity representation method to learn a fixed-size vocabulary based on the collapsed KG, which is built by the selected anchor entities, concept semantics respect to these anchors and relation types. This work can learn entity embeddings on large-scale KGs with fewer anchor entities (and their corresponding concepts) embeddings and then enhance the generalization performance of KGC model. Therefore, this work can enhance the implementation and application capabilities of KG representation learning technology.

Acknowledgements. We thank anonymous reviewers for valuable comments. This work is funded by: (i) the National Natural Science Foundation of China (No. 62106243, U19B2026, U22B2601).

References

1. Bordes, A., Usunier, N., Garcia-Duran, A., Weston, J., Yakhnenko, O.: Translating embeddings for modeling multi-relational data. In: NIPS'13, pp. 2787–2795 (2013)
2. Cao, Y., Ji, X., Lv, X., Li, J.Z., Wen, Y., Zhang, H.: Are missing links predictable? an inferential benchmark for knowledge graph completion. ArXiv abs/2108.01387 (2021)
3. Chen, X., et al.: Hybrid transformer with multi-level fusion for multimodal knowledge graph completion. In: Proceedings of the 45th International ACM SIGIR Conference on Research and Development in Information Retrieval (2022)
4. Dettmers, T., Minervini, P., Stenetorp, P., Riedel, S.: Convolutional 2D knowledge graph embeddings. In: AAAI (2018)
5. Fu, J., Wang, J., Rui, Y., Wang, X.J., Mei, T., Lu, H.: Image tag refinement with view-dependent concept representations. IEEE Trans. Circuits Syst. Video Technol. **25**, 1409–1422 (2015)
6. Galkin, M., Wu, J., Denis, E., Hamilton, W.L.: NodePiece: compositional and parameter-efficient representations of large knowledge graphs. ArXiv abs/2106.12144 (2021)
7. Gao, H., Yang, K., Yang, Y., Zakari, R.Y., Owusu, J.W., Qin, K.: QuatDE: dynamic quaternion embedding for knowledge graph completion. ArXiv abs/2105.09002 (2021)
8. Guo, L., Wang, W., Sun, Z., Liu, C., Hu, W.: Decentralized knowledge graph representation learning. ArXiv abs/2010.08114 (2020)
9. Hiraoka, T.: MaxMatch-Dropout: subword regularization for wordpiece. In: International Conference on Computational Linguistics (2022)
10. Huang, H., Wang, Y., Feng, C., Liu, Z., Zhou, Q.: Leveraging conceptualization for short-text embedding. IEEE Trans. Knowl. Data Eng. **30**(7), 1282–1295 (2018)
11. Ji, G., He, S., Xu, L., Liu, K., Zhao, J.: Knowledge graph embedding via dynamic mapping matrix. In: ACL (2015)
12. Ji, S., Pan, S., Cambria, E., Marttinen, P., Yu, P.S.: A survey on knowledge graphs: representation, acquisition and applications. IEEE Trans. Neural Netw. Learn. Syst. (2021)
13. Lin, Y., Liu, Z., Sun, M., Liu, Y., Zhu, X.: Learning entity and relation embeddings for knowledge graph completion. In: AAAI'15, pp. 2181–2187 (2015)
14. Lv, X., et al.: Do pre-trained models benefit knowledge graph completion? a reliable evaluation and a reasonable approach. In: FINDINGS (2022)

15. Ma, S., Ding, J., Jia, W., Wang, K., Guo, M.: TransT: type-based multiple embedding representations for knowledge graph completion. In: Joint European Conference on Machine Learning and Knowledge Discovery in Databases, pp. 717–733 (2017)
16. Niu, G., Li, B., Zhang, Y., Pu, S., Li, J.: AutoETER: automated entity type representation for knowledge graph embedding. ArXiv abs/2009.12030 (2020)
17. Park, J.W., Hwang, S.W., Wang, H.: Fine-grained semantic conceptualization of FrameNet. In: AAAI, pp. 2638–2644 (2016)
18. Shi, B., Weninger, T.: ProjE: embedding projection for knowledge graph completion. In: AAAI (2016)
19. Song, Y., Wang, H., Wang, Z., Li, H., Chen, W.: Short text conceptualization using a probabilistic knowledgebase. In: Proceedings of the Twenty-Second International Joint Conference on Artificial Intelligence - Volume Volume Three, pp. 2330–2336 (2011)
20. Song, Y., Wang, S., Wang, H.: Open domain short text conceptualization: a generative + descriptive modeling approach. In: Proceedings of the 24th International Conference on Artificial Intelligence (2015)
21. Song, Y., Wang, S., Wang, H.: Open domain short text conceptualization: a generative + descriptive modeling approach. In: International Conference on Artificial Intelligence, pp. 3820–3826 (2015)
22. Sun, Z., Deng, Z., Nie, J.Y., Tang, J.: RotatE: knowledge graph embedding by relational rotation in complex space. ArXiv abs/1902.10197 (2019)
23. Tang, J., Shu, X., Li, Z., Jiang, Y.G., Tian, Q.: Social anchor-unit graph regularized tensor completion for large-scale image retagging. IEEE Trans. Pattern Anal. Mach. Intell. **41**, 2027–2034 (2019)
24. Tang, J., et al.: Tri-clustered tensor completion for social-aware image tag refinement. IEEE Trans. Pattern Anal. Mach. Intell. **39**, 1662–1674 (2017)
25. Toutanova, K., Chen, D.: Observed versus latent features for knowledge base and text inference. In: Proceedings of the 3rd Workshop on Continuous Vector Space Models and their Compositionality (2015)
26. Wang, F., Wang, Z., Li, Z., Wen, J.R.: Concept-based short text classification and ranking. In: The ACM International Conference, pp. 1069–1078 (2014)
27. Wang, Y., Huang, H., Feng, C.: Query expansion based on a feedback concept model for microblog retrieval. In: International Conference on World Wide Web, pp. 559–568 (2017)
28. Wang, Y., Huang, H., Feng, C., Zhou, Q., Gu, J., Gao, X.: CSE: conceptual sentence embeddings based on attention model. In: 54th Annual Meeting of the Association for Computational Linguistics, pp. 505–515 (2016)
29. Wang, Y., Liu, Y., Zhang, H., Xie, H.: Leveraging lexical semantic information for learning concept-based multiple embedding representations for knowledge graph completion. In: APWeb/WAIM (2019)
30. Wang, Y., Ouyang, X., Zhu, X., Zhang, H.: Concept commons enhanced knowledge graph representation. In: Memmi, G., Yang, B., Kong, L., Zhang, T., Qiu, M. (eds.) Knowledge Science, Engineering and Management - 15th International Conference, KSEM 2022, Singapore, August 6–8, 2022, Proceedings, Part I. Lecture Notes in Computer Science, vol. 13368, pp. 413–424. Springer (2022). https://doi.org/10.1007/978-3-031-10983-6_32
31. Wang, Y., Wang, Z., Zhang, H., Liu, Z.: Microblog retrieval based on concept-enhanced pre-training model. ACM Trans. Knowl. Discov. Data **17**, 1–32 (2022)
32. Wang, Y., Zhang, H.: HARP: a novel hierarchical attention model for relation prediction. ACM Trans. Knowl. Discov. Data **15**, 171–1722 (2021)
33. Wang, Y., Zhang, H., Li, Y., Xie, H.: Simplified representation learning model based on parameter-sharing for knowledge graph completion. In: CCIR (2019)
34. Wang, Z., Zhang, J., Feng, J., Chen, Z.: Knowledge graph embedding by translating on hyperplanes. In: AAAI'14, pp. 1112–1119 (2014)

35. Wang, Z., Zhao, K., Wang, H., Meng, X., Wen, J.R.: Query understanding through knowledge-based conceptualization. In: International Conference on Artificial Intelligence, pp. 3264–3270 (2015)
36. Wu, L., Hua, X., Yu, N., Ma, W.Y., Li, S.: Flickr Distance: a relationship measure for visual concepts. IEEE Trans. Pattern Anal. Mach. Intell. **34**, 863–875 (2012)
37. Wu, W., Li, H., Wang, H., Zhu, K.Q.: Probase: a probabilistic taxonomy for text understanding. In: Proceedings of the 2012 ACM SIGMOD International Conference on Management of Data, pp. 481–492 (2012)
38. Xiao, H., Huang, M., Zhu, X.: TransG: a generative model for knowledge graph embedding. In: Meeting of the Association for Computational Linguistics, pp. 2316–2325 (2016)
39. Xie, R., Liu, Z., Jia, J.J., Luan, H., Sun, M.: Representation learning of knowledge graphs with entity descriptions. In: AAAI (2016)
40. Xie, R., Liu, Z., Sun, M.: Representation learning of knowledge graphs with hierarchical types. In: International Joint Conference on Artificial Intelligence, pp. 2965–2971 (2016)
41. Yao, L., Mao, C., Luo, Y.: KG-BERT: BERT for knowledge graph completion. ArXiv abs/1909.03193 (2019)

A Novel Convolutional Neural Network Architecture with a Continuous Symmetry

Yao Liu[1], Hang Shao[2], and Bing Bai[1(✉)]

[1] Tsinghua University, Beijing, China
baibing12321@163.com
[2] Zhejiang Future Technology Institute, Jiaxing, China
shaohang@zfti.org.cn

Abstract. This paper introduces a new Convolutional Neural Network (ConvNet) architecture inspired by a class of partial differential equations (PDEs) called quasi-linear hyperbolic systems. With comparable performance on image classification task, it allows for the modification of the weights via a continuous group of symmetry. This is a significant shift from traditional models where the architecture and weights are essentially fixed. We wish to promote the (internal) symmetry as a new desirable property for a neural network, and to draw attention to the PDE perspective in analyzing and interpreting ConvNets in the broader Deep Learning community.

Keywords: Convolutional Neural Networks · Partial Differential Equations · Continuous Symmetry

1 Introduction

With the tremendous success of Deep Learning in diverse fields from computer vision [10] to natural language processing [18], the model invariably acts as a black box of numerical computation [2], with the architecture and the weights largely fixed, i.e., they can not be modified without changing the output (for a fixed input), except by permuting neurons or units from the same layer. One would say that the **symmetry** of the neural network, the set of transformations that do not affect the model's prediction on any input, is

$$\mathrm{Sym(model)} = \prod_i S_{n_i}\,,$$

the product of symmetric groups on n_i "letters," where n_i is the number of *interchangeable* neurons in the i-th layer.

Although quite large as a group, it does not permit us to modify the model in any meaningful way. In the case of Convolutional Neural Networks (ConvNets), the channels are essentially fixed and frozen in place, due to the presence of coordinate-wise activation functions [21] (such as ReLU), which arguably build certain semantic contents in the channels [13], thus mixing them would destroy

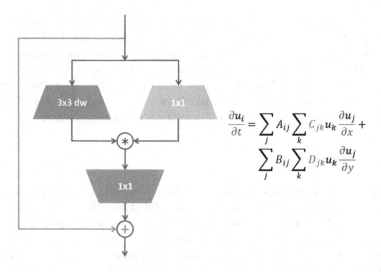

$$\frac{\partial u_i}{\partial t} = \sum_j A_{ij} \sum_k C_{jk} u_k \frac{\partial u_j}{\partial x} +$$
$$\sum_j B_{ij} \sum_k D_{jk} u_k \frac{\partial u_j}{\partial y}$$

Fig. 1. Schematic of a single block of our ConvNet architecture based on Eq. (3), to replace the bottleneck block of ResNet50. The trapezoids represent the increase/decrease in the number of channels. The corresponding components in PDE are illustrated (best viewed in color).

the model. The *nonlinear* activation unit is generally thought to be an essential component for the neural network to fit arbitrary *nonlinear* functions.

With inspiration and guidance from partial differential equations (PDEs), specifically *first-order quasi-linear hyperbolic systems* [1], we introduce a new architecture of ConvNet with a different type of nonlinearity, which allows us to remove most of the activations without degrading performance. As a result, the new architecture admits a *continuous* group of symmetry (i.e., a Lie group, as opposed to a discrete group) that allows mixing of the channels; in one version, it's the full *general linear* (GL) *group*, the set of all invertible $n_i \times n_i$ matrices:

$$\text{Sym(model)} = \prod_i GL(n_i, \mathbb{R}).$$

With a judicious choice of transformations, we may alter the weights so that the connections become more sparse, resulting in a smaller model (a kind of lossless pruning). Since the group is continuous, one might use the method of gradient descent to search for it. In addition, it may also lead to a better understanding of the inner workings of the neural network, much like how matrix diagonalization leads to the decoupling of a system of (linear) differential equations into different "modes", which are easier to interpret.

We primarily present the simplest version of our model based on ResNet50, illustrated in Fig. 1 alongside the corresponding PDE. The nonlinearity is at the element-wise multiplication of the two branches (3×3 and 1×1 conv), and we apply activations at the end of only *four* of the 16 blocks. See §5.1 for details.

This is a preliminary report on our new architecture, and the relevant parts of partial differential equations behind it. It is our hope that the research community builds upon and analyzes the properties of this new architecture, and to take the PDE perspective seriously in designing, analyzing, or simply describing *all* aspects of ConvNets. Moreover, given that the Transformer architecture [18] also involves a similar nonlinearity apart from softmax and activation functions, it could potentially be made to admit a continuous symmetry as well.

2   Related Work

The link with **differential equations** has been recognized, and well exploited [4, 8, 9, 12, 16], soon after the introduction of ResNet [10] by He et al. in 2015, if not known implicitly before; see also [17] for a more recent analysis. With a few exceptions [12, 16], most discussions do not make an emphasis on *partial* differential equations, and to the best of our knowledge, little activity has been devoted to *designing* new architecture from this perspective. Even though a PDE can be regarded as an ODE for which the state space is "infinite-dimensional", or upon discretization, a finite but large system with interactions only between neighboring pixels, we find the PDE perspective, specifically of hyperbolic systems, more illuminating and *fruitful* (albeit limited to ConvNets), and deserves more attention and further study in the broader Deep Learning community.

A related but distinct field of research is using Deep Learning methods to solve various PDEs of interests to physicists, applied mathematicians, and engineers. We shall only mention two pioneering works that have attracted the most attention: Fourier Neural Operators (FNO) [11] and Physics-informed Neural Networks (PINN) [15].

Symmetry and **equivariance** often appear in the theoretical discussions of ConvNets and Graph Neural Networks (GNN) [3, 6], though it's worth pointing out the distinction from our usage: More often, we say a network has a (translational or permutation) *symmetry*, or is *equivariant* or *invariant* (under translation or permutation), if when we transform the input in a certain way, the output is also transformed accordingly, or does not change at all; the model itself remains fixed. In our scenario, we are directly transforming the weights of the model, which incidentally does not need to be trained. Nevertheless, much of our work involves finding a good enough model that achieves comparable performance on standard training sets. (It shall be apparent that, like conventional ConvNets, our model is also equivariant under translations.)

One type of operation, known under the term **structural reparametrization** [7], can claim to modify the model after training. However, it can only merge consecutive layers or operations that are both linear; the basic example is conv followed by batchnorm. As such, it is better regarded as a trick in training: for whatever reason, it is better to train with a deeper and more complicated network than is necessary for the model, and is fundamentally different from the kind of symmetry that our model has.

3   Designing ConvNets from the PDE Perspective

Given that ResNet is numerically solving a particular system of PDEs,

$$\frac{\partial u_i}{\partial t} = \sigma\left(\sum_j L_{ij} u_j\right), \quad \text{for} \quad i = 1, \ldots, n$$

$$L_{ij} := \alpha_{ij}\frac{\partial}{\partial x} + \beta_{ij}\frac{\partial}{\partial y} + \gamma_{ij}\frac{\partial^2}{\partial x \partial y} + \cdots$$

of n unknowns $u_i \equiv u_i(t, x, y)$, with initial condition at $t = 0$, wherein the coefficients are learned (for background, see Appendix), it is natural to take inspiration from other PDEs as found in mathematics and physics, and see what new ConvNet architecture would come out. Here are some natural changes that one could make:

- Change the constant coefficients to be variables (of x and y), simply as linear or polynomial functions. The equation would still be linear, but now the space of PDEs would include, e.g., this special equation ($n = 1$)

$$\frac{\partial u}{\partial t} = -y\frac{\partial u}{\partial x} + x\frac{\partial u}{\partial y},$$

which is solved by simply *rotating* the initial data $f(x, y)$ by angle t (around the origin):

$$u(t, x, y) = f(x \cos t - y \sin t, x \sin t + y \cos t),$$

as can be readily verified. It is reasonable to expect that such a variation on ResNet would allow the model to make rotations and dilations — in addition to translations — on the input image.
- One might think that the standard "zero padding" of conv layer corresponds to the Dirichlet boundary condition: the value of u on the boundary being fixed at a prescribed value for all time. On closer inspection, it is a little different. One could also experiment with other boundary conditions: the other most common one is the Neumann condition, that the "normal derivative" of u on the boundary is prescribed. The different conditions have the effects that the signals would "bounce back" off the boundary differently.
- In a typical PDE, the matrix of coefficients is constant or slowly varying with time, while in neural networks the weights from different layers are initialized independently, drawn from a (normal) distribution. One could try to force the weights from neighboring layers to correlate, either by weight-sharing or by introducing a term in the loss function that penalizes large variations between layers.

Having experimented with some of these ideas on small datasets, we have not found a specific variation that yields convincing results on the full ImageNet. We then looked into ways that the coefficients may depend on u itself, which makes the equation nonlinear (apart from the activation functions). It may be viewed as a kind of "dynamic kernel", but we draw inspiration from a class of PDEs called *quasi-linear hyperbolic systems*, which may be the simplest, well-studied nonlinear systems for which the number of equations can be arbitrary.

In two spatial and one time dimensions, a first-order **quasi-linear system** is typically of the form

$$\frac{\partial u_i}{\partial t} = \sum_j \mathcal{A}_{ij}(u)\frac{\partial u_j}{\partial x} + \sum_j \mathcal{B}_{ij}(u)\frac{\partial u_j}{\partial y}, \tag{1}$$

where the coefficient matrices may depend on u (but not derivatives of u), and it is **hyperbolic**[1] if $\mathcal{A}$ and $\mathcal{B}$ are diagonalizable with *only* real eigenvalues $\lambda_i(u)$, e.g., when $\mathcal{A}$ and $\mathcal{B}$ are symmetric, for *any* u. Leaving aside the latter condition, the simplest example is to make each entry a linear function of u:

$$\mathcal{A}_{ij}(u) = \sum_k \mathcal{A}_{ijk}u_k, \tag{2}$$

and similarly for $\mathcal{B}$. By dimension count, such a tensor would be very large (for large n), and it would deviate too much from typical ConvNets. Instead, we shall restrict to

$$\frac{\partial u_i}{\partial t} = \sum_j A_{ij} \sum_k C_{jk}u_k\frac{\partial u_j}{\partial x} + \sum_j B_{ij} \sum_k D_{jk}u_k\frac{\partial u_j}{\partial y}, \tag{3}$$

which is straightforward to turn into a ConvNet (see §5.1 and Fig. 1 for details), and the number of parameters is kept at a reasonable level. Since nonlinearity is already built-in, we thought it would not be necessary to introduce activations, at least not at every turn; and much to our surprise the model trains just as well, if not better. With this simple change, the model now has a *continuous* symmetry, from mixing of the channels, that is not present in conventional ConvNets with coordinate-wise activation after every conv layer, and we believe it is a more significant contribution than matching or breaking state of the art. It is likely that, with enough compute, ingenuity, and techniques from Neural Architecture Search, variations of this architecture could compete with the best image models of comparable size. (We have not tried to incorporate the modifications listed earlier, as they are all linear and we wish to explore the new nonlinearity on its own.)

It is observed that, once we remove all the activations, or use only ReLU, the training is prone to break down: it would fail at a particular epoch, with one or more samples (either from the training or validation set) causing the network to output NaN, and the model could not recover from it. To mitigate this, we add activations such as hardtanh that clip off large values, once every few blocks.

---

[1] The designation may appear cryptic. It originates from the classic wave equation, which has some semblance in form with the equation of a hyperbola or hyperboloid. See Appendix §A.3.

It is also observed that resuming training with a smaller learning rate may get around the "bad regions" of the parameter space. More analyses are needed to determine the precise nature and cause of this phenomenon (it might be related to the formation of "shock waves" in nonlinear hyperbolic equations [1]), and perhaps other ways to avoid it.

We provide here the details of the activation functions. In standard PyTorch [14], nn.Hardtanh is implemented as

$$\mathrm{hardtanh}(x) := \begin{cases} \mathrm{max\_val} & x > \mathrm{max\_val} \\ \mathrm{min\_val} & x < \mathrm{min\_val} \\ x & \mathrm{otherwise}. \end{cases}$$

We typically use ± 1 as the clip-off values. We also introduce two multi-dimensional variants that we call "hardball" and "softball," which appear to give the best performance. Hardball is so defined that it takes a vector $\mathbf{x} \in \mathbb{R}^n$ and maps it into the ball of radius R

$$\mathrm{hardball}(\mathbf{x}) := \begin{cases} \mathbf{x} & |\mathbf{x}| < R \\ R\mathbf{x}/|\mathbf{x}| & |\mathbf{x}| \geq R, \end{cases}$$

where $|\mathbf{x}|$ is the Euclidean norm. We set R to be the square root of n (the number of channels), though other choices may be better. Softball is a soft version,

$$\mathrm{softball}(\mathbf{x}) := \frac{\mathbf{x}}{\sqrt{1 + |\mathbf{x}|^2/R^2}},$$

and they both have the property of being equivariant under rotation. (One may perhaps regard them as normalization layers rather than activation functions.)

4   Symmetry of the Model

The symmetry of our model would depend on the specific implementation, and may be more complicated than one would naively expect. We shall first consider it on the level of the PDE.

With a change of coordinates $\tilde{u}_i = \sum_j T_{ij} u_j$ for an invertible matrix T, the general equation (1) with (2)

$$\frac{\partial u_i}{\partial t} = \sum_{j,k} \mathcal{A}_{ijk} u_k \frac{\partial u_j}{\partial x} + \sum_{j,k} \mathcal{B}_{ijk} u_k \frac{\partial u_j}{\partial y}$$

would transform *only* in the coefficient tensors $\mathcal{A}_{ijk}$ and $\mathcal{B}_{ijk}$. Indeed,

(For clarity, we omit the second half involving $\mathcal{B}$ and $\frac{\partial}{\partial y} \cdot$)

$$\frac{\partial \tilde{u}_i}{\partial t} = \sum_j T_{ij} \frac{\partial u_j}{\partial t}$$

$$= \sum_j T_{ij} \sum_{k,l} \mathcal{A}_{jkl} u_l \frac{\partial u_k}{\partial x}$$

$$= \sum_j T_{ij} \sum_{k,l} \mathcal{A}_{jkl} \sum_m T_{lm}^{-1} \tilde{u}_m \sum_r T_{kr}^{-1} \frac{\partial \tilde{u}_r}{\partial x}$$

$$= \sum_{m,r} \underbrace{\left(\sum_{j,k,l} T_{ij} \mathcal{A}_{jkl} T_{lm}^{-1} T_{kr}^{-1} \right)}_{\tilde{\mathcal{A}}_{irm}} \tilde{u}_m \frac{\partial \tilde{u}_r}{\partial x} \cdot$$

We note in passing that similar calculations are commonplace in *classical* differential geometry when making a change of coordinates *on the base space*. Here, we are making a change of coordinates on the "dependent" variables. From a more abstract point of view, this is the induced representation of $GL(V)$ on the tensor product $V \otimes V^* \otimes V^* \cong \mathrm{Hom}(V \otimes V, V)$ for a vector space $V \cong \mathbb{R}^n$.

On the level of the neural network, we only need to make sure that the T^{-1} comes from the previous layer, i.e., it is the inverse of the T that appears in transforming the previous block. With such a transformation at each block, we find that the overall symmetry of the model is

$$\mathrm{Sym(model)} = \prod_i G_i \,,$$

with([4]The group $O(n) := \{M, n \times n \text{ matrix} \mid MM^T = I_n\}$ is known as the orthogonal group, and it preserves the Euclidean norm of $\mathbb{R}^n$.)

$$G_i = \begin{cases} S_n & \sigma = \text{relu, or any element-wise activation} \\ O(n) & \sigma = \text{hardball, softball, etc.}^4 \\ GL(n, \mathbb{R}) & \sigma = \text{identity} \,. \end{cases}$$

As noted before, this "fully connected" block would be too costly to train, if we are to match ResNet in which the last stage uses as many as $n = 512$ channels. One simple way to reduce the number of parameters is to make the tensor "block diagonal", and the transformation would only mix channels from the same block. The bottleneck block of ResNet addresses this by shrinking the number of channels before applying the 3×3 conv, but a similar approach would introduce additional layers that shield the main operations that we would like to transform.

If we are to take $\mathcal{A}_{ijk}$ to be of the special form as in Eq. (3), i.e., as the product of two matrices $A_{ij}C_{jk}$ (no summation), then it is not guaranteed that the transformed tensor would still factorize in the same way, for a generic T. By simple dimension count, the set of tensors that are factorizable in the prescribed way is a subvariety (submanifold) of dimension at most $2n^2$, and one wishes to find a $T \in GL(n, \mathbb{R})$ that keeps the resulting $\tilde{\mathcal{A}}_{ijk}$ on the subvariety. Given that the dimension of $GL(n, \mathbb{R})$ is n^2 and that $2n^2 + n^2 \ll n^3$, it is not *a prior* obvious that such transformations exist, apart from simple scalings and S_n, that are universal for all $\mathcal{A}_{ijk}$.

What one may hope for is that, for a specific $\mathcal{A}_{ijk}$ that factors, we can find such a T. For example, if for some pair of indices j, j', we have $A_{ij} C_{jk} = A_{ij'} C_{j'k}$ for all i, k, then we can perform a rotation in the plane of the j and j' directions:

$$\begin{cases} \tilde{u}_j = \alpha u_j + \beta u_{j'} \\ \tilde{u}_{j'} = \gamma u_j + \delta u_{j'} \end{cases} \qquad \alpha\delta - \beta\gamma \neq 0\,.$$

Further investigation, either theoretical or numerical, may be needed to answer this question satisfactorily. It may be the case that there exists a symmetry that preserves the output *not* for all inputs, but only those inputs that are "similar" to the dataset. It would be a weaker form of symmetry, but no less useful in practice.

Lastly, it should be remarked that there is a trivial "symmetry" in the tensor $\mathcal{A}_{ijk}$ in the last two indices, i.e., $\mathcal{A}_{ijk}$ and $\mathcal{A}_{ikj}$ can be interchanged (so long as their sum is fixed). One may regard this as a redundancy in the parameter space, for we can force $\mathcal{A}_{ijk} = \mathcal{A}_{ikj}$, or $\mathcal{A}_{ijk} = 0$ for $j < k$ (and reduce the dimension roughly by half), and not due to mixing of the channels. We have not exploited this in the present work.

5 Experimental Results

5.1 Details of the Architecture

How do we turn Eq. (3) into a ConvNet? We first make the differential operators $\partial/\partial x$ and $\partial/\partial y$ into 3×3 convolutional kernels, but we allow the weights to be trainable instead of fixed. Each incoming channel (each u_j) would split into 2 — or better, 4 — channels, and this is conveniently implemented in nn.Conv2d by setting groups to equal the number of input channels, as in "depthwise convolution" [5]. The matrices are simply 1×1 conv layers, with A and B stacked into one, and C and D stacked into one. Batchnorm is applied after the 3×3 and at the end. As with standard ResNet, the time derivative turns into the skip connection, and we arrive at the architecture of a single block as illustrated in Fig. 1. We do not enforce symmetry of these matrices to make the equation hyperbolic in the technical sense. As a general rule, we need not strictly follow the equation, but take the liberty in relaxing the weights whenever convenient.

One novelty is to make the weights of the 3×3 conv shared across the groups, which would make the transformations easier to implement. One may achieve this by making an nn.Conv2d with 1 input channel and 4 output channels, and at forward pass, we "repeat" the $4 \times 1 \times 3 \times 3$ weights in the zeroth dimension before acting on the input tensor. Most of our experiments are with this "minimalist" 3×3 conv, except the ones marked "no ws" (no weight-sharing) in Table 1.

It may be possible to implement this kind of "variable-coefficient convolution" *natively*, instead of using the existing PyTorch layer which is tailored to conventional convolutions.

For the full design of the neural network, we simply take the classic ResNet50 with [3,4,6,3] as the numbers of blocks in the four stages. No activation is applied except once only in each stage (e.g., at each downsampling), and we use nn.Hardtanh or our variants, hardball and softball (see §3 for definitions), instead of ReLU, lest the training would fail completely or the resulting model would not work as well.

5.2 Experiments

As is standard in computer vision since the 2012 Deep Learning revolution, we trained our model as an image classification task, on the ImageNet dataset. The ImageNet-1k contains 1000 classes of labeled images, and for the sake of faster iterations, we primarily trained on a 100-class subset on a single GPU, while maintaining the standard image size of 224×224.

We use the timm library of PyTorch image models [19] for best practices in implementing ResNet and its training [20]. We took the official training script for ResNeXt-50 (SGD, cosine learning rate, with a warmup of 5 epochs, etc.) except that the peak learning rate is set to 0.3 instead of 0.6, and the total number of epochs is set to 50. The results are in Table 1, where we mainly record two ways of modifying the model: changing only the activation function, and altering the placements of the conv layers within the block, corresponding to modifying Eq. (3) into Eqs. (4)–(7):

$$\frac{\partial u_i}{\partial t} = \sum_k C_{ik} u_k \sum_j A_{ij} \frac{\partial u_j}{\partial x} + \sum_k D_{ik} u_k \sum_j B_{ij} \frac{\partial u_j}{\partial y} \tag{4}$$

$$\frac{\partial u_i}{\partial t} = \sum_k C_{ik} u_k \frac{\partial u_i}{\partial x} + \sum_k D_{ik} u_k \frac{\partial u_i}{\partial y} \tag{5}$$

$$\frac{\partial u_i}{\partial t} = \sum_j A_{ij} \frac{\partial}{\partial x} \sum_k C_{jk} u_j u_k + \sum_j B_{ij} \frac{\partial}{\partial y} \sum_k C_{jk} u_j u_k \tag{6}$$

$$\frac{\partial u_i}{\partial t} = \sum_j A_{ij} \frac{\partial}{\partial x} \sum_k C_{jk} u_j u_k + \sum_j B_{ij} \frac{\partial}{\partial y} \sum_k D_{jk} u_j u_k \tag{7}$$

Note that Eq. (5) only has each u_i depending on its own derivatives, and the model is thus smaller and limited in expressivity or capacity. Equations (6) and (7) have the derivative acting on the product $u_j u_k$, and it is often called a system of *conservation laws*. They can easily be rewritten in the form of Eq. (3), and the difference in performance may be attributable simply to the model size.

Table 1. Performance on a 100-class subset of ImageNet-1k, trained for 50 epochs with identical training strategy. For our model, activation is applied either at the end of each block (@all), or only at downsampling (@ds). Inside the block, the number of channels increases by a factor of 4, except when indicated with "x6".

model	# parameters	top-1 acc	activation
ResNet50	23.7M	84.52	
MobileNet_v3_large	4.33M	82.91	
Eq. (3)	8.61M	82.06	relu@all
Eq. (3)	8.61M	82.34	hardtanh@ds
Eq. (3)	8.61M	83.50	hardball@ds
Eq. (3)	8.61M	83.66	softball@ds
Eq. (3) (no ws)	8.73M	84.24	softball@ds
Eq. (3) (no ws, x6)	13.0M	84.58	softball@ds
Eq. (4)	5.70M	81.88	softball@ds
Eq. (5)	4.26M	78.64	
Eq. (6)	5.61M	82.52	
Eq. (7) (no ws, x6)	13.0M	**84.96**	

It is expected that, when going to the full ImageNet-1k, and allowing longer training and hyperparameter tuning, the best-performing model may be different from the ones we found in Table 1.

We refrain from making assertions on *why* — or *if* — this class of PDEs is a better base model for ConvNets, or how the theory of PDE can provide the ultimate answer to the *effectiveness* of neural networks as universal function approximators, using gradient descent. Whatever mechanisms that make ResNet work, also make our model work.

6 Conclusion

We present a new Convolutional Neural Network architecture inspired by a class of PDEs called quasi-linear hyperbolic systems, and with preliminary experiments, we found a simple implementation that showed promising results. Even though it is known, within small circles, the close connection between PDE and ConvNet, we made the first architecture design directly based on a nonlinear PDE, and as a result, we are able to remove most of the activation functions which are generally regarded as indispensable. The new architecture admits a continuous symmetry that could be exploited, hopefully in future works. We expect that this opens up a new direction in neural architectural design, demonstrates the power of the PDE perspective for ConvNets, and opens the door for other concepts and techniques in nonlinear PDE, both theoretical and numerical, for improved understanding of deep neural networks.

Acknowledgements. This work is supported in part by Provincial Key R&D Program of Zhejiang under contract No. 2021C01016, in part by Young Elite Scientists Sponsorship Program by CAST under contract No. 2022QNRC001.

References

1. Alinhac, S.: Hyperbolic partial differential equations. Springer Science & Business Media (2009). https://doi.org/10.1007/b13382_10
2. Bai, B., Liang, J., Zhang, G., Li, H., Bai, K., Wang, F.: Why attentions may not be interpretable? In: Proceedings of the 27th ACM SIGKDD Conference on Knowledge Discovery & Data Mining, pp. 25–34 (2021)
3. Bronstein, M.M., Bruna, J., Cohen, T., Veličković, P.: Geometric deep learning: grids, groups, graphs, geodesics, and gauges. arXiv preprint arXiv:2104.13478 (2021)
4. Chen, R.T., Rubanova, Y., Bettencourt, J., Duvenaud, D.K.: Neural ordinary differential equations. In: Advances in Neural Information Processing Systems 31 (2018)
5. Chollet, F.: Xception: deep learning with depthwise separable convolutions. In: Proceedings of the IEEE Conference on Computer Vision and Pattern Recognition, pp. 1251–1258 (2017)
6. Cohen, T., Welling, M.: Group equivariant convolutional networks. In: International Conference on Machine Learning, pp. 2990–2999. PMLR (2016)
7. Ding, X., Zhang, X., Han, J., Ding, G.: Diverse branch block: building a convolution as an inception-like unit. In: Proceedings of the IEEE/CVF Conference on Computer Vision and Pattern Recognition, pp. 10886–10895 (2021)
8. E, W.: A proposal on machine learning via dynamical systems. Commun. Math. Stat. **1**(5), 1–11 (2017)
9. E, W., Han, J., Li, Q.: Dynamical systems and optimal control approach to deep learning. Math. Aspects Deep Learn., 422 (2022)
10. He, K., Zhang, X., Ren, S., Sun, J.: Deep residual learning for image recognition. In: Proceedings of the IEEE Conference on Computer Vision and Pattern Recognition, pp. 770–778 (2016)
11. Li, Z., et al.: Fourier neural operator for parametric partial differential equations. arXiv preprint arXiv:2010.08895 (2020)
12. Long, Z., Lu, Y., Ma, X., Dong, B.: PDE-Net: Learning PDEs from data. In: International Conference on Machine Learning, pp. 3208–3216. PMLR (2018)
13. Olah, C., Cammarata, N., Schubert, L., Goh, G., Petrov, M., Carter, S.: Zoom. In: An Introduction to Circuits. Distill **5**(3), e00024–001 (2020)
14. Paszke, A., et al.: PyTorch: an imperative style, high-performance deep learning library. In: Advances in Neural Information Processing Systems 32 (2019)
15. Raissi, M., Perdikaris, P., Karniadakis, G.E.: Physics-informed neural networks: a deep learning framework for solving forward and inverse problems involving nonlinear partial differential equations. J. Comput. Phys. **378**, 686–707 (2019)
16. Ruthotto, L., Haber, E.: Deep neural networks motivated by partial differential equations. J. Math. Imaging Vis. **62**, 352–364 (2020)
17. Sander, M., Ablin, P., Peyré, G.: Do residual neural networks discretize neural ordinary differential equations? Adv. Neural. Inf. Process. Syst. **35**, 36520–36532 (2022)
18. Vaswani, A., et al.: Attention is all you need. In: Advances in Neural Information Processing Systems 30 (2017)

19. Wightman, R.: PyTorch image models (2019). https://github.com/rwightman/pytorch-image-models, https://doi.org/10.5281/zenodo.4414861
20. Wightman, R., Touvron, H., Jégou, H.: ResNet strikes back: an improved training procedure in TIMM. arXiv preprint arXiv:2110.00476 (2021)
21. Zhang, H., Weng, T.W., Chen, P.Y., Hsieh, C.J., Daniel, L.: Efficient neural network robustness certification with general activation functions. In: Advances in Neural Information Processing Systems 31 (2018)

Dynamic Task Subspace Ensemble for Class-Incremental Learning

Weile Zhang, Yuanjian He, and Yulai Cong[✉]

Sun Yat-sen University, Shenzhen 518107, China
heyj39@mail2.sysu.edu.cn, yulaicong@gmail.com

Abstract. Deep-learning models are expected to continually learn new concepts without forgetting old ones in real-world applications with shifting data distributions. However, the notorious catastrophic forgetting often occurs. Recently, methods based on task subspace modeling have been developed to address this issue by gradually adding new subspaces to learn new concepts. In this paper, we reveal that such task-subspace-modeling methods may suffer from the inter-task confusion issue, leading to degraded performance in challenging class-incremental learning settings. Concerning addressing the forgetting issue of deep learning models, we propose a two-stage framework called *Dynamic tAsk Subspace Ensemble* (DASE), the first stage of which involves the dynamic expansion of the extractor network for memory efficiency, while the second stage delivers dynamic learning and aggregation of diverse features. To further enhance the discriminative capacity of the aggregated features for both historical and new classes, we also introduce new feature-enhancement techniques. Experimental results demonstrate that our method achieves state-of-the-art CIL performance on natural image datasets (CIFAR-100 and ImageNet) and Synthetic Aperture Radar image datasets (MSTAR and OpenSARShip).

Keywords: class-incremental learning · inter-task confusion · dynamic task subspace ensemble · memory-efficient

1 Introduction

Most deep-learning models, trained with static training data, cannot adapt to the data-distribution drifting that is widespread in real-world applications. When given a dynamically changing task stream, these models often suffer from *catastrophic forgetting* [20,21], where a model learns new task skills by modifying its parameters that contain old task skills, leading to catastrophic forgetting of its old skills during that task stream. Concerning addressing the forgetting issue of deep learning models, many researchers [25,27,31,35] have concentrated on the setup of class-incremental learning (CIL), where the goal is to sequentially master the discrimination capabilities of new classes without forgetting those of historical ones.

L. Fang et al. (Eds.): CICAI 2023, LNAI 14474, pp. 322–334, 2024.
https://doi.org/10.1007/978-981-99-9119-8_29

A promising direction that delivers state-of-the-art performance for CIL lies in dynamic architectures, where the network architecture is dynamically expanded to deal with the *stability-plasticity dilemma* [22]. Task subspace modeling [2,23], a dynamic-architecture-based method, suggests that low dimensional subspaces within the whole neural network parameters may be manifested as capabilities for different tasks, allowing tasks to be represented as latent basis tasks and their linear combinations [14]. These methods are particularly memory-efficient due to their minimal demand for new parameters to construct a new sub-network [2,23] and often employ two steps to solve a CIL task. Step I deals with the task-ID prediction, *i.e.*, using a minimal-entropy criterion to predict which task is a data sample from. Step II solves within-task prediction, predicting a special class within the task from the last step [11]. Although they work well on some natural image datasets like CIFAR-100 [23], we empirically reveal that the task-ID prediction based on the minimal-entropy criterion is not generic for Synthetic Aperture Radar (SAR) CIL tasks with a high degree of sample similarity between tasks [16]. Due to the unavailability of historical task data, the features learned in each sub-network, referred to as local features, only distinguish classes within the task in the task-incremental (TIL) setting, where the task-id is given. However, we empirically reveal that in the case of CIL, where task-ID is not provided, there is a high likelihood of incorrect task-ID prediction with the commonly used minimal-entropy criterion [11,23]. Therefore, the task subspace model often fails to deliver features that discriminate among all classes after two steps, a phenomenon known as inter-task confusion [19].

To address the inter-task confusion issue and improve task subspace modeling for CIL, we propose a novel two-stage method called *Dynamic tAsk Subspace Ensemble* (DASE), which involves dynamic expansion of task subspace hybrid network and dynamic learning and aggregation of features. The presented DASE is developed based on empirical experiences [17,35], which show that shallow layers of a convolutional (conv) network tend to acquire generalized features, while deep layers specialize in high-level features tailored to specific tasks. Accordingly, DASE leverages a memory-efficient hybrid network structure that expands shallow conv layers using a low-rank reconstruction method and fully expands deep layers, as illustrated in Fig. 1(a). To preserve the old task skills learned from previous tasks while also ensuring that the new skills contain concepts from both the old and new tasks, the presented DASE preserves old feature extractors for stability and adds the new feature extractor for plasticity, ensuring new features to be discriminative across all observed classes, thus reducing inter-task confusion as depicted in Fig. 1(b).

The main contributions of this paper are summarized as follows.

- We empirically analyze that existing CIL methods based on task subspace modeling may suffer from severe inter-task confusion, which contributes valuable insights for future research in developing improved CIL algorithms.
- We propose a novel two-stage framework called DASE that learns the global discriminative features for CIL, addressing the inter-task confusion and achieving state-of-the-art performance on a wide range of CIL scenarios.

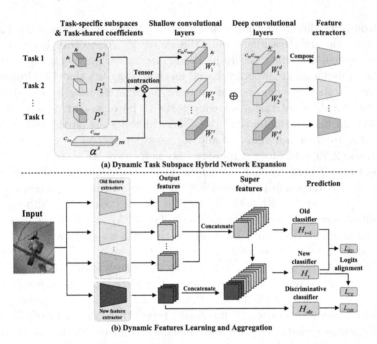

(a) Dynamic Task Subspace Hybrid Network Expansion

(b) Dynamic Features Learning and Aggregation

Fig. 1. The framework of our method. (a) Step I expands a feature extractor dynamically for each task. When new task data arrives, shallow convolutional layers share coefficients, while we initialize a new set of task-specific subspaces and deep convolutional layers framed in green for each task. (b) Step II dynamically learns and aggregates the features for CIL. $\mathcal{L}_{KD}$ signifies knowledge distillation loss at the logit level, $\mathcal{L}_{CE}$ refers to the classification loss, and $\mathcal{L}_{DIS}$ is the discriminative loss that guides the training of the feature extractors. (Color figure online)

2   Related Work

Existing CIL methods can be roughly divided into three categories, *i.e.,* replay-based [1,3,9,25], dynamic-architectures-based [5,31,35], and regularization-based methods [10,12]. Below we briefly introduce the first two categories that are most relevant to this paper.

Replay-based methods rely either on exemplar-stored historical data [1,9,25] or on pseudo-replay via *e.g.,* a GAN generator [3,29], to alleviate catastrophic forgetting. However, limited storage space for examplar can lead to an imbalance issue in class samples between historical tasks and the new one [7]. To relieve that issue, one may resort to the techniques presented in [27,28,30,33] to balance the old and new predictions. On the other hand, pseudo-replay methods [3,29] often have high training complexity and are likely computationally expensive to retain many generators.

Dynamic-architectures-based methods [31,32,35] incrementally enhance network capacity to accommodate new tasks while preserving previously acquired skills. Although these methods might prevent catastrophic forgetting, they rely on an additional task-ID to ensure the network accurately assigns data to the correct module. This dependency on task-ID makes these methods impractical for CIL. Existing methods [23,26] employ entropy to forecast the task-ID, but we empirically reveal that such a strategy is not generally applicable, as revealed in Sect. 3.1. Recently, a series of methods [27,31,34,35] preserve old feature extractors to maintain knowledge for old categories. When new tasks come, DER [31] expands a new backbone per incremental task and concatenates it with old feature extractors to form a higher dimensional feature space, FOSTER [27] adds an extra model compression stage to maintain limited model storage and MEMO [35] decouples the network structure and only expands deep layers. Compared to MEMO [35], we propose a strategy where shallow layers share coefficients and gradually expand task-specific subspaces to modify the features from the shallow layers with a slight shift, enhancing their generalizability across different tasks.

3   Method

We consider the problem of learning T tasks sequentially in CIL. Each task t encompasses data $(\mathcal{X}_t, \mathcal{Y}_t)$ sampled from the respective training data D_t. Here, $\mathcal{X}_t$ denotes N input data samples for task t, and $\mathcal{Y}_t$ denotes the matching ground truth labels. We assume no class overlap between tasks. The objective is to minimize the statistical risk of all observed tasks, given limited or no access to prior task data. Unlike TIL, task-ID isn't provided for inference, necessitating feature learning capable of task-ID prediction. In this context, we employ data replay, where the t-th training data $\tilde{\mathcal{D}}_t$ contains new task data from $\mathcal{D}_t$ and selected representative old task data from $\tilde{\mathcal{D}}_{t-1}$.

Below, we first analyze the inter-task confusion issue in task subspace modeling in Sect. 3.1. Then, we discuss the presented Dynamic tAsk Subspace Ensemble (DASE), with the dynamic task subspace hybrid network expansion presented in Sect. 3.2 and the dynamic features learning and aggregation for CIL in Sect. 3.3.

3.1   Inter-task Confusion in Task Subspace Modeling

We experimentally reveal that task subspace modeling methods mentioned in Sect. 1, which gradually add new sub-networks to learn the new tasks, are likely to suffer from severe inter-task confusion in practical applications.

We first experiment on MSTAR, which is a Synthetic Aperture Radar (SAR) image dataset consisting of ten classes. The dataset exhibits a high degree of similarity between samples across classes, as demonstrated in the supplementary material. We design a five-task CIL experiment on the MSTAR dataset, with each task encompassing two classes. After training, we have five sub-networks. Then, we input the data from the first task to these sub-networks, respectively, visualize every feature space with t-SNE [18] and use the minimal-entropy criterion to find the task-ID of each sample for the following analysis.

It is expected that all data points originating from task 1 are determined as task-ID $= 1$, *i.e.,* all points in Fig. 2(a) should be represented in a dark color,

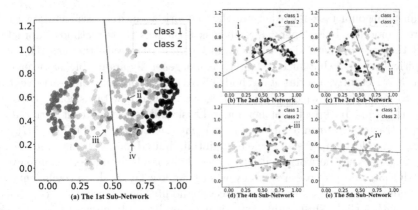

Fig. 2. t-SNE visualization of the output features of each sub-network. The data of the first task (*i.e.*, the data with task-ID = 1) are fed into the five sub-networks, and the task-ID of each sample is determined with the minimal-entropy criterion. If the task-ID of a specific sample is determined to be t, it will be presented in a dark color in the figure of the t-th sub-network and in a light color in other figures. It's expected that all points in (a) should be represented in a dark color. However, most samples with task-ID = 1 are incorrectly determined, such as the i-iv samples, which are incorrectly determined as the other task-IDs as illustrated in (b) to (e), meaning the minimal entropy criterion exhibits a high rate of misidentification for the task-ID.

while those in Figs. 2(b) to (e) should be represented in a light color. However, as depicted in Fig. 2(a), many samples near the decision boundary, indicated in a light color in the feature space, are given the incorrect task-ID, because they typically exhibit higher entropy values compared to those from other sub-networks as illustrated in Fig. 2(b) to (e). Consequently, these samples are prone to task-ID misidentification, causing substantial inter-task confusion.

Moreover, we observe similar results on CIFAR-10 [13] and MNIST [15]. The accuracy of task-ID is low, which restricts the accuracy of all classes in CIL, manifesting different degrees of inter-task confusion in various datasets. More experimental details are presented in the supplementary material and further experiments will be presented in Table 1 and Fig. 4.

In CIL, as the task with new class comes in sequentially, the model cannot learn the global features that are discriminative among all historical data classes and non-observed future classes. Accordingly, it's likely that the cause of inter-task confusion is, *i.e.*, multiple local features sequentially learned are only discriminative for the classes within the task, and there is a lack of a mechanism to merge these local features into global ones that can discriminate all observed classes.

3.2 Dynamic Task Subspace Hybrid Network Expansion

Motivated by the reason for inter-task confusion, we first introduce an efficient method for sub-network expansion, where each new feature extractor is constructed by a task subspace hybrid network. It establishes the network foundation for dynamic features learning and aggregation in Sect. 3.3.

Tensor Contraction. As presented in Fig. 1(a), given a conv neural network (CNN), tensor contraction decomposes the weight of each layer, $W^i \in R^{c_{in} \times c_{out} \times k \times k}$ of the i-th layer to two parts of parameters according to its shape as [23,24] in a low-rank manner. The first part is called the task-shared coefficient $\alpha^i \in R^{c_{in} \times c_{out} \times m}$, and the second part is called the task-specific subspace $P^i \in R^{m \times k \times k}$, where c_{in} and c_{out} are the number of input and output channels, k is the kernel size and m is the number of bases. Specifically, $W^i = \alpha^i \otimes P^i$.

Task Subspace Hybrid Network. As shown in Fig. 1(a), upon the arrival of the t-th task, we employ tensor contraction to reconstruct the shallow conv layer as $W_t^s = \alpha^s \otimes P_t^s$; For the deep conv layer W_t^d, we initialize a new one for each task, where $t \in \{1, 2, ..., T\}$. α^s is only initialized when $t = 1$ and P_t^s increases with the addition of tasks. Therefore, the shallow extractor Φ_t^s is composed by multiple shallow conv layers $\{W_t^s\}_S$ and the deep extractor Φ_t^d is composed by multiple deep conv layers $\{W_t^d\}_D$.

Accordingly, the output of the newly added extractor or called task subspace hybrid network Φ_t is

$$\Phi_t(x) = \Phi_t^d(\Phi_t^s(x)). \tag{1}$$

In the training phase of task 1, the task-shared coefficient α^s, the shallow subspace P_1^s and the deep layer W_1^d are all trainable parameters. In the following task t, only the parameters of the newly added shallow subspace P_t^s and that of the deep layer W_t^d are optimized to learn a new task.

3.3 Dynamic Features Learning and Aggregation

To learn global features for CIL that can discriminate among all observed task classes, we first introduce the pipeline of dynamic features learning and aggregation after task subspace hybrid network expansion. Moreover, we propose additional feature-enhancement techniques for further feature learning.

The Pipeline. Following the training process introduced in Sect. 3.2, at each task $t \in \{2, ..., T\}$, we add a new feature extractor Φ_t while keeping the parameters of previous extractors $\{\Phi_1, ..., \Phi_{t-1}\}$ and previous classifier $\mathcal{H}_{t-1}$ frozen to preserve the old skills learned from previous tasks. Simultaneously, the parameters for the classes from the old tasks of $\mathcal{H}_t$ are initialized with $\mathcal{H}_{t-1}$. As shown in Fig. 1(b), considering an image x, drawn from the set of observed data, we concatenate the extracted features to the super features u_t as follows

$$u_t = [\Phi_1(x), ..., \Phi_t(x)]. \tag{2}$$

Then we feed the super features u_t into the new classifier $\mathcal{H}_t$. Since the classes of the old and new tasks are highly unbalanced, we use Logits Alignment [27] to correct the task bias and get the output logits $F_t(x)$ as follows

$$F_t(x) = \gamma \mathcal{H}_t(u_t), \tag{3}$$

where γ is a scale factor vector hyperparameter of Logits Alignment. We employ the softmax cross entropy as the classification loss as follows

$$\mathcal{L}_{CE} = - \sum_{(x,y) \in \tilde{\mathcal{D}}_t} \sum_{i=1}^{|\mathcal{Y}_t|} \delta_{y=i} \log \left[\sigma(F_t(x))_i \right], \tag{4}$$

where $\sigma(\cdot)$ is softmax function, $\sigma(F_t(x))_i$ is the prediction of the i-th class and $|\mathcal{Y}_t|$ denotes all observed classes.

Feature Enhancement. Feature enhancement consists of the logit-level distillation loss $\mathcal{L}_{KD}$ and the discriminative loss $\mathcal{L}_{DIS}$. We employ the former for avoiding catastrophic forgetting and the latter for further alleviating inter-task confusion.

For each training image x, we can get $F_t(x)$ and $F_{t-1}(x)$ from the outputs of classifiers $\mathcal{H}_t$ and $\mathcal{H}_{t-1}$, respectively. As shown in Fig. 1(b), the logit-level knowledge distillation loss is calculated as follows

$$\mathcal{L}_{KD} = \sum_{(x,y) \in \tilde{\mathcal{D}}_t} \sum_{j=1}^{c_{old}} \mathrm{KD} \left(\sigma(F_{t-1}(x))_j, \sigma(F_t(x))_j \right), \tag{5}$$

where $\mathrm{KD}(\cdot, \cdot)$ is the standard logit distillation loss as [6] and c_{old} denotes the classes from old tasks.

Since the parameters of the old feature extractors remain fixed, the extracted features for the same input sample by the old feature extractors remain unchanged. We aim for the features outputted by the new feature extractor to be capable of distinguishing between the classes of both the new and old tasks, which allows for more extensive learning of task-specific features for the new task, leading to improved discrimination between the classes of the new and old tasks. We employ the softmax cross entropy between the output logit of the new feature extractor $\mathcal{F}_t(x)$ and true labels $y \in \mathcal{Y}_t$ as follows

$$\mathcal{L}_{DIS} = - \sum_{(x,y) \in \tilde{\mathcal{D}}_t} \sum_{i=1}^{|\mathcal{Y}_t|} \delta_{y=i} \log \left[\sigma(\mathcal{F}_t(x))_i \right], \tag{6}$$

where $\mathcal{F}_t(x) = \mathcal{H}_{dis}(\Phi_t(x))$ and $\mathcal{H}_{dis}$ is the discriminative classifier as shown in Fig. 1(b). $|\mathcal{Y}_t|$ denotes all observed classes as in Eq. (4).

As illustrated in Fig. 1(b), the total loss is

$$\mathcal{L} = \mathcal{L}_{CE} + \mathcal{L}_{KD} + \mathcal{L}_{DIS}. \tag{7}$$

DASE is trained with a uniform combination of the above-mentioned three losses, i.e., the classification loss $\mathcal{L}_{CE}$ calculated by cross-entropy, the logit-level knowledge distillation loss $\mathcal{L}_{KD}$ given by Eq. 5 and a discriminative loss $\mathcal{L}_{DIS}$ to maximum the discrepancy between old-new classes.

Table 1. Average incremental accuracy on CIFAR-100.

Method	Average accuracy of all sessions(%)					
	B0 5 steps	B0 10 steps	B0 20 steps	B50 5 steps	B50 10 steps	B50 25 steps
Bound	**80.38**	**80.40**	**80.41**	**81.39**	**81.49**	**81.74**
iCaRL [25]	65.51	64.42	63.50	54.54	53.78	50.60
BiC [30]	66.60	65.08	62.37	54.13	53.21	48.96
WA [33]	67.14	67.08	64.64	62.72	57.57	54.10
Filter Atom [23]	60.27	56.97	51.01	65.44	62.48	-
DER [31]	71.63	69.74	67.98	67.53	66.36	-
MEMO [35]	72.01	70.20	68.10	68.35	66.94	66.12
FOSTER [27]	74.00	72.90	70.65	70.30	67.95	63.83
Ours	**75.35**	**74.12**	**70.89**	**72.23**	**69.86**	**66.63**

4 Experiments

We compare our method with state-of-the-art methods on benchmark CIL datasets. Ablation experiments are conducted to empirically reveal the role of each objective function of our method, the robustness of hyperparameters and the influence of the number of exemplars.

4.1 Experimental Settings

Datasets. We verify the effectiveness of our method on the widely used CIL benchmark datasets: (i) natural image datasets CIFAR-100 [13], ImageNet-100 and ImageNet-1000 [4]; (ii) Synthetic Aperture Radar (SAR) image datasets MSTAR and OpenSARShip [8]. Each dataset and more implementation details are specifically introduced in the supplementary material.

Protocol. (i) CIFAR-100 or ImageNet-100 B0 (base 0): In this protocol, we train the model gradually with class steps of 5, 10, and 20 for CIFAR-100 and only 10 for ImageNet-100 with a fixed memory size of 2,000 exemplars. (ii) CIFAR-100 or ImageNet-100 B50 (base 50): We also train the model with 50 classes in the first task and the other 50 classes are gradually divided with class steps of 2, 5, and 10 for CIFAR-100 and only 10 for ImageNet-100 with 20 exemplars per class. (iii) ImageNet-1000 B0 10 steps: We train all 1000 classes with 100 classes per step with a fixed memory size of 20,000 exemplars. (iv) For MSTAR and OpenSARShip, we design three scenarios shown in Table 2.

4.2 Quantitative Results

Evaluation on CIFAR-100. Table 1 summarizes the results of the CIFAR-100 benchmark. Our method consistently outperforms other methods in different incremental settings. Particularly, we achieve **1.93%**, **1.91%** and **0.51%** improvement under the incremental learning setting of base 50 with 5 steps, 10 steps and 25 steps, respectively. We also surpass the state-of-the-art method

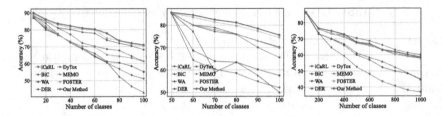

Fig. 3. Incremental Accuracy on ImageNet. Our method achieves comparable performance in the settings of B0 10 steps (left), B50 5 steps (middle) of ImageNet-100 and B0 10 steps (right) of ImageNet-1000.

Table 2. Three experiment scenarios of MSTAR and OpenSARShip. Class 1–13 are ZIL131, D7, BTR70, T72, BMP2, BRDM2, T62, BTR60, 2S1, ZSU23/4, Container Ship, Bulk Carrier and Tanker.

Class	1	2	3	4	5	6	7	8	9	10	11	12	13
Scenario 1	task 1	task 2	task 3	task 4	task 5	task 6	task 7	task 8	task 9	task 10	task 11	task 12
Scenario 2	task 1	task 2		task 3		task 4		task 5		task 6			task 7
Scenario 3	task 1					task 2					task 3		

by **1.35%**, **1.22%** and **0.24%** under the incremental learning setting of base 0 with 5 steps, 10 steps and 20 steps, respectively. Filter Atom [23] fails to learn sufficiently generalized features and performs badly in the B0 configuration with fewer classes in the first task because it updates coefficients from all layers with larger parameters only during the first task. In addition, our method, as shown in Table 1 and 3, achieves both high performance and memory efficiency.

Evaluation on ImageNet. Figure 3 summarizes the results of the ImageNet benchmark. Our method still outperforms other methods in most settings. Specifically, our method surpasses the state-of-the-art by about **0.85%** and **1.02%** in two settings of ImageNet-100. Moreover, we achieve performance comparable to the state-of-the-art method DyTox [5] under the ImageNet-1000 setting. It illustrates that our method is also successful for larger-scale incremental learning.

Evaluation on MSTAR and OpenSARShip. Figure 4 summarizes the experimental results for the MSTAR and OpenSARShip datasets. We can see that our method consistently surpasses other methods in three scenarios. Specifically, our method surpass the state-of-the-art with about **1.29%**, **0.95%** and **0.25%** in scenario 1, 2 and 3, respectively. It demonstrates that our method has good adaptability in the field of SAR remote sensing. However, as demonstrated in Sect. 3.1, Filter Atom [23] suffers from high inter-task confusion, which significantly lowers performance.

4.3 Ablation Study

The Effect of Each Objective Function. Table 4 summarizes the results of our ablative experiments on CIFAR-100 B50 with 5 steps. First, equipping

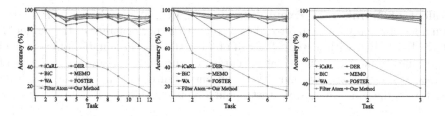

Fig. 4. Incremental Accuracy of scenario 1 (Left), 2 (Middle), and 3 (Right). Our method achieves state-of-the-art performance in three scenarios.

the CE loss gets 65.17% accuracy. In the following, applying KD loss and Logits Alignment further gives 1.08% and 2.38% accuracy improvements. Logits Alignment (LA) better distinguishes between old-new classes and gives marginal improvements. Finally, DIS loss also reports a large improvement of 3.6%, indicating that this loss effectively alleviates inter-task confusion.

Table 3. The number of parameters (million) in CIFAR-100 settings.

Method	B0 10 steps	B50 10 steps
DER	4.60	2.76
MEMO	3.62	2.22
Our Method	**3.63**	**2.23**

Table 4. Ablation on each objective function.

CE loss	KD loss	LA	DIS loss	Avg
✓				65.17
✓	✓			66.25
✓	✓	✓		68.63
✓	✓	✓	✓	**72.23**

Sensitive Study of Hyperparameters. To evaluate the robustness of our method, we perform experiments on CIFAR-100 B50 5 steps with varied hyperparameters m. We test $m = 3, 6, 9, 12, 15$ and 18 respectively. The experimental results are shown in Fig. 5(a). We choose the hyperparameter $m = 12$ with the best performance. At this point, the shallow layers simply add minimal parameters as shown in Table 3.

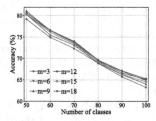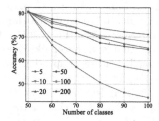

Fig. 5. Robustness Testing. Sensitive study of the number of bases m (left) and influence of the number of exemplars (right).

Effect of Number of Exemplars. In Fig. 5(b), We gradually increase the number of exemplars from 5 to 200 and report the performance of the model on CIFAR-100 B50 with 5 steps. We can observe that as the number of exemplars continues to increase, the average accuracy of our method rises in a majority of cases. It illustrates that our method can make full use of the information from historical data and mitigate catastrophic forgetting well.

5   Conclusion

In this study, we empirically reveal that existing task subspace methods may suffer from severe inter-task confusion under the challenging setup of class-incremental learning. To deal with that, we present a novel two-stage framework called DASE. Our method dynamically expands the extraction network for improved memory efficiency, and enables dynamic learning and aggregation of diverse features, thereby addressing the challenge of inter-task confusion.

References

1. Buzzega, P., Boschini, M., Porrello, A., Abati, D., Calderara, S.: Dark experience for general continual learning: a strong, simple baseline. Adv. Neural. Inf. Process. Syst. **33**, 15920–15930 (2020)
2. Chaudhry, A., Khan, N., Dokania, P., Torr, P.: Continual learning in low-rank orthogonal subspaces. Adv. Neural. Inf. Process. Syst. **33**, 9900–9911 (2020)
3. Cong, Y., Zhao, M., Li, J., Wang, S., Carin, L.: Gan memory with no forgetting. Adv. Neural. Inf. Process. Syst. **33**, 16481–16494 (2020)
4. Deng, J., Dong, W., Socher, R., Li, L.J., Li, K., Fei-Fei, L.: Imagenet: a large-scale hierarchical image database. In: 2009 IEEE Conference on Computer Vision and Pattern Recognition, pp. 248–255. IEEE (2009)
5. Douillard, A., Ramé, A., Couairon, G., Cord, M.: Dytox: Transformers for continual learning with dynamic token expansion. In: Proceedings of the IEEE/CVF Conference on Computer Vision and Pattern Recognition, pp. 9285–9295 (2022)
6. Hinton, G., Vinyals, O., Dean, J., et al.: Distilling the knowledge in a neural network. arXiv preprint arXiv:1503.02531 2(7) (2015)
7. Hou, S., Pan, X., Loy, C.C., Wang, Z., Lin, D.: Learning a unified classifier incrementally via rebalancing. In: Proceedings of the IEEE/CVF Conference on Computer Vision and Pattern Recognition, pp. 831–839 (2019)
8. Huang, L., Liu, B., Li, B., Guo, W., Yu, W., Zhang, Z., Yu, W.: Opensarship: a dataset dedicated to sentinel-1 ship interpretation. IEEE J. Sel. Top. Appl. Earth Observ. Remote Sens. **11**(1), 195–208 (2017)
9. Iscen, A., Zhang, J., Lazebnik, S., Schmid, C.: Memory-efficient incremental learning through feature adaptation. In: Vedaldi, A., Bischof, H., Brox, T., Frahm, J.-M. (eds.) ECCV 2020. LNCS, vol. 12361, pp. 699–715. Springer, Cham (2020). https://doi.org/10.1007/978-3-030-58517-4_41
10. Joseph, K., Khan, S., Khan, F.S., Anwer, R.M., Balasubramanian, V.N.: Energy-based latent aligner for incremental learning. In: Proceedings of the IEEE/CVF Conference on Computer Vision and Pattern Recognition, pp. 7452–7461 (2022)
11. Kim, G., Xiao, C., Konishi, T., Ke, Z., Liu, B.: A theoretical study on solving continual learning. arXiv preprint arXiv:2211.02633 (2022)

12. Kirkpatrick, J., Pascanu, R., Rabinowitz, N., Veness, J., Desjardins, G., Rusu, A.A., Milan, K., Quan, J., Ramalho, T., Grabska-Barwinska, A., et al.: Overcoming catastrophic forgetting in neural networks. Proc. Natl. Acad. Sci. **114**(13), 3521–3526 (2017)

13. Krizhevsky, A., Hinton, G., et al.: Learning multiple layers of features from tiny images (2009)

14. Kumar, A., Daume III, H.: Learning task grouping and overlap in multi-task learning (2012)

15. LeCun, Y.: The mnist database of handwritten digits (1998). http://yann.lecun.com/exdb/mnist/

16. Li, B., Cui, Z., Cao, Z., Yang, J.: Incremental learning based on anchored class centers for sar automatic target recognition. IEEE Trans. Geosci. Remote Sens. **60**, 1–13 (2022)

17. Long, M., Cao, Y., Wang, J., Jordan, M.: Learning transferable features with deep adaptation networks. In: International Conference on Machine Learning, pp. 97–105. PMLR (2015)

18. Van der Maaten, L., Hinton, G.: Visualizing data using t-sne. J. Mach. Learn. Res. **9**(11) (2008)

19. Masana, M., Liu, X., Twardowski, B., Menta, M., Bagdanov, A.D., van de Weijer, J.: Class-incremental learning: survey and performance evaluation on image classification. arXiv preprint arXiv:2010.15277 (2020)

20. McClelland, J.L., McNaughton, B.L., O'Reilly, R.C.: Why there are complementary learning systems in the hippocampus and neocortex: insights from the successes and failures of connectionist models of learning and memory. Psychol. Rev. **102**(3), 419 (1995)

21. McCloskey, M., Cohen, N.J.: Catastrophic interference in connectionist networks: The sequential learning problem. In: Psychology of Learning and Motivation, vol. 24, pp. 109–165. Elsevier (1989)

22. Mermillod, M., Bugaiska, A., Bonin, P.: The stability-plasticity dilemma: investigating the continuum from catastrophic forgetting to age-limited learning effects. Front. Psychol. **4**, 504 (2013)

23. Miao, Z., Wang, Z., Chen, W., Qiu, Q.: Continual learning with filter atom swapping. In: International Conference on Learning Representations (2022)

24. Qiu, Q., Cheng, X., Sapiro, G., et al.: Dcfnet: Deep neural network with decomposed convolutional filters. In: International Conference on Machine Learning, pp. 4198–4207. PMLR (2018)

25. Rebuffi, S.A., Kolesnikov, A., Sperl, G., Lampert, C.H.: icarl: incremental classifier and representation learning. In: Proceedings of the IEEE conference on Computer Vision and Pattern Recognition, pp. 2001–2010 (2017)

26. Von Oswald, J., Henning, C., Sacramento, J., Grewe, B.F.: Continual learning with hypernetworks. arXiv preprint arXiv:1906.00695 (2019)

27. Wang, F.Y., Zhou, D.W., Ye, H.J., Zhan, D.C.: Foster: Feature boosting and compression for class-incremental learning. arXiv preprint arXiv:2204.04662 (2022)

28. Wang, Y., Ma, Z., Huang, Z., Wang, Y., Su, Z., Hong, X.: Isolation and impartial aggregation: A paradigm of incremental learning without interference. arXiv preprint arXiv:2211.15969 (2022)

29. Wu, C., Herranz, L., Liu, X., van de Weijer, J., Raducanu, B., et al.: Memory replay gans: Learning to generate new categories without forgetting. Advances in Neural Information Processing Systems 31 (2018)

30. Wu, Y., Chen, Y., Wang, L., Ye, Y., Liu, Z., Guo, Y., Fu, Y.: Large scale incremental learning. In: Proceedings of the IEEE/CVF Conference on Computer Vision and Pattern Recognition, pp. 374–382 (2019)
31. Yan, S., Xie, J., He, X.: Der: Dynamically expandable representation for class incremental learning. In: Proceedings of the IEEE/CVF Conference on Computer Vision and Pattern Recognition, pp. 3014–3023 (2021)
32. Yoon, J., Yang, E., Lee, J., Hwang, S.J.: Lifelong learning with dynamically expandable networks (2017)
33. Zhao, B., Xiao, X., Gan, G., Zhang, B., Xia, S.T.: Maintaining discrimination and fairness in class incremental learning. In: Proceedings of the IEEE/CVF Conference on Computer Vision and Pattern Recognition, pp. 13208–13217 (2020)
34. Zhou, D.W., Wang, Q.W., Qi, Z.H., Ye, H.J., Zhan, D.C., Liu, Z.: Deep class-incremental learning: A survey. arXiv preprint arXiv:2302.03648 (2023)
35. Zhou, D.W., Wang, Q.W., Ye, H.J., Zhan, D.C.: A model or 603 exemplars: towards memory-efficient class-incremental learning. arXiv preprint arXiv:2205.13218 (2022)

Energy-Based Policy Constraint
for Offline Reinforcement Learning

Zhiyong Peng, Changlin Han, Yadong Liu, and Zongtan Zhou[✉]

National University of Defense Technology, Changsha, China
narcz@163.com

Abstract. Offline RL suffers from the distribution shift problem. One way to address this issue is to constrain the divergence between the target policy and the behavior policy. However, directly using the behavior policy-based constraint has two drawbacks: first, it does not directly distinguish the in-distribution samples and OOD samples, possibly overly constraining the target policy and limiting performance. Second, the practical datasets may be collected from multiple different behavior policies, which results in a multi-modal distribution, making it hard to represent the behavior policy. To address this problem, we propose a policy constraint method based on the energy-based model. On the one hand, the energy-based model constrains the target policy by energy function rather than directly constraining it to the dataset actions, making it suitable for multi-modal distribution. On the other hand, the energy-based model can effectively detect OOD samples, avoiding over-constraint of the target policy and improving the ceiling of the algorithm's performance. The proposed algorithm is evaluated on the D4RL datasets. Experimental results show that compared to the behavior policy-based constraint methods, the energy-based policy constant significantly improves the performance and outperforms existing offline RL baselines.

Keywords: Reinforcement learning · Offline reinforcement learning · Energy-based Model

1 Introduction

RL has made impressive achievements in games [29,31,32] and robotics [2,8,16]. RL learns by trial and error, which requires lots of interactions with environments. For both games and robotics, it usually first builds a high-fidelity simulator and then trial-and-error in the simulated environment. Compared to trial-and-error in the real environment, interacting with the simulator is much cheaper, faster and safer. However, it is challenging to build precision simulators for many real-world tasks such as healthcare and autonomous driving. In addition, learning in an imperfect simulated environment can raise the sim2real gap [21] when employing the policy in the real world. Some studies tried to directly learn robotic skills in the real environment [19], but it is expensive and takes several months to collect enough experience. Offline RL aims to learn policy from existing history datasets, which neither requires a simulator nor direct trial-and-error in the real environment. As a fully data-driven policy learning method, offline RL has great potential to solve a wide range of real-world control and decision tasks.

One of the previously widely used offline policy learning methods is behavior cloning imitation learning, which imitates expert behavior via a supervised learning paradigm. However, behavior cloning needs expert demonstrations and is hard to outperform the behavior policy (the policy used for collecting the dataset). In the real world, large amounts of datasets are non-expert or suboptimal, and the cost is much lower compared to the elaborated expert datasets. Offline RL can learn from these suboptimal datasets and outperform the behavior policy. Offline RL suffers from the distribution shift problem [26], which is raised when the target policy deviates from the behavior policy. When the distribution shift occurs, the agent will visit out-of-distribution (OOD) samples and the value function estimation of these OOD samples is inaccurate.

To avoid value estimation on OOD samples, a line of studies in offline RL focus on constraining the divergence of the learned policy and the behavior policy [11,12,23,35]. However, directly using the behavior policy-based constraint has several disadvantages. Firstly, the practical datasets may be collected from multiple different behavior policies, which results in a multi-modal distribution, making it hard to effectively constrain the target policy. For example, if there are two different actions conditioned on the same state, constraining the target policy to both actions will make it imitate the average action, which is unexpected. Secondly, the behavior policy-based constraint methods do not directly distinguish the in-distribution samples and OOD samples, possibly overly constraining the target policy and limiting its performance. The implicit inductive bias of behavior policy-based constraint methods is that the further away from the dataset, the more likely to be an OOD sample. It is well-known that neural networks can generalize well, this kind of inductive bias does not take advantage of the generalization ability of the value function. To address the aforementioned issues, we propose a policy constraint method based on the energy-based model (EBM) [25]. EBM is widely used in the deep learning community for OOD detection [6,28], the model is trained to allocate low energy for in-distribution samples and high energy for OOD samples. In this paper, we first train the energy model on the offline datasets, then apply the trained model as an OOD detector to constrain the policy to avoid visiting OOD actions. Similar to the OOD detection methods, there are several studies on the uncertainty-based offline RL [1,3]. These approaches make use of the Q ensembles to measure uncertainty and conservatively estimate the OOD value function. Compared to the uncertainty-based methods, our energy-based policy constraint method applies the pre-trained energy model in policy improvement rather than learning Q ensembles in policy evaluation. Another advantage of EBM is the ability to represent complex multi-modal distribution, this characteristic is exploited in learning more representative RL policy [15] or implicit behavior cloning [9]. The multi-modal characteristic of EBM makes it suitable for modeling the monolithic behavior policy when the offline datasets are collected from different individual behavior policies. We evaluate the proposed algorithm on the D4RL Gym tasks, experimental results show that the energy-based policy constraint method significantly improves the performance compared to these direct behavior policy-based policy constraint counterparts. We utilize the energy surface visualization to qualitatively analyze the learned EBM and reveal the different patterns between the energy-based policy constraint loss function and the behavior cloning-based one.

2   Related Work

Though EBM is widely studied in the deep learning community, to our best knowledge, it is the first that the EBM be employed in offline RL.

Offline RL. Offline RL can be divided into model-free and model-based methods. Model-based methods like MOPO [36] and MOReL [20] learn multiple environment models simultaneously and utilize the prediction differences between these models as an uncertainty measurement. Among model-free methods, BCQ [12] employs a generative model to constrain the policy actions near datasets, BEAR [23] proposes support set constraints to regularize the learned policy. TD3BC [11] directly adds a behavior cloning loss item during the policy improvement steps. CQL [24] penalizes the value functions of OOD samples to obtain conservative value function estimation. SAC-N [1] shows that extending double Q to a larger size of Q function ensembles can be sufficiently conservative and outperforms many previous classical offline RL algorithms. EDAC [1] reduces ensemble numbers required on SAC-N by increasing ensemble diversity. PBRL [3] penalizes the uncertainty for the in-distribution target as well as additional OOD sampling to regularize the learned Q function. IQL [22] employs the SARSA-like updates and expectile regression to learn the Q function. DT [5] and TT [18] introduce the transformer network into offline RL and generate actions autoregressively. Compared to the autoregressive paradigm, Diffuser [17] and Diffusion-QL [34] employ powerful diffusion probabilistic models to sample the whole trajectory. The Diffusion-QL learns a more expressive target policy via diffusion models while our method obtains better behavior policy constraints by energy-based models, these two methods all use generative models but are orthogonal to each other.

Energy-Based Model. EBM [25] captures dependencies between variables by associating scalar energy to each configuration of the variables. JEM [14] reinterpret a standard discriminative classifier of $p(y|x)$ as an energy-based model for the joint distribution $p(x, y)$, which improves calibration, robustness, and out-of-distribution detection. Liu et al. [28] propose a unified framework using an energy score for OOD detection. Elflein et al. [6] propose the energy-prior network which enables the estimation of various uncertainties within an EBM for classification. SQL [15] employs the EBM to learn expressive energy-based policies for continuous states and actions, improving exploration and compositionality that allows transferring skills between tasks.

3   Preliminaries

3.1   Offline RL

RL aims to solve the sequential decision problem modeled by a Markov decision process (MDP) $(S, A, r, P, \rho, \gamma)$, with state space S, action space A, reward function r, dynamic P, initial state distribution ρ and the discount factor γ. RL learns the policy $\pi(a|s)$ to maximize the cumulative discount reward and the optimization objective can be represented as:

$$J(\pi) = \mathbb{E}_{s_0 \sim \rho, a_t \sim \pi(s_t), s_{t+1} \sim P(\cdot|s_t, a_t)} [\sum_{t=0}^{\infty} \gamma^t r_t(s_t, a_t)], \tag{1}$$

where the cumulative discount reward is usually approximated by the value function $Q(s_0, a_0) = \sum_{t=0}^{\infty} \gamma^t r_t(s_t, a_t)$.

The overestimated Q values can be exploited by the greedy policy. In the offline RL setting, inaccurate Q values can not be corrected by collecting new experiences like online RL. Moreover, the overestimated Q values will propagate via bootstrap updating and damage the whole Q values estimation. For a given dataset, these in-distribution Q values can be estimated well while OOD Q values are inaccurate. Therefore, a line of work in offline RL is to constrain the policy not to visit OOD actions by constraining the divergence between the learned policy and the behavior policy like:

$$\pi_\theta := \arg\max_{\pi_\theta} \mathbb{E}_{s \sim B} [Q(s, \pi_\theta(s)) - \beta D(\pi_\theta(s), \pi_b(s))], \tag{2}$$

where D is the divergence criterion and β controls the strength of divergence constraint and B refers to the dataset batch.

The minimalist offline RL method TD3BC [11] adopts the mean square error (MSE) between the policy action and the corresponding action in the dataset as the divergence metric:

$$\pi_\theta := \arg\max_{\pi_\theta} \mathbb{E}_{s, a \sim B} [\lambda Q(s, \pi_\theta(s)) - (\pi_\theta(s) - a)^2]. \tag{3}$$

3.2   Energy-Based Model

The energy-based model maps a sample (x, y) to a single scalar called energy $E_\omega(x, y) : R^{m+n} \to R$. The energy values can be converted to a conditional probability density $p(y|x)$ through the Gibbs distribution:

$$p(y|x) = \frac{e^{-E_\omega(x,y)/T}}{Z_\omega(x)}, \tag{4}$$

where $E_\omega(x, y)$ is the energy model, T is the temperature parameter and $Z_\omega(x)$ is the partition function.

4 Method

To avoid visiting OOD samples, the policy constants methods restrain the divergence between the target policy and behavior policy, which do not directly distinguish OOD samples and make it hard to deal with the multi-modal distribution. In this paper, we propose an energy-based policy constraint method to address the above problems. The method pipeline consists of two steps, firstly, an energy model is trained on the datasets to assign low energy for the datasets samples and high energy for OOD samples, then, the learned energy model is plugged into the RL agent as an additional component to constrain the target policy.

4.1 Learning the Energy-Based Model

For a dataset of M samples $\{s_i, a_i\}_{i=1}^M$, we aim to learn an energy model to assign low energy $E(s_i, a_i)$ on those samples in the dataset and high energy $E(s_i, \tilde{a}_i)$ to OOD samples, where the OOD samples consist of in-distribution states s_i and OOD actions $\tilde{a}_i$. Similar to the implicit BC method [9], we adopt an InfoNCE-style [30] loss function to train the model, which is defined as follows:

$$
\begin{aligned}
\mathcal{L}_{EBM} &= -\frac{1}{N} \sum_{i=1}^N \log p(a_i|s_i) = -\frac{1}{N} \sum_{i=1}^N \log \frac{e^{-E_\omega(s_i, a_i)}}{Z_\omega(s_i)} \\
&= -\frac{1}{N} \sum_{i=1}^N \log \frac{e^{-E_\omega(s_i, a_i)}}{e^{-E_\omega(s_i, a_i)} + \sum_{j=1}^{N^{neg}} e^{-E_\omega(s_i, \tilde{a}_i^j)}}
\end{aligned}
\tag{5}
$$

where N is the batch size, N^{neg} is the number of negative samples, note that the temperature T is Eq. 4 is set to be 1. Since we do not have ground truth OOD samples, the negative samples are used to surrogate the OOD ones. Specifically, for each state s_i in the dataset, we uniform sampling N^{neg} actions from the range of $[a_{min}, a_{max}]$ as negative samples. By minimizing the InfoNCE-style loss function, the energy of dataset samples is pushed down to increase the conditional probability. At the same time, the energy of negative samples is pushed up since the loss function will maximize the partition function in Eq. 5.

4.2 Energy-Based Policy Constraint

After training the energy-based model $E(s, a)$, the conditional probability Eq. 4 can be used to constrain the target policy via maximizing this equation. However, the conditional probability includes partition function $Z(s)$, which needs negative sampling and can be time-consuming. Instead, we directly plug the learned energy function $E(s, a)$ as an auxiliary item into the policy optimization objective:

$$
\pi_\theta := \arg\max_{\pi_\theta} \mathbb{E}_{s \sim B}[\alpha Q(s, \pi_\theta(s)) - E(s, \pi_\theta(s))],
\tag{6}
$$

where α is a hyperparameter to control the policy constraint degree.

Compared to the explicit policy constraint method represented in Eq. 2, the energy-based policy constraint is an implicit one since the behavior policy is implicitly modeled by the energy-based model $E(s, a)$. The energy-based model not only can represent complex multi-modal behavior policy but also act as an OOD detector. The energy minimum auxiliary item forces the target policy to stay in the low energy area, i.e. the in-distribution. Compared to the conservative Q-learning (CQL) [24], CQL pushes down the OOD Q values during the whole policy learning process while our method pushes up the energy of OOD samples only at the energy-based model training state, after that, the parameters of the energy-based model fixed and needs no more updates. Compared to the uncertainty-based methods [1,3], we adopt the energy-based model to detect OOD samples rather than uncertainty.

4.3 Algorithm Summary

The energy-based policy constraint (EBPC) algorithm consists of two states, i.e., training an EBM neural network first and then learning the energy-constrained RL policy. For training the EBM, we utilize the InfoNCE-like loss function in Eq. 5. For each state in the dataset, we sample 200 negative samples from the uniform distribution $[a_{min}, a_{max}]$. In the policy learning state, we build our EBPC offline RL algorithm based on the TD3 online RL method and modify the policy optimization objective by adding the energy minimization item. We summarize the EBPC algorithm[1] in Algorithm 1 and Algorithm 2.

Algorithm 1: Energy-based Policy Constraint (EBPC)

Initialize: Initialize value networks Q_{ϕ_1}, Q_{ϕ_2}, target networks $\overline{\phi}_1 \leftarrow \phi_1$, $\overline{\phi}_2 \leftarrow \phi_2$, actor π_θ, target actor $\overline{\theta} \leftarrow \theta$, and replay buffer $\mathcal{D}$
Setting hyperparameters $\{c, \sigma, \gamma, \tau\}$, α
for $i{=}1$ to max_steps **do**

    Sample batch $B = (s, a, r, s', d)$ from dataset $\mathcal{D}$
    $a'(s') = clip(\pi_{\bar{\theta}}(s') + clip(\epsilon, -c, c), a_{min}, a_{max}), \epsilon \sim \mathcal{N}(0, \sigma)$
    $y = r + \gamma(1 - d) \min_{\phi_{1,2}} Q_{\bar{\phi}_{1,2}}(s', a'(s'))$
    $\phi_{1,2} \leftarrow \arg\min_{\phi_{1,2}} \frac{1}{|B|} \sum_B (Q_{\phi_{1,2}}(s, a) - y)^2$
    if i mod policy_update_frequency $== 0$ **then**

        $\theta \leftarrow \arg\max_\theta \frac{1}{|B|} \sum_B \alpha Q_{\phi_1}(s, \pi_\theta(s)) - E(s, \pi_\theta(s))$
        $\overline{\phi}_{1,2} \leftarrow (1 - \tau)\overline{\phi}_{1,2} + \tau\phi_{1,2}$
        $\overline{\theta} \leftarrow (1 - \tau)\overline{\theta} + \tau\theta$

    end

end

---

[1] Our implementation is available at https://github.com/qsa-fox/EBPC.

Algorithm 2: Learning EBM

Initialize: Initialize energy networks E_ω and replay buffer $\mathcal{D}$
Setting hyperparameter N^{neg}
for $i=1$ *to* max_steps **do**
    Sample batch $B = \{s_i, a_i\}_{i=1}^{N}$ from dataset $\mathcal{D}$
    Sample N^{neg} negative actions $\tilde{a}_i$ for each state s_i from $U(a_{min}, a_{max})$
    Calculate $\mathcal{L}_{EBM}$ loss using Equation 5
    $\omega \leftarrow \arg\min_\omega \mathcal{L}_{EBM}$
end

---

5   Experiments

In the experiments section, the performance of EBPC is compared with several previous offline RL baseline algorithms on the D4RL gym datasets, including behavior clone (BC), TD3BC [11], CQL [24], OneStep [4], DT [5], and Rvs-R [7]. We also study whether the learned EBM can effectively distinguish in-distribution samples and OOD ones. To answer these questions, we qualitatively analyze the learned model by visualizing the energy surface and compare the difference between the energy-based policy constraint and the behavior cloning-based policy constraint.

5.1   Performance on Offline RL Benchmarks

The D4RL [10] Gym datasets consist of three different environments, i.e., the halfcheetah, the hopper, and the walker2d, each environment contains several different datasets. The "random" datasets are collected by a random behavior policy, the "medium" datasets are collected by a partially-trained policy, and the "medium-replay" datasets consist of recording all samples in the replay buffer observed during training until the policy reaches "medium" level of performance and the "medium-expert" datasets mix up equal amounts of expert demonstrations and suboptimal data.

The most similar counterpart of EBPC is the minimalist TD3BC algorithm, which adds a behavior cloning penalty in the policy optimization objective while our method adds the energy function loss item. Compared to the TD3BC, EBPC significantly improves the performance on most tasks, especially on the "hopper-medium" and "hopper-medium-replay" tasks. Compared to the CQL algorithm, our method also outperforms it while keeping simplicity like TD3BC at the same time.

Experimental Details. For the comparison results in Table 1, the scores for BC, 10%BC, RvS-R, OneStep and DT are referred from the RvS paper and the scores for IQL and CQL are referred from their official paper. We run 5 seeds on

Table 1. Normalized scores on the D4RL gym datasets(bold indicates the highest score). The variances of EBPC are also reported. EBPC receives the highest total score.

Task name (-v2)	BC	10%BC	DT	RvS-R	OneStep	CQL	TD3BC	EBPC
halfcheetah-random	2.3	2.0	2.2	3.9	6.9	18.6	11.7	**29.2± 2.6**
hopper-random	4.8	4.1	7.5	0.2	7.8	**9.1**	8.6	8.2± 4.1
walker2d-random	1.7	1.7	2.0	**7.7**	6.1	2.5	0.9	6.6± 5.9
halfcheetah-medium	42.6	42.5	42.6	41.6	**55.6**	49.1	48.2	52.9±0.3
hopper-medium	52.9	56.9	67.6	60.2	83.3	64.6	57.7	**99.8±1.5**
walker2d-medium	75.3	75.0	74.0	71.7	**85.6**	82.9	83.2	81.5±5.5
halfcheetah-medium-replay	36.6	40.6	36.6	38.0	42.4	**47.3**	44.6	45.7±0.6
hopper-medium-replay	18.1	75.9	82.7	73.5	71.0	97.8	67.4	**100.5±4.3**
walker2d-medium-replay	26.0	62.5	66.6	60.6	71.6	86.1	83.7	**91.4±3.1**
halfcheetah-medium-expert	55.2	92.9	86.8	92.2	**93.5**	85.8	90.7	78.2± 5.5
hopper-medium-expert	52.5	**110.9**	107.6	101.7	106.1	102.1	102.0	91.3±12.2
walker2d-medium-expert	107.5	109.0	108.1	106	**110.9**	109.5	110.1	**110.4±0.8**
Total	475.5	674.0	684.3	657.3	736.8	755.5	712.9	**795.7**

the TD3BC and the EBPC algorithms and report the average normalized score. For each task, we train the algorithm with 2 million steps and evaluate the performance every 5000 steps. We train the EBM for 500 thousand steps. The evaluation results are averaged on 10 episodes, and the final scores are reported in the table. The EBPC algorithm is implemented on top of the TD3. The only new hyperparameter introduced is the α in Eq. 6, we sweep this hyperparameter in $\{0.1, 1.0, 10.0\}$ and find $\alpha = 1.0$ works well for all tasks.

5.2 Qualitative Analyzation of the EBM

Loss surface visualization technology is widely used to qualitatively analyze the training characteristic of deep neural networks [13,27]. The reward surface visualization is used to study the stability of different RL algorithms [33]. In this work, we visualize the energy surface of the learned EBM to qualitatively analyze whether the learned model can effectively distinguish in-distribution samples and OOD ones.

To generate the energy surface, we choose the dataset sample (s_i, a_i) as the center point, and choose two random directions δ and η. We define the energy surface as the following function:

$$f_i(\alpha, \beta) = E(s_i, a_i + \alpha \frac{\delta}{|\delta|} + \beta \frac{\eta}{|\eta|}). \tag{7}$$

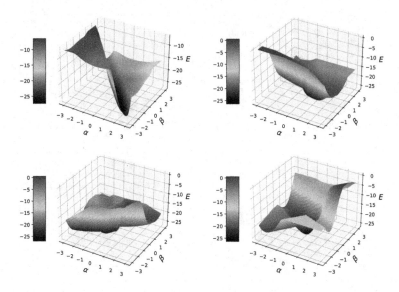

Fig. 1. Energy surface of the learned EBM on the hopper-medium-replay dataset.

The energy change of high dimensions samples can be visualized by the energy surface function, Fig. 1 shows the energy surface of the learned EBM on the hopper-medium-replay dataset, each sub-figure is generated on the same center point (s_i, a_i) but different random directions. From the energy surface, it can seem that the center point (s_i, a_i) indeed has the lowest energy. For the behavior cloning (BC) constraint in TD3BC, the BC penalty is proportional to the distance between the target policy action and the dataset action, so the corresponding loss surface is a bowl-like paraboloid. However, the energy surface of the learned EBM is more complex. For example, the first sub-figure in Fig. 1 shows a narrow low-energy furrow, which means that the energy is not always proportional to the distance to the center point in some directions. Figure 2 shows the relationship between distance and energy. In general, greater distance tends to have greater energy, but there is no strict positive proportion. Some samples, though far from the dataset samples, still have low energy. This phenomenon can be explained by the generalization ability of neural networks, in some directions, the network generalizes well, and these samples still stay in dataset distribution though relatively far from the dataset.

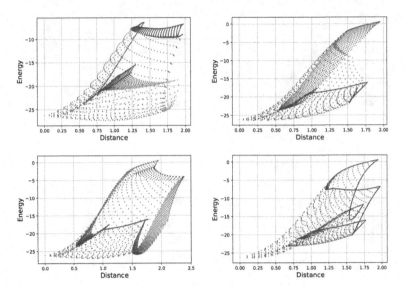

Fig. 2. Energy surface of the learned EBM on the hopper-medium-replay dataset.

6   Conclusion

One of the key challenges of offline RL is the value overestimation on OOD samples, thus avoiding visiting OOD samples is critical. Previously policy constraint methods constrain the target policy to be close to the behavior policy, which indirectly reduces OOD samples access. In this paper, we adopt the energy-based model to directly detect OOD samples by assigning low energy to dataset samples and high energy to negative samples. Experimental results show that the proposed energy-based policy constraint method not only significantly improves the performance compared to the behavior policy-based constraint ones, but also outperforms the previous offline RL baselines. The energy surface visualization reveals the different patterns between the behavior cloning loss function and the energy loss function.

Acknowledgments. This work was supported by the National Natural Science Foundation of China under Grant U19A2083.

References

1. An, G., Moon, S., Kim, J.H., Song, H.O.: Uncertainty-based offline reinforcement learning with diversified Q-ensemble. Adv. Neural Inf. Process. Syst. **34**, 7436–7447 (2021)
2. Andrychowicz, O.M., et al.: Learning dexterous in-hand manipulation. Int. J. Robot. Res. **39**(1), 3–20 (2020)
3. Bai, C., et al.: Pessimistic bootstrapping for uncertainty-driven offline reinforcement learning. In: International Conference on Learning Representations (2022). https://openreview.net/forum?id=Y4cs1Z3HnqL

4. Brandfonbrener, D., Whitney, W., Ranganath, R., Bruna, J.: Offline RL without off-policy evaluation. Adv. Neural Inf. Process. Syst. **34**, 4933–4946 (2021)
5. Chen, L., et al.: Decision transformer: reinforcement learning via sequence modeling. Adv. Neural Inf. Process. Syst. **34**, 15084–15097 (2021)
6. Elflein, S.: Out-of-distribution detection with energy-based models. arXiv preprint arXiv:2302.12002 (2023)
7. Emmons, S., Eysenbach, B., Kostrikov, I., Levine, S.: RVS: what is essential for offline RL via supervised learning? arXiv preprint arXiv:2112.10751 (2021)
8. Fan, T., Cheng, X., Pan, J., Manocha, D., Yang, R.: CrowdMove: autonomous mapless navigation in crowded scenarios. arXiv preprint arXiv:1807.07870 (2018)
9. Florence, P., et al.: Implicit behavioral cloning. In: Conference on Robot Learning, pp. 158–168. PMLR (2022)
10. Fu, J., Kumar, A., Nachum, O., Tucker, G., Levine, S.: D4RL: datasets for deep data-driven reinforcement learning. arXiv preprint arXiv:2004.07219 (2020)
11. Fujimoto, S., Gu, S.S.: A minimalist approach to offline reinforcement learning. Adv. Neural Inf. Process. Syst. **34**, 20132–20145 (2021)
12. Fujimoto, S., Meger, D., Precup, D.: Off-policy deep reinforcement learning without exploration. In: International Conference on Machine Learning, pp. 2052–2062. PMLR (2019)
13. Goodfellow, I.J., Vinyals, O., Saxe, A.M.: Qualitatively characterizing neural network optimization problems. arXiv preprint arXiv:1412.6544 (2014)
14. Grathwohl, W., Wang, K.C., Jacobsen, J.H., Duvenaud, D., Norouzi, M., Swersky, K.: Your classifier is secretly an energy based model and you should treat it like one. In: International Conference on Learning Representations (2020). https://openreview.net/forum?id=Hkxzx0NtDB
15. Haarnoja, T., Tang, H., Abbeel, P., Levine, S.: Reinforcement learning with deep energy-based policies. In: International Conference on Machine Learning, pp. 1352–1361. PMLR (2017)
16. Hwangbo, J., et al.: Learning agile and dynamic motor skills for legged robots. Sci. Robot. **4**(26), eaau5872 (2019)
17. Janner, M., Du, Y., Tenenbaum, J., Levine, S.: Planning with diffusion for flexible behavior synthesis. In: International Conference on Machine Learning, pp. 9902–9915. PMLR (2022)
18. Janner, M., Li, Q., Levine, S.: Offline reinforcement learning as one big sequence modeling problem. Adv. Neural. Inf. Process. Syst. **34**, 1273–1286 (2021)
19. Kalashnikov, D., et al.: Scalable deep reinforcement learning for vision-based robotic manipulation. In: Conference on Robot Learning, pp. 651–673. PMLR (2018)
20. Kidambi, R., Rajeswaran, A., Netrapalli, P., Joachims, T.: MOReL: model-based offline reinforcement learning. Adv. Neural Inf. Process. Syst. **33**, 21810–21823 (2020)
21. Koos, S., Mouret, J.B., Doncieux, S.: Crossing the reality gap in evolutionary robotics by promoting transferable controllers. In: Proceedings of the 12th Annual Conference on Genetic and Evolutionary Computation, pp. 119–126 (2010)
22. Kostrikov, I., Nair, A., Levine, S.: Offline reinforcement learning with implicit Q-learning. In: International Conference on Learning Representations (2022). https://openreview.net/forum?id=68n2s9ZJWF8
23. Kumar, A., Fu, J., Soh, M., Tucker, G., Levine, S.: Stabilizing off-policy Q-learning via bootstrapping error reduction. In: Advances in Neural Information Processing Systems, vol. 32 (2019)

24. Kumar, A., Zhou, A., Tucker, G., Levine, S.: Conservative Q-learning for offline reinforcement learning. Adv. Neural Inf. Process. Syst. **33**, 1179–1191 (2020)
25. LeCun, Y., Chopra, S., Hadsell, R., Ranzato, M., Huang, F.: A tutorial on energy-based learning. Predicting Struct. Data **1**(0) (2006)
26. Levine, S., Kumar, A., Tucker, G., Fu, J.: Offline reinforcement learning: tutorial, review, and perspectives on open problems. arXiv preprint arXiv:2005.01643 (2020)
27. Li, H., Xu, Z., Taylor, G., Studer, C., Goldstein, T.: Visualizing the loss landscape of neural nets. In: Advances in Neural Information Processing Systems, vol. 31 (2018)
28. Liu, W., Wang, X., Owens, J., Li, Y.: Energy-based out-of-distribution detection. Adv. Neural. Inf. Process. Syst. **33**, 21464–21475 (2020)
29. Mnih, V., et al.: Human-level control through deep reinforcement learning. Nature **518**(7540), 529–533 (2015)
30. Oord, A.v.d., Li, Y., Vinyals, O.: Representation learning with contrastive predictive coding. arXiv preprint arXiv:1807.03748 (2018)
31. Silver, D., et al.: Mastering the game of go with deep neural networks and tree search. Nature **529**(7587), 484–489 (2016)
32. Silver, D., et al.: Mastering the game of go without human knowledge. Nature **550**(7676), 354–359 (2017)
33. Sullivan, R., Terry, J.K., Black, B., Dickerson, J.P.: Cliff diving: exploring reward surfaces in reinforcement learning environments. arXiv preprint arXiv:2205.07015 (2022)
34. Wang, Z., Hunt, J.J., Zhou, M.: Diffusion policies as an expressive policy class for offline reinforcement learning. arXiv preprint arXiv:2208.06193 (2022)
35. Wu, Y., Tucker, G., Nachum, O.: Behavior regularized offline reinforcement learning. arXiv preprint arXiv:1911.11361 (2019)
36. Yu, T., et al.: MOPO: model-based offline policy optimization. Adv. Neural. Inf. Process. Syst. **33**, 14129–14142 (2020)

Lithology Identification Method Based on CNN-LSTM-Attention: A Case Study of Huizhou Block in South China Sea

Zhikun Liu[1(✉)], Xuedong Yan[1], Yanhong She[2], Fan Zhang[1], Chongdong Shi[3], and Liupeng Wang[1]

[1] College of Petroleum Engineering, Xi'an Shiyou University, Xi'an 7100652, China
lzk12431@xsyu.edu.cn
[2] School of Science, Xi'an Shiyou University, Xi'an 7100652, China
[3] Chuanqing Drilling Changqing Drilling Company, Xi'an 7100212, China

Abstract. Lithology identification of rock is one of the main bases for stratigraphic division in geology and plays a very important role in oil and gas exploration. In recent years, with the increasing amount of data obtained by MWD and other methods, it is possible to use artificial intelligence method to dynamically identify lithology based on these data. This paper establishes a formation lithology prediction model based on CNN-LSTM-Attention, predicts formation lithology through drilling parameters and logging data, and verifies the drilling data of a block in Huizhou, South China Sea. Three artificial intelligence methods, convolutional neural network - Long short-term memory neural network -Attention mechanism (CNN-LSTM-Attention), convolutional neural network - long short-term memory neural network (CNN-LSTM) and long short-term memory neural network (LSTM), are compared and analyzed. The results show that the lithology prediction model proposed in this paper has good accuracy and low error, and has certain reliability and practicability.

Keywords: Convolutional neural network · Long short-term memory neural network · Attention mechanism · The lithology prediction

1 Introduction

The upper shale in Huizhou area of the east South China Sea has high clay content, which is prone to shrinkage and collapse after hydration expansion. The Paleogene strata have high degree of compaction and high structural stress, and are prone to shaft wall falling. The lower part is pre-paleogene igneous rock, which is characterized by deep burial, dense rock, high temperature and low porosity and permeability, and requires high reservoir protection [19, 20]. Due to the complex lithology in Huizhou block, accurate and rapid lithology identification and prediction in this area can help drilling

Funding: Supported by the Postgraduate Innovation and Practice Ability Development Fund of Xi'an Shiyou University (YCS22215305).

L. Fang et al. (Eds.): CICAI 2023, LNAI 14474, pp. 347–358, 2024.
https://doi.org/10.1007/978-981-99-9119-8_31

more efficiently, reduce the complexity of fractured formations, avoid drilling accidents such as well collapse and sticking to a certain extent, reduce development costs and improve development efficiency. However, the nonlinear interference among many variables affecting the drilling process leads to a certain complexity in the identification of formation lithology. Traditional formation lithology identification methods have certain limitations, low accuracy and high error, and are difficult to meet the current requirements [1].

At present, the main methods for stratigraphic lithology identification are probabilistic statistical lithology identification, cluster analysis lithology identification and neural network lithology identification. Neural network is widely used in lithology identification because of its strong adaptability, fault tolerance and associative memory function. In 2019, CAI Huihui et al. [2] proposed to replace traditional manual calculation by using one-dimensional convolutional neural network, to mine comprehensive metallogenic information by training geochemical and geophysical element data of metal ores, and to classify four types of metallogenic prospect areas according to the training results. In 2020, Duan Youxiang et al. [3] pointed out that a single machine learning method had a low fault tolerance rate in porosity prediction. The shortcomings of overfitting and so on put forward the method of building prediction model by selective ensemble learning based on lithology classification. This method takes into account the influence of lithology on porosity, overcomes the shortcomings of a single model, and the model has strong generalization ability. In 2021, Xu Ting; In view of the cost problem of lithology identification based on supervised learning, et al. [4] proposed an active learning method for lithology identification. This method can reduce the cost significantly while ensuring the classification accuracy. In 2022, Lei Mingfeng et al. [6] selected the Mask R-CNN model and made targeted improvements, proposing an intelligent detection method for rock flake mineral identification and quantitative content statistics. The accuracy and error of the method for rock classification meet expectations, but the model has high data requirements and requires a large number of original images to carry out model deep learning training. Otherwise, the prediction accuracy of the model will be affected.

Although the above scholars have established various formation lithology prediction models by using artificial intelligence algorithms, the above methods have certain limitations and deficiencies in terms of low accuracy, cost problems and generalization ability. Therefore, on the basis of the characteristic value of the extracted data of traditional CNN, this paper combined with the LSTM network to find out the characteristics of drilling parameters with drilling depth (time series), and added an AM module to assign weights according to the importance of the correlation between drilling parameters and formation, and built a CNN-LSTM-Attention lithology prediction model. In this paper, multiple models are adopted for lithology identification. CNN and LSTM in the model have different functional tasks. Compared with a single model for data processing and prediction, they can overcome the limitations of input feature redundancy, low precision and higher data set requirements. Experiments show that this model has the best prediction accuracy, has certain reliability and practicability, and its identification results and model can provide certain reference for formation lithology identification [5, 7].

2 CNN-LSTM-Attention Formation Lithology Identification Model

Based on the data of a well in Huizhou area of the South China Sea, this paper conducts experiments to identify the underlying lithology to be drilled. In view of the limitations of traditional lithology prediction methods and CNN, combined with the advantages of LSTM network, a CNN-LSTM-ATTENTION [8, 9] model is proposed for lithology identification to provide decision-making basis for guiding drilling and improving drilling efficiency.

2.1 Sample Set Data Analysis and Processing

Since the acquired drilling data inevitably has some redundancy, it is very necessary to carry out dimensionality reduction processing of input vectors based on the importance of drilling parameters. In addition, data cleaning and data labeling are the basic links of supervised learning and prediction classification model. Finally, the sample set is divided into training set and test set according to reasonable proportion.

Taking the integrated drilling parameters as the input vector established by the neural network model [10], the probability of different formation lithology in the output layer is obtained through a series of nonlinear transformations of the hidden layer. Among all the predicted formation lithology, the formation lithology with the highest probability is determined to be the current predicted formation lithology. Due to the large number of drilling parameters, the neural network system will be huge if all parameters are treated as independent input feature vectors (input feature redundancy). Compared with other models, this paper combines two kinds of neural networks. Compared with the single LSTM [11, 12] model, the sample set data first reduces the types of input parameters through CNN to avoid input feature redundancy.

By analyzing the correlation between drilling parameters, data with high correlation is selected as the input parameters of the model to improve the accuracy of the model and reduce the redundancy of the established formation lithology prediction model [16].

Since most drilling parameters are non-linear correlated with the target, mutual information is used to measure the correlation between drilling data. When $(X, Y) \sim p (x, y)$, the mutual information between variables X and Y is defined as

$$MI(X; Y) = H(X) + H(Y) - H(X, Y) \tag{1}$$

where: H (X, Y) is the joint entropy of variables X and Y; H (X) and H (Y) are the unconditional entropy of X and Y respectively.

The thermal map calculated by mutual information values of drilling parameters is shown in Fig. 1. It can be seen that the classification of formation lithology is affected by these parameters. Among them, formation lithology has strong correlation with drilling parameters such as WOB (weight on bit), WOH (Weight on Hook), RPM (revolutions per minute), ROP (rate of penetration), SSP (pumping pressure), etc., and has low mutual information value with displacement and drilling time. Therefore, parameters with strong correlation of drilling parameters are selected as input variables of formation lithology prediction model.

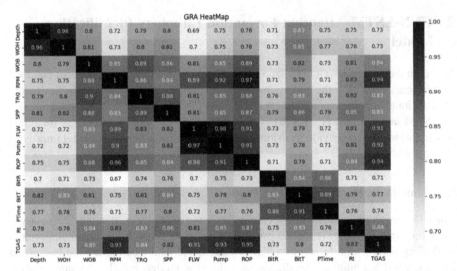

Fig. 1. Mutual information heat map of drilling data.

Table 1. Statistical analysis of sample input data set

Drilling Parameter	Minimum	Maximum	Mean	Standard Deviation
Depth (m)	1018.00	3586.00	2302.10	741.73
WHO (k/bs)	211.80	390.00	309.43	47.43
WOB (k/bs)	0.40	35.50	15.17	6.11
RPM (l/min)	20.00	101.00	63.13	6.66
SSP (n/m²)	488.00	3248.00	2648.03	440.36
ROP (min/m)	0.36	139.19	4.26	8.18

2.2 Model Input Parameter Preprocessing – Pavelet Filter Processing

Due to the complex conditions in downhole drilling, there is a certain error between the measured value received while drilling and the actual value under the interference of many factors. Therefore, wavelet filtering is applied for data noise reduction [18].

The analysis signal S (t) is transformed into a wavelet, i.e.

$$W_f(\tau, a) = \frac{1}{\sqrt{a}} \int\limits_{-\infty}^{+\infty} X(t) \cdot \varphi\left(\frac{t-\tau}{a}\right) dt \qquad (2)$$

where: a is the scale factor, and a > 0, the basic wavelet scaling transformation is realized; Is the translation factor to realize the translation transformation of the basic wavelet on the time axis.

This method determines the optimal block size and threshold, and uses the block threshold to de-noise the original drilling data. The structure before and after de-noising is shown in Fig. 2. The original data curve contains many spikes and abrupt changes, and the curve after forced de-noising by wavelet de-noising is smooth, but its curve trend is consistent with the trend of the original curve. Keep the changing nature of the original data.

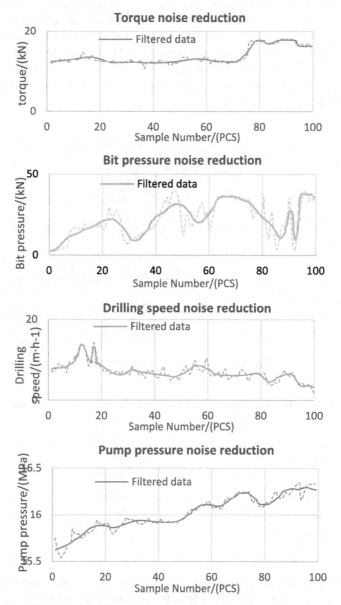

Fig. 2. Comparison of the effect of wavelet noise reduction

2.3   CNN-LSTM-Attention Modeling of Formation Lithology Identification

In the network structure of the model [14, 15], the convolutional network consists of two convolutional layers and one pooled layer. The convolutional filter of the two layers is 128 and 64 respectively, the kernel is 1, the activation function is relu, and the pooled layer selects the maximum pooled operation to extract features and explore hidden rules in the lithology data. The features extracted by the convolutional network are input into the timing rule between the extracted data in the LSTM network, where the inpu_DIM is 4 and the number of time steps is 4. After the LSTM network, AM module is added to screen the state at different times. Finally, the retained features are output to the fully connected layer and combined with Softmax activation function to realize lithology classification (Fig. 3 and Table 2).

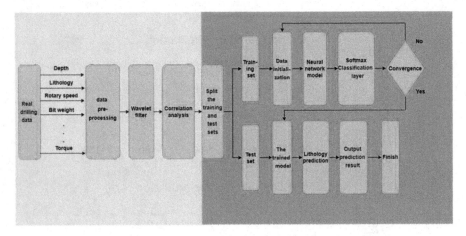

Fig. 3. Lithology identification flow chart

Table 2. Hyperparameters and network structure of the optimal model

Options	Parameter
Maximum number of passes	100
Train/test set	80%/20%
Optimization algorithm	Adam
Number of hidden layers	20
Number of nodes per layer	70

LSTM [17] has a Gate mechanism, in which the input gate selects the current information to input, and the forget gate selects the past information to forget. In fact, it is also a certain degree of Attention. However, if you want to solve the problem of long-term dependence, LSTM actually needs to further capture sequence information, and the performance in long text will slowly decay with the increase of step, and it is difficult to retain all useful information. The Attention mechanism in this paper weights the hidden state of all steps, and focuses attention on the more important hidden state information in the whole text. Making it easy to visualize those steps is important, but it can lead to overfitting and increase computation.

The Attention mechanism is actually a learning mechanism that continuously adjusts the weight of each encoder by measuring the contribution of the JTH hidden state of the encoder and the previous decoder state pair. Thus, it pays more attention to the similar parts of the input elements and suppresses other useless information. This model assigns more weight to the main influencing parameters such as WOH, WOB and RPM, and reduces the weight of parameters such as T GAS main and Return, which improves the lithology prediction effect. The other two models predict the lithology data with equal weight, and produce too much weight for other useless information. As a result, the accuracy of lithology prediction decreases.

3   CNN-LSTM-Attention Model Training and Performance Evaluation

3.1   CNN-LSTM-Attention Model Training Process

The dataset in this paper is divided into a training set and a test set according to the ratio of 8:2, with 2054 samples in the training set and 514 samples in the test set. The number of samples for each training input is 32, the maximum number of iterations is chosen to be 100, and the selected optimization model algorithm is Adam (Fig. 4).

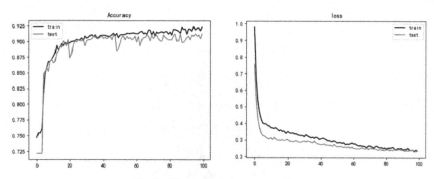

Fig. 4. Changes of accuracy and loss

The CNN-LSTM-Attention intelligent lithology prediction model has the highest accuracy and the lowest loss function (the gap between the predicted value and the actual value is the lowest), and the final accuracy of the test set and training set are 91.6% and 92.57% respectively. Therefore, the intelligent lithology prediction model proposed in this paper can more accurately predict the three lithology conditions (mudstone, sandstone and magmatic rock), and the loss function training set and test set gradually tend to 0.2, indicating that the prediction effect gradually reaches the actual situation.

3.2 Performance Evaluation of Neural Network Model for Formation Lithology Identification

The primary task of realizing the intelligent discrimination of drilling conditions is to select the best neural network algorithm according to the data characteristics of the sample set, and then select the evaluation indicators such as accuracy, precision, recall rate and score to evaluate the classification and discrimination performance of the machine learning model.

It is often necessary to evaluate the generalization error of machine learning models and select the model with the smallest generalization error. Therefore, the test set should be used to test the classification discrimination ability of the model, and the test error of the test set should be used as an approximation of the generalization error. Four evaluation metrics, Accuracy, Precision, Recall and -score, are used in this paper [13]. Combining the actual category and the model predicted category for classification, the confusion matrix of binary classification is shown in Table 1, and the confusion matrix of multi-classification is shown in Fig. 5 where TP is true positive, FP is false positive, TN is true negative, and FN is false negative (Table 3).

Table 3. Confusion matrix of binary classification results

Actual value	Predicted value	
	Positive	Negative
True	TP	FN
False	FP	TN

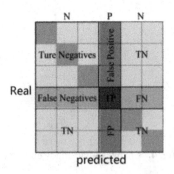

predicted

Fig. 5. Confusion matrix for a multiclass classification problem

Different indexes directly reflect the performance of classification discrimination. The most common metric is "accuracy," which is the number of correctly classified samples divided by the total number of samples. For balanced classification problems, generally a higher correct rate indicates a better classifier. "Precision" and "recall" are contradictory measures. "Precision" can reflect how many predictions are correct and how many are incorrect for a certain class of test samples. The "recall" shows how many predictions of a certain class of prediction results are correct. The "score" is the harmonic mean of precision and recall.

According to Table 4, P, A, TPR and Score of CNN-LSTM-Attention model are 0.80, 0.92, 0.88 and 0.84, respectively. These four indicators all tend to 1, which indicates that the model has high prediction accuracy for lithology prediction. Secondary indicators A, P, TPR through the confusion matrix based on the extended indicators. P is accuracy, where bigger is better, and is a measure of how likely a classifier is to actually classify the positive class as positive. TPR is recall and measures how well a class finds all the positive classes.

Table 4. Three model evaluation indicators

Algorithm model	F_1 Sorce	P	TPR	Acc
LSTM	0.7937	0.7669	0.8225	0.8966
CNN-LSTM	0.7762	0.7391	0.8173	0.9042
CNN-LSTM-Attention	0.8403	0.8000	0.8849	0.9257

3.3 Confusion Matrix and Results of CNN-LSTM-Attention Model

The confusion matrix is used to show the overall performance index of the model. It can be seen from Table 4 and Fig. 6 that the LSTM network and CNN-LSTM network have strong feature learning ability, and the LSTM lithology prediction rate reaches 89.66%. After adding and CNN network, the lithology prediction rate of CNN-LSTM reaches 90.42%. After adding the AM module, the lithology prediction rate reaches 92.57%, which is 2.91% higher than that of the LSTM model and 2.51% higher than that of the CNN-LSTM model. It can be seen that Acc, F1 Score, P and TPR of CNN-LSTM-Attention model are better than those of CNN-LSTM model and CNN-LSTM model. On the one hand, this is because CNN may convolutional features, and on the other hand, AM model increases the weight of key parameters.

Figure 7 shows the lithology of part of a well section in Huizhou, South China Sea. Taking the 50-m well section from 3000 m to 3050 m as an example, the actual formation lithology is compared with the recognition results of CNN-LSTM-Attention model. In this well section, only the recognition results of adjacent sandstone and mudstone are different.

confusion_matrix
mud stone
sand stone
magmatite
True label
370
93.67%
25
6.33%
0
0.00%
13
11.40%
100
87.72%
1
0.88%
1
20.00%
0
0.00%
4
80.00%
mudstone
sandstone
magmatite
Predict label
0.8
0.6
0.4
0.2
0.0

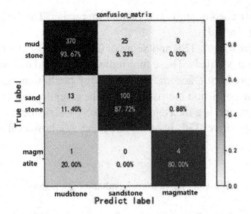

Fig. 6. CNN-LSTM-Attention model recognition results

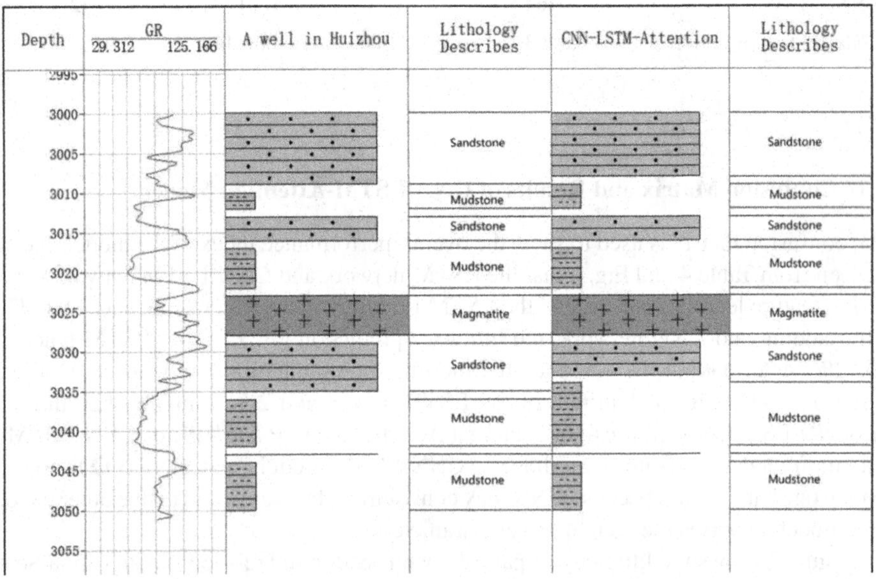

Depth
GR
29.312
125.166
A well in Huizhou
Lithology Describes
CNN-LSTM-Attention
Lithology Describes
2995
3000
3005
3010
3015
3020
3025
3030
3035
3040
3045
3050
3055
Sandstone
Mudstone
Sandstone
Mudstone
Magmatite
Sandstone
Mudstone
Mudstone
Sandstone
Mudstone
Sandstone
Mudstone
Magmatite
Sandstone
Mudstone
Mudstone

Fig. 7. Partial actual lithology of a well in Huizhou, South China Sea and lithology identified by CNN-LSTM-Attention model

4 Conclusion

This paper builds a deep lithology prediction method based on CNN-LSTM-Attention is constructed to fully mine the spatial characteristics of the data.. From the model evaluation index reference, it shows that CNN-LSTM-Attention model can achieve better results in lithology prediction, and further infer the advantages of Attention mechanism. For the prediction of too long text, it is necessary to assign a large proportion of weight to important information. By comparing the analysis results, it shows that it has a better effect on prediction accuracy. Firstly, it extracts the input parameters with high correlation in drilling parameters through CNN, and then redistributes the weights through the Attention mechanism, and then performs lithology identification through the LSTM model. Compared with a single model, it has higher accuracy and avoids input feature redundancy. The model has reference value in the application of lithology identification, which can quickly and accurately identify the lithology of the target layer, and the accuracy of lithology identification is 92.57%.

References

1. Xu, Z., et al.: Lithology identification: method, current situation and intelligent development trend. Geolog. Rev. **68**(06), 2290–2304 (2022)
2. Cai, H., et al.: Classification of metallogenic prospect areas based on convolutional neural network model: a case study of gold polymetallic ore field in Daqiao Area, Gansu Province. Geolog. Bull. China **38**(12), 1999–2009 (2019)
3. Duan, Y., Wang, Y., Sun, Q.: Application of selective ensemble learning model in lithology-porosity prediction. Sci. Technol. Eng. **20**(03), 1001–1008 (2020)
4. Xu, T., et al.: Evaluation of active learning algorithms for formation lithology identification. J. Petrol. Sci. Eng. **206**, 108999 (2021)
5. Arnø, M., Morten, J., Morten Aamo, O.: Real-time classification of drilled lithology from drilling data using deep learning with online calibration. In: SPE/IADC International Drilling Conference and Exhibition (2021)
6. Lei, M., et al.: Research on intelligent recognition method and application of Rock lithology Mask R-CNN. J. Railway Sci. Eng. **19**(11), 3372–3382 (2022)
7. Yue, Z., et al.: Research progress of machine learning algorithms for lithology identification based on LWD data. Sci. Technol. Eng. **23**(10), 4044–4057 (2023)
8. Wei, Y., Gong, J.: Rolling bearing fault diagnosis based on CNN-LSTM-attention. J. Shenyang Univ. Technol. (08) (2022)
9. Liu, W., Liu, W., Gu, J.: Prediction of daily oil production of oil Wells based on machine learning method. Oil Drill. Prod. Technol. **421**, 70–75 (2020)
10. Ma, Z., Ma, L., Li, K., Yao, W., Wang, P., Wang, X.: Multi-scale lithology recognition based on deep learning of rock images. Geolog. Sci. Technol. Bull. **41**(06), 316–322 (2022)
11. Yang, J., Zhang, H.: Research on neural network method of formation lithology identification while drilling. Nat. Gas. Ind. **26**(12), 109–111 (2006)
12. Hochreiters, S.J.: Longshort-termmemory. Neural Comput. **8**, 1735–1780 (1997)
13. Yin, Q.S., Yang, J., Hou, X.X., et al.: Drilling performance improvement in offshore batch Wells based on rig state classification using machine learning. J. Petrol. Sci. Eng. **192**, 107306 (2020)
14. Zhu, H., Ning, Q., Lei, Y.J., et al.: Fault classification of rolling bearing based on attention mechanism-inception CNN model. J. Vib. Shock **39**(19), 84–93 (2020)
15. Roger, Z.L., et al.: Research on construction of deep prospecting prediction model based on PSO-CNN. J. Chengdu Univ. Technol. (Nat. Sci. Edn.) (09) (2020)
16. Zhang, J.F., Zhu, Y., Zhang, X.P., et al.: Developing a long short-term memory (LSTM) based model for predicting water table depth in agricultural areas. J. Hydrol. **6**(561), 918–929 (2018)
17. Wu, Z., Zhang, X., Zhang, C., et al.: Lithologic reservoir identification method based on LSTM recurrent neural network. Litholog. Reserv. **33**(3), 120–128 (2021)
18. Arps, J.J., Arps, J.L.: The subsurface telemetry problem - a practical solution. Soc. Petrol. Eng. (1964)

19. Tong, W., Zhao, R., Guo, C.: Limestone slurry density prediction based on grey relational analysis and mutual information theory. China Testing **06**, 1–7 (2022)
20. Yan, D., Chen, B., Song, L., Wang, B.: Drilling fluid system of deep Wells in Huizhou area, South China Sea. Chem. Eng. Equip. **06**, 109–110 (2021)
21. Wang, A., et al.: Characteristics and controlling factors of physical properties of deep tight sandstone reservoirs: a case study of the second lower Member of Xuerang Formation in Yuanba West area, Northeast Sichuan Basin. Acta Sedimentol. Sinica **40**(02), 410–421 (2012)

Deployment and Comparison of Large Language Models Based on Virtual Cluster

Kai Li[1,2], Rongqiang Cao[1,2(✉)], Meng Wan[1,2], Xiaoguang Wang[1,2],
Zongguo Wang[1,2], Jue Wang[1,2], and Yangang Wang[1,2]

[1] Computer Network Information Center, Chinese Academy of Sciences, Beijing 100083, China
caorq@sccas.cn
[2] School of Computer Science and Technology, Chinese Academy of Sciences, Beijing 100049,
China

Abstract. **[Objective]** Currently, large language model (LLM) is one of research
highlights in the field of natural language processing. This paper selected some
open-source LLMs for deployment and comparison from the perspective of
consumer-grade GPU and support for Chinese and English. **[Coverage]** This
paper uses keywords search and citation secondary search to collect papers and
information from international computer journals, conferences and open source
code warehouse. **[Methods]** From the perspective of supporting both Chinese and
English, we selected LLaMA, MOSS, ChatGLM-6B, ColossalChat, and Chinese-
LLaMA-Alpaca for deployment at the same virtual task, on the virtual cluster with
consumer-grade GPU. Furthermore, we made horizontal comparisons on semantic understanding, logical reasoning, code programming, ancient poetry, and legal
questions, and then, discuss the advantages and disadvantages of these models.
[Results] Limited parameters scale, most of them are not very friendly to support
Chinese, have weak Chinese understanding abilities, and have varying abilities in
logical reasoning. At present, researchers have paid less attention to issues such
as Chinese support and resource consumption. They generally focus on increasing the scale of model parameters and using higher graphics card resources for
model training and inference. **[Conclusions]** Although the development of LLMs
is rapid, many models do not fully support Chinese. Understanding Chinese ability
needs to be further improved, and more efforts need to be made in logical reasoning. It is believed that in the future, there will be more large language models that
consume lower resources and support stronger Chinese.

Keywords: Large Language Model · General Language Model ·
consumer-grade acceleration card · virtual cluster

Fund project: Chinese Academy of Sciences strategic leading science and technology project.
(XDB38050200).

L. Fang et al. (Eds.): CICAI 2023, LNAI 14474, pp. 359–365, 2024.
https://doi.org/10.1007/978-981-99-9119-8_32

1 Introduction

With the rapid development of artificial intelligence, the scale of obtaining and process- ing massive text data has rapidly increased, providing sufficient training data for the development of large language models (LLMs). At the same time, deep learning tech- nologies such as convolutional neural network and recurrent neural network have made rapid progress in recent years, providing technical support for the development of LLMs. In particular, the birth of ChatGPT [1] developed by OpenAI [2] has set off a new wave of research in the field of natural language processing, demonstrated the ability of artificial general intelligence (AGI), and attracted extensive attention in the industry. However, due to the extremely expensive training and deployment of large language models, it has created certain obstacles for building transparent and open academic research and application promotion.

In order to promote the open research and application promotion of LLMs in the Chinese NLP [3–5] community and various fields, this paper started from the perspective of consumer-grade GPU graphics cards and support for Chinese and English, selected some open source LLMs and deployed them in virtual cluster. Moreover, horizontal comparisons will be made on semantic understanding, logical reasoning, code program- ming, ancient poetry, and legal questions, mainly comparing the loading efficiency of model parameters and resource utilization, and then discussing the advantages and dis- advantages of these models. Through this approach, we hope to promote the research and application of large models, and contribute to the implementation of academic transparency and openness.

2 Related Work

In order to verify the performance and effectiveness of the universal open source large lan- guage models in semantic understanding, logical reasoning, code programming, ancient poetry, and legal questions, we selected several typical open source models, includ- ing LLaMA [6, 7], MOSS [8], ChatGLM-6B [9–12], ColossalChat [13, 14], Chinese- LLaMA-Alpaca [15, 16], and chose some common questions in the above fields to test and verify in the virtual cluster [17, 18] of consumer-grade GPU graphics cards.

2.1 Selection of LLMs

This paper focuses on deploying and comparing several open source LLMs [19– 29], enabling them to complete inference tasks in a consumer-grade acceleration card environment. These models are open source, highly focused and support for Chinese.

2.2 Selection of Questions

We selected five types of questions covering semantic understanding, logical reasoning, code programming, ancient poetry, and legal question, including Chinese and English, to test the response time, GPU and CPU resources consumption, and response effectiveness of these models. Each question was given a corresponding simple rating standard.

3   Virtual Cluster and Deployment

The virtual cluster used in this paper is developed by Computer Information Network Center, Chinese Academy of Sciences. It is based on supercomputing hardware resources, and realizes the construction of heterogeneous computing resources into configurable, dynamically applied and monitorable virtualization computing resources to meet the needs of AI applications in various fields for adapting to the running environment of various software required. It has a high degree of customizability.

3.1   Deployment of Virtualization Task

This section will introduce how to build a LLM runtime environment in virtual cluster [30] (see Fig. 1).

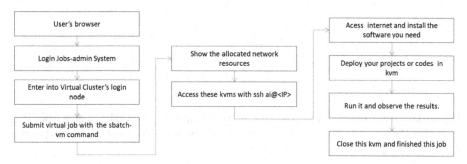

Fig. 1. Virtualization Task Process. We enter into virtual cluster, submit virtualization task, configure base image or perform inference tasks for LLMs.

We use the *sbatch-vm* command to submit a virtualization task, which has *kvm-edit* parameter at first time. Then the cluster allocates accelerator card resources and network resources, and assigns a job ID for this task. We can view the network resource information assigned to this task through the *showip* command. Importantly, we can access this task with SSH, install the required software from internet, and program. Furthermore, we deploy the above LLMs in the same task, run and monitor the results.

3.2   Deployment of the Models

After starting the virtualization task, we enter the virtualization task to customize the basic image. Firstly, we configured the base image (*Ubuntu 20.04, NVIDIA GPU Driver 520.61, and CUDA 11.8*). Secondly, we installed the CONDA environment, PyTorch2.0, and other running environment for LLMs. Then, we deployed step by step and write inference test code according to the description of the above model. Moreover, we passed the above questions as parameters to test programs, run the test codes multiple times, obtained data such as model parameter loading time, response time of each question, and GPU resource utilization etc. Finally, we normalized and organized these data to prepare for the comparison.

4   Result Analysis

In this section, we compared the loading time, response time, model effects and resource consumption of the above models.

4.1   Comparison of Model Loading and Problem Response Time

In order to visually compare the performance of these models in terms of model parameter loading time and Chinese English problem response time, we visualized the data (see Fig. 2).

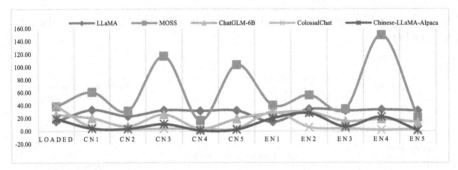

Fig. 2. Response Time (Seconds). The value *LOADED* represents the model loading time. The value *CN1,CN2,...,EN5* represent Chinese and English questions. The ordinate values represent the inference response time of these questions.

From the above figure (see Fig. 2), we can see that MOSS has the longest response time when answering the questions, and there is a significant deviation in response time for different questions. Even for the same question, there is a significant difference in language (such as CN4 and EN4). ColossalChat has the shortest response time. ChatGLM-6B model has a relatively moderate response time when dealing with Chinese and English questions. However, the length of problem inference response time does not represent the effectiveness of the models, because there may be a situation that one model provides search results for certain problems rather than generate results randomly.

4.2   Comparison of Model Effects

Through questionnaire survey, we obtained the average score of each model answering (see Fig. 3).

ChatGLM-6B has the highest scores and performs well in both Chinese and English questions. Next is MOSS, but its logical reasoning ability is weak. LLaMA performs significantly better in dealing with English problems than Chinese problems, and its understanding of Chinese semantics is also relatively weak. ColossalChat generally has low scores and needs to be strengthened in terms of Chinese semantic understanding, logical reasoning, and ancient poetry. Chinese-LLaMA-Alpaca has made significant progress on the basis of LLaMA, but there is still a certain gap compared to ChatGLM-6B and MOSS.

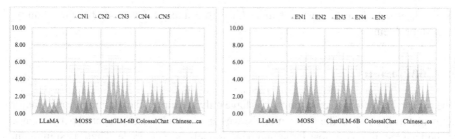

Fig. 3. Comparison of model effects. Each question has a maximum score of 10 points. The left figure shows the Chinese question scores of each model. The right figure shows the English question scores of each model.

4.3 Comparison of Resource Consumption

The resource consumption of inference question refers to the amount of GPU or CPU resources occupied by the relevant problem during the model response period as below.

$$ResCS_{GPU} = T_{response} * (MEM_{GPU1} + MEM_{GPU2}) \tag{1}$$

$$ResCS_{CPU} = T_{response} * MEM_{CPU} \tag{2}$$

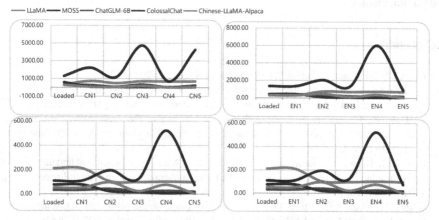

Fig. 4. Resource consumption situation (GPU and CPU). The above two figures represent GPU resources consumption (ResCSGPU) for Chinese and English problems, respectively. The below two figures are for CPU resources consumption (ResCSCPU).

It is basically the same in response to Chinese and English problems for each model from the perspective of CPU resource consumption (below two figures), (see Fig. 4). However, there are slight differences in CPU resource consumption among different models, but the difference is not very obvious.

From the perspective of GPU resource consumption (above tow figures), there is a significant difference in resource consumption between Chinese and English problems.

Especially for MOSS, it consumes more GPU resources on Chinese problems than on English problems.

5 Conclusion

Due to time and resource constraints, we have only attempted the aforementioned major language models. From the above models, we can conclude that the overall performance of the basic abilities of most large language models is acceptable. There is significant room for improvement in logical reasoning, code programming, and contextual understanding. LLaMA and its derived models exhibit relatively weak performance in logical reasoning, code generation, and ancient poetry abilities. Both semantic understanding and keyword extraction in Chinese and English need to be further strengthened. At the level of basic abilities, ChatGLM-6B and MOSS demonstrate excellent Chinese writing skills. However, MOSS consumes too much resources and cannot achieve democratization. ChatGLM-6B performs better in Chinese question answering, in terms of semantic understanding, poetry creation, and simple logical reasoning. It also has relatively less resource consumption and performs well.

In the future, with the improvement of accuracy, large language models will need to be applied in more fields while reducing resource consumption to facilitate the promotion and application.

We have created a video demonstration about ChatGLM-6B which deployed in the virtual cluster [31].

References

1. ChatGPT, Introducing ChatGPT, OpenAI Blog, November 2022
2. OpenAI, Our approach to alignment research, OpenAI Blog, August 2022
3. Thede, S.M., Harper, M.P.: A second-order hidden Markov model for part-of-speech tagging. In: Dale, R., Church, K.W. (eds.) 27th Annual Meeting of the Association for Computational Linguistics, University of Maryland, College Park, Maryland, USA, 20–26 June 1999, pp. 175–182. ACL (1999)
4. Bahl, L.R., Brown, P.F., de Souza, P.V., Mercer, R.L.: A tree-based statistical language model for natural language speech recognition. IEEE Trans. Acoust. Speech Signal Process. **37**(7), 1001–1008 (1989)
5. Brants, T., Popat, A.C., Xu, P., Och, F.J., Dean, J.: Large language models in machine translation. In: Eisner, J. (ed.) EMNLP-CoNLL 2007, Proceedings of the 2007 Joint Conference on Empirical Methods in Natural Language Processing and Computational Natural Language Learning, 28–30 June 2007, Prague, Czech Republic, pp. 858–867. ACL (2007)
6. Touvron, H., et al.: LLaMA: open and efficient foundation language models. CoRR (2023)
7. Facebookresearch/llama [open source]. https://github.com/facebookresearch/llama. Accessed 6 June 2023
8. OpenLMLab/MOSS [open source]. https://github.com/OpenLMLab/MOSS. Accessed 6 June 2023
9. Zeng, A., et al.: GLM-130B: an open bilingual pre-trained model, vol. abs/2210.02414 (2022)
10. Du, Z., et al.: GLM: General Language Model Pretraining with Autoregressive Blank Infilling (2021). https://doi.org/10.48550/arXiv.2103.10360

11. THUDM/GLM [open source]. https://github.com/THUDM/GLM. Accessed 6 June 2023
12. THUDM/ChatGLM-6B [open source]. https://github.com/THUDM/ChatGLM-6B. Accessed 6 June 2023
13. Bian, Z., et al.: Colossal-AI: a unified deep learning system for large-scale parallel training. CoRR, vol. abs/2110.14883 (2021)
14. Hpcaitech/ColossalAI [open source]. https://github.com/hpcaitech/ColossalAI. Accessed 7 June 2023
15. Cui, Y., Yang, Z., Yao, X.: Efficient and effective text encoding for Chinese LLaMA and Alpaca. abs/2304.08177 (2023)
16. Ymci/Chinese-LLaMA-Alpaca [open source]. https://github.com/ymcui/Chinese-LLaMA-Alpaca. Accessed 7 June 2023
17. Yao, T., et al.: VenusAI: an artificial intelligence platform for scientific discovery on supercomputers. abs/pii/S1383762122001059 (2022)
18. Virtual Cluster User's Manual. http://www.aicnic.cn/jobs-admin/static/manual.pdf. Accessed 3 June 2023
19. Shanahan, M.: Talking about large language models. CoRR, vol. abs/2212.03551 (2022)
20. Liu, P., Yuan, W., Fu, J., Jiang, Z., Hayashi, H., Neubig, G.: Pre-train, prompt, and predict: a systematic survey of prompting methods in natural language processing. ACM Comput. Surv. **55**, 195:1–195:35 (2023)
21. Han, X., et al.: Pretrained models: past, present and future. AI Open **2**, 225–250 (2021)
22. Qiu, X., Sun, T., Xu, Y., Shao, Y., Dai, N., Huang, X.: Pre-trained models for natural language processing: a survey. CoRR, vol. abs/2003.08271 (2020)
23. Li, J., Tang, T., Zhao, W.X., Wen, J.: Pretrained language model for text generation: a survey. In: Zhou, Z. (ed.) Proceedings of the Thirtieth International Joint Conference on Artificial Intelligence, IJCAI 2021, Virtual Event/Montreal, Canada, 19–27 August 2021. ijcai.org, pp. 4492–4499 (2021)
24. Lu, P., Qiu, L., Yu, W., Welleck, S., Chang, K.: A survey of deep learning for mathematical reasoning. CoRR, vol. abs/2212.10535 (2022)
25. Dong, Q., et al.: A survey for in-context learning. CoRR, vol. abs/2301.00234 (2023)
26. Huang, J., Chang, K.C.: Towards reasoning in large language models: a survey. CoRR, vol. abs/2212.10403 (2022)
27. Qiao, S., et al.: Reasoning with language model prompting: a survey. CoRR, vol. abs/2212.09597 (2022)
28. Zhou, J., Ke, P., Qiu, X., Huang, M., Zhang, J.: ChatGPT: potential, prospects, and limitations. Front. Inf. Technol. Electron. Eng. 1–6 (2023)
29. Zhao, W.X., Liu, J., Ren, R., Wen, J.: Dense text retrieval based on pretrained language models: a survey. CoRR, vol. abs/2211.14876 (2022)
30. Virtual Cluster. http://www.aicnic.cn/jobs-admin. Accessed 26 June 2023
31. Demo Video. https://pan.baidu.com/s/1Qral_mni6iKa80y6LGbjtg?pwd=imam

Social Network Community Detection Based on Textual Content Similarity and Sentimental Tendency

Jie Gao, Junping Du$^{(\boxtimes)}$, Zhe Xue, and Zeli Guan

Beijing Key Laboratory of Intelligent Telecommunication Software and Multimedia, School of Computer Science (National Pilot School of Software Engineering), Beijing University of Posts and Telecommunications, Beijing 100876, China
junpingdu@126.com

Abstract. Shared travel has gradually become one of the hot topics discussed on social networking platforms such as Micro Blog. In a timely manner, deeper network community detection on the evaluation content of shared travel in social networks can effectively conduct research and analysis on the public opinion orientation related to shared travel, which has great application prospects. The existing community detection algorithms generally measure the similarity of nodes in the network from the perspective of spatial distance. This paper proposes a community detection algorithm based on textual content similarity and sentimental tendency (CTST), considering the network structure and node attributes at the same time. The content similarity and sentimental tendency of network community users are taken as node attributes, and based on this, an undirected weighted network is constructed for community detection. This paper conducts experiments with actual data and analyzes the experimental results. It is found that the modularity of the community detection results is high and the effect is good.

Keywords: Social Network · Community Detection · Shared Travel · Content Similarity · Sentimental Tendency · Undirected Weighted Network

1 Introduction

With the development of the Internet and network communication, electronic social networks such as Facebook, Twitter, WeChat, blogs, etc., have gradually become an indispensable social channel and way in the daily life of the public [1–4]. Expressing personal opinions and thoughts on the Internet and sharing daily personal life are the main ways people utilize such electronic social networks. In this cyberspace, a large number of Internet users continuously publish massive amounts of information on various topics, such as posts, pictures, videos, comments and likes. Due to their different knowledge systems, hobbies, and discussion topics, network users will have different

This work was supported by the Program of the National Natural Science Foundation of China (62192784, U22B2038, 62172056).

information preferences. Users with similar preferences often form communities, where individuals within the same community act as network nodes, expressing similar opinions in the network space, so the connections within the community will be closer than the connections between different communities [5–7], which constitutes the so-called social network.

Micro Blog is one of the most commonly used social networking platforms among internet users in our country [9–12]. The research on the community structure of Micro Blog network is a prominent topic in the field of community detection research. At the same time, the research on it helps to find the behavior rules of user groups within social networks [5]. It can also facilitate the effective grouping of Micro Blog network users based on different specific requirements, allowing for the rapid targeting of desired user groups. This research holds important theoretical and practical significance [1, 13]. The user posting and commenting mechanism of Micro Blog makes the user relationship abstract into an undirected network, in which the nodes of the network represent each user, and the relationships between users can be represented as edge between the commenters and the posters of blog posts.

The existing traditional community detection algorithms generally interpret the results of community detection from the perspective of "distance", which only considers the topology of the network, that is, the topological relationship between nodes, while ignoring the attributes of nodes, making the results of community detection lack semantics and inconvenient to understand and explain [7, 14–16]. In this paper, we propose a community detection algorithm based on textual content similarity and sentimental tendency (CTST), aiming at the lack of semantics in the results of some traditional community detection algorithms, and considering the network topology and node attributes at the same time [17].

Our contributions are summarized as follows:

- This paper proposes a social network community detection algorithm based on textual content similarity and sentimental tendency (CTST). In this algorithm, we utilize the text content as a measure of similarity for node attributes. By combining the network structure with node attributes, we transform the network into an undirected weighted graph.
- In addition, we introduce the concept of "sentiment bias value" to represent the sentimental tendency of text content. We construct the sentiment vector based on the sentiment analysis of the text content, and combine the sentiment vectors of two distinct users to obtain the final composite sentiment vector. By further transforming and calculating the composite sentiment vector, we obtain the sentiment bias value between two nodes in the network.
- We evaluate our algorithm on two real-world datasets, and the experimental results verify that our algorithm performs better on undirected weighted networks, with higher modularity and improved the effect of community detecting.

The rest of the paper is organized as follows. In Sect. 2 we give an overview of the related work. In Sect. 3 we present and analyze the algorithm. In Sect. 4, we show experimental results on real-world datasets. Finally, we give conclusion in Sect. 5.

2 Related Work

Within the research process of social network analysis, the clustering analysis of node clusters in the network is often referred to as "graph clustering" [18, 19], which is essentially synonymous with the term "community detection". Sociological theory holds that members of an online community often share similar views and cognitive understandings regarding specific topic or aspects. Then, when comments on a specific topic are generated, due to the commonalities among members of the network community, the entire community will show a certain degree of consistency for these comments [20–23]. Based on this, the consensus reached by community members on these topics follows certain rules [5, 6], which helps us to study and discover the behavior rules of user groups within social networks.

With the continuous advancement of research on community structure detection algorithms in complex networks [24–30], various community detection algorithms have been continuously proposed and improved. Classic community detection algorithms mainly include graph segmentation-based methods, hierarchical clustering methods, modularity-based methods [31–34], and spectral clustering-based methods. The traditional method based on graph segmentation involves conducting community detection by iteratively removing edges connecting different communities until there no remaining inter-community connections in the network. The main idea of the hierarchical clustering method is to identify the vertex sets with high similarity, which can be divided into two types: aggregation algorithm and splitting algorithm according to the different identification process. One of the most famous splitting algorithm is the GN algorithm proposed by Newman and Girvan in 2002 [6, 36], which iteratively divides the structure of the community by continuously removing the edge with the highest Edge Betweenness Centrality (EBC) in the network. Based on the idea of GN algorithm, Newman proposed an improved GN algorithm, also known as Fast Newman algorithm [37], aiming at the shortcomings of GN algorithm in 2004. It is also in this algorithm that the concept of "Modularity" was first proposed. The method based on spectral clustering [38, 39] involves analyzing the eigenvectors of the Laplacian operator of the social network as its fundamental approach. The Fast Unfolding algorithm is another Modularity-based community detection algorithm [33]. The algorithm emphasizes the concept of "community folding", and iteratively analyzes the entire community that has been discovered as a new node, resulting in the formation of a community structure with a hierarchical organization. Many domestic scholars have also made a lot of contributions to the research field of community detection. Z XIE et al. optimized the Fast Unfolding algorithm and proposed a community detection algorithm with lower time complexity [40]. YW Jiang, CY Jia and other scholars used the Jaccard distance formula to measure the similarity between nodes in a social network [41]. They also applied various clustering methods such as K-means and hierarchical clustering to effectively conduct community detection in complex networks, yielding positive results.

Most of the traditional community detection algorithms primarily measure node similarity based on spatial distance and solely focus on the network structure, disregarding the attributes between nodes, which makes the results lack of semantics and inconvenient to interpret. In this paper, we present the CTST model, which takes into account both the network structure and node attributes simultaneously. The CTST model utilizes text similarity and sentiment tendency as measures of node attribute similarity. By incorporating these aspects, the CTST model addresses the challenge of sparse attribute data and facilitates community division. Consequently, the obtained community detection results possess meaningful semantics and improved interpretability, thereby enhancing the overall effectiveness of the community detection process.

3 Methodology

In this section, we first describe the overall situation of the CTST model, and give the mathematical calculation of the content similarity and propose the concept of the sentiment bias value, and then show the details about our algorithm.

3.1 Overview

Figure 1 shows the overall structure of our CTST model. It can be seen that the CTST model is generally divided into two routes for processing user comment text. On the one hand, the upper part of Fig. 1 shows the content similarity calculation route. Firstly, the CTST model conducts word segmentation on the textual content. Then, it calculates the $tf - idf$ value of the word entry after word segmentation, and extracts the text information eigenvector based on this. Subsequently, the CTST model employs the proposed formulas to calculate the cosine similarity between the two eigenvectors. This cosine similarity value serves as a measure of the similarity between the two user nodes. On the other hand, the lower part of Fig. 1 demonstrates the calculation process for sentiment bias values. In this segment, the model begins by constructing the fundamental sentiment vectors for the two users, which are subsequently combined. Following a series of transformations, the sentiment bias value is computed. Lastly, the model integrates the outcomes of both parts, constructs the undirected weighted network, and performs community detection on the network based on this framework [42, 43].

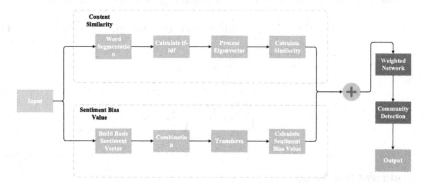

Fig. 1. The overall logical architecture of our algorithm. It consists of two separate parts. On the top, it calculates the *Content Similarity*, and it calculates the *Sentiment Bias Value* on the bottom. These two values are then processed into weights, and *Community Detection* is performed on the *Weighted Network*.

3.2 Content Similarity Calculation

First, we conduct word segmentation on the comment information text content of the user nodes, and then calculate the $tf - idf$ value of the entry. The tf value represents the frequency of a certain word in the text. The calculation formula is shown in Formula (1), where the numerator $n_{i,j}$ represents the frequency of a word in the text d_j, and the denominator represents the total number of words in the entire text, and the tf value is used to indicate the significance of a word in the entire text content.

$$tf_{i,j} = \frac{n_{i,j}}{\sum_k n_{k,j}}. \tag{1}$$

idf represents the reverse document frequency, and its calculation formula is shown in Formula (2). The numerator $|D|$ represents the total number of documents in the corpus, and the denominator $\{j : t_i \in d_j\}$ represents the number of documents containing the word t_i, and then takes the logarithm of the quotient to get idf value.

$$idf_{i,j} = \log \frac{|D|}{\{j : t_i \in d_j\}}. \tag{2}$$

Multiply the obtained tf value by the idf value, as shown in Formula (3), to get the $tf - idf$ value of the term:

$$tf - idf_{i,j} = tf_{i,j} \times idf_{i,j}. \tag{3}$$

After the $tf - idf$ value of the entry is obtained, the user text information is processed to obtain the information text eigenvector between any two network user nodes $V1$ and $V2$, and the cosine similarity between the two eigenvectors is then calculated as the similarity measure s between the two user nodes. The specific calculations are shown in Formula (4) and Formula (5). Among them, $tf - idf_{V1}$ represents the text eigenvectors of user node $V1$, and $tf - idf_{V2}$ represents the text eigenvectors of user node $V2$.

$$s(V1, V2) = \cos(tf - idf_{V1}, tf - idf_{V2}). \tag{4}$$

$$\cos \theta = \frac{\sum_{i=1}^{n} (V1_i \times V2_i)}{\sqrt{\sum_{i=1}^{n} V1_i^2} \times \sqrt{\sum_{i=1}^{n} V2_i^2}}. \tag{5}$$

3.3 Sentiment Bias Value

Each user node in the network has corresponding text information content. Correspondingly, the text information contains the user's opinions, emotions and attitudes about the topic. There are two traditional methods for sentiment analysis, one is based on sentiment dictionary and the other is based on machine learning [44, 45].

Micro Blog has a mechanism for users to post comments. From an emotional stand-point, the poster expresses their emotions in the post, and this emotion can have an impact on the emotional expression of the comment users who reply to the blog post to some extent. From the perspective of user relationships, these two users not only share a connection through mutual comments but also have an emotional connection, which we refer to as "sentiment bias". To capture this sentiment bias, we first construct a polar coordinate system. Within this system, the emotional state of each user is represented by a sentiment vector e_i, as shown in Formula (6):

$$e_i = (\rho_i, \omega_i). \tag{6}$$

In the above formula, ρ_i and ω_i represent the emotional intensity and the corresponding weight assigned to it, respectively. Each piece of text information exhibits varying degrees of emotional tendency. To quantify the intensity of emotion, the polar diameter $\rho_i \in [0, 1]$ is introduced, which represents the emotional score. By transforming and combining the two sentiment vectors, a composite sentiment vector can be generated (Fig. 2).

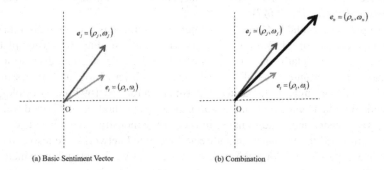

(a) Basic Sentiment Vector   (b) Combination

Fig. 2. The basic sentiment vectors of the two users e_i and e_j are combined to obtain a composite sentiment vector e_n.

The composite sentiment vector can be further transformed and calculated to obtain the final sentiment bias value. The calculation of the sentiment bias value is shown in Formula (7).

$$sv = \rho_n \times \omega_n. \tag{7}$$

3.4 Weighted Network

We map network users to the graph, where $V = \{V_1, V_2, \ldots, V_i, \ldots, V_n\}$ represents a set of network user nodes, and (V_i, V_j) represents an edge between two nodes. W_{ij} is the weight of the edge, which comprises the text content similarity value s and the sentiment bias value between user V_i and user V_j, as shown in Formula (8). Then $WG(V, E, W)$ represents an undirected weighted graph with V as the user node set,

$E \in \{(V_i, V_j)|V_i, V_j \in V\}$ as the edge set, and $W = \{W_{ij} : (V_i, V_j) \in E\}$ as the weight set. On this basis, the algorithm is applied to partition the graph into communities, ultimately yielding the community detection result for the social network user communities.

$$W_{ij} = s \times 0.5 + sv \times 0.5. \tag{8}$$

4 Experiments

4.1 Datasets

In recent years, shared travel relying on the mobile Internet has become a prominent topic on social networking platforms like Micro Blog [46]. Therefore, conducting a comprehensive social network community detection based on the Micro Blog social network and the comment content of shared travel can significantly enhance our understanding of user groups with diverse preferences. Such research holds great value and practical significance in studying the characteristics and patterns of user groups and their opinions in the realm of shared travel [47].

As no publicly available Chinese sentiment analysis dataset specifically related to the comment content of "shared travel" exists, this paper uses the Micro Blog platform as the data collection platform. It involves crawling the links of some discussion blog posts under the relevant entries of "shared travel" on Micro Blog, and processing the acquired data accordingly. In the actual experiment, a total of 1693 pieces of blog data were finally obtained, and based on the aforementioned method, an undirected weighted network was constructed. Additionally, in order to demonstrate the advantages of the algorithm proposed in this paper on undirected weighted networks, we also conducted comparative experiments using the well-known Zachary's Karate Club, which is an undirected and unweighted network with 34 nodes and 156 edges.

4.2 Experimental Procedure

We initially processed the data and obtained 85 actual users from the blog post dataset. According to the method described above, an undirected graph with 85 nodes and 396 edges was established. The text information of the users is then calculated using the approach outlined in Sect. 3.2, resulting in the generation of a similarity matrix, as depicted in Table 1.

Table 1. Textual content similarity matrix.

Nodes	1	2	3	...	85
1	0	0.24	0.53	...	0.05
2	0.24	0	0.62	...	0.12
3	0.53	0.62	0	...	0.33
...	...	...	...	...	...
85	0.05	0.12	0.33	...	0

Then, according to the method described in Sect. 3.3, the sentiment bias value matrix is calculated and shown in Table 2.

Table 2. Sentiment bias value matrix.

Nodes	1	2	3	...	85
1	0	0.37	0.6	...	0.22
2	0.37	0	0.81	...	0.25
3	0.6	0.81	0	...	0.59
...	...	...	...	...	...
85	0.22	0.25	0.59	...	0

Finally, a weighted network is constructed according to the contents described in Sect. 3.4. And based on the aforementioned results, we augment the undirected network by adding weighted edges and subsequently conduct community detection to analyze the network structure in greater detail. The process of community detection is divided into two steps. The first step involves selecting appropriate central nodes as the center of the clusters. In the second step, the correlation between adjacent nodes is examined to determine whether to add a node to the cluster. In the experiment, we set the number of initial central nodes to 2, 3, and 4, respectively, to partition the undirected weighted network into communities.

4.3 Experimental Results

We employ modularity as a metric to evaluate the results of community detection, which is a value within the range of [0, 1]. A higher modularity indicates a stronger internal connection within the community, signifying better overall connectivity in the community detection results. Its specific calculation is shown in Formula (9). In the following formula, L_n represents the edges within the community, L represents the number of all edges in the network, and D_n represents the sum of the degrees of all nodes in the community.

$$Q = \sum_{n=1}^{m} [\frac{L_n}{L} - \left(\frac{D_n}{2L}\right)^2].$$ (9)

For the three experiments on our dataset, we calculated their modularity separately, and the results are shown in Table 3.

Table 3. Modularity comparison.

Number of initial central nodes	modularity
2	0.298
3	0.329
4	0.443

It can be seen from Table 3 that the modularity increases as the number of initial central nodes increases, and the best community detection result is achieved when the number of initial central nodes is 4. Additionally, we also compared the algorithm on the Zachary's Karate Club dataset, and found that the modularity of the CTST model on the undirected weighted network is higher than that on the undirected unweighted network. This indicates that the community detection effect on the undirected weighted graph is superior (Fig. 3).

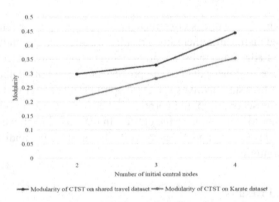

Fig. 3. The comparison results of our CTST model on *shared travel dataset* and *Karate dataset*.

5 Conclusion

In this paper, we propose a social network community detection algorithm based on textual content similarity and sentimental tendency (CTST). We measure the similarity of user nodes by using the similarity and sentimental tendency of the relevant comment texts shared by users on the topic of "shared travel" within the social network as node attributes. It is worth mentioning that we introduce the concept of sentiment bias value as a numerical representation of sentiment bias in the emotional orientation part of the text content. Experiments have shown that our CTST model, which capitalizes on the wealth of information present in the text content while also considering the network structure and node attributes, has the capacity to imbue the results of community detection with semantics. Moreover, it effectively characterizes similar users within the same community. It is easier to understand and explain, and to a certain extent solves the problem of sparse attribute data that may be caused by only considering a single situation. Indeed, the CTST model enhances the similarity and stability among user nodes within the community, thereby enhancing the overall quality of community detection. In future work, we will focus on researching the results of community detection based on shared travel comment content, and analyze its impact on public opinion orientation, and utilize this knowledge to develop online public opinion monitoring and real-time early warning systems.

References

1. Sun, Y.F., Li, S.: Similarity-based community detection in social network of microblog. J. Comput. Res. Dev. **51**(12), 2797–2807 (2004)
2. Li, A., Li, Y., Shao, Y., Liu, B.: Multi-view scholar clustering with dynamic interest tracking. IEEE Trans. Knowl. Data Eng. **35**, 1–14 (2023)
3. Wei, X., Du, J., Liang, M., Ye, L.: Boosting deep attribute learning via support vector regression for fast moving crowd counting. Pattern Recogn. Lett. **119**, 12–23 (2019)
4. Li, A., et al.: Scientific and technological information oriented semantics-adversarial and media-adversarial cross-media retrieval. arXiv preprint arXiv:2203.08615 (2022)
5. Newman, M.E.J.: The structure and function of complex networks. SIAM Rev. **45**(2), 167–256 (2003)
6. Newman, M.E.J., Girvan, M.: Finding and evaluating community structure in networks. Phys. Rev. E **69**(2), 026113 (2004)
7. Chunaev, P.: Community detection in node-attributed social networks: a survey. Comput. Sci. Rev. **37**, 100286 (2020)
8. Huang, J., et al.: HGAMN: heterogeneous graph attention matching network for multilingual POI retrieval at Baidu maps. In: Proceedings of the 27th ACM SIGKDD Conference on Knowledge Discovery & Data Mining, KDD 2021, pp. 3032–3040 (2021)
9. Kou, F., et al.: Hashtag recommendation based on multi-features of microblogs. J. Comput. Sci. Technol. **33**(4), 711–726 (2018)
10. Meng, D., Jia, Y., Du, J., Yu, F.: Tracking algorithms for multiagent systems. IEEE Trans. Neural Networks Learn. Syst. **24**(10), 1660–1676 (2013)
11. Guan, Z., Li, Y., Xue, Z., Liu, Y., Gao, H., Shao, Y.: Federated graph neural network for cross-graph node classification. In: 2021 IEEE 7th International Conference on Cloud Computing and Intelligent Systems, CCIS 2021, pp. 418–422 (2021)
12. Cao, T., et al.: Reliable and efficient multimedia service optimization for edge computing-based 5G networks: game theoretic approaches. IEEE Trans. Netw. Serv. Manage. **17**(3), 1610–1625 (2020)
13. Hong, L., Davison, B.D.: Empirical study of topic modeling in Twitter. In: Proceedings of the First Workshop on Social Media Analytics, pp. 80–88 (2010)

14. Parthasarathy, S., Ruan, Y., Satuluri, V.: Community discovery in social networks: applications, methods and emerging trends. In: Aggarwal, C. (eds.) Social Network Data Analytics, pp. 79–113. Springer, Cham (2011). https://doi.org/10.1007/978-1-4419-8462-3_4
15. Xin, Y., Yang, J., Tang, C.H.: An overlapping semantic community detection algorithm based on local semantic cluster. J. Comput. Res. Dev. **52**(7), 1510–1521 (2015)
16. Bedi, P., Sharma, C.: Community detection in social networks. Wiley Interdisc. Rev. Data Min. Knowl. Discovery. **6**(3), 115–135 (2016)
17. Li, Z., Liu, J., Wu, K.: A multiobjective evolutionary algorithm based on structural and attribute similarities for community detection in attributed networks. IEEE Trans. Cybern. **48**(7), 1963–1976 (2017)
18. Fortunato, S.: Community detection in graphs. Phys. Rep. **486**(3–5), 75–174 (2010)
19. Zhou, Y., Cheng, H., Yu, J.X.: Graph clustering based on structural/attribute similarities. Proc. VLDB Endow. **2**(1), 718–729 (2009)
20. Li, W., Jia, Y., Du, J.: Tobit Kalman filter with time-correlated multiplicative measurement noise. IET Control Theory Appl. **11**(1), 122–128 (2017)
21. Li, W., Sun, J., Jia, Y., Du, J., Fu, X.: Variance-constrained state estimation for nonlinear complex networks with uncertain coupling strength. Digital Signal Process. **67**, 107–115 (2017)
22. Li, Y., et al.: Heterogeneous latent topic discovery for semantic text mining. IEEE Trans. Knowl. Data Eng. **35**(1), 533–544 (2023)
23. Xiao, S., Shao, Y., Li, Y., Yin, H., Shen, Y., Cui, B.: LECF: recommendation via learnable edge collaborative filtering. Sci. China Inf. Sci. **65**(1), 1–15 (2022)
24. Cao, J., Mao, D., Cai, Q., Li, H., Du, J.: A review of object representation based on local features. J. Zhejiang Univ. Sci. C **14**(7), 495–504 (2013)
25. Li, W., Jia, Y., Du, J.: Distributed consensus extended Kalman filter: a variance-constrained approach. IET Control Theory Appl. **11**(3), 382–389 (2017)
26. Shao, Y., Huang, S., Li, Y., Miao, X., Cui, B., Chen, L.: Memory-aware framework for fast and scalable second-order random walk over billion-edge natural graphs. VLDB J. **30**(5), 769–797 (2021)
27. Li, W., Jia, Y., Du, J.: Resilient filtering for nonlinear complex networks with multiplicative noise. IEEE Trans. Autom. Control **64**(6), 2522–2528 (2018)
28. Li, Y., Li, W., Xue, Z.: Federated learning with stochastic quantization. Int. J. Intell. Syst. **37**, 11600–11621 (2022)
29. Meng, D., Jia, Y., Du, J.: Robust iterative learning protocols for finite-time consensus of multi-agent systems with interval uncertain topologies. Int. J. Syst. Sci. **46**(5), 857–871 (2015)
30. Wang, P., Huang, Y., Tang, F., et al.: Overlapping community detection based on node importance and adjacency information. Secur. Commun. Netw. **2021**, 1–17 (2021)
31. Shang, R., Bai, J., Jiao, L., et al.: Community detection based on modularity and an improved genetic algorithm. Physica A **392**(5), 1215–1231 (2013)
32. Dang, T.A., Viennet, E.: Community detection based on structural and attribute similarities. In: International Conference on Digital Society (ICDS). vol. 659, pp. 7–12 (2012)
33. Blondel, V.D., Guillaume, J.L., Lambiotte, R., et al.: Fast unfolding of communities in large networks. J. Stat. Mech: Theory Exp. **2008**(10), P10008 (2008)
34. Li, Y., Zeng, I.Y., Niu, Z., Shi, J., Wang, Z., Guan, Z.: Predicting vehicle fuel consumption based on multi-view deep neural network. Neurocomputing **502**, 140–147 (2022)
35. Li, Y., Yuan, Y., Wang, Y., Lian, X., Ma, Y., Wang, G.: Distributed multimodal path queries. IEEE Trans. Knowl. Data Eng. **34**(7), 3196–3321 (2022)
36. Girvan, M., Newman, M.E.J.: Community structure in social and biological networks. Proc. Natl. Acad. Sci. **99**(12), 7821–7826 (2002)
37. Newman, M.E.J.: Fast algorithm for detecting community structure in networks. Phys. Rev. E **69**(6), 066133 (2004)

38. Verma, D., Meila, M.: A comparison of spectral clustering algorithms. University of Washington Technical report UWCSE030501, vol. 1, pp. 1–18 (2003)

39. Jia, H., Ding, S., Xu, X., et al.: The latest research progress on spectral clustering. Neural Comput. Appl. **24**(7), 1477–1486 (2014)

40. Wang, X.F.: A fast algorithm for detecting local community structure in complex networks. Comput. Simul. **24**(11), 82–85 (2007)

41. Jiang, Y.W., Jia, C.Y., Yu, J.: Community detection in complex networks based on vertex similarities. Comput. Sci. **38**(7), 185 (2011)

42. Li, S., Jiang, L., Wu, X., et al.: A weighted network community detection algorithm based on deep learning. Appl. Math. Comput. **401**, 126012 (2021)

43. Midoun, M.A., Wang, X., Talhaoui, M.Z.: A jungle community detection algorithm based on new weighted similarity. Arab. J. Sci. Eng. **46**(9), 8493–8507 (2021)

44. Yadollahi, A., Shahraki, A.G., Zaiane, O.R.: Current state of text sentiment analysis from opinion to emotion mining. ACM Comput. Surv. (CSUR). **50**(2), 1–33 (2017)

45. Thelwall, M., Buckley, K., Paltoglou, G., et al.: Sentiment strength detection in short informal text. J. Am. Soc. Inform. Sci. Technol. **61**(12), 2544–2558 (2010)

46. Du, M., Cheng, L., Li, X., et al.: Investigating the influential factors of shared travel behavior: comparison between app-based third taxi service and free-floating bike sharing in Nanjing, China. Sustainability **11**(16), 4318 (2019)

47. Zhang, Y.: Investigation on the status quo of shared bicycles in Wenzhou and analysis of travel characteristics. Acad. J. Sci. Technol. **1**(3), 60–65 (2022)

IvyGPT: InteractiVe Chinese Pathway Language Model in Medical Domain

Rongsheng Wang[1], Yaofei Duan[1], ChanTong Lam[1], Jiexin Chen[1],
Jiangsheng Xu[2], Haoming Chen[1], Xiaohong Liu[3], Patrick Cheong-Iao Pang[1],
and Tao Tan[1(✉)]

[1] Faculty of Applied Sciences, Macao Polytechnic University, Rua de Luís Gonzaga
Gomes, Macao 999078, China
taotanjs@gmail.com
[2] Opera inc, Beijing, China
[3] John Hopcroft Center, Shanghai Jiao Tong University, Shanghai 200240, China

Abstract. General large language models (LLMs) such as ChatGPT
have shown remarkable success. However, such LLMs have not been
widely adopted for medical purposes, due to poor accuracy and inability
to provide medical advice. We propose IvyGPT, an LLM based
on LLaMA that is trained and fine-tuned with high-quality medical
question-answer (QA) instances and Reinforcement Learning from
Human Feedback (RLHF). In the training, we used QLoRA to handle 33
billion parameters on a small number of NVIDIA A100 (80 GB) GPUs.
Experimental results show that IvyGPT has outperformed other medical
GPT models. The online demo is available at http://81.71.71.157:52022.
Our demo video can be found at https://youtu.be/O4D74pQh8Is.

Keywords: Large language models · Medical · Reinforcement
Learning

1 Introduction

Large language models (LLMs) have seen rapid growth after the introduction
of ChatGPT [5]. Nevertheless, LLMs such as ChatGPT and GPT-4 are general-
purpose models that are not trained for medical or clinical purposes. Therefore,
the use of LLMs in the medical domain may lead to incorrect or misleading
information, and may result in harmful consequences.

Using LLMs in the medical sector has many advantages. They can potentially
provide a variety of answers like health advice. The applications of LLMs can also
relieve the shortage of healthcare resources, which is a major factor to deteriorate
doctor-patient relationships [4].

Several LLMs for the medical domain have been proposed for the Chinese
language, including HuaTuo [7], ChatMed [9], ZhenNong-TCM [10], and MedicalGPT (zh) [2]. However, these models have shortcomings such as small parameter sizes (like 6B or 7B) and the lack of high-quality training data conforming
to real doctor-patient scenarios. These limitations restrict their generalization
abilities in medicine. To address these issues, we propose IvyGPT for the medical
settings in China and our contributions in this paper include:

L. Fang et al. (Eds.): CICAI 2023, LNAI 14474, pp. 378–382, 2024.
https://doi.org/10.1007/978-981-99-9119-8_34

1. We propose a comprehensive training method for medical LLMs which comprises three parts: supervised training, reward model training, and reinforcement learning using the QLoRA method to train large models with 33B parameters on devices with low computing power.
2. We contribute a high-quality dataset containing verified and realistic scenarios of doctor-patient conversations.
3. We evaluate the IvyGPT against other LLMs in the medical domain.

2   Methodology

Our approach focuses on integrating high-quality data with human feedback to enhance the quality of responses in medical consultations. This is achieved through a two-stage training strategy: Supervised Fine-tuning (SFT) utilizing mixed data and Reinforcement Learning (RL) derived from human feedback, as illustrated in Fig. 1.

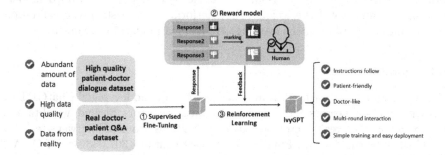

Fig. 1. Process strategy for IvyGPT training

2.1   Supervised Fine-Tuning

The training of large language models with full parameters can be resource-intensive. To address this, we employ LoRA [3], which can significantly reduces the number of trainable parameters required for downstream tasks by freezing the weights of pre-trained models and injecting trainable rank decomposition matrices into each layer of the Transformer architecture. Furthermore, QLoRA [1] performs a 4-bit quantization of the base model to ensure that the model can be fine-tuned for Large Language Models (LLMs) with extensive parameters, even in a low memory footprint.

2.2   Reinforcement Learning from Human Feedback

Reward Model. We generate multiple responses from our supervised fine-tuned model. These responses are then evaluated by humans, taking into account factors like informativeness, coherence, adherence to human preferences, and factual accuracy based on real doctors' diagnoses. We utilize this paired data for the training of the reward model.

Reinforcement Learning. During the reinforcement learning process, we feed the top k responses generated by the model for the same question into the reward model for evaluation, thereby obtaining a score value. To avert the risk of the model, trained via reinforcement learning, deviating excessively from the model in the Supervised Fine-Tuning (SFT) stage, we introduce a KL divergence constraint during the training process. This ensures that the model attains improved results without deviating from the intended path.

3   Experiments

3.1   Training Details

Our model was implemented in PyTorch using the Accelerate, PEFT, and transformers packages, with LLaMA-33B [6] as the base model. We trained models for 3 epochs and saved the weights that performed the best on the validation set. During the reinforcement learning, We trained models for 2 epochs.

3.2   Dataset

First we used dataset released from HuaTuoGPT [8], and leveraged ChatGPT to verify grammatical errors and remove common sense errors. Second, we obtained a large number of real doctor-patient conversations from public websites and added them to the dataset. Finally, our dataset contains 307,038 sets of Q&As.

As shown in Fig. 2(a), we used a smaller amount of data to complete the training process of the model. In Fig. 2(b), Comparison of the length of the model's response generated for the query is shown. IvyGPT generates an average word count of 271.05, which is higher than other datasets in the chinese medical domain. Higher word count of responses indicates that the model can cover richer information in the response.

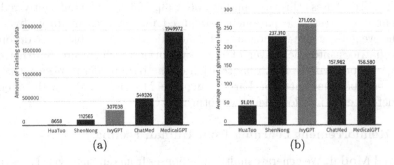

(a)   (b)

Fig. 2. (a) The total amount of data in the training set, (b) The average number of words in the generated responses

3.3   Evaluation Results

We used the word2vec-based method of computing cosine similarity to evaluate semantic similarity of AI answers and human answers. We used the trained 64-dimensional word2vec model to embed answers. We compared IvyGPT with four proposed baseline models: HuaTuo, ShenNong, ChatMed, and MedicalGPT for analysis. We used 100 query pairs to evaluate the semantic similarity between ChatGPT and the Chinese medical domain model, respectively. The results are shown in Table 1, and our model obtains the highest semantic similarity. This implies that our model is better able to capture and express semantic relatedness, resulting in more accurate and consistent language generation.

Table 1. Semantic similarity comparison with real doctor

Model	Score
ShenNong [10]	77.71
HuaTuo [7]	71.20
ChatMed [9]	84.51
MedicalGPT [2]	83.73
ChatGPT	89.13
IvyGPT (ours)	**93.58**

We compare the output word count of the model with different training methods. As shown in the Fig. 3(a), the model with reinforcement learning has richer answers to users' questions. Figure 3(b) shows that QLoRA has a shorter training time and is more efficient compared to LoRA.

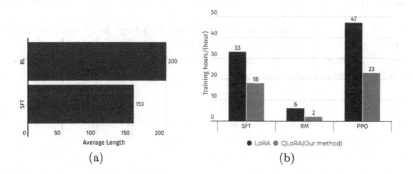

(a)   (b)

Fig. 3. (a) The average word count of responses produced by the model after reinforcement learning. (b) The training duration under different fine-tuning methods

4   Conclusion and Discussion

We propose IvyGPT, a medical LLM trained through supervised fine-tuning using high-quality medical question-and-answer instances and RLHF. With QLoRA, IvyGPT is able to load and train a model with 33 billion parameters, even with limited computational resources. The results show that the responses generated by IvyGPT have a higher similarity to the Q&A in real doctor-patient scenarios, indicating its potential in promoting self-service healthcare and supporting healthcare professionals.

Funding Information. This work was funded by the Science and Technology Development Fund of Macau SAR (Grant Number 0105/2022/A and File Number 0041/2023/RIB2).

References

1. Dettmers, T., Pagnoni, A., Holtzman, A., Zettlemoyer, L.: QLORA: efficient fine-tuning of quantized LLMs. arXiv preprint arXiv:2305.14314 (2023)
2. Liu, H., Liao, Y., Meng, Y., Wang, Y., Wang, Y.: MedicalGPT-zh (2023). https://github.com/MediaBrain-SJTU/MedicalGPT-zh
3. Hu, E.J., et al.: LoRA: low-rank adaptation of large language models. arXiv preprint arXiv:2106.09685 (2021)
4. Li, J., Pang, P.C.I., Xiao, Y., Wong, D.: Changes in doctor-patient relationships in china during COVID-19: a text mining analysis. Int. J. Environ. Res. Public Health **19**(20), 13446 (2022)
5. Ouyang, L., et al.: Training language models to follow instructions with human feedback (2022)
6. Touvron, H., et al.: LLaMA: open and efficient foundation language models (2023)
7. Wang, H., et al.: HuaTuo: tuning LLaMA model with Chinese medical knowledge (2023)
8. Zhang, H., et al.: HuatuoGPT, towards taming language model to be a doctor (2023)
9. Zhu, W., Wang, X.: ChatMed: a Chinese medical large language model (2023). https://github.com/michael-wzhu/ChatMed
10. Zhu, W., Wang, X.: ShenNong-TCM-LLM (2023). https://github.com/michael-wzhu/ShenNong-TCM-LLM

Network Pruning via Explicit Information Migration

Jincheng Wu[1,2], Dongfang Hu[1(✉)], and Zhitong Zheng[1]

[1] OPPO, Dongguan, China
hudongfang2017@gmail.com, {hudongfang,liam}@oppo.com
[2] University of Science and Technology of China, Hefei, China

Abstract. Neural network pruning is a widely used approach for reducing the inference cost of deep models in order to deploy on resource-limited settings. However, current pruning works lack attention to information migration from pruned to the remaining part of the Deep Neural Network, and on balancing model performance and compression rate. On these two issues, in this paper, we propose a novel **E**xplicit **I**nformation **M**igration network **P**runing (EIMP) algorithm. Specifically (1) the constrained gradient update method transfers valid information from redundant networks to the preserved, and (2) the newly designed λ-decay regularization method learns the trade-off between the performance and penalty item. Experiments show that our EIMP algorithm achieves state-of-the-art performance on several datasets with various benchmark network architectures. Notably, EIMP achieves +1.54% better than SOTA on ImageNet.

Keywords: Model compression · Neural network pruning · Explicit information migration · Constrained gradient · Regularization

1 Introduction

With the advancement of deep neural networks (DNN) in recent years [20], accompanied by the effectiveness, there has been a rising amount of storage, memory, computation resources, and energy cost [25]. Therefore, neural network pruning has attracted lots of attention for its capability of reducing model complexity and speeding up inference which is achieved by discarding some redundant structures, such as channels, filters, layers, *etc* [20,30]. Huge efforts have been devoted by researchers over years [3,27], which can be roughly divided into pruning based on criteria [1,2,6,16,21] and learn-able pruning methods [3,18].

The former focuses on directly proposing some theoretically feasible redundancy criteria, like magnitude [20], Hessian [8] and Group Fisher [21], to prune unimportant weights once and for all, followed by a fine-tuning of the pruned network. The latter learnable pruning usually automatically selects unimportant

---

Work done during the internship at OPPO.

© The Author(s), under exclusive license to Springer Nature Singapore Pte Ltd. 2024
L. Fang et al. (Eds.): CICAI 2023, LNAI 14474, pp. 383–395, 2024.
https://doi.org/10.1007/978-981-99-9119-8_35

weights during training. In this line, a variety of approaches add a regularization term to the origin loss [20], and we also name them regularization-based methods. Regularization-based pruning is dedicated to impelling unimportant weights close to zero so that subsequent shearing operation will affect as little as possible performance even followed by a simple weight magnitude-based criteria [29]. And we conduct research on the latter path.

In pruning, [29] has pointed out that, high-redundancy parts in a network can have helpful information. Directly trimming is detrimental to the model. From this perspective, learnable regularization-based pruning attempts to automatically transfer useful information from the redundant part to that to be preserved during training. But there are still problematic hurdles: (1) the process of impelling unimportant weights close to zero is too subtle to be perceptible which deserves further research [3]. (2) previous works, such as L_1, L_2, $L_{2,1}$ regularization, *etc.*, ignore the balance between model convergence and pruning [18]. For the first concern, we present a novel constrained gradient update method to better transfer the model capability to the remaining part. In pruning applications, we find a practical situation: when the magnitude of the model parameter is less than a certain threshold τ, the impact of pruning on the performance can be minimized. We utilize this phenomenon and repel the information of networks to be pruned moves towards the threshold τ during training. Different from the commonly used Stochastic Gradient Descent (SGD), the constrained gradient update method can proactively prompt redundant parameters to move below the threshold. The process undergoes information migration so that when pruning occurs, the model effectiveness suffers as little as possible.

Considering the second issue, we propose the adapted λ-decay regularization. Pruning in the training procedure causes the model to go from initial model overfitting to subnet underfitting and then to pruned net fitting again [24]. During the underfitting phase, the high-frequency information tends to be lost, so it is necessary to migrate it from the pruned part to the remaining. Our adapted λ-decay method regards regularization as a function. When the model has not yet been converged, the optimization is more performance-oriented. And when the model is close to the global optimum, the regularization can increase to speed up the convergence.

The commonality between constrained gradient update and adapted λ-decay methods is characteristic of adaptive learning. In constrained gradient update, both S^0 and S^1 are learning concurrently, but different gradient update methods allow the pruned conv kernels to gradually decay and knowledge progressively flows to the retained parts S^1. And meanwhile, the structures of the S^0 and S^1 sets are dynamically adjusting, so the pruned but potential structures have chances to be resurrected. Furthermore, the adapted λ-decay method allows the model to learn performance when it should improve it, and learn convergence when it should improve it, dynamically balancing this process.

To summarize, our main contributions are as follows:

- We have further explored the explicit information migration for pruning and propose Explicit Information Migration network Pruning (**EIMP**) consists of a constrained gradient update and an adapted λ-decay method.

- The proposed constrained gradient update method is designed to transfer the model expressivity to the remaining part during pruning. It ensures the model compression rate while enhancing its ability as much as possible.
- The adapted λ-decay regularization scheme is simple yet effective, which can help improve model performance during pruning training.
- Extensive experiments on various networks (*i.e.* ResNet-32, ResNet-50, ResNet-110, MobileNet-V1, *etc.*) and datasets (*i.e.* CIFAR-10, ImageNet) verify the effectiveness of the two methods. Combining them has delivered the best performance against the state-of-the-art. Besides, we find some interesting phenomena and provide some experience for pruning.

2   Related Work

The pruning algorithm is a commonly used model compression algorithm that can effectively improve the efficiency of model operations [20]. It can be divided into three paradigms: pruning before training, pruning after training, and pruning during training.

Pruning before training Some work [4,22] direct randomization network, followed by randomization of the structure of the network and finally training of the model parameters. [22] think that fine-tuning a pruned model only gives comparable or worse performance than training that model with randomly initialized weights.

Pruning after training The classic training paradigm is pre-trained a well-trained model, then prune the unimportant parts by human heuristic algorithm, and finally fine-tuning, it can be performed once or multiple times. The heuristic algorithms include based on magnitude [7], Hessian information [21], *etc.* Some work introduces trainable structures that adaptively learn the importance of each computing unit.

Pruning during training Pruning during training is another pruning paradigm that does not require fine-tuning, as it allows for pruning to be performed concurrently with training. It includes soft pruning strategy [3,10,25] and hard pruning [20,32]. Hard pruning is performed by ADMM [36] algorithm and the projection operator [32] to directly prune unimportant conv kernels. And the Soft pruning strategy [10] inducing structure sparsity is most relevant to the algorithm proposed in this paper, pruned conv kernels continue to participate in training iterations, which means that they are not directly discarded.

3   Method

The overall of our proposed EIMP algorithm is shown in Fig. 1. In this section, we first introduce the widely-used formulations of channel pruning for deep neural networks (DNNs) and floating-point operations ($FLOPs$), which are used in our paper and presented in Sect. 3.1. Next, we describe the proposed constrained gradient update method in Sect. 3.2, and finally present our method of adapted λ-decay in Sect. 3.3.

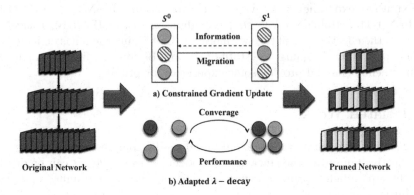

Fig. 1. Overall workflow of the proposed approach EIMP. The entire algorithm EIMP is a pruning-in-training algorithm and it automatically learns how to prune better by the two designed methods. (a). The constrained gradient update method helps information flow between networks S^0 and S^1 and repels valid information to S^1. (b) The adapted λ-decay method automatically balances convergence and performance.

3.1   Preliminaries

Formulation Denote the dataset with N samples as $\mathbb{X} = \{x_i\}_{i=1}^N$, and $\mathbb{Y} = \{y_i\}_{i=1}^N$ are the corresponding labels, a well-trained model with total L layers weights as $\mathbb{W} = \{\mathbf{W}_l\}_{l=1}^L$ where $\mathbf{W}_l \in \mathbb{R}^{c_l \times c_{l-1} \times k_l \times k_l}$ represents the weight parameter of the l-th layer, $l \in \{1, 2, ..., L\}$. c_l is the channel number of l-th layer and k_l is the kernel size in l-th layer.

Some researchers often insert new layers with parameters $\mathbb{W}' = \{\mathbf{W}'_l\}_{l=1}^L$ after each convolution kernel in the original model [18,20]. We denote this layer as the auxiliary pruning layer, which can be vectors obeying some distribution, crafted matrices, or even conv layers that are introduced to facilitate the pruning of the network architecture. That is, if the L_2 norm of the auxiliary pruning layer is approximately or equal to 0, the convolution kernel of the previous layer to which it is attached is equivalent to having no effect on the model.

$$\min_{\mathbb{W}, \mathbb{W}'} \sum_{i=1}^N \mathcal{L}(x_i, \mathbb{W}, \mathbb{W}'; y_i) + \lambda \cdot \sum_{l=1}^L ||\mathbf{W}'_l||_{2,1} \qquad (1)$$

$$\text{s. t.} \quad R_{FLOPs} \leq R_{budget} \qquad (2)$$

The optimization of previous learn-able regularization-based pruning can be formulated as Eq. 1−2, where $\mathcal{L}$ is the original loss of model, for instance, Cross

Entropy Loss. $\|\cdot\|_{2,1}$ represents $l_{2,1}$-norm and $\|\mathbf{W}_l'\|_{2,1} = \sum_{j=1}^{c_l} \|\mathbf{W}_{l,j,:,:,:}'\|_2$. λ is a scalar coefficient, which balances the origin model loss and sparsification penalty term, and a larger λ urges a sparser convolution kernel and thus obtains a more compact network, whereas a smaller λ tends to improve model effectiveness. R_{FLOPs} means $FLOPs$ of sheared model during pruning training, and R_{budget} is the upper limit of resource constraint.

We consider the stochastic gradient update algorithm, the one-step update formula for $\mathbf{W}'^l$ as shown in Eq. 3, where η' is the learning rate for auxiliary pruning layers training.

$$\mathbf{W}_l'^{(t+1)} = \mathbf{W}_l'^{(t)} - \eta' \cdot \nabla_{\mathbf{W}_l'} \mathcal{L} - \eta' \cdot \lambda \cdot \frac{\mathbf{W}_l'}{\|\mathbf{W}_l'\|_{2,1}} \tag{3}$$

FLOPs During Pruning. We find when the conv kernel weight is less than the threshold τ_1, pruning it will affect little. Therefore, we only calculate the networks satisfying this condition.

3.2   Constrained Gradient Update

We define S^0 as the set of structures to be pruned of $\mathbb{W}'$ and S^1 as the set of remaining structures. The selection process is not done all at once. Initially, all parameters of $\mathbb{W}'$ belong to S^1 and S^0 is an empty set. As pruning progresses, every few epochs, sort $\mathbb{W}'$ according to the L_2 norm, and when $Epoch/u == 0$, calculate $Q = m * Epoch/u$, where u represents the number of times selection is executed and m indicates how many convolutional kernels are selected in each selection. It should be noticed that the Q value will not increase until the $FLOPs$ value of the S^1 set meets the predefined value. After that, put the top Q minimum structures from S^1 into S^0. Before each selection, the S^0 set is reset to an empty set, and all parameters of $\mathbb{W}'$ are again placed in S^1. Then, the selected elements are removed from S^1 and added to S^0.

To update the parameter weights in the S^1 set, we use the SGD method, which is the same as Eq. 3. For the parameters of network structures to be pruned in S^0, we define the information learning intensity function $f(\mathbf{W}_{l,j}')$ as Eq. 4, which indicates how much the gradient weight for the loss function contributes to the update of the convolution kernel weight parameters in the S^0 set. Thus, the parameter weights in the S^0 set are updated according to Eq. 5.

$$f(\mathbf{W}_j'^l) = \lambda_2 \cdot max(0, \tau_2 - \|\mathbf{W}_j'^l\|_2) \tag{4}$$

$$\mathbf{W}_{j,t+1}'^l = \mathbf{W}_{j,t}'^l - \eta' \cdot \nabla_{\mathbf{W}_j'^l} \mathcal{L} \cdot f(\mathbf{W}_j'^l) - \eta' \cdot \lambda \cdot \frac{\mathbf{W}_j'^l}{\|\mathbf{W}'^l\|_{2,1}} \tag{5}$$

where λ_2 is the coefficient and τ_2 is the threshold. By employing the proposed CGU algorithm, the weights of the S^0 set can continuously learn and transfer information until it reaches a stage where the information is either fully depleted or repartitioned into the S^1 space. The updated weight of a conv kernel in the S^0 set is monitored by Eq. 5. If the updated weight surpasses the decay rate of the gradient of the loss function, it implies that the conv kernel is gradually increasing its significance and will therefore be reclassified into the S^1 set.

3.3 Adapted λ-Decay Regularization

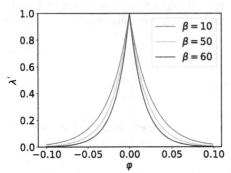

Fig. 2. λ' **function.** Different lines in the graph represent the function under various values of β.

The convergence of the model is represented via gradients φ of the last layer in the model. Set the number of each batch size is B, thus the average convergence of every batch is depicted as:

$$\varphi = \nabla \mathcal{L} = \frac{1}{B \times K} \sum_{j=1}^{B} \sum_{i=1}^{K} |\frac{\partial \mathcal{L}}{\partial z_i^j}| \tag{6}$$

$$\lambda' = \lambda \cdot e^{-\beta * \varphi} \tag{7}$$

where z_i^j represents the logit for the i-th class and j-th input data among K total classes. In our paper, the regularization coefficients λ are reparameterized as a function λ' in Eq. 7, which can automatically balance the effect and convergence during different pruning training phases. Figure 2 shows that the parameter β controls the elasticity of the regularization term.

$$\min_{\mathbf{W}, \mathbf{W}'} \sum_{i=1}^{N} \mathcal{L}(x_i, \mathbf{W}, \mathbf{W}'; y_i) + \lambda' \cdot \sum_{l=1}^{L} ||\mathbf{W}_l'||_{2,1} \tag{8}$$

Therefore, with the constraint Eq. 2, the model optimization Eq. 1 can be transformed into Eq. 8. It can be seen a rapid decrease in the regularization term enhances the effectiveness of the model, whereas an increase in the norm term accelerates the model's convergence.

Pseudo-code in Algorithm 1 shows the implementation procedure of our proposed pruning methods. In initialization, all structure parameters are regarded as belonging to S^1 set. And during the training process, in every u iterations, a few structure parameters with the least L_2 norm are chosen and added to the S^0 set, where they undergo constrained gradient update rule.

4   Experiments and Results

4.1   Datasets, Networks and Pruning Traning Settings

Datasets and Networks We evaluate the performance of the proposed algorithm on the CIFAR-10 [15] and ImageNet [26] datasets. (1) On the CIFAR-10 dataset, we have investigated three models ResNet-32, ResNet-56 and ResNet-110. All the models are trained from scratch with initialization, using Cosine Annealing LR with initial learning rate $lr = 0.1$, and decayed between 120 epochs and 180 epochs, for a total of 240 epochs. (2) The ImageNet dataset is a widely used large dataset in the classification task. We have conducted experiments on the ImageNet dataset using MobileNet-V1 [14] and ResNet-50 models [9] with a total of 180 epochs, respectively. MobileNet-V1 i+s trained from scratch with a learning rate of 0.1, batch size of 256, and a decay scheme: cosine learning rate with initialization $lr = 0.01$, $cosine minimum = 0$, and before pruning, its Top-1 accuracy is 69.26%. In addition, ResNet-50 is the official website implementation integrated into torchvision with Top-1 accuracy 76.15%.

Algorithm 1: The pseudo-code for proposed EIMP.

Input: Pre-trained model $\mathbb{W}$, auxiliary pruning layer $\mathbb{W}'$, Target Flops R_{budget}, layer l, regularization coefficient λ, select interval u.
Output: Pruned model $\mathbb{W}, \mathbb{W}'$.

1   Initialization $\mathbb{W}'$ with identity matrix;
2   **while** $Epoch < Epoch\_max$ **do**
3   **if** $Epoch\%u == 0$ and $Epoch > 0$ **then**
4   Set $S^0 = \varnothing$, $\mathbb{W}' \in S^1$;
5   Sort $\mathbb{W}'$ according to L_2 norm;
6   **if** $R_{budget} < R_{FLOPs}$ **then**
7   Calculate $Q = m * Epoch\%u$;
8   Put the top Q minimum structures from S^1 into S^0;
9   **end**
10   **end**
11   Update $\mathbb{W}$ by SGD; **if** $\mathbb{W}' \in S^1$ **then**
12   update weight by SGD;
13   **else if** $\mathbb{W}' \in S^0$ **then**
14   calculate information learning intensity function by Eq. 4;
15   update weight by Eq. 5;
16   **end**
17   calculate $\varphi = \nabla \mathcal{L} = \frac{1}{B \times K} \sum_{j=1}^{B} \sum_{i=1}^{K} |\frac{\partial \mathcal{L}}{\partial z_i^j}|$;
18   update $\lambda' = \lambda \cdot e^{-\beta*\varphi}$;
19 **end**

Pruning Traning Settings. As [3] has done, we insert architecture parameters in the basic block first (3×3) conv layers. For ResNet-50 and MobileNet-V1 pruning, we use architecture parameters in the first and second layers in the bottleneck layer. The pruning process uses consine learning rate with an initial value of 0.01 and cosine minimum set to 0. The optimizer selects the SGD algorithm and momentum is set to 0.99. η' is optimized by Cosine learning rate and initialized with 0.01. Threshold $\tau_1 = 10^{-5}$, $\tau_2 = 10^{-2}$, $\lambda_1 = 10^{-4}$, $\lambda_2 = 1$, $m = 4$ and set select interval $u = 5$. So we obtain the S^0 set by selecting the Top-4 smallest parameters of every 5 epochs.

4.2   Pruning Results on CIFAR-10 and ImageNet

CIFAR-10. Experiments on benchmark models of ResNet-32, ResNet-56 and ResNet-110 and conducted on CIFAR-10 data and as the Table 1 shows our pruning obtains the best performance overall comparative experiments. Specifically speaking, for the ResNet-56 model, we compare our approach with several recent methods. As the pruning ratio descends to 53.87%, EIMP performs best, +0.55% Top-1 accuracy. In addition, for ResNet-110, our method gets +0.24% increase than baseline while the second best only obtains +0.06%. In a word, all these results validate the effectiveness of EIMP on CIFAR-10 dataset.

ImageNet. On ImageNet data, benchmark networks ResNet-50 and MobileNet-V1 are tried on, the results of which are shown in Table 2 and verify the superiority of our pruning. As far as we know, with 54.5% compression rate, we achieve an accuracy improvement of 0.07%, which surpasses the previous SOTA. Meanwhile, in MobileNet-V1 experiments, at the same compression rate, our EIMP surpasses previous by 1.54%.

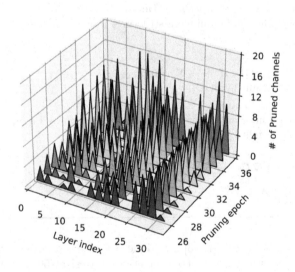

Fig. 3. Number of Pruned channels per layer.

4.3   Information Migration Between S^0 and S^1

As shown in Fig. 3, when the iteration increases, more and more channels are cut. It can be noticed that between *epochs* $= 32 - 36$, some pruned channels

Table 1. CIFAR-10 dataset experiments compared with state-of-the-art. We conduct experiments on three models: ResNet-32, ResNet-56 and ResNet-110 with 50–65% *FLOPs* drops. Our proposed pruning method achieves obvious gains compared with others. $\Delta = (Pruned - Baseline)/Baseline$.

Model	Method	Top-1 Accuracy(%)			Flops↓ (%)
		Baseline	Pruned	Δ	
ResNet-32	LFPC [12]	92.63	92.12	−0.51	52.60
	WML50 [34]	92.63	92.10	−0.53	50.00
	EIMP	93.16	**93.40**	**+0.14**	**53.40**
ResNet-56	HRank [19]	93.26	93.17	−0.09	50.00
	LFPC [12]	93.59	93.24	−0.35	52.90
	SRR-GR [31]	93.38	93.75	+0.37	53.80
	GDP [6]	93.90	93.97	+0.07	53.35
	ResRep [3]	93.71	93.71	+0.00	52.91
	CC [17]	93.33	93.64	+0.31	48.20
	AGMG [33]	93.39	92.76	−0.63	50.00
	NPPM [5]	93.04	93.40	+0.36	50.00
	EIMP	93.80	**94.35**	**+0.55**	**53.87**
	SCOP [28]	93.70	93.64	−0.06	56.00
	Zhang et al. [35]	93.62	93.68	+0.08	56.00
	EIMP	93.80	**93.91**	**+0.11**	**56.00**
ResNet-110	FPGM [11]	93.68	93.74	+0.06	52.30
	AGMC [33]	93.68	93.08	−0.60	50.00
	ResRep [3]	93.64	93.62	−0.02	58.21
	EIMP	94.24	**94.46**	**+0.24**	**58.21**

are revived. That is to say, in our algorithm, information flows bidirectionally between the pruned and unpruned networks, which enables the proposed algorithm more flexible. And the results highlight the effectiveness of the CGU algorithm in preserving important convolution kernels while pruning.

4.4 Ablation Studies

Overall of EIMP. As a baseline, we use the method in [18], and select Top-4 smallest L_2 norm elements of the auxiliary pruning layer from S^1 to S^0 set every 5 epochs and then prune them. For a fair comparison, we use the same base pruning framework and under the situation of pruning 60% of *Flops*, compare the pruned model accuracies of ablation experiments. The impacts of whether using CGU or adapted λ-decay is investigated in Table 3.

Effectiveness of CGU. To inspect the information migration effectiveness of CGU, we compare it with other learning-based pruning methods. For example,

Table 2. ImageNet dataset experiments compared with state-of-the-art.
Results of our method on two models with 54% and 74% *FLOPs* drops respectively.
Our proposed pruning method achieves obvious gains compared with others.

Model	Method	Top-1 Accuracy(%)			Flops↓ (%)
		Baseline	Pruned	Δ	
ResNet-50	MetaPruning [23]	76.60	75.40	−1.20	51.10
	FPGM [11]	76.15	74.83	−1.32	53.50
	HRank [19]	76.15	74.98	−1.17	42.40
	CHEX [13]	**77.80**	**77.40**	−0.40	48.78
	SRR-GR [31]	76.13	75.76	−0.37	44.10
	GroupFisher [21]	76.79	76.42	−0.37	50.00
	ResRep [3]	76.15	76.15	−0.00	54.50
	EIMP	76.15	76.22	**+0.07**	**54.50**
MobileNet-V1	MetaPruning [23]	70.60	66.10	−4.50	73.81
	ResRep [3]	**70.78**	68.02	−2.76	73.91
	EIMP	69.26	**68.04**	**−1.22**	**74.00**

Table 3. Effectiveness of EIMP. We train on CIFAR-10 with ResNet-56. λ-decay is short for adapted λ-decay regularization. The Top-1 accuracy of the pruned networks and the gaps from the base networks are reported.

Model	λ-decay	CGU	Top-1 Accuracy(%)	Δ(%)
ResNet-56	✗	✗	93.21%	0%
	✓	✗	93.36%	+0.15%
	✗	✓	93.55%	+0.34%
	✓	✓	93.76%	+0.55%

in [3], after its division of the conv kernel into S^0 and S^1 sets, only attenuation is performed on the S^0 set without information migration. Results in Fig. 4 left and middle demonstrate the superiority of our CGU method over Res in ResRep. We conduct the experiments on ResNet-56 on the CIFAR-10 and the CGU method consistently outperforms the Res method in terms of accuracy, because when the L_2 norm of the conv kernels in the S^0 set is less than the threshold τ_2, they migrate information to the S^1 space to allow the model achieves better performance.

Robustness of Adapted λ-decay Method. To demonstrate the generalizability of the adapted λ-decay regularization method, we have transferred the method to the ResRep framework and verified its performance. Specifically, we trained the ResNet-32 benchmark network on the CIFAR-10 dataset for a total of 400 epochs, using the same optimizer settings and learning rate as in the Sect. 4.2. Figure 4 (right) shows the accuracy variation from the 160th epoch to the 480th

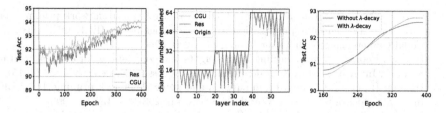

Fig. 4. Ablation Study. Left and Middle: the change in accuracy of the ResNet-56 model on the CIFAR-10 using the Res or CGU algorithms, respectively, along with the number of channels retained in each layer after pruning. The FLOPs reduction in this case is 52.9%. Right: the change in accuracy with the use or non-use of the adapt λ-decay in ResNet-32 model on the CIFAR-10 with $FLOPs$ reduced by 52.94%.

epoch during the pruning training process, with $FLOPs$ reduced by 52.94%. We can observe that after implementing the λ-decay method, the accuracy can be improved while reducing the same $FLOPs$ value. This implies that λ-decay can help the model converge closer to the global optimum.

5   Conclusions

In this paper, we examine the problem of model pruning from the perspective of information migration and propose an EIMP schema. Our key contribution is the introduction of a new gradient update rule, referred to as CGU, and an adapted λ-decay regularization method. The CGU lets the pruned network transfer its valid information to the retained portion of the model, potentially offering a new approach for future pruning research. And the adapted λ-decay regularization enables the automatic trade-off between performance and regularization item, via assigning varying regularization intensities. Two methods of EIMP can be easily applied to other DNN frameworks.

References

1. Chen, T.Y., et al.: Only train once: a one-shot neural network training and pruning framework. In: Advances in Neural Information Processing Systems, vol. 34 (2021)
2. Chin, T.W., Ding, R.Z., Zhang, C., Marculescu, D.: Towards efficient model compression via learned global ranking. In: Proceedings of the IEEE/CVF Conference on Computer Vision and Pattern Recognition, pp. 1518–1528 (2020)
3. Ding, X.H., Hao, T.X., et al.: Resrep: Lossless CNN pruning via decoupling remembering and forgetting. In: Proceedings of the IEEE/CVF International Conference on Computer Vision, pp. 4510–4520 (2021)
4. Frankle, J., Carbin, M.: The lottery ticket hypothesis: finding sparse, trainable neural networks. arXiv preprint arXiv:1803.03635 (2018)
5. Gao, S.Q., Huang, F.H., Cai, W.D., Huang, H.: Network pruning via performance maximization. In: Proceedings of the IEEE/CVF Conference on Computer Vision and Pattern Recognition, pp. 9270–9280 (2021)

6. Guo, Y., Yuan, H., Tan, J.C., et al.: GDP: stabilized neural network pruning via gates with differentiable polarization. In: Proceedings of the IEEE/CVF International Conference on Computer Vision, pp. 5239–5250 (2021)
7. Han, S., Pool, J., et al.: Learning both weights and connections for efficient neural network. In: Advances in Neural Information Processing Systems, vol. 28 (2015)
8. Hassibi, B., et al.: Optimal brain surgeon and general network pruning. In: IEEE International Conference on Neural Networks, pp. 293–299. IEEE (1993)
9. He, K.M., Zhang, X.Y., Ren, S.Q., Sun, J.: Deep residual learning for image recognition. In: Proceedings of the IEEE Conference on Computer Vision and Pattern Recognition, pp. 770–778 (2016)
10. He, Y., Kang, G.L., et al.: Soft filter pruning for accelerating deep convolutional neural networks. arXiv preprint arXiv:1808.06866 (2018)
11. He, Y., Liu, P., Wang, Z.W., Hu, Z.L., Yang, Y.: Filter pruning via geometric median for deep convolutional neural networks acceleration. In: Proceedings of the IEEE/CVF Conference on Computer Vision and Pattern Recognition, pp. 4340–4349 (2019)
12. He, Y., et al.: Learning filter pruning criteria for deep convolutional neural networks acceleration. In: Proceedings of the IEEE/CVF Conference on Computer Vision and Pattern Recognition, pp. 2009–2018 (2020)
13. Hou, Z.J., Qin, M.H., et al.: Chex: channel exploration for CNN model compression. arXiv preprint arXiv:2203.15794 (2022)
14. Howard, A.G., Zhu, M.L., et al.: Mobilenets: efficient convolutional neural networks for mobile vision applications. arXiv preprint arXiv:1704.04861 (2017)
15. Krizhevsky, A., Hinton, G.: Learning multiple layers of features from tiny images. Technical report 0, University of Toronto, Toronto, Ontario (2009)
16. Li, B., Wu, B., Su, J., Wang, G.: EagleEye: fast sub-net evaluation for efficient neural network pruning. In: Vedaldi, A., Bischof, H., Brox, T., Frahm, J.-M. (eds.) ECCV 2020. LNCS, vol. 12347, pp. 639–654. Springer, Cham (2020). https://doi.org/10.1007/978-3-030-58536-5_38
17. Li, Y.C., Lin, S.H., et al.: Towards compact CNNs via collaborative compression. In: Proceedings of the IEEE/CVF Conference on Computer Vision and Pattern Recognition, pp. 6438–6447 (2021)
18. Li, Y.W., Gu, S.H., et al.: Group sparsity: the hinge between filter pruning and decomposition for network compression. In: Proceedings of the IEEE/CVF Conference on Computer Vision and Pattern Recognition, pp. 8018–8027 (2020)
19. Lin, M.B., Ji, R.R., et al.: Hrank: filter pruning using high-rank feature map. In: Proceedings of the IEEE/CVF Conference on Computer Vision and Pattern Recognition, pp. 1529–1538 (2020)
20. Liu, Z., Li, J.G., et al.: Learning efficient convolutional networks through network slimming. In: Proceedings of the IEEE International Conference on Computer Vision, pp. 2736–2744 (2017)
21. Liu, L.Y., et al.: Group fisher pruning for practical network compression. In: International Conference on Machine Learning, pp. 7021–7032. PMLR (2021)
22. Liu, Z., et al.: Rethinking the value of network pruning. arXiv preprint arXiv:1810.05270 (2018)
23. Liu, Z.C., et al.: MetaPruning: meta learning for automatic neural network channel pruning. In: Proceedings of the IEEE/CVF International Conference on Computer Vision, pp. 3296–3305 (2019)
24. Miao, L., et al.: Learning pruning-friendly networks via frank-Wolfe: one-shot, any-sparsity, and no retraining. In: International Conference on Learning Representations (2021)

25. Park, J.H., et al.: Dynamic structure pruning for compressing CNNs. arXiv:2303.09736 (2023)
26. Russakovsky, O., Deng, J., et al.: Imagenet large scale visual recognition challenge. Int. J. Comput. Vision **115**(3), 211–252 (2015)
27. Shen, M.Y., Yin, H.X., et al.: Structural pruning via latency-saliency knapsack. arXiv preprint arXiv:2210.06659 (2022)
28. Tang, Y.H., et al.: Scientific control for reliable neural network pruning. Scop. Adv. Neural. Inf. Process. Syst. **33**, 10936–10947 (2020)
29. Wang, H., Qin, C., et al.: Neural pruning via growing regularization. arXiv preprint arXiv:2012.09243 (2020)
30. Wang, N.G., Liu, C.C., et al.: Deep compression of pre-trained transformer models. Adv. Neural. Inf. Process. Syst. **35**, 14140–14154 (2022)
31. Wang, Z., Li, C.C., Wang, X.Y.: Convolutional neural network pruning with structural redundancy reduction. In: Proceedings of the IEEE/CVF Conference on Computer Vision and Pattern Recognition, pp. 14913–14922 (2021)
32. Yang, H.C., Gui, S.P., Zhu, Y.H., Liu, J.: Automatic neural network compression by sparsity-quantization joint learning: A constrained optimization-based approach. In: Proceedings of the IEEE/CVF Conference on Computer Vision and Pattern Recognition, pp. 2178–2188 (2020)
33. Yu, S.X., Mazaheri, A., Jannesari, A.: Auto graph encoder-decoder for neural network pruning. In: Proceedings of the IEEE/CVF International Conference on Computer Vision (ICCV), pp. 6362–6372 (2021)
34. Zhang, M., et al.: Weighted mutual learning with diversity-driven model compression. In: Advances in Neural Information Processing Systems (2022)
35. Zhang, Y.F., et al.: Exploration and estimation for model compression. In: Proceedings of the IEEE/CVF International Conference on Computer Vision, pp. 487–496 (2021)
36. Zheng, S., et al.: Fast-and-light stochastic ADMM. In: IJCAI, pp. 2407–2613 (2016)

Multidisciplinary Research with AI

Attention-Based RNA Secondary Structure Prediction

Liya Hu[1], Xinyi Yang[1], Yuxuan Si[1], Jingyuan Chen[1,2(✉)],
Xinhai Ye[1,2], Zhihua Wang[1,2(✉)], and Fei Wu[1,2]

[1] Zhejiang University, Hangzhou, China
{liyahu,syx_sue,jingyuanchen,yexinhai,zhihua.wang,wufei}@zju.edu.cn,
xinyiy.21@intl.zju.edu.cn
[2] Shanghai Institute for Advanced Study of Zhejiang University, Shanghai, China

Abstract. RNA is a molecule composed of ribonucleotides and plays a crucial role in biological activities. The computational prediction of RNA secondary structures has been a long-standing issue in computational biology. Traditional methods for this problem are based on free energy minimization, but the performance of these methods has reached an upper limit. In recent years, various deep learning-based methods have been proposed, but these models are still primitive and prone to overfitting, resulting in poor performance across RNA families. In this paper, we propose two methods, AttnUFold and TransUFold, which utilize the attention mechanism to enhance the model's learning ability for the global features of RNA sequences. Additionally, we modify the loss function to cope with sample distribution imbalances and attempt to introduce relevant constraints for RNA folding. Compared with the baseline, the two models have brought improvements in both within- and cross-family tasks. AttnUFold achieved a high F1 score of 0.852 on the ArchiveII dataset, surpassing all traditional and most deep learning methods.

Keywords: RNA secondary structure prediction · Deep learning · Attention

1 Introduction

RNA is a common biomolecule that plays a key role in the encoding, decoding, regulation, and expression of genes. It is a linear molecule composed of ribonucleotides linked by phosphodiester bonds, with four bases in its nucleotides: adenine (A), cytosine (C), guanine (G), and uracil (U). The spatial structure of RNA has a greater impact on its function than its ribonucleotides sequence. However, due to the high experimental cost and limited resolution of RNA measurement, there are significant difficulties in determining the tertiary structure of RNA through experimental methods such as nuclear magnetic resonance and X-ray crystallography. Many studies have focused on determining the secondary structure of RNA, which is defined as the group of nucleotide pairs with hydrogen bonds [9].

The traditional methods of RNA secondary structure prediction minimize the free energy to find the thermodynamically stable state, using a dynamic programming algorithm to determine the secondary structure that only contains nested base pairs [1,13,21]. However, due to RNA not completely following the Watson-Crick base pairing rules similar to DNA [8], and the existence of non-nested pseudoknot structures, existing models are not accurate. Even if a perfect prediction model exists, finding the optimal solution is still NP-complete. The performance of traditional thermodynamic models has reached an upper limit [23].

In recent years, deep learning has shown great potential in RNA secondary structure prediction [4,10,20,22]. Compared with traditional methods, deep learning methods can predict RNA secondary structures with pseudoknots and can learn richer features from training data to improve prediction accuracy. Latest deep learning-based RNA secondary structure prediction method UFold [10] transformed RNA sequences into image inputs and designed relevant channels to represent base pairing probabilities. However, some issues exist with UFold: (1) the U-net model UFold used is a pure convolutional architecture, which can only capture the interactions between bases within the convolution window, leading to the limited ability to capture the relationships between distant bases; (2) the input matrix is sparse, indicating a significantly imbalanced distribution in the input samples; (3) UFold is purely data-driven, with no relevant constraints introduced in the training stage.

In this paper, we first propose two new methods, AttnUFold and TransU-Fold, which utilize the attention mechanism [27] to enhance the model's learning ability for the global features of RNA sequences. We integrate attention gates and transformer layers into the traditional U-net architecture to address the limitations of UFold. Second, we modify the loss function to cope with sample distribution imbalances and attempt to introduce relevant constraints for RNA folding. Experiments on various common datasets demonstrate the effectiveness of our methods.

2   Methods

2.1   Problem Definition

The RNA secondary structure prediction problem is defined as determining the base pairings for a given RNA sequence [9]. The input RNA sequence with length L is denoted as $x = (x_1, x_2, \ldots, x_L)$, where $x_i \in \{A, U, C, G\}$. The result of RNA secondary structure prediction is denoted by an adjacency matrix $A \in \{0, 1\}^{L \times L}$, where $A_{ij} = 1$ if x_i and x_j is paired, and 0 otherwise.

2.2   Overview

In our work, the input sequence will be converted into an tensor of size $L \times L$ with 17 channels (see Sect. 2.3). For model design, we add Attention Gate and

Transformer module to the U-net framework, allowing a larger range of related nucleotides to be learned from the data. The output of the model is a $L \times L$ adjacency matrix Y, representing the probability score of each nucleotide base pairing. Subsequently, a post-processing network is employed to convert it into the 0–1 adjacency matrix $\hat{Y}^*$ representing the final results of actual pairing.

2.3 Data Preprocessing

We follow the steps used in UFold [10] that convert the input sequence to an tensor of size $L \times L$ with 17 channels, in which 16 channels represent 16 distinct types of base-pairing. Such channels are Kronecker products between the one-hot representation and itself, hence allowing the model to learn all of the possible base-pairing, including non-canonical patterns that cause failure in traditional models. The other channel represents the unique feature of the sequence itself, which is obtained by the algorithm adopted from CDPFold [30]. It captures the influence from the adjacent base-pairing and meanwhile overcoming the sparsity of the input matrix.

2.4 Models

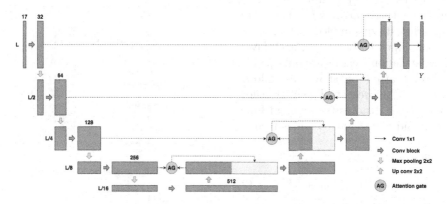

Fig. 1. AttnUFold architecture. The Attention Gate [17] module is added, with an input $17 \times L \times L$ matrix and an output $1 \times L \times L$ pairing probability matrix Y.

AttnUFold. AttnUFold model, as illustrated in Fig. 1, is based on U-net and incorporates the Attention Gate module. The goal of introducing the Attention Gate Module in this model is to capture the mutual influence among nucleotides along the entire RNA sequence, overcoming the constraints of the limited feature extraction range of convolutional kernels. The detailed structure of the Attention Gate module is depicted in Fig. 2, where F, H and W represent the number of feature channels, height and width of each matrix respectively. The integration of the gating matrix provides contextual information extracted from the

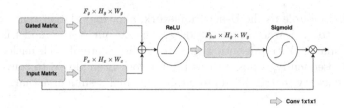

Fig. 2. The Attention Gate architecture. The input matrix x^l is scaled with the attention weights $\alpha \in [0,1]$, which can then adjust the output $\hat{x}^l_k$ based on the spatial information of the gating matrix. The output is formulated by $\hat{x}^l_k = \alpha^l_k \cdot x^l_k$.

coarse scale, thus disambiguating irrelevant responses in the skip connection of the U-net model. It focuses on convolutional operations and the relevant computations of the Attention mechanism, while still retaining the capability to handle variable-length RNA sequences.

The output of the Attention Gate Module is the element-wise multiplication between the input matrix x^l and attention weights α^l in each layer l. We use the up-sampled matrix g as the gating matrix, while the corresponding down-sampled matrix serves as the input matrix x^l for each layer. The attention weights are obtained by additive attention shown as follows, where W represents linear transformations, σ denotes the sigmoid function, and b^g, b^{int} denotes bias terms:

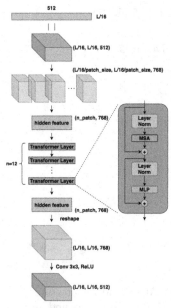

$$\hat{g}_k = \sigma_1(W^g g^l_k + W^x x^l_k + b^g) \quad (1)$$

$$\alpha^l_k = \sigma_2(W^{int} \hat{g}_k + b^{int}) \quad (2)$$

During the training process, we employ the Adam algorithm to optimize the loss function between the output and the ground truth pairing matrix. Detailed information of the loss function can be found in Sect. 3.4. It is important to note that

Fig. 3. The sub-module of TransUFold. TransUFold adds a Transformer structure to the hidden layer of UFold.

the model's output needs to be converted into the final prediction result by a post-processing network, as described in Sect. 2.5.

TransUFold. TransUFold is another approach to incorporating the attention mechanism by introducing the Transformer module into the hidden layer of

UFold, as shown in Fig. 3. Similar to AttnUFold, TransUFold can handle different lengths of RNA sequences effectively. In our experiment, the input RNA sequences are within a length of 600, and it is sufficient to align the tensor size L with 608 in the Transformer module. TransUFold employs the feature map generated by CNN as the input.

Similar to Chen et al. [3], for the tensor $\mathbf{h}$ after U-net encoder layers, we split $\mathbf{h}$ into a series of size $P \times P$ patches $\{\mathbf{h}_p^1, \ldots, \mathbf{h}_p^N\}$ (N is the number of patches), and reassemble them into a latent D-dimensional embedding $\mathbf{z}_0$. Adding a positional encoding lets the model learn embedding representations to distinguish patches at different positions:

$$\mathbf{z}_0 = \text{Concat}(\mathbf{h}_p^1 \mathbf{E}, \ldots, \mathbf{h}_p^N \mathbf{E}) + \mathbf{P} \tag{3}$$

where $\mathbf{E} \in \mathbb{R}^{(P^2 \cdot C) \times D}$, $\mathbf{P} \in \mathbb{R}^{N \times D}$ denotes the patch embedding projection and the position embedding, respectively, and C is the number of $\mathbf{h}$'s channels. Next, the latent vector $\mathbf{z}$ is further encoded by the Transformer encoder to obtain the hidden feature.

In the Transformer encoder, Multihead Self-Attention (MSA) and Multi-Layer Perceptron (MLP) are applied, allowing the model to attend to information in different positions. The output of the l-th layer can be formulated as follows:

$$\mathbf{z}_l' = \text{MSA}(\text{LN}(\mathbf{z}_{l-1})) + \mathbf{z}_{l-1} \tag{4}$$

$$\mathbf{z}_l = \text{MLP}(\text{LN}(\mathbf{z}_l')) + \mathbf{z}_l' \tag{5}$$

where $\text{LN}(\cdot)$ corresponds to the layer normalization operator and $\mathbf{z}_L$ is the encoded tensor representation.

This approach enables feature learning for the entire spatial positions in the hidden layers, enhancing the model's ability to learn global features of RNA sequences.

2.5 Post-processing

The post-processing procedure is needed to convert the scoring matrix Y into the final RNA secondary structure. We adopt the post-processing method from E2Efold [4], incorporating hard constraints into RNA secondary structure, details can be found in Appendix A.

3 Experiment

3.1 Experiment Setup

Datasets. Several benchmark datasets are used in this study: (1) ArchiveII [25], a widely used dataset for RNA secondary structure prediction which contains 3,975 sequences from 10 RNA families; (2) RNAStralign [26], which contains 37,149 sequences from 8 RNA families, and has 30,451 remaining sequences after eliminating redundant sequences and structures; (3) bpRNA-1m [5], one of the most comprehensive datasets of RNA structures, contains 102,318 sequences from 2,588 families; (4) bpRNA-new, a dataset derived from Rfam 14.2 [12, 22], contains sequences from 1,500 new RNA families. As every RNA family that appears in bpRNA-1m or other datasets is excluded from bpRNA-new, it's commonly used for evaluating cross-family model generalization.

Regarding the limited training resources, our study is focused on RNA sequences within a length of 600 only. Redundant sequences are removed in the same way as E2Efold [4]. Stratified sampling is used to balance training samples from each RNA family in all test sets. For the RNAStralign dataset, we use the CD-HIT program [14] to remove redundant sequences and then select sequences shorter than 600. For the ArchiveII dataset, we follow the same approach as E2Efold [4], extracting RNA sequences shorter than 600 and excluding sequences present in RNAStralign, in order to form the test set. For the bpRNA-1m, we apply the training and test sets the same as MXfold2 [22], denoted as TR0 and TS0, respectively. The non-redundant bpRNA-new dataset is specifically used to test the model's cross-family prediction capability. The datasets used in our study are shown in Table 1.

Table 1. Datasets

Name	Size	Description
RNAStralign_600	20881	The training set in RNAStralign consisted of sequences with lengths up to 600
test_600	1790	The test set in ArchiveII consisted of sequences with lengths up to 600
TR0	5038	The bpRNA training set
TS0	1305	The bpRNA test set
bpnew	5401	Cross-family training set

Description. In this study, three experiments are conducted to evaluate the performance of the AttnUFold and TransUFold models:

1. To assess the within-family prediction capability of the AttnUFold and TransUFold models, we run tests on the RNAStralign, ArchiveII, and bpRNA datasets.
2. To assess the cross-family prediction capability of the AttnUFold and TransUFold models, we run tests on the bpRNA and bpnew datasets.
3. We compare various loss functions by conducting experiments on the AttnUFold model using the bpRNA dataset, as illustrated in Sect. 3.4.

For TransUFold, we set hidden_size D as 768, patch_size P as 1 and the number of Transformer Layer is 12. All models are trained using DiceLoss [16] (see Sect. 3.4) and Adam optimizer with 1×10^{-3} learning rate and 1×10^{-4}

weight_decay. All experiments were run on an Intel(R) Xeon(R) Gold 6330 CPU @ 2.00GHz, and an GeForce RTX 3090 GPU with 24G memory.

Evaluation. The evaluation metrics for RNA secondary structure prediction consist of precision, recall, and F1 score, as illustrated in equation 6, for evaluating all predicted base pairs in each RNA.

$$Prec = \frac{TP}{TP + FP}, \quad Recall = \frac{TP}{TP + FN}, \quad F_1 = 2 \cdot \frac{Prec \cdot Recall}{Prec + Recall} \quad (6)$$

Table 2. Results on ArchiveII Dataset

Method	Precision	Recall	F1 Score
baseline	0.864	0.802	0.820
AttnUFold	**0.936**	**0.820**	**0.852**
TransUFold	**0.873**	0.803	0.830
ContexFold	**0.873**	0.821	**0.842**
MXfold2	0.788	0.760	0.768
SPOT-RNA	0.743	0.726	0.711
E2Efold	0.734	0.660	0.686
Linearfold	0.724	0.605	0.647
Mfold	0.668	0.590	0.621
Eternafold	0.622	0.802	0.636
RNAfold	0.665	0.594	0.622
RNAsoft	0.664	0.606	0.628
RNAStructure	0.613	0.802	0.631
Contrafold	0.651	0.802	0.665

Table 3. Results on TS0 Dataset

Method	Precision	Recall	F1 Score
baseline	0.514	0.600	0.528
AttnUFold	**0.560**	**0.604**	**0.554**
TransUFold	0.555	**0.604**	0.547
CNNFold	**0.640**	0.566	0.582
SPOT-RNA	0.594	**0.693**	**0.619**
MXfold2	0.519	0.646	0.558
E2Efold	0.140	0.129	0.130
Mfold	0.501	0.627	0.538
Linearfold	0.561	0.581	0.550
Contrafold	0.528	0.644	0.567
Eternafold	0.516	0.666	0.563
ContexFold	0.529	0.607	0.546
RNAfold	0.494	0.631	0.536
RNAsoft	0.497	0.626	0.535
RNAStructure	0.494	0.622	0.533

3.2 Results on Within-Family Datasets

To assess the secondary structure prediction ability of AttnUFold and TransU-Fold within RNA families, we evaluate their performance on the RNAStralign-ArchiveII dataset and the bpRNA dataset. The evaluation results of the reproduced UFold (i.e. baseline), AttnUFold, and TransUFold on the ArchiveII dataset are presented in Table 2. We compare their performance with several deep learning-based methods, including MXfold2 [22], SPOT-RNA [24], and E2Efold [4]. Moreover, we compare them with traditional methods such as Con-textfold [29], Contrafold [7], Linearfold [11], Eternafold [28], RNAfold [15], RNAS-tructure (Fold) [18], RNAsoft [2], and Mfold [31]. The two highest values for each metric are indicated in bold. As seen in Table 2, except that ContextFold achieves an impressive F1 score of 0.842, other traditional methods only reach a maximum F1 score of 0.665. In contrast, all three deep learning methods surpass the 0.665, with MXfold2 performing the best at 0.768. The baseline model achieves an F1 score of 0.820, showing a 5.2% improvement over MXfold2. Additionally, both AttnUFold and TransUFold outperform the baseline across all three metrics. TransUFold shows a 1.0% increase in F1 score, while AttnUFold exhibits an impressive improvement of 3.2%, surpassing ContextFold. Notably, AttnUFold achieves a prediction accuracy of 93.6%, which is 6.3% higher than the second-highest accuracy, highlighting a significant enhancement in prediction precision.

In Table 3, we present the experimental results on the TS0 test set. Overall, the performance of UFold-based methods on this dataset is average. Among the four deep learning methods, AttnUFold outperforms E2Efold by 42.4% and achieves a comparable level to MXfold2. Its F1 score of 0.554 surpasses the majority of traditional prediction methods. Both AttnUFold and TransUFold consistently outperform the baseline across all three evaluation metrics. TransU-Fold shows a 1.9% improvement in the F1 score, while AttnUFold demonstrates a greater improvement of 2.6%.

We noticed that the performance of TransUFold is not as good as AttnUFold, which may have the following reasons: (1) Transformer is a harder structure to train, which requires a large amount of data, and the number of known RNA with secondary structure is limited; (2) TransUFold only learns global features in one hidden layer, unlike AttnUFold, which can obtain global information in each skip connection; (3) In the patch_embedding process, a padding operation is used, and the actual effective patch range is from 16 to 1444. Therefore, this padding process may affect the model's performance.

3.3 Results on Cross-Family Datasets

The experimental results of cross-family prediction are presented in Table 4. The table provides a comparison with deep learning-based methods, including CNN-Fold [20], MXfold2 [22], SPOT-RNA [24], and E2Efold [4], as well as traditional methods such as Contrafold [7], TORNADO [19], Contextfold [29], and RNAfold [15]. The deep learning methods shown in the table were all trained on the TR0 training set, and tested on the bpnew test set.

Compared to traditional methods, the cross-family prediction ability of UFold-based methods is not particularly outstand-

Table 4. Results on bpRNA-new Dataset

Method	Precision	Recall	F1 Score
baseline	0.541	0.600	0.552
AttnUFold	0.537	0.635	0.568
TransUFold	0.600	0.564	0.559
CNNFold	–	–	0.496
MXfold2	0.585	0.710	0.632
SPOT-RNA	0.599	0.619	0.596
E2Efold	0.047	0.031	0.036
Contrafold	0.620	0.736	0.661
TORNADO	0.636	0.638	0.620
ContexFold	0.595	0.539	0.554
RNAfold	0.552	0.720	0.617

ing. When compared to deep learning models, the data-driven UFold model exhibits lower generalization ability than MXfold2, which incorporates thermodynamic constraints into its loss function. However, it is worth noting that the baseline model achieves an F1 score 5.6% higher than CNNFold, which also transforms RNA sequences into image inputs. In comparison to the baseline, both AttnUFold and TransUFold show slight improvements in their F1 scores. TransUFold has a 0.6% increase, while AttnUFold shows a 1.6% increase. Furthermore, TransUFold achieves an accuracy of 0.600, which is higher than all other deep-learning methods.

3.4 Comparison of Loss Functions

Due to the highly imbalanced distribution between 0 and 1 in the input data, and no relevant constraints being introduced in UFold's training stage. To overcome these two limitations, we apply DiceLoss and $DiagLoss$ as loss functions, respectively. The training and testing sets used are TR0 and TS0.

Table 5. Different Loss Fuctions' Results on TS0 Dataset

Method	Precision	Recall	F1 Score
AttnUFold_bce	0.501	**0.624**	0.536
AttnUFold_dice	**0.560**	0.604	**0.554**
AttnUFold_bce+diag	0.538	0.572	0.527
AttnUFold_dice+diag	0.526	0.612	0.536

Cross-entropy loss is used for comparison. As shown below, we incorporate a positive weight ω of 300 from UFold into the cross-entropy calculation, where A represents the ground truth adjacency matrix.

$$Loss(Y, A; \theta) = -\sum_{ij}[A_{ij}log(Y_{ij}) + (1 - A_{ij})log(1 - Y_{ij})] \tag{7}$$

DiceLoss. Given the prediction matrix Y and the ground truth matrix Y^T, the smaller value of $DiceLoss$ represents the greater similarity between the two matrices. $DiceLoss$ is suitable for the imbalanced sample distribution in this study.

$$DiceLoss = 1 - \frac{2|Y \cap Y^T|}{|Y| + |Y^T|} \tag{8}$$

DiagLoss. Regarding to the RNA folding constraint, the output matrix should have zero diagonal values to ensure that each nucleotide is not paired with itself. Hence, $DiagLoss$ is applied as follows, where $Loss$ can be Cross-entropy loss or DiceLoss as mentioned above.

$$DiagLoss = Loss + c\sum |x_i|^2 \tag{9}$$

The results are presented in Table 5, where the best performance is achieved by DiceLoss, with an F1 score of 0.554, 1.8% higher than the cross-entropy loss with positive weight. We observed that incorporating the diagonal constraint in the loss function does not improve the model's prediction results and leads to a decrease in all three metrics instead. It could be attributed to the fact that the post-processing stage and model training stage are separate. Therefore, the effect of diagonal constraints during the training process is not reflected in the final evaluation results. On the other hand, the excellent performance of DiceLoss indicates that calculating the similarity between the predicted pairing matrix and the ground truth matrix effectively complements the U-net model architecture. Other experiments conducted in this study also demonstrate the improvement brought by DiceLoss.

3.5 Visualization

Prediction Results. We visually showcased the partial prediction results of the AttnUFold model on the RNAStralign dataset [26]. We convert the predictions into dot-bracket notation and generate structure diagrams using the

VARNA [6] tool. Figure 4 highlights the model's remarkable prediction ability, as it accurately captures most of the structure in longer RNA sequences, with only minor discrepancies in certain details.

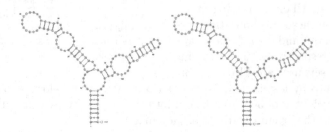

Fig. 4. The predicted structure of RNA with the ID B05038. The left figure shows the ground-truth structure, and the right figure shows the predicted structure.

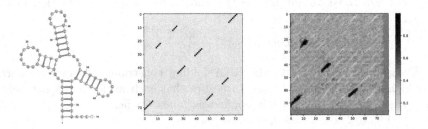

Fig. 5. The attention weight of RNA with the ID tdbD00008575. The left figure shows its actual secondary structure. The middle figure shows its pairing matrix. The right figure shows the attention weight coefficient of the last layer attention gate of AttnU-Fold.

Attention Weights. We visualize the attention weight coefficients in the Attention Gate module of AttnUFold, as shown in Fig. 5. The sample used is the RNA sequence with the identifier tdbD00008575 from the RNAStralign dataset, which consists of 76 base pairs and is perfectly predicted by our model. Specifically, the visualization process involves extracting the attention weight matrix from the last layer of the Attention Gate in AttnUFold and summing the 32 channels to obtain a 2-dimensional coefficient matrix.

The figure provides some insights of our model. First, due to the alignment operation performed to ensure compatibility with the U-net architecture, indicated by the light green borders on the right and bottom sides of the image, the model effectively recognizes the padded regions in the input data. Second, we observe that regions with high attention coefficients align closely with the true pairing regions, such as the lower-left triangle in the figure, which reflects the symmetry of the actual pairings. This indicates that the model correctly identifies potential pairing regions. Interestingly, similar patterns appear in the visualization results of other examples as well, where the attention coefficient matrix tends to reflect pairing information through the lower-left triangle rather than the upper-right triangle. The underlying reason for this phenomenon remains to be explored.

4 Conclusion

In this study, we propose AttenUFold and TransUFold for RNA secondary structure prediction, two novel models developed based on attention mechanism. These models incorporate the Attention Gate module and draw inspiration from the ViT model, marking the first integration of attention mechanism into the field of RNA secondary structure prediction. Additionally, different loss functions are explored to overcome the challenge posed by the highly imbalanced 0–1 sample distribution in the input data. The results demonstrate our method's significant performance enhancement across various tasks.

Acknowledgements. This work was supported by Shanghai Rising-Star Program (23QA1409000), the Starry Night Science Fund at Shanghai Institute for Advanced Study (SN-ZJU-SIAS-0010), and CEIEC-2022-ZM02-0246.

References

1. Allali, J., Sagot, M.F.: A new distance for high level rna secondary structure comparison. IEEE/ACM Trans. Comput. Biol. Bioinf. **2**(1), 3–14 (2005)
2. Andronescu, M., Aguirre-Hernandez, R., Condon, A., Hoos, H.H.: Rnasoft: a suite of rna secondary structure prediction and design software tools. Nucleic Acids Res. **31**(13), 3416–3422 (2003)
3. Chen, J., et al.: Transunet: transformers make strong encoders for medical image segmentation. arXiv preprint arXiv:2102.04306 (2021)
4. Chen, X., Li, Y., Umarov, R., Gao, X., Song, L.: RNA secondary structure prediction by learning unrolled algorithms. arXiv preprint arXiv:2002.05810 (2020)
5. Danaee, P., Rouches, M., Wiley, M., Deng, D., Huang, L., Hendrix, D.: bpRNA: large-scale automated annotation and analysis of RNA secondary structure. Nucl. Acids Res. **46**(11), 5381–5394 (2018). https://doi.org/10.1093/nar/gky285
6. Darty, K., Denise, A., Ponty, Y.: Varna: interactive drawing and editing of the RNA secondary structure. Bioinformatics **25**(15), 1974 (2009)
7. Do, C.B., Woods, D.A., Batzoglou, S.: Contrafold: RNA secondary structure prediction without physics-based models. Bioinformatics **22**(14), e90–e98 (2006)
8. Fallmann, J., Will, S., Engelhardt, J., Grüning, B., Backofen, R., Stadler, P.F.: Recent advances in rna folding. J. Biotechnol. **261**, 97–104 (2017)
9. Fox, G.E., Woese, C.R.: 5s RNA secondary structure. Nature **256**(5517), 505–507 (1975)
10. Fu, L., Cao, Y., Wu, J., Peng, Q., Nie, Q., Xie, X.: Ufold: fast and accurate RNA secondary structure prediction with deep learning. Nucl. Acids Res. **50**(3), e14 (2022)

11. Huang, L., et al.: Linearfold: linear-time approximate RNA folding by 5'-to-3'dynamic programming and beam search. Bioinformatics **35**(14), i295–i304 (2019)
12. Kalvari, I., et al.: RFAM 14: expanded coverage of metagenomic, viral and microrna families. Nucl. Acids Res. **49**(D1), D192–D200 (2021)
13. Knudsen, B., Hein, J.: Using stochastic context free grammars and molecular evolution to predict RNA secondary structure. Bioinformatics **15**(6), 446–454 (1999)
14. Li, W., Godzik, A.: Cd-hit: a fast program for clustering and comparing large sets of protein or nucleotide sequences. Bioinformatics **22**(13), 1658–1659 (2006)
15. Lorenz, R., et al.: Viennarna package 2.0. Algorithms Molecul. Biol. **6**, 1–14 (2011)
16. Milletari, F., Navab, N., Ahmadi, S.A.: V-net: fully convolutional neural networks for volumetric medical image segmentation. In: 2016 Fourth International Conference on 3D Vision (3DV), pp. 565–571. IEEE (2016)
17. Oktay, O., et al.: Attention u-net: learning where to look for the pancreas. arXiv preprint arXiv:1804.03999 (2018)
18. Reuter, J.S., Mathews, D.H.: RNAstructure: software for RNA secondary structure prediction and analysis. BMC Bioinformatics **11**(1), 1–9 (2010)
19. Rivas, E., Lang, R., Eddy, S.R.: A range of complex probabilistic models for rna secondary structure prediction that includes the nearest-neighbor model and more. RNA **18**(2), 193–212 (2012)
20. Saman Booy, M., Ilin, A., Orponen, P.: RNA secondary structure prediction with convolutional neural networks. BMC Bioinformatics **23**(1), 1–15 (2022)
21. Sankoff, D.: Simultaneous solution of the RNA folding, alignment and protosequence problems. SIAM J. Appl. Math. **45**(5), 810–825 (1985)
22. Sato, K., Akiyama, M., Sakakibara, Y.: Rna secondary structure prediction using deep learning with thermodynamic integration. Nat. Commun. **12**(1), 941 (2021)
23. Seetin, M.G., Mathews, D.H.: RNA structure prediction: an overview of methods. In: Bacterial Regulatory RNA: Methods and Protocols, pp. 99–122 (2012)
24. Singh, J., Hanson, J., Paliwal, K., Zhou, Y.: RNA secondary structure prediction using an ensemble of two-dimensional deep neural networks and transfer learning. Nat. Commun. **10**(1), 5407 (2019)
25. Sloma, M.F., Mathews, D.H.: Exact calculation of loop formation probability identifies folding motifs in rna secondary structures. RNA **22**(12), 1808–1818 (2016)
26. Tan, Z., Fu, Y., Sharma, G., Mathews, D.H.: Turbofold II: RNA structural alignment and secondary structure prediction informed by multiple homologs. Nucl. Acids Res. **45**(20), 11570–11581 (2017)
27. Vaswani, A., et al.: Attention is all you need. Adv. Neural Inf. Process. Syst. **30** (2017)
28. Wayment-Steele, H., Kladwang, W., Participants, E., Das, R.: RNA secondary structure packages ranked and improved by high-throughput experiments. In: biorxiv (p. 2020.05. 29.124511) (2020)
29. Zakov, S., Goldberg, Y., Elhadad, M., Ziv-Ukelson, M.: Rich parameterization improves rna structure prediction. J. Comput. Biol. **18**(11), 1525–1542 (2011)
30. Zhang, H., et al.: A new method of RNA secondary structure prediction based on convolutional neural network and dynamic programming. Front. Genet. **10**, 467 (2019)
31. Zuker, M.: Mfold web server for nucleic acid folding and hybridization prediction. Nucl. Acids Res. **31**(13), 3406–3415 (2003)

A Velocity Controller for Quadrotors Based on Reinforcement Learning

Yu Hu[1,2], Jie Luo[1,2], Zhiyan Dong[1,2]([✉]), and Lihua Zhang[1,2]

[1] Academy for Engineering and Technology, Fudan University, Shanghai, China
21210860050@m.fudan.edu.cn
[2] Institute of AI and Robotics, Fudan University, Shanghai, China
{19210860032,dongzhiyan,lihuazhang}@fudan.edu.cn

Abstract. UAVs based on PID controllers are having increasing difficulties in handling complex tasks. Whereas, reinforcement learning-based high-dimensional models provide an important entry point for flight control to handle complex and high-dimensional tasks. In this paper, a neural network controller training framework for outer-loop control is proposed, which is used as a base platform for velocity controller training. Also, a reinforcement learning-based quadrotor neural network speed controller is proposed which maps the state of the UAV to the throttle commands of the rotor for stable control of speed. In addition, this paper employs the idea of curriculum learning to help the UAV adapt to a larger speed tracking range and improve its overall performance. We demonstrate the performance of the trained neural network controller by comparing it with a conventional PI controller in simulations, achieving improvements in both steady-state response time and tracking performance.

Keywords: Reinforcement Learning · Quadrotor · Velocity Controller

1 Introduction

Unmanned Aerial Vehicles are extensively utilized in military exploration, agriculture, transportation, and other fields due to their agile maneuverability. In these complex application scenarios, precise execution of position commands is crucial, leading to an increasing demand for advanced intelligence in UAV flight controllers. Traditional UAV controllers typically consist of an outer-loop position control loop and an inner-loop attitude control loop. The outer-loop position control loop generates attitude and thrust commands to track specific trajectories or speeds, while the inner-loop attitude control loop generates actuation commands to the rotors to accurately track the target attitude state. However, these conventional controller architectures face several challenges: (1) classical PID control algorithms are commonly used for tracking expected values in both loops. Although PID control can be effective, it relies on small angle assumptions that approximate the nonlinear dynamics of UAV as linear. Additionally, the process of simultaneously adjusting PID parameters is also can be complex and challenging to generalize. (2) Differences in the modal periods of the

L. Fang et al. (Eds.): CICAI 2023, LNAI 14474, pp. 411–421, 2024.
https://doi.org/10.1007/978-981-99-9119-8_37

two control loops result in delays between control signals and longer periods for upper-level position control. Recent research has shown that intelligent control algorithms based on reinforcement learning offer a promising solution to these challenges.

For attitude loop control, using reinforcement learning, an optimal attitude control strategy for UAVs can be achieved without assuming aircraft dynamics, ensuring that UAVs maintain good attitude stability despite unknown dynamics or unexpected conditions. Deep reinforcement learning uses neural network models instead of traditional control strategies. It has the advantage that Deep Reinforcement Learning (DRL) has the ability to adapt to any non-linear input/output mapping. Training data is collected iteratively in a simulation environment to adapt the neural network model to the ideal controller distribution. In [5], a simulation platform for reinforcement learning attitude controllers is built, which simulates a realistic UAV maneuvering dynamics model and motor model, and on which the Proximal Policy Optimization (PPO) algorithm-based attitude controller is trained. Adversarial reinforcement learning algorithms were proposed by [15] to enhance the capabilities of Sim2Real and based on [5] the attitude controller was migrated to a real environment for deployment, on top of which [10] proposed domain randomisation and course adversarial reinforcement learning to further improve the robustness of the attitude controller.

For position loop control, which is more challenging due to the incorporation of UAV's high-level dynamics attributes, Neural network controllers have been applied to solve specific problems. For instance, [12] addresses the accurate landing of quadrotors and introduces neural network controllers for quadrotor landing. In [7], a deep Q-network hierarchy is designed for the autonomous landing of a quadrotor in various simulated environments, demonstrating superior performance compared to human-controlled aerial vehicles in the presence of environmental disturbances. [11] applies the Deep Deterministic Policy Gradient (DDPG) algorithm for outer-loop control of aircraft landing and showcases the capability and potential of the reinforcement learning approach, validating its robustness under diverse wind disturbance conditions. In [14], a simulation platform for a hybrid UAV velocity controller is developed, and a velocity control strategy is trained on this platform for the hybrid UAV model. While [4] proposes a reinforcement learning-based position controller, it only addresses recovery from an arbitrary initial state to a stationary state, limiting its applicability to all cases of UAV position control.

In this work, a simulation platform for training a quadrotor velocity controller based on the gym-pybullet-drone project [6] was proposed. On top of this, a DRL-based quadrotor velocity controller was proposed, which maps the target speed and the quadrotor's state as inputs directly to the rotor commands. A PPO-based reinforcement learning strategy is used to train the controller, and a curriculum learning approach is added on top of this to extend further the velocity tracking range and robustness of the controller.

2   Background

2.1   Reinforcement Learning

The RL-based algorithm learns the optimal strategy by allowing agents to inter-act with the environment continuously so as to maximize returns or to achieve specific goals [1]. The interaction process between an agent and the environment is modeled as a Markov Decision Process (MDP). Namely, at a certain time t, an agent obtains the state value S_t from the environment, takes the corresponding action A_t according to the current state value, changes the environment, and obtains the state value S_{t+1} at the next time. The state transition is defined as a probability of transition to state s'. Assume that the current state and action are denoted as s and a, respectively. Then, the probability of transition to state s' can be expressed as $p_r\{S_{t+1} = s'|S_t = s, A_t = a\}$. The behavior of an agent is defined by its policy, which is essentially a mapping of actions to be taken for a specific state, as shown in Fig. 1.

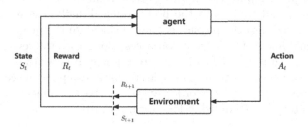

Fig. 1. Schematic diagram of the RL framework

2.2   Curriculum Learning

Curriculum learning, introduced by [3] in 2009, is regarded as a training strategy and has been widely referenced in the field of machine learning. It aims to mimic the human learning process by initially training on simple tasks and gradually increasing the training difficulty, enabling the model to accomplish more chal-lenging tasks. Through curriculum learning, the intelligent agent can organize a systematic exploration of the state space to address the issue of sparse rewards and accelerate the convergence rate of the model. Therefore, we incorporate the concept of curriculum learning into the training process to further enhance the convergence efficiency and control effectiveness of the model.

3   Simulation Platform

3.1   Framework

We have designed the training framework shown in Fig. 2 as the basis for this work, which is suitable for training not only velocity controllers but also supports the training of inner and outer loop controllers such as attitude and position. In each interaction, the Pybullet physics engine will give a 12-dimensional variable O representing the state of the UAV, including the position, attitude angle, angular velocity, and linear velocity of the quadrotor, which can be expressed as

$$O = [x, y, z, \varphi, \theta, \psi, v_x, v_y, v_z, \omega_x, \omega_y, \omega_z] \qquad (1)$$

Then, the input state $S(t)$ of the neural network controller was calculated based on the state of the UAV and the target state given by the environment. Based on $S(t)$ the neural network controller selects an optimal action $A(t)$ based on its own policy distribution, which corresponds to the throttle commands of the four motors, denoted as

$$A(t) = [\sigma_1, \sigma_2, \sigma_3, \sigma_4], \sigma_i \in [0, 1] \qquad (2)$$

Subsequently, the throttle commands are then fed into the rotor model to compute the revolutions per minute (RPM) of the propellers. These RPMs are further supplied as inputs to the quadrotor's rigid body force model, yielding the total thrust and torques in the body frame. Finally, the total thrust and torques will be fed into the physics engine to calculate the quadrotor's state at the subsequent time step. The state variables, along with the rewards computed based on these states, are fed as input into the neural network controller.

The above process constitutes a single simulation step for the controller, with the neural network controller's objective being the attainment of the optimal execution strategy by maximizing long-term rewards.

3.2   Rotor Model

The RPM of the four rotors of a quadrotor is commonly controlled by a throttle command, denoted as $\sigma \in [0, 1]$. It takes a certain amount of time for the rotor to reach a steady-state speed, which is referred to as the rotor's dynamic response time, denoted as T_m. The dynamic response process of a rotor can be approximated as a first-order low-pass filter, represented by:

$$T_m \frac{d\varpi(t)}{t} + \varpi(t) = C_m \sigma(t) + \varpi_0 \qquad (3)$$

C_m is the revolution rate of a rotor. ϖ_0 denotes the rotor revolution when the throttle command is 0. Equation (1) can be further represented as:

$$\dot{\varpi}(t) = \frac{d\varpi(t)}{dt} = \frac{1}{T_m}(C_m \sigma(t) + \varpi_m - \varpi(t)) \qquad (4)$$

$\dot{\varpi}(t)$ denotes the change rate of rotor revolution. Finally, the rotor revolution in the next timestep can be represented as:

$$\varpi(t+1) = \dot{\varpi}(t) * d_t + \varpi(t) \qquad (5)$$

d_t denotes the simulation timestep of the physical engine, and T_m denotes the response time of the rotor.

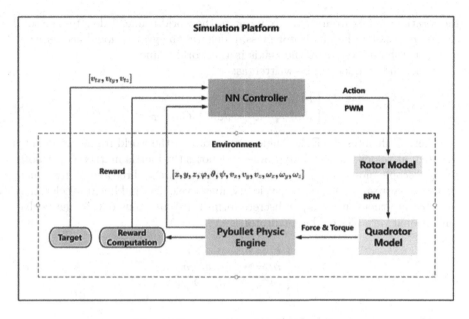

Fig. 2. Controller training Framework

3.3 Quadrotor Model

Depending on the rotational speed of the rotor, we can obtain the total thrust and the corresponding torque for each axis:

$$
\begin{bmatrix} f \\ \tau_y \\ \tau_y \\ \tau_z \end{bmatrix} = \begin{bmatrix} c_T & c_T & c_T & c_T \\ -\frac{\sqrt{2}}{2}dc_T & \frac{\sqrt{2}}{2}dc_T & \frac{\sqrt{2}}{2}dc_T & -\frac{\sqrt{2}}{2}dc_T \\ \frac{\sqrt{2}}{2}dc_T & -\frac{\sqrt{2}}{2}dc_T & \frac{\sqrt{2}}{2}dc_T & -\frac{\sqrt{2}}{2}dc_T \\ c_M & c_M & c_M & c_M \end{bmatrix} \begin{bmatrix} w_1^2 \\ w_2^2 \\ w_3^2 \\ w_4^2 \end{bmatrix} \tag{6}
$$

where c_T and c_M denote the drag force factor and torque factor, which depend mainly on the geometry of the propeller. And w_i denotes the RPM of the propellers.

3.4 Dynamic Model

The quadrotor is considered as a rigid body and its total thrust is always parallel to the z-axis of the body frame. We can calculate the positional and rotational dynamics of the quadrotor based on the total thrust and torques on the three axes in the body frame obtained in (6).

$$
\begin{cases} \dot{\mathbf{p}}^e = \mathbf{v}^e \\ \dot{\mathbf{v}}^e = g\mathbf{e}_3 - \frac{f}{m}\mathbf{R}_b^w \mathbf{e}_3 \end{cases} \tag{7}
$$

where p denotes position, v denotes velocity, the right superscript w denotes the world frame, and the right superscript b denotes the body frame. R_b^e denotes the rotation matrix from the body frame to the world frame, f denotes the total thrust, m denotes the quadrotor mass, g denotes the gravitational acceleration, and e_3 is the unit vector of the z-axis in the world frame.

Its attitude model can be written as:

$$\begin{cases} \dot{\boldsymbol{\Theta}} = \boldsymbol{R}_b^w \cdot \boldsymbol{\omega}^b \\ J\dot{\boldsymbol{\omega}}^b = -\boldsymbol{\omega}^b \times J\boldsymbol{\omega}^b + \boldsymbol{G}_a + \boldsymbol{\tau} \end{cases} \tag{8}$$

where $\dot{\boldsymbol{\Theta}}$ denotes the Euler angle change rate in the world frame, $\boldsymbol{\omega}^b$ denotes the angular velocity in the body frame, J denotes the inertia matrix, $\boldsymbol{G}_a$ denotes the gyroscopic moment, and $\boldsymbol{\tau}$ denotes the torque in the three axes. After calculating the corresponding angular velocity, linear velocity, and linear acceleration, we integrate them in a simple discrete form in (9) and thus obtain the twelve-dimensional variables of the quadrotor UAV.

$$\begin{cases} \boldsymbol{v}_i = \boldsymbol{v}_{i-1} + d_t \cdot \dot{\boldsymbol{v}}_i \\ \boldsymbol{p}_i = \boldsymbol{p}_{i-1} + d_t \cdot \boldsymbol{v}_i \\ \boldsymbol{\Theta} = \boldsymbol{\Theta} + d_t \cdot \dot{\boldsymbol{\Theta}} \end{cases} \tag{9}$$

4   Neural Network Velocity Controller

4.1   Network Structure

Based on Sect. 3, a neural network velocity controller was trained using the PPO algorithm. The controller adopts the network structure of the classical PPO algorithm, shown in Fig. 3. The input $S(t) \in R^{12}$ to the network comes from the state of the quadrotor and the target velocity, which can be represented as:

$$S(t) = [\varphi, \theta, \psi, \alpha, \beta, \gamma, P_{v_x}, P_{v_y}, P_{v_z}, I_{v_x}, I_{v_y}, I_{v_z}] \in R^{12} \tag{10}$$

where $[\varphi, \theta, \psi]$ respectively denote the quadrotor's roll, pitch, and yaw angles. $[\alpha, \beta, \gamma]$ respectively denote the angular velocity of the three axes in the body frame of the quadrotor. $[P_{v_x}, P_{v_y}, P_{v_z}]$ denote the proportional term for the difference between the line velocity of the quadrotor and the desired velocity. $[I_{v_x}, I_{v_y}, I_{v_z}]$ denote the integral term. The output of the network is the throttle commands σ_i for the four rotors.

Given that the model of the controller does not incorporate temporal networks like RNN or LSTM, the incorporation of an integration module will enhance the tracking performance of the controller [14]. Therefore, an integral module is incorporated into the model, and its discrete form is shown below:

$$I_n = \eta I_{n-1} + D_n \tag{11}$$

where D_n denotes the velocity error sum in the current timestep. I_{n-1} denotes cumulative velocity integral in the last timestep. And η is the constant decay coefficient.

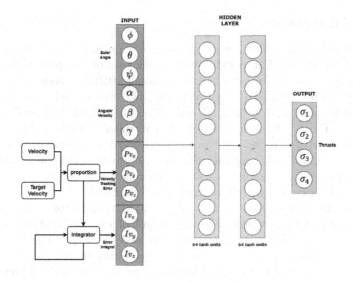

Fig. 3. Neural Network Controller

4.2 Reward Function

The reward function consists of the following components: (1)Survival reward g_s, which is used to keep the flight stable at the beginning. (2)Velocity error penalty g_v, which is used to ensure the quadrotor tracks the desired velocity as closely as possible. (3)Angular velocity penalty g_ω, which is used to ensure that the quadrotor remains as stable as possible during the tracking and to prevent constant oscillations.

$$R(t) = g_s - w_v g_v - w_\omega g_\omega \tag{12}$$

where w_v, w_ω are the corresponding weights and g_s is defined as a constant.

4.3 Normalization

Normalization is a significant part of the training process, as there are gaps in the magnitudes of the original inputs. If normalization is not performed, the training process will not converge and result in the accumulative reward oscillating repeatedly. To address this issue, a normalization step for the state variables of the quadrotor is adopted.

In our approach, we normalize the state quantities of the UAV by constraining the velocity and desired velocity within the range of $[-3, 3]$ and the angular velocity within the range of $[-50, 50]$. Subsequently, we apply a max-min normalization to ensure that all the state variables fall within a consistent range.

By performing input normalization, we aim to mitigate the adverse effects of input disparities and facilitate the training process of the controller. This normalization step helps to establish a more stable and effective learning environment, allowing the controller to converge toward optimal performance.

4.4 Training Setup

We implemented the PPO algorithm using Python and based on the Stable Baselines3 library [8]. The training process consists of two phases. In the first phase, the quadrotor is initialized with a random attitude (pitch and roll angles randomly selected from the range of $[-\pi/2, \pi/2]$ rad/s) and a random target velocity along each axis (with single-axis velocities randomly selected from the range of $[-1, 1]$ m/s). At each time step, the controller executes motor commands within the range of $[0,1]$ and receives a reward based on the current state. When any of the pitch, roll, and angular velocities exceeds the predefined threshold, it indicates that the quadrotor has lost stability and is no longer in a controlled state. In such cases, the early termination is considered.

The simulation operates at a frequency of 240 Hz, and a total of eight million steps are trained in this phase. In the second phase, we fine-tune the model trained in the first phase by increasing the task difficulty. The target speed range is expanded to a single-axis target speed within the range of $[-3,3]$ m/s, allowing for a larger speed range. We train a total of two million steps in this stage.

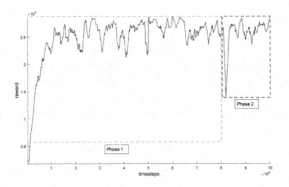

Fig. 4. The mean reward during the training process

The average reward curve for each episode during the final convergence is illustrated in Fig. 4. It's worth noting that at the beginning of the second stage, there is a significant drop in the reward. This reduction is due to the larger speed variation range, which results in a higher penalty term for tracking speeds, impacting the overall reward.

In order to enhance the robustness of the controller in real-world control scenarios, we incorporated noise during the training process [10,15,16]. Specifically, we added Gaussian noise with a mean of 0 and a variance of 0.1 to several state variables, including velocity, angular velocity, and attitude of the UAV, during the training phase. The inclusion of noise helps the controller to adapt and handle variations and uncertainties that may arise during actual control tasks, thereby improving its overall robustness.

5   Result

Since in practice quadrotors do not have an accurate estimation of acceleration, conventional PID controllers generally rely on proportional and integral terms in the velocity loop. A comparative analysis is conducted between our controller and the PI controller. It is worth noting that the NN controller also relies explicitly only on the proportional and integral terms as input, and its knowledge of the derivative terms will probably come from its derivation of the pose. Moreover, it is important to consider that the velocity control loop exhibits a longer steady-state response time delay in comparison to the attitude control loop, typically ranging from 0.5 to 1 s, depending on the tracked speed.

To evaluate their performance, both the NN Controller and PI Controller are initialized from a static state and tasked with tracking four velocity changes within 12 s. The velocities are randomly altered every 3 s, with the third change being of greater magnitude. The step response performance of the controllers with different algorithm implementations under the interference of random noise is illustrated in Fig. 5. Intuitively, the NN Controller has an advantage in response latency and exhibits a smoother response curve when tracking larger velocity changes (6–9 s) in comparison to the PI Controller.

In order to effectively evaluate the tracking performance and emphasize the impact of high errors, we utilize the root mean square error (RMSE) as a performance metric. The RMSE amplifies errors and provides a more comprehensive measure of the actual tracking error. The formula of RMSE is as follows:

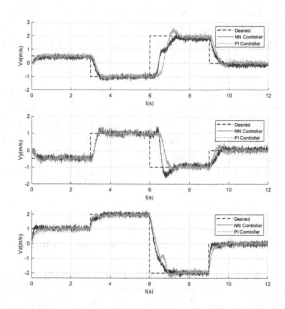

Fig. 5. Comparison of two controllers in the simulation of velocity tracking

$$RMSE(x,y,f) = \sqrt{\frac{1}{m}\sum_{i=1}^{m}[f(x_i)-y_i]^2} \qquad (13)$$

The results of the RMSE are presented in the table, clearly demonstrating that our algorithm outperforms the PI controller in terms of smaller RMSE values across all three axes. This indicates that our algorithm excels in velocity tracking for the flight task, achieving superior performance.

Table 1. The RMSE of PI Controller and NN Controller

	$RMSE_x$ (m/s)	$RMSE_y$ (m/s)	$RMSE_z$ (m/s)
PI Controller	0.7942	0.5366	0.6233
NN Controller	**0.6773**	**0.4943**	**0.5207**

6   Conclusion

In this paper, we propose a simulation framework used to train the neural network controller for quadrotors and a novel neural network velocity controller for quadrotors based on the PPO algorithm. Additionally, we extend the tracking speed range of the controller by incorporating the concept of curriculum learning. The obtained controller demonstrates stable tracking control of the quadrotor's upper velocity, exhibiting comparable or superior tracking performance compared to the conventional PI controller.

In our future work, we aim to further enhance the algorithm model to improve the tracking performance of the controller. Additionally, we plan to incorporate control of the yaw angle of the quadrotor. Furthermore, we intend to deploy the developed controller to real aircraft and utilize it as the underlying controller for path planning. By pursuing these avenues, we anticipate achieving significant advancements in the tracking capabilities of the controller, expanding its potential applications, and enhancing its performance in real-world scenarios.

References

1. Andrew, G., Barto, R.S.S.: Reinforcement Learning: An Introduction. MIT Press (1998)
2. Bagnell, J., Schneider, J.: Autonomous helicopter control using reinforcement learning policy search methods. In: Proceedings 2001 ICRA. IEEE International Conference on Robotics and Automation (Cat. No. 01CH37164). vol. 2, pp. 1615–1620 (2001). https://doi.org/10.1109/ROBOT.2001.932842

3. Bengio, Y., Louradour, J., Collobert, R., Weston, J.: Curriculum learning. In: Proceedings of the 26th Annual International Conference on Machine Learning (ICML 2009), pp. 41–48. Association for Computing Machinery, New York (2009). https://doi.org/10.1145/1553374.1553380
4. Hwangbo, J., Sa, I., Siegwart, R., Hutter, M.: Control of a quadrotor with reinforcement learning. IEEE Robot. Automat. Lett. **2**(4), 2096–2103 (2017). https://doi.org/10.1109/LRA.2017.2720851
5. Koch, W., Mancuso, R., Bestavros, A.: Neuroflight: next generation flight control firmware (2019)
6. Panerati, J., Zheng, H., Zhou, S., Xu, J., Prorok, A., Schoellig, A.P.: Learning to fly - a gym environment with pybullet physics for reinforcement learning of multi-agent quadcopter control. arXiv preprint arXiv:2103.02142 (2021)
7. Polvara, R., et al.: Toward End-to-End Control for UAV Autonomous Landing via Deep Reinforcement Learning, pp. 115–123 (2018). https://doi.org/10.1109/ICUAS.2018.8453449
8. Raffin, A., Hill, A., Gleave, A., Kanervisto, A., Ernestus, M., Dormann, N.: Stable-baselines3: reliable reinforcement learning implementations. J. Mach. Learn. Res. **22**(268), 1–8 (2021). http://jmlr.org/papers/v22/20-1364.html
9. Barros dos Santos, S.R., Givigi, S.N., Júnior, C.L.N.: An experimental validation of reinforcement learning applied to the position control of UAVs. In: 2012 IEEE International Conference on Systems, Man, and Cybernetics (SMC), pp. 2796–2802 (2012). https://doi.org/10.1109/ICSMC.2012.6378172
10. Sheng, J., Zhai, P., Dong, Z., Kang, X., Chen, C., Zhang, L.: Curriculum Adversarial Training for Robust Reinforcement Learning, pp. 1–8 (2022). https://doi.org/10.1109/IJCNN55064.2022.9892908
11. Tang, C., Lai, Y.C.: Deep reinforcement learning automatic landing control of fixed-wing aircraft using deep deterministic policy gradient. In: 2020 International Conference on Unmanned Aircraft Systems (ICUAS), pp. 1–9 (2020). https://doi.org/10.1109/ICUAS48674.2020.9213987
12. Vankadari, M.B., Das, K., Shinde, C., Kumar, S.: A reinforcement learning approach for autonomous control and landing of a quadrotor. In: 2018 International Conference on Unmanned Aircraft Systems (ICUAS), pp. 676–683 (2018). https://doi.org/10.1109/ICUAS.2018.8453468
13. Williams, J.D., Young, S.: Partially observable Markov decision processes for spoken dialog systems. Comput. Speech Lang. **21**(2), 393–422 (2007). https://doi.org/10.1016/j.csl.2006.06.008
14. Xu, J., et al.: Learning to fly: computational controller design for hybrid UAVs with reinforcement learning. ACM Trans. Graph. **38**(4), 1–12 (2019). https://doi.org/10.1145/3306346.3322940
15. Zhai, P., Hou, T., Ji, X., Dong, Z., Zhang, L.: Robust adaptive ensemble adversary reinforcement learning. IEEE Robot. Automat. Lett. **7**(4), 12562–12568 (2022). https://doi.org/10.1109/LRA.2022.3220531
16. Zhang, W., Song, K., Rong, X., Li, Y.: Coarse-to-fine UAV target tracking with deep reinforcement learning. **16**, 1522–1530 (2019). https://doi.org/10.1109/TASE.2018.2877499
17. Zhen, Y., Hao, M., Sun, W.: Deep Reinforcement Learning Attitude Control of Fixed-Wing UAVs (2020). https://doi.org/10.1109/ICUS50048.2020.9274875

DACTransNet: A Hybrid CNN-Transformer Network for Histopathological Image Classification of Pancreatic Cancer

Yongqing Kou[1], Cong Xia[2], Yiping Jiao[3(✉)], Daoqiang Zhang[1],
and Rongjun Ge[4(✉)]

[1] Key Laboratory of Brain-Machine Intelligence Technology, Ministry of Education,
Nanjing University of Aeronautics and Astronautics, Nanjing 211106, China
[2] Jiangsu Key Laboratory of Molecular and Functional Imaging,
Department of Radiology, Zhongda Hospital, Medical School of Southeast University,
Nanjing, China
[3] Nanjing University of Information Science and Technology, Nanjing, China
ping@nuist.edu.cn
[4] School of Instrument Science and Engineering,
Southeast University, Nanjing, China
rongjun_ge@126.com

Abstract. Automated and accurate classification of histopathological images of pancreatic cancer can lead to higher survival rates for
more pancreatic cancer patients in the clinic. However, there are very
scarce existing studies for pancreatic cancer, and the diagnosis of
pancreatic cancer remains a challenge for pathologists, especially for
well-differentiated pancreatic cancer with a clinical histological pattern
similar to that of chronic pancreatitis. We propose a hybrid CNN-
Transformer model incorporating deformable atrous spatial pyramids
(DACTransNet) to perform automated and accurate classification of
histopathological images of pancreatic cancer. We elegantly integrate the
powerful local feature extraction capability of CNN for spatial features
and the global modeling capability of transformer for abstract patterns.
Moreover, we imitate pathologists in the clinic by better integrating
deformable convolution and multiscale methods to review histopathology slides in pyramidal format. In addition, a migration learning approach was used to improve the classification accuracy of pancreatic cancer
histopathology images. The experimental results show that the proposed
method not only has a high classification accuracy (up to 96%), but
also its good robustness and generalizability as validated by real clinical datasets from multiple centers. Consequently, we provide an effective
tool for the clinical diagnosis of pancreatic cancer.

Keywords: Pancreatic cancer · Histopathological image · Transformer

---

Y. Kou and C. Xia—Contribute equally to this work.

L. Fang et al. (Eds.): CICAI 2023, LNAI 14474, pp. 422–434, 2024.
https://doi.org/10.1007/978-981-99-9119-8_38

1   Introduction

Pancreatic cancer is a highly fatal malignant tumor, known as the "king of cancers", with a low 5-year survival rate of only 10% [1]. Clinically, many patients with pancreatic cancer are mistaken for pancreatitis when early symptoms appear and miss the most optimal time for treatment, resulting in a terrible prognosis. That is why pancreatic cancer has the highest mortality rate among all malignant tumors [2]. Thus, the accurate classification of pancreatic cancer plays an important role in the diagnosis and treatment process.

The gold standard for clinical medical diagnosis is the histopathological image evaluation by pathologists. Currently, there are fewer studies on automated analysis of histopathological images of pancreatic cancer. One reason may be due to the scarcity of resources and lack of high-quality annotation because of the low rate of early diagnosis. Another important reason may be that classification of pancreatic cancer is challenging because early stage pancreatic cancer is clinically very similar to pancreatitis. Most of the existing models used in studies of histopathological images are fine-tuned models [3–6] pre-trained on large natural image datasets (e.g., ImageNet datasets). However, due to the distinctiveness of histopathological images, such as the differences in data structure between histopathological images and natural images, as well as the heterogeneity of tumor cells, these models often result in suboptimal performance.

Recently, there has been significant progress in the accuracy of medical image analysis facilitated by methods based on Vision Transformers (ViT). Advanced approaches [7,8] for medical image analysis tasks rely on the ViT framework, leveraging its remarkable achievements in computer vision tasks. However, compared to methods based on convolutional neural networks, ViT-based models have certain limitations on their performance. Firstly, the serialization operation of ViT results in the loss of spatial information modeling. Secondly, ViT exhibits a higher dependency on large-scale datasets.

To address the aforementioned issues, we propose a novel and efficient hybrid network architecture called the CNN-Transformer hybrid model for Deformable Atrous Spatial Pyramids (DACTransNet). Which combines of local features of convolutional neural networks (CNN) and global features of ViT-based model. This model incorporates a lightweight transformer block at each layer of the CNN, allowing for the extraction of local features while considering global contextual information. Additionally, we employ a novel Deformable Atrous Spatial Pyramids (DC-ASPP) module to capture information from multi-scale irregular objects. Our method offers three primary contributions compared to existing approaches:

- We propose an integrated model that elegantly combines the local information of convolutional neural networks (CNN) and the long-range characteristics of Transformers, allowing for the simultaneous utilization of their strengths to enhance the model's ability to extract distinctive features from pancreatic cancer histopathological images.
- We incorporate deformable convolution into ASPP to extract multi-scale target information as well as irregular target information via multiple atrous

convolution layers with different scales of dilatation rate and deformable convolution in parallel with the features extracted by the encoder.

- We conducted extensive experiments on multiple central datasets, including training on the publicly available TCGA dataset annotated by multiple pathologists and testing on actual clinical datasets from three different regions. This comprehensive evaluation validated the generalization performance and clinical value of our model.

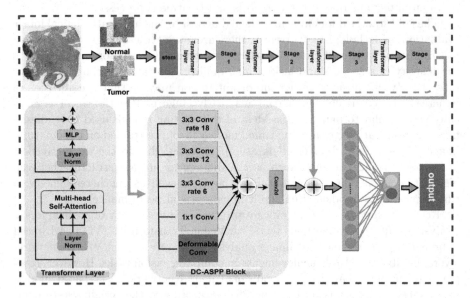

Fig. 1. Illustration of the DACTransNet. DACTransNet contains a CNN-Transformer backbone and a DC-ASPP block incorporating deformable convolution and atrous spatial pyramid pooling (ASPP).

2   Related Works

2.1   CNNs for Histopathological Images Classification

Convolutional Neural Networks (CNNs) have considerably contributed to the development of computational pathology with the development of CNN models due to their robust feature representation capabilities. Most histopathology image classification models [9–15] are derived from the prevalent natural image classification backbone. However, histopathological images are different from other medical images due to their inherent characteristics, such as extremely high resolution images, insufficient labeling, and multi-scale information, trinh et al. [16] developed a multi-scale binary-type coding network to enhance cancer classification by using binary pattern codes to capture and exploit patterns at different scales, which are further converted to decimal numbers. Zhang et al. [17] proposed the concept of a "virtual package" to classify histopathology whole slide images through a multi-instance learning (MIL) architecture with a two-layer feature distillation. However, since resizing the original image causes information loss when processing high-resolution images, hou et al. [18] proposed a novel spatial hierarchical graph neural network framework to improve the classification accuracy of histopathological images by adding a dynamic structure learning to obtain the spatial topology and hierarchical dependencies of entities.

2.2 Vision Transformers for Histopathological Images Classification

More recently, transformers, originally proposed for natural language processing (NLP), have rapidly become the main architecture in computer vision [7, 8, 19] and they are considered as alternatives to their CNN counterparts. Several works have been proposed for the processing of medical images [20–22] because the self-attention mechanism of the visual transformer is able to directly capture long-range dependencies. However, due to the limited number of medical images, especially histopathology images, such methods are difficult to optimize and computationally expensive, so most existing studies have focused on creating hybrid CNN-transformer models for feature processing. Zhang et al. [23] proposed a multi-stage hybrid transformer combining the CNN and transformer, achieving high accuracy on the ROSE image dataset. Zheng et al. [24] proposed a graph transformer classifier that fuses graph neural networks and transformers to predict disease grades.

Discussion: Although the mentioned methods (CNNs-based as well as ViT-based methods) have achieved good results, they still ignore some inherent characteristics of histopathological images, such as the heterogeneity and heterogeneity of tumor cells, so they lead to challenging classification tasks of histopathological images. Therefore, the problem is how to elegantly combine the advantages of CNN and transformer, yet reduce the model complexity and classify well for targets with large shape differences and irregularities.

3 Methodology

For the features of histopathology images, the hybrid architecture of DAC-transNet is designed to combine the robust local feature extraction capability of CNN for spatial features and the global modeling capability of transformer for abstract patterns. The overall architecture is shown in Fig. 1. DACtransNet consists of two main modules: a hybrid CNN-Transformer network as backbone and an ASPP module that incorporates deformable convolution. And then the ASPP module based on deformable convolution is designed to acquire multi-scale information, and the more robust deformable convolution is used to extract information from irregular targets. Our proposed network DACTransNet is optimized using the standard cross-entropy function as the loss function:

$$L = -[y log \; \hat{y} + (1 - y) log(1 - \hat{y})] \tag{1}$$

where y is the true label value (positive class value is 1 and negative class value is 0) and $\hat{y}$ is the predicted label value ($\hat{y} \in (0, 1)$).

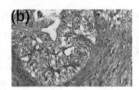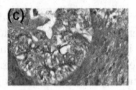

Fig. 2. Color normalization visualization, (a) is the image of the training dataset, (b) is the result of coloring and normalizing the images in the multicenter clinical dataset, and (c) is the original image in the multicenter clinical dataset, after coloring and normalizing the images in the test and training sets are closer in style.

3.1   CNN-Transformer Hybrid Network

We propose an integrated backbone that elegantly combines the local information of a convolutional neural network (CNN) and the long-range properties of transformer. It consists of convolutional and transformer blocks in an alternating superposition.

Convolution Block. To better encode spatial location information, we use convolutional blocks to extract local spatial features. First, fine-tuning of VGG19-Net pre-trained in ImageNet is designed by stacking a stem module and a CNN bottleneck module in four stages, where the blocks downsample the image $I \in R^{3 \times H \times W}$ with edge size of H and W into abstractive features. The convolutional blocks in the four stages are the same as $Conv_2$, $Conv_3$, $Conv_4$, $Conv_5$. The modeling process for each convolutional block is shown as follows:

$$F_i^C = ConvBlock_i(F_{i-1}^T), i \in \{1, 2, 3, 4\} \qquad (2)$$

where $F_i^C \in R^{\frac{H}{2^{i+1}} \times \frac{W}{2^{i+1}} \times C_i}$ is the local feature obtained from the $i - th$ block.

Transformer Block. The structure of the Transformer layer is shown in the lower left corner of Fig. 1, which contains a Multi-Headed Self-Attention (MHSA) layer to model long-range dependencies, two Layer Normalization layers, and a Multilayer Perceptron (MLP). Traditional ViT-based models use linear position projections for MHSA computation, which results in the loss of spatial information in the transformer, but this is crucial for medical image processing. Existing methods would alleviate this problem by adding positional encoding, however this would add additional computational cost and lead to poor optimization of the model. Therefore, inspired by [25], we replace the position-wise linear projection before each MHSA in the transformer module with a convolutional projection operation that employs $s \times s$ depth-separable convolution on a two-dimensional reshaped token mapping. Such an operation allows the model to further capture local information in the attention mechanism and can remove the original position embedding, simplifying the computational effort. The modeling process of the Transformer block is shown below:

Formally, the two-dimensional feature form of the stage i of the convolutional block output $F_i^C \in R^{\frac{H}{2^{i+1}} \times \frac{W}{2^{i+1}} \times C_i}$ is given as input, we learn a 2D convolution operation of kernel size $s \times s$, stride $s - o$ and p padding as a function $f(\cdot)$. The tokens maps $f(x_i)$ obtained after convolutional mapping. Next, the convolutional projection is implemented using a deeply separable convolutional layer with kernel size s. Finally, the projected tokens are panned to one dimension, i.e., $Query\ Q, Key\ K, Value\ V$ is used as the input for multi-headed self-attention. The modeling process of this process can be formulated as:

$$x_i^{q/k/v} = Flatten(DepthConv2d(Reshape(x_i), s)) \tag{3}$$

where $x_i^{q/k/v}$ is the token input for $Q/K/V$ matrics at layer i, $DepthConv2d$ is a deep-wise separable convolution, x_i is the token prior to the convolutional projection, and s refers to the convolution kernel size. Hereinafter, applying a MHSA, the output is obtained as follows:

$$SA_i = \sigma(\frac{Q \times K^T}{\sqrt{d}})V \tag{4}$$

where σ is the softmax function, d is the dimension of the input token. After applying a residual operation and MLP, this process can be expressed as follows:

$$F_i^T = MLP(SA_i + F_i^C) + SA_i \tag{5}$$

We omitted the layer normalization (LN) in the equation for simplicity. Finally, the output of $Transformer\ Block$ is as the input of $Convolutional\ Block$ at stage $i + 1$.

3.2 DC-ASPP Block

To combine information at multiple scales, we introduce an *astrous spatial pyramid pooling* (ASPP) block to detect incoming features by using various filters or pooling operations under multiple perceptual fields and multiple dilation, although using astrous convolution can obtain larger perceptual fields without increasing the computation. Since deformable convolution [26] focuses on adding adaptive 2D spatial offsets to enhance the flexibility of the convolutional sampling locations and to keep the channel dimensions unchanged, we propose a new feature aggregation module that, by introducing deformable convolution instead of normal convolution, the sampling locations of the convolution are no longer limited to fixed sampling locations, making the sampling locations more flexible and capable of producing more accurate localizations. Thus our proposed $DC - ASPP\ Block$ enables the network to focus on both overall features and detailed features and obtain more detailed localization for better feature extraction capability. Our improved $DC - ASPP\ Block$ consists of one 1×1 convolution, three 3×3 convolutions with $rates = (6, 12, 18)$ (all with 256 filters and Batch Normalization) and $Deformable\ Convolution$, and then the features of all branches are then concatenated and pass through another 1×1 convolution.

428   Y. Kou et al.

Table 1. Performance comparison for previous and DACTransNet models on TCGA datasets. Boldfaced results indicate better results.

Method	Accuracy ↑	AUC ↑	F1 Score ↑	Precision ↑		Recall ↑	
				Normal	Tumor	Normal	Tumor
VGGNet	0.8963	0.9765	0.8381	0.9802	0.7332	0.8663	0.9584
ResNet	0.9451	0.9740	0.8919	0.9734	0.9742	0.9921	0.8224
DenseNet	0.9207	0.9785	0.8439	0.9212	0.9216	0.9754	0.7782
ViT	0.8232	0.9451	0.7293	0.9412	0.6293	0.8073	0.8672
SwinTransformer	0.8719	0.9625	0.7836	0.9381	0.7312	0.8823	0.8442
MobileViT	0.8537	0.9387	0.7545	0.9201	0.7061	0.8742	0.8100
DACTransNet (Ours)	**0.9634**	**0.9894**	**0.9791**	**0.9821**	**1.000**	**1.000**	**0.9591**

Table 2. Performance comparison for previous and DACTransNet models on center A datasets. Boldfaced results indicate better results.

Method	Accuracy ↑	AUC ↑	F1 Score ↑	Precision ↑		Recall ↑	
				Normal	Tumor	Normal	Tumor
VGGNet	0.8797	0.9493	0.8692	0.9031	0.8535	0.8764	0.8854
ResNet	0.8822	0.9530	0.8664	0.8822	0.8853	0.9120	0.8482
DenseNet	0.8940	0.9539	0.8758	0.8514	0.8643	0.8833	0.8876
ViT	0.8240	0.9121	0.8166	0.8792	0.7723	0.7893	0.8662
SwinTransformer	0.8541	0.9381	0.8437	0.8852	0.8234	0.8452	0.8651
MobileViT	0.8499	0.9296	0.8372	0.8752	0.8216	0.8482	0.8533
DACTransNet (Ours)	**0.8973**	**0.9714**	**0.8831**	**0.8731**	**0.9344**	**0.9522**	**0.8933**

4 Experiment

4.1 Datasets and Details

Public Datasets (TCGA). In this study, we utilized the TCGA (The Cancer Genome Atlas) dataset [27]. Due to the limited availability of pancreatic cancer histopathological image resources, we opted to use H&E stained tissue slides from 190 pancreatic cancer patients from the TCGA dataset as the sole training dataset for our experiments. The entire image is magnified up to a resolution of 160k×160k pixels at 40× zoom. To facilitate computation and achieve better classification accuracy, we downscaled the images to a 4x zoom level. Moreover, because the whole slide images (WSI) are too large to be loaded into memory, and because the TCGA dataset didn't contain annotations due to the difficulty of annotating pancreatic cancer histopathology images, a group of pathologists collaborated to select annotated portions of the pancreatic cancer WSIs containing both tumor and normal tissue, while maintaining a balanced ratio of negative and positive samples. To facilitate training, we cropped each WSIs into non-overlapping patches of 256×256 pixels. 1336 patches were used as the training dataset, and 164 patches were used as the test dataset.

Table 3. Performance comparison for previous and DACTransNet models on center B datasets. Boldfaced results indicate better results.

Method	Accuracy ↑	AUC ↑	F1 Score ↑	precision ↑		Recall↑	
				Normal	Tumor	Normal	Tumor
VGGNet	0.8426	0.9171	0.8443	0.8441	0.8415	0.8382	0.8472
ResNet	0.8594	0.9362	0.8585	0.8413	0.8830	0.8851	0.8354
DenseNet	0.8605	0.9465	0.8534	0.8222	0.9092	0.9184	0.8041
ViT	0.7945	0.8738	0.7963	0.7923	0.7971	0.7942	0.7956
SwinTransformer	0.8417	0.9250	0.8407	0.8312	0.8524	0.8556	0.8293
MobileViT	0.7814	0.9631	0.9171	0.8592	0.9423	0.9511	0.8933
DACTransNet (Ours)	**0.8714**	**0.9631**	**0.9171**	**0.8592**	**0.9243**	**0.9511**	**0.8933**

Table 4. Performance comparison for previous and DACTransNet models on center C datasets. Boldfaced results indicate better results.

Method	Accuracy ↑	AUC ↑	F1 Score ↑	Precision ↑		Recall↑	
				Normal	Tumor	Normal	Tumor
VGGNet	0.8655	0.9357	0.8660	0.8661	0.8652	0.8644	0.8672
ResNet	0.9087	0.9738	0.8984	0.8834	0.9154	0.9141	0.8821
DenseNet	0.9039	0.9630	0.9013	0.8793	0.9324	0.9365	0.8722
ViT	0.8444	0.9264	0.8447	0.8412	0.8482	0.8483	0.8412
Swin Transformer	0.8544	0.9536	0.8727	0.8563	0.8920	0.8534	0.8543
MobileViT	0.8522	0.9497	0.8538	0.8632	0.8543	0.8432	0.8543
DACTransNet (Ours)	**0.9113**	**0.9801**	**0.9091**	**0.8881**	**0.9374**	**0.9432**	**0.8824**

Multicenter Clinical Dataset for External Validation. In order to evaluate the generalization performance of our proposed model, we applied it to clinical datasets from three different centers, which comprehensively encompassed different types of pancreatic cancer. To ensure patient privacy, the clinical datasets from the three centers were anonymized as Center 1, Center 2, and Center 3, consisting of 30, 35, and 38 H&E stained histopathological slides, respectively. Since the staining of data from different centers varies widely, color normalization becomes an essential step in preprocessing. We adopted the method proposed by Jiao et al. [28], the results after color normalization are shown in Fig2.

Implementation Details. We use Pytorch and the adam optimizer with a learning rate of 1e-4 to run all our experiments. We used the pre-trained weights of VGG19-Net pre-trained on imagenet to train our proposed DACTransNet and ran 300 epochs. To avoid overfitting, our data were enhanced as follows: rotation (90°), horizontal, vertical flip and color disturbance. For the TCGA training dataset, we set the batch size to 4. Appropriate test values, including recall, precision, F1-score, accuracy, and AUC are calculated to quantify and compare the model performance of these four test cohorts.

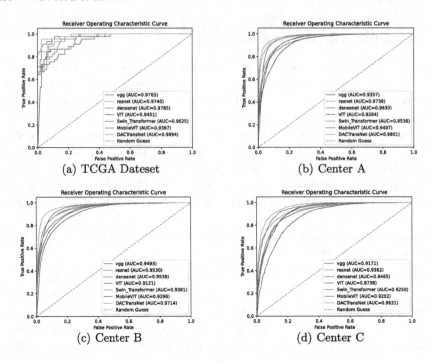

Fig. 3. Acceptance operating characteristic (ROC) curves.

Table 5. Ablation study on TCGA dataset, Trans denotes transformer layer, DC denotes deformable convolution. The baseline model we used was VGG19Net.

Models	Baseline	Trans	ASPP	DC	Accuracy ↑	F1 Score ↑	Precision ↑	Recall ↑
E.1	✓				0.8963	0.8308	0.7332	0.9584
E.2		✓			0.8444	0.8447	0.8482	0.8412
E.3	✓	✓			0.9123	0.9116	0.8829	0.9422
E.4	✓	✓	✓		0.9328	0.9521	0.9533	0.9511
E.5	✓	✓	✓	✓	0.9634	0.9891	1.0000	0.9591

4.2 Comparison with State-of-the-Art Methods

To demonstrate the effectiveness of our proposed method, we compare our approach with state-of-the-art classification methods, including transformer-based models and CNN-based models on the ImageNet dataset.

Results on TCGA Dataset. Compared to these models, our approach largely outperforms both the pre-trained transformer-based model and the CNN-based model. More specifically, our DACTransNet achieves 96.34% accuracy. From the results in Table 1, we found that DACTransNet has a relatively significant advantage in cancer recall compared to the transformer-based and CNN-based

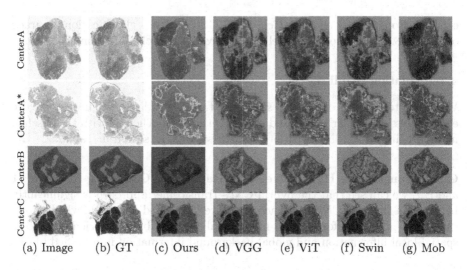

Fig. 4. Visualization results on an multicenter clinical dataset. Where DAT stands for our proposed DACTransNet, Swin stands for Swin Transformer, and Mob stands for MobileViT. The greater the color indicates a higher probability of lesions and conversely a higher probability of normal tissue.

models pre-trained on the ImageNet dataset, a performance that is consistent with the requirements of clinical diagnosis, as pathologists must scrutinize and not overlook any patches that may be cancerous.

Results on Multicenter Clinical Datasets. Then, in order to verify the portability and robustness of the model, we performed the same tests on multi-center clinical datasets. The results are shown in Tables 2, 3 and 4. It can be seen from the results that our model also achieves better results on the multicenter clinical dataset, which is sufficient to prove the good generalization performance of our model.

Visualization Results. In addition, we analyze the performance of our proposed DACTransNet model by means of receiver operating characteristic curve (ROC) curves, as shown in Fig. 3. In the results, our model achieves better performance both on the internal training dataset and the multicenter clinical dataset. Finally, we plotted cancer probability heatmaps on all four datasets, as shown in Fig. 4 and what can be seen is that for the slice shown in CenterA, our model DACTransNet can achieve a classification accuracy of 98.32%, while the classification accuracy of VGG-Net is only 89.45%. For more challenging cases, such as CenterA* and CenterB slices, which are very difficult to classify, because these two slices show a very rare type of cancer in pancreatic cancer, our model can achieve a classification accuracy of 64.42%, while VGG-Net only has a classification accuracy of 56.23%. Finally, for the more common type of pancreatic cancer histopathology slides in CenterC, VGG-Net has a classification accuracy of about 98.67%, while our model can achieve 99.45% classification accuracy.

4.3 Ablation Studies

We conducted a series of ablation studies on the TCGA dataset to investigate the effectiveness of DACTransNet and to justify the design choices, with the baseline method being VGG19Net with pre-trained weights, as shown in Table 5.

CNN-Transformer Hybrid Network Backbone. Comparing E.3 with E.1 and E.2, we can see that the hybrid CNN-Transformer approach is significantly better than the pure CNN approach and pure ViT-based approach, which shows that CNN and Transformer can indeed make it possible to efficiently process spatial local information and global background information in a unified block.

DC-ASPP Block. Comparing E.3 and E.4, we can see that the ASPP module can bring better performance, which indicates that the multi-scale approach is important for histopathology image classification since pathologists read films are operating at multiple resolutions. And by for E.5 and E.4, we can see that deformable convolution can also bring better performance because of the heterogeneity of tumor cells, and by adding deformable convolution can be a good learning effect for irregular cancer types.

5 Conclusion

In this work, to address some of the challenging classification tasks for histopathological images. We propose a DACTransNet network for pancreatic cancer classification, which elegantly combines the advantages of CNN and transformer to improve the model's ability to model local information and long-distance dependencies, and to classify targets with large differences in shape and irregularities well. It outperforms pure CNN methods pre-trained on ImageNet or pure Transformer methods, and can show better performance on small datasets.

Acknowledgements. This study was supported by the Natural Science Foundation of Jiangsu Province (No. BK20210291), the National Natural Science Foundation (No. 62101249 and No. 62136004), and the China Postdoctoral Science Foundation (No. 2021TQ0149 and No. 2022M721611).

References

1. Siegel, R.L., Miller, K.D., Wagle, N.S., Jemal, A.: Cancer statistics, 2023. CA: Cancer J. Clinicians **73**(1), 17–48 (2023)
2. Pereira, S.P., et al.: Early detection of pancreatic cancer. Lancet. Gastroenterol. Hepatol. (2020)

3. Krizhevsky, A., Sutskever, I., Hinton, G.E.: Imagenet classification with deep convolutional neural networks. Commun. ACM **60**(6), 84–90 (2017)
4. Simonyan, K., Zisserman, A.: Very deep convolutional networks for large-scale image recognition. CoRR, abs/1409.1556 (2014)
5. He, K., Zhang, X., Ren, S., Sun, J.: Deep residual learning for image recognition. In: 2016 IEEE Conference on Computer Vision and Pattern Recognition (CVPR), pp. 770–778 (2015)
6. Huang, G., Liu, Z., Van Der Maaten, L., Weinberger, K.Q.: Densely connected convolutional networks. In: 2017 IEEE Conference on Computer Vision and Pattern Recognition (CVPR), pp. 2261–2269 (2017)
7. Dosovitskiy, A., et al.: An image is worth 16x16 words: transformers for image recognition at scale. arXiv, abs/2010.11929 (2020)
8. Liu, Z., et al.: Swin transformer: hierarchical vision transformer using shifted windows. In: 2021 IEEE/CVF International Conference on Computer Vision (ICCV), pp. 9992–10002 (2021)
9. Chen, H., Dou, Q., Wang, X., Qin, J., Heng, P.-A.: Mitosis detection in breast cancer histology images via deep cascaded networks. In: AAAI Conference on Artificial Intelligence (2016)
10. Tian, Y., et al.: Computer-aided detection of squamous carcinoma of the cervix in whole slide images. arXiv, abs/1905.10959 (2019)
11. Fu, H., et al.: Automatic pancreatic ductal adenocarcinoma detection in whole slide images using deep convolutional neural networks. Front. Oncol. **11** (2021)
12. Yang, H., et al.: Deep learning-based six-type classifier for lung cancer and mimics from histopathological whole slide images: a retrospective study. BMC Med. **19** (2021)
13. Coudray, N., et al.: Classification and mutation prediction from non-small cell lung cancer histopathology images using deep learning. Nat. Med. **24**, 1559–1567 (2018)
14. Ianni, J.D., et al.: Tailored for real-world: a whole slide image classification system validated on uncurated multi-site data emulating the prospective pathology workload. Sci. Rep. **10** (2020)
15. Liu, M., Lanlan, H., Tang, Y., Chu Wang, Yu., He, C.Z., et al.: A deep learning method for breast cancer classification in the pathology images. IEEE J. Biomed. Health Inform. **26**, 5025–5032 (2022)
16. Vuong, T.T.L., Song, B., Kim, K., Cho, Y.M., Kwak, J.T.: Multi-scale binary pattern encoding network for cancer classification in pathology images. IEEE J. Biomed. Health Inform. **26**, 1152–1163 (2021)
17. Zhang, H., et al.: DTFD-mil: double-tier feature distillation multiple instance learning for histopathology whole slide image classification. In: 2022 IEEE/CVF Conference on Computer Vision and Pattern Recognition (CVPR), pp. 18780–18790 (2022)
18. Hou, W., Huang, H., Peng, Q., Yu, R., Yu, L., Wang, L.: Spatial-hierarchical graph neural network with dynamic structure learning for histological image classification. In: International Conference on Medical Image Computing and Computer-Assisted Intervention (2022)
19. Touvron, H., Cord, M., Douze, M., Massa, F., Sablayrolles, A., J'egou, H.: Training data-efficient image transformers & distillation through attention. arXiv, abs/2012.12877 (2020)
20. Chen, H., et al.: Gashis-transformer: a multi-scale visual transformer approach for gastric histopathological image detection. Pattern Recogn. **130**, 108827 (2021)

21. Shao, Z., Bian, H., Chen, Y., Wang, Y., Zhang, J., et al.: TransMil: transformer based correlated multiple instance learning for whole slide image classication. In: Neural Information Processing Systems (2021)
22. Xiong, Y., et al.: Nyströmformer: a nyström-based algorithm for approximating self-attention. In: AAAI Conference on Artificial Intelligence, vol. 35, pp. 16:14138–16:14148 (2021)
23. Zhang, T., Yunlu Feng, Yu., Zhao, G.F., Yang, A., Lyu, S., et al.: MSHT: multi-stage hybrid transformer for the rose image analysis of pancreatic cancer. IEEE J. Biomed. Health Inform. **27**, 1946–1957 (2021)
24. Zheng, Y., Gindra, R., Green, E., Burks, E.J., Betke, M., Beane, J.E., et al.: A graph-transformer for whole slide image classification. IEEE Trans. Med. Imaging **41**, 3003–3015 (2022)
25. Wu, H., et al.: CVT: introducing convolutions to vision transformers. In: 2021 IEEE/CVF International Conference on Computer Vision (ICCV), pp. 22–31 (2021)
26. Dai, J., et al.: Deformable convolutional networks. In: 2017 IEEE International Conference on Computer Vision (ICCV), pp. 764–773 (2017)
27. The cancer genome atlas (TCGA) (2016). http://cancergenome.nih.gov/
28. Jiao, Y., Li, J., Fei, S.M.: Staining condition visualization in digital histopatholog-ical whole-slide images. Multimedia Tools Appl. **81**, 17831–17847 (2022)

YueGraph: A Prototype for Yue Opera Lineage Review Based on Knowledge Graph

Songjin Yang[1], Fuxiang Fu[1], Chenxi Zhu[2], Hao Zeng[1(✉)], Youbing Zhao[1(✉)],
Hao Xie[1], Xuxue Sun[1], Xi Guo[1], Bin Han[3], Guofen Tao[3], and Shengyou Lin[1]

[1] Communication University of Zhejiang, Hangzhou 310018, China
{hao.zeng,zyb}@cuz.edu.cn
[2] Zhejiang University, Hangzhou 310058, China
[3] Zhejiang Xiaobaihua Yue Opera Theatre, Hangzhou 310005, China

Abstract. Yue opera, as one of the representatives of China's intangible cultural heritage, embodies a profound regional history and folk art. This paper focuses on utilizing knowledge graphs to promote research and preservation efforts in the field of Yue Opera's lineage. By mining the characteristics of the lineage domain, designing an ontology model, and employing ChatGPT for knowledge extraction, a knowledge graph specific to Yue Opera's lineage is constructed to facilitate knowledge integration in the domain. Furthermore, this study also develops an application prototype named YueGraph (A video is shown in https://github. com/Ani-li/YueGraph/blob/main/demo.mp4) that explores the knowledge graph of the Yue Opera lineage through question-answering and visualization. YueGraph has been deployed at the Zhejiang Xiaobaihua Yue Opera Theatre to provide strong support for the preservation of the Yue Opera lineage.

Keywords: Yue Opera · Knowledge Graphs · ChatGPT

1 Introduction

Yue opera is renowned as the "second-largest opera genre" in China. Lineage is a crucial component in the field of Yue opera, as it preserves traditional artistic techniques and performance methods through the generations via the master-apprentice system. This practice also allows apprentices to incorporate their own styles and expressions within the traditional framework, resulting in the development of various Yue opera genres.

In recent years, the growing recognition of intangible cultural heritage has propelled the advancement of digital technologies in the field of cultural preservation. Leveraging digital tools and methodologies, the analysis and exploration of Yue opera lineage can effectively facilitate the integration and study of Yue opera cultural knowledge, thus injecting new vitality into the development and inheritance of the Yue opera domain.

L. Fang et al. (Eds.): CICAI 2023, LNAI 14474, pp. 435–441, 2024.
https://doi.org/10.1007/978-981-99-9119-8_39

This study focuses on a number of Yue opera performers as research subjects. By constructing a prototype YueGraph, a knowledge graph for the Yue opera lineage, the transmission chains between Yue opera performers are documented and investigated. By employing a novel approach to analyze knowledge in the domain of Yue opera lineage, this research not only enables the discovery and inference of unknown or implicit semantic relationships, but also lays the foundation for filling gaps in existing knowledge [1].

2   Related Work

The existing historical and cultural resources in the field of Yue opera lineage are characterized by diverse and heterogeneous content, often with loosely structured organizational frameworks, which pose challenges to the utilization of cultural resources related to Yue opera lineage [2]. Traditional methods of information organization and retrieval may not adequately address the diversity and heterogeneity of these resources.

The knowledge graph, proposed by Google, is a graph-based structure that is used to establish semantic relationships between entities and allow reasoning and associative analysis [3]. Knowledge graphs have been widely applied in the field of historical and humanistic studies. For example, Zhou et al. collected and organized a huge amount of data on Tang poetry to build a knowledge graph focused on Tang poems and poets [4]. Yang et al. constructed a knowledge graph of academic lineages in the Song Dynasty based on data from the Chinese Biographical Database of the History of the Chinese Academy [5]. Carriero et al. built the ArCo knowledge graph of Italian cultural heritage based on the official general catalog and related coding rules of the Italian Ministry of Cultural Heritage and Activities [6,7].

3   Methodology

We employ a top-down approach to construct a knowledge graph for the domain of the Yue opera lineage. Starting from high-level concepts and relationships within the Yue Opera lineage domain, the construction process gradually refines and elaborates on these concepts. By defining the top-level concepts and relationships of the Yue opera lineage, an overall framework is established. Subsequently, relevant entity, relationship, and attribute information are extracted from the data sources. Finally, by employing entity linking, a complete knowledge graph of the Yue opera lineage is formed.

3.1   Data Sources

The primary data source for this study is the "Encyclopedia of Chinese Yue Opera" [8]. Considered the most comprehensive and authoritative encyclopedia on Yue opera, the "Encyclopedia of Chinese Yue Opera" documents the development of Yue opera from the mid-19th century to 2005. This compendium includes a wealth of representative content related to notable artists, repertoire, and lyrics in the field of Yue opera. Furthermore, additional data was collected from web pages and existing databases.

3.2 Ontology Construction

Based on the cultural characteristics of the Yue opera lineage, a feature analysis is conducted to uncover distinctive attributes and relationships within the set of core concepts. Under expert guidance, the rules for the ontology model are defined.

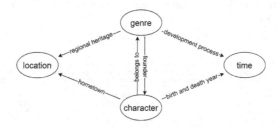

Fig. 1. Ontology concept relationship for Yue opera lineage domain

The feature analysis of Yue opera lineage culture reveals that temporal, regional, and diverse aspects are significant features. The construction process adheres to the principles of ontology independence and reusability while minimizing the number of categories [9]. As a result, four top-level concepts are derived for the Yue opera lineage domain: "character", "time", "location", and "genre", along with their interrelationships, as depicted in Fig. 1. After the overall framework of the ontology is established, further steps are taken to describe the features and relationships between concepts. This involves defining the attributes of the concepts, categorizing the attributes in a specific manner, and imposing restrictions on the number of possible attribute values.

3.3 Knowledge Extraction

Large-scale pre-trained language models (LLMs) such as ChatGPT [10] have shown impressive performance with few or zero-shot learning capabilities, where they can perform well given appropriate prompts [11]. Utilizing a ChatGPT-based knowledge extraction approach offers flexibility, generality, and transferability, making it adaptable to various knowledge domains.

We divide the knowledge extraction task into two steps. (1) Define and explain the professional terms and concepts in the domain of the Yue opera lineage to ChatGPT, providing additional contextual information to aid in its understanding of knowledge within this domain as it may not fully understand domain-specific terminologies [12]. (2) Extract knowledge by querying ChatGPT via OpenAI APIs. Descriptive prompt fields encode the data request to provide contextual cues and the obtained results are parsed and extracted for further analysis.

Appropriate prompt templates can significantly improve the performance of ChatGPT. We design prompt templates based on the types of information to be extracted (entities, relationships, attributes) and the ontology model of Yue opera lineage. The prompt templates consist primarily of input text, output format, and target information type, as shown in Fig. 2. Since entities and attributes share a nominal relationship, attribute extraction is considered a type of relationship extraction, and the same set of templates can be used. The format for the entity extraction results is defined as "entity type: specific entity," while the format for the relationship extraction results is "specific entity-relationship type-specific entity."

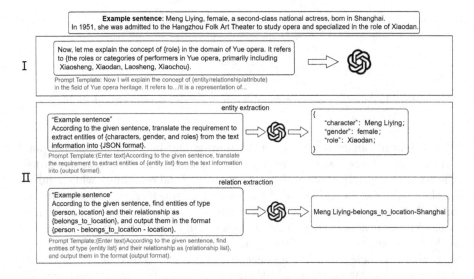

Fig. 2. Knowledge extraction workflow and prompt templates based on ChatGPT

3.4   Knowledge Storage

We employ the Neo4j graph database to store and query the knowledge graph of the Yue opera lineage. Data are structured as a graph model consisting of nodes and relationships in Neo4j. The nodes are made up of labels and properties, while relationships are made up of types and properties.

To develop the YueGraph prototype, we extracted 987 entities and 542 relationships in total. These data were organized as CSV files for easy import into Neo4j. On the basis of different types of information, we prepare two types of CSV files: one for storing entities and their attributes and the other for storing relationships between entities.

4   Applications

4.1   Question-Answering System

Leveraging a deep understanding of semantic relationships in the domain of the Yue Opera lineage, a question-answering system has been developed to answer complex and in-depth questions using semantic reasoning. As illustrated in Fig. 3, it employs techniques such as tokenization and entity recognition to analyze user input questions, then matches with the corresponding question templates, and converts into Cypher commands to query relevant entities or relationships in Neo4j. This process ultimately generates logically coherent answers to the questions.

Fig. 3. Yue opera lineage question-answering system

4.2   Visualization

YueGraph also provides visualization of the Yue Opera lineage with d3.js, as shown in Fig. 4. Users can filter by genres, Cypher commands are used to search for corresponding entities and relationships in Neo4j. The query results are presented to users in visual forms, effectively showcasing the lineage relationships and developmental changes among different factions in Yue Opera.

440   S. Yang et al.

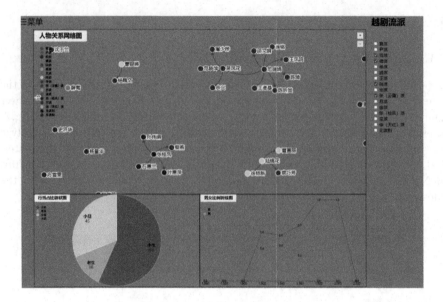

Fig. 4. Visualization of Yue opera lineage domain

Specifically focusing on the master-apprentice relationships, a person relationship network graph is constructed, where different genres are distinguished by colors, facilitating the study of lineage paths and relationships between genres. Additionally, line charts and pie charts are utilized to visualize the male-to-female ratio and distribution of roles, revealing the specific gender distribution and expertise of genres in Yue Opera.

5   Conclusion

This study investigates the construction and application of a knowledge graph in the domain of the Yue Opera heritage. The cultural characteristics of the Yue Opera heritage were explored to build an ontology model. Subsequently, knowledge was extracted through interactions with ChatGPT, and the workflow for knowledge extraction using ChatGPT was summarized. Finally, based on the constructed knowledge graph, a question-answering system is implemented, and visual representations of the Yue Opera heritage are explored. This research provides a fresh perspective on the preservation, inheritance and development of the Yue Opera heritage. In the future, it is possible to enrich the knowledge graph by incorporating information on repertoire, music, and other aspects, thus enhancing its comprehensiveness. Furthermore, integrating the Yue Opera heritage knowledge graph with knowledge graphs from other domains can facilitate interdisciplinary knowledge discovery and cross-domain research.

References

1. Tian, L., Zhang, J., Zhang, J.: A survey on knowledge graphs: representation, construction, reasoning, and hypergraph theory. J. Comput. Appl. **41**(08), 2161–2186 (2021)
2. Xu, Z., Sheng, Y., He, L.: A comprehensive survey on knowledge graph technologies. J. Univ. Electron. Sci. Technol. China **45**(04), 589–606 (2016)
3. Huang, H., Yu, J., Liao, X.: A comprehensive review of knowledge graph research. J. Comput. Syst. Appl. **28**(06), 1–12 (2019)
4. Zhou, L., Hong, L., Gao, Z.: Construction of the Tang poetry knowledge graph and design of intelligent knowledge services. Libr. Inf. Serv. **63**(02), 24–33 (2019)
5. Yang, H., Wang, J.: Construction and visualization of knowledge graph on academic lineages in the Song Dynasty. Data Anal. Knowl. Discov. **3**(06), 109–116 (2019)
6. Carriero, V.A., Gangemi, A., Mancinelli, M.L.: ArCo: the Italian cultural heritage knowledge graph. In: Ghidini, C., et al. (eds.) The Semantic Web-ISWC 2019: 18th International Semantic Web Conference, Auckland, New Zealand, 26–30 October 2019, Proceedings, Part II 18, pp. 36–52. Springer, Cham (2019). https://doi.org/10.1007/978-3-030-30796-7_3
7. ArCo [EB/OL]. http://wit.istc.cnr.it/arco/?lang=en
8. Encyclopedia of Yue Opera in China. Zhejiang Literature, Art, and Audiovisual Publishing House, Beijing (2006)
9. Liu, Y.S.: Research of approaches and development tools in constructing ontology. J. Mod. Inf. **29**(9), 17–24 (2009)
10. OpenAI. Introducing ChatGPT. [EB/OL]. https://openai.com/blog/ChatGPT
11. Han, R., Peng, T., Yang, C.: Is information extraction solved by ChatGPT? An analysis of performance, evaluation criteria, robustness and errors. arXiv preprint arXiv:2305.14450 (2023)
12. Bao, T., Zhang, C.: ChatGPT's Chinese information extraction capability evaluation: taking three typical extraction tasks as examples. In: Data Analysis and Knowledge Discovery, pp. 1–16 (2023)

An Integrated All-Optical Multimodal Learning Engine Built by Reconfigurable Phase-Change Meta-Atoms

Yuhao Wang[1,2], Jingkai Song[1,2], Penghui Shen[1,2], Qisheng Yang[1,2], Yi Yang[1,2], and Tian-ling Ren[1,2(✉)]

[1] School of Integrated Circuits, Tsinghua University, Beijing 100084, China
RenTL@tsinghua.edu.cn
[2] Beijing National Research Center for Information Science and Technology, Tsinghua University, Beijing 100084, China

Abstract. Optical computing is regarded as one of the most promising computing paradigms for solving the computational bottleneck and accelerating artificial intelligence in the post-Moore age. While reconfigurable optical processors make artificial general intelligence (AGI) possible, they often cannot process multimodal signals. Here, we propose an integrated all-optical multimodal learning engine (AOMLE) built by reconfigurable phase-change meta-atoms. The engine architecture can be mapped to different optical neural networks by laser direct writing for phase-change materials, enabling more efficient processing of visual and auditory information at the speed of light. The AOMLE provides a cutting-edge idea for reconfigurable optical processors with increasing demands for complicated AI models.

Keywords: All-Optical Computing · Reconfigurable Chip · Multimodal Learning

1 Introduction

The thriving development of photonics has paved the way for faster and more energy-efficient AI computing. Optical processors are considered one of the most promising solutions for accelerating AI [1–7], leveraging the unique advantages of light speed, ultralow power consumption, and multiplexing. As the complexity of AI models continues to increase, the development of reconfigurable optical processors becomes increasingly important. There is a need to design new devices and explore suitable materials to make progress [8–12]. Chalcogenide phase-change materials play a crucial role in the field of reconfigurable photonics [13–16]. The non-volatile materials can transition between crystalline and amorphous phases under external excitation, exhibiting significant differences in optical properties [17], which find widespread applications in light field modulation. Undeniably, reconfigurable optical computing hardware enabled by phase-change materials has fruitful achievements [18–20]. However, current advanced optical processors can only demonstrate particular types of information, such as visual

© The Author(s), under exclusive license to Springer Nature Singapore Pte Ltd. 2024
L. Fang et al. (Eds.): CICAI 2023, LNAI 14474, pp. 442–451, 2024.
https://doi.org/10.1007/978-981-99-9119-8_40

signals like images and videos or sequential signals represented by audio. Due to the monotony of optical computing architecture and coupling signals, there are still limitations in multimodal signals processing, which prevents optical processors from progressing toward general computers.

Herein, we propose an integrated all-optical multimode learning engine (AOMLE) built by reconfigurable phase-change meta-atoms. By arranging phase-change meta-atoms covering an individual optical waveguide, we map them to different optical neural networks, enabling light-speed multimodal learning. We successfully reconstruct the all-optical computing architecture by taking advantage of the excellent properties of chalcogenide phase-change materials: disordered metasurface corresponds to the optical scattered neural network suitable for auditory signals, whereas layer-by-layer metalines corresponds to the optical diffractive neural network that is better for visual signals. We unify the training models for both optical neural network architectures by solving Maxwell's equations, and the adjoint method is used for backpropagation to update the medium gradient. Finally, we obtain 95.83% accuracy in vowel recognition and 96.34% accuracy in handwritten digit recognition, both of which are comparable to state-of-the-art electronic platforms and with a boost in energy efficiency. In conclusion, our proposed all-optical computing engine can efficiently perform multimodal learning, providing promise for general AI processors.

2 Architecture of AOMLE

Figure 1 depicts the architecture of AOMLE, which is actually a physical neural network built by phase-change materials. The red and purple marks represent the directions of data propagation in the feed-forward neural network and recurrent neural network, respectively, in the artificial neural network model shown in Fig. 1(a), and AOMLE is mapped to two neural networks by programming the pattern of phase-change materials covered on the waveguide, as illustrated in Fig. 1(b). The amplitude of light is pre-coded at the left input port of AOMLE. The optical path is obviously changed as light propagates through the intermediate training region due to the modulation of phase-change materials at the top of the waveguide, and the light is eventually coupled out at the right port, and the intensity is detected by photodetectors to obtain the classification result. This is the entire inference procedure of AOMLE.

It should be noted that the chalcogenide phase-change material used in AOMLE is Sb_2Se_3, and its extinction coefficient in the telecommunication wavelength tends to be negligible, meaning that intrinsic loss to the propagating light will be minimal. The refractive index of crystalline Sb_2Se_3 is around 4.0, it has a stronger effect on the phase modulation of light than that of amorphous Sb_2Se_3, so we regard the crystalline phase-change meta-atom as an effective neuron, which function is to sum the input light and then transmit them to the next effective neuron. In the nonlinear activation function of the neural network, we use the Kerr effect of silicon itself to establish a nonlinear connection between output and input optical power.

Considering the wave characteristics of light itself, when it propagates through the disordered dielectric layer, it will continue to scatter in all directions, which will introduce a feedback loop. This process is equivalent to the recurrent neural network, which is more suitable for processing data with time series information. Previous work has also proved this in theory [21]. Therefore, we use the metasurface formed by the random distribution of meta-atoms in different crystal phases to map the optical scattered neural network to AOMLE. It should be pointed out that our preprocessing of time series information only involves basic operations such as windowing and sampling, and we will not use spectrogram and other methods to make it into a matrix for subsequent calculation so that the time step of data will be preserved and the recurrent neural network will be driven to compute when scattered light propagates backward.

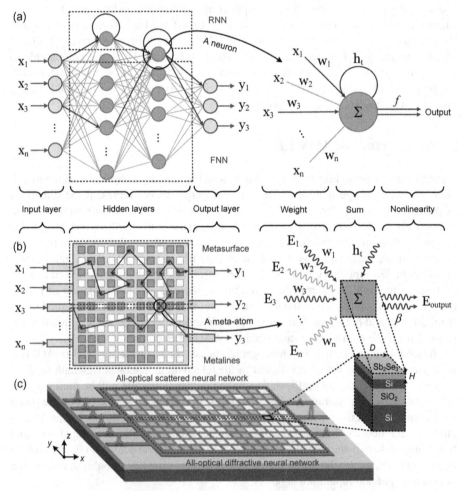

Fig. 1. Architecture of AMOLE. (a) Artificial neural network model. (b) Reconfigurable principle of AOMLE. (c) A phase-change meta-atom and hardware implementation of AOMLE.

For the on-chip diffractive neural network, its mathematical model is based on the Huygens-Fresnel diffraction principle, which reveals that diffractive neurons are essentially secondary wave sources, so their diffractive characteristics are more similar to convolution operation, even though this is a one-dimensional situation. In this way, we use crystalline effective neurons to form layers of diffractive metalines as hidden layers of feedforward neural networks. Furthermore, while the layered optical diffractive architecture is a subset of the bulk scattered architecture, the modulation mechanism of different architectures for the propagation light field determines which artificial neural network model they correspond to and which modal data computing scenarios are better suitable for.

In the experiment, the waveguide pattern of AOMLE architecture is realized by optically programming the phase-change materials, and its experimental platform is shown in Fig. 2. The experimental platform is mainly divided into two types of optical paths, propagation computation part and laser programming part. For the image input of the first kind of optical path, the 1550nm CW laser passes through the beam expander, and the image information is programmed by the digital micro-mirror. It is input to a single-mode fiber through the objective lens after passing through a 4f system. The acoustic-optical modulator provides waveform information of voice signals to the light, which is subsequently fed to AOMLE through the fiber. Finally, the photodetector receives the classified optical signal. The second type of optical path is used to implement the programming of AOMLE. We reconstruct the device using optical pulses generated by a 638nm laser diode, and the piezo stage accomplishes the movement required for array programming.

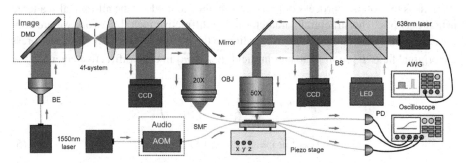

Fig. 2. Reconfigurable experimental schematic diagram of AOMLE

3   Training Algorithm of AOMLE

Light propagation in the AOMLE training region follows Maxwell's electromagnetic theory, and the primary light field distribution can be obtained by solving Maxwell's equations in the frequency domain. We describe their training methods in relation to the differences between the all-optical scattered and diffractive neural networks, respectively.

The following loss function Γ is defined.

$$\Gamma = \frac{1}{2}\sum_{i=1}^{N}(I_i - y_i)^2 \tag{1}$$

where I_i denotes the light intensity detected at the i th output port, and y_i is the ground truth as one-hot encoding.

First, we discuss the training process of the AOMLE scattered neural network model. At this moment, the structure of all phase-change meta-atoms in the training region is noticed. We use the FDTD method in the frequency domain to solve the primary light field $\vec{E}_{pri}(r)$ of any point r.

$$\left(\nabla^2 - \omega^2\mu_0\varepsilon(r)\right)\vec{E}_{pri}(r) = -i\omega\mu_0\vec{J}_s \tag{2}$$

where $\varepsilon(r)$ is the complex relative dielectric constant at r, μ_0 is the permeability of vacuum, and $\vec{J}_s$ is the current source density of the input light field distribution. We then determine the derivative $\partial\Gamma/\partial\vec{E}_{pri}(r)$ and use it as the excitation source of the adjoint field $\vec{E}_{adj}(r)$. Consequently, $\vec{E}_{adj}(r)$ can correspond to any r in the training region.

$$\left(\nabla^2 - \omega^2\mu_0\varepsilon(r)\right)\vec{E}_{adj}(r) = -\frac{\partial\Gamma}{\partial\vec{E}_{pri}(r)} \tag{3}$$

This is the solution process of two electromagnetic fields in the all-optical scattered neural network. The structural parameter in the diffractive neural network we are concerned about is a certain layer $\vec{m}$. According to the previous work [22], we can express the original field $\vec{E}_{pri}(\vec{m})$ and adjoint field $\vec{E}_{adj}(\vec{m})$ in each diffractive layer.

$$\vec{E}_{pri}(\vec{m}) = \left(\prod_{m=1}^{M} F^{-1}P_m F\Phi_m\right)\vec{E}_s \tag{4}$$

$$\vec{E}_{adj}(\vec{m}) = \vec{E}_{pri}(\vec{m}) \otimes (I_i - y_i) \tag{5}$$

where F and F^{-1} denote the discrete Fourier transform and inverse form, P_m and Φ_m express the diagonal matrix including the light propagation from m th layer to $m+1$ th layer, and phase shifts of m th layer, respectively.

Combining the adjoint field $\vec{E}_{adj}$ with the original field $\vec{E}_{pri}$, we get the gradient of AOMLE's structural parameters.

$$\frac{\partial\Gamma}{\partial\varpi} \propto \text{Re}\left\{\vec{E}_{adj} \cdot \vec{E}_{pri}\right\} \tag{6}$$

where $\partial\Gamma/\partial\varpi$ denotes the gradient value, and ϖ here represents the structural parameter r and $\vec{m}$ corresponding scattered and diffractive neural network, respectively, and the gradient has a linear relationship with the real component of $\left\{ \vec{E}_{adj} \cdot \vec{E}_{pri} \right\}$. We can update the state $\Delta \varpi$ of phase-change meta-atoms.

$$\Delta \varpi = \varpi_{t+1} - \varpi_t \propto \frac{\partial \Gamma}{\partial \varpi} \tag{7}$$

The above process represents the training algorithm of AOMLE. It is crucial to acknowledge that training a scattered neural network requires an inverse design method rooted in photonics. Genetic algorithms, the adjoint method, generative adversarial networks, and reinforcement learning are all common methods for inverse design. For the proposed on-chip diffractive neural network, the primary modeling approach is based on the Rayleigh-Sommerfeld diffraction equation, although with substantial computational complexity. We build a unified model for all-optical scattered and diffractive neural networks in this work by solving the original and adjoint electromagnetic fields within the training domain for forward propagation and using the adjoint method for backpropagation, which enables the realization of a more efficient training algorithm. The training algorithm flow chart of AOMLE is presented in Fig. 3.

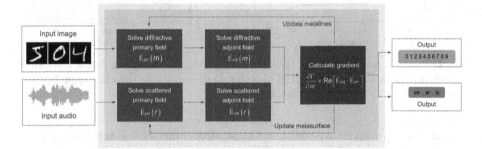

Fig. 3. Training algorithm flow chart of AOMLE

4 Multimodal Learning of AOMLE

According to the reconfigurable properties of AOMLE, we separate multimodal learning into scattered computing mode (SCM) and diffractive computing mode (DCM). In this work, we classify vowels and handwritten digital images using disctinct strategies.

In the face of SCM, we conduct vowel recognition tasks using the dataset [23]. This dataset consists of 270 audio messages from individuals of different genders and covers a variety of pronunciations, including ae, ei and ow. The training epoch of SCM is set to 30. As shown in Fig. 4(a-b), AOMLE achieves rapid convergence in the vowel recognition, with the training dataset reaching 96%, and the testing dataset likewise reaching 95.83%. Figure 4(c) shows the confusion matrix for both the training and testing dataset. Additionally, the effect of varying the length of the training region on recognition accuracy is also investigated as depicted in Fig. 4(d). By keeping the width of

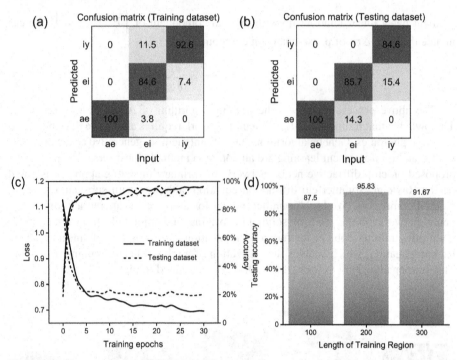

Fig. 4. Vowel recognition of SCM. (a-b) The confusion matrix of training dataset and testing dataset. (c) The loss and accuracy of SCM with training epochs. (d) The relationship between recognition accuracy and training region length.

the training area constant in AOMLE at 100, we identify that the testing dataset achieves the greatest accuracy (95.83%) when the training area length is set to 200.

We employ the classical MNIST dataset to assess its performance for DCM. Following 30 epochs of training, we achieve a recognition accuracy of 96.82% on the training dataset and 96.34% on the testing dataset. The confusion matrix is shown in Fig. 5(a-b), while Fig. 5(c) illustrates the performance of loss function and accuracy in handwritten digital classification. It is evident that AOMLE and other models have comparable accuracy. We further explore the scalability of DCM by changing the number of layers in the diffractive neural network, as presented in Fig. 5(d). By maintaining a fixed interval between metalines, we show that increasing the number of diffractive layers improves accuracy. The rate of improvement, however, becomes limited as the number of layers increases, indicating some redundancy within the neural network. We have a total of 2000 effective neurons per diffractive layer, which can be further optimized through pruning. Nevertheless, we decide to use five diffractive layers of meta lines for handwritten digital image classification, reaching a remarkable accuracy of 96.34% on the testing dataset.

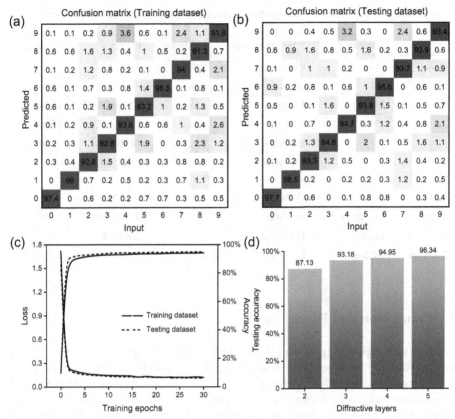

Fig. 5. Handwritten digital recognition of DCM. (a-b) The confusion matrix of training dataset and testing dataset. (c) The loss and accuracy of DCM with training epochs. (d) The scalablity of diffractive layers.

We conduct a comparative analysis between our proposed AOMLE and several processor chips used for intelligent classification tasks in the fields of natural language processing and machine vision, as shown in Table 1. The current optical processors primarily rely on devices or architectures such as Mach-Zehnder interferometer, optical diffraction, wavelength-division multiplexing, and optical scattering. Upon comparing AOMLE with other advanced optical processors, it becomes evident that AOMLE outperforms them in various indicators, including programmability, processible modality, among others. Furthermore, AOMLE achieves a remarkable increase in computing energy-efficiency, surpassing commercial electric processors by several orders of magnitude, while retaining outstanding recognition accuracy. These results highlight AOMLE's exceptional competitive edge over both optical and electric processors.

Table 1. Comparison of state-of-the-art integrated optical and electronic AI chips.

	Processor	Programmability	Modality	Energy	Latency
Optical processor	[1]	Electrical	Audios	30fJ/MAC	< 100ps
	[3]	Optical	Images	–	< 1ns
	[4]	Electrical	Images	1.58pJ/MAC	110ns
	[5]	Optical	Images	17fJ/MAC	250ps
	[7]	Electrical	Images	345fJ/MAC	< 60ps
	[9]	Electrical	Images & Videos	0.82fJ/MAC	–
	[12]	Optical	Audios	20pJ/MAC	40ps
Electronic processor	Google TPU	Electrical	–	0.43pJ/MAC	1.4ns
	Flash	Electrical	–	7fJ/MAC	15ns
Our work	AOMLE	Optical	Images & Audios	< 5fJ/MAC	< 200ps

5   Conclusion and Discussion

We propose a highly integrated all-optical multimodal learning engine called AOMLE, which effectively switches tasks based on input data modality and achieve array programming using externally modulated laser pulses. By leveraging the tunable property of phase-change materials, we successfully implement the reconfigurability of all-optical scattered and diffractive neural network on a single chip. We update the neural network's parameters using the unified form of the adjoint method, resulting in exceptional performance in both vowel recognition and handwritten digit recognition multimodal tasks.

It is worth mentioning the adjoint method has been widely employed in the inverse design of photonic devices. However, practical device fabrication often necessitates the binarization of trained medium parameters. The obtained device parameters of AOMLE can be quasi-continuous, with the level of discretization depending on the programming ability of laser pulses. This approach effectively overcomes the constraints imposed by binarization during fabrication, thereby further enhancing the computing performance of the optical processor.

Additionally, we utilize externally modulated laser pulses to program the phase-change materials, enabling precise control and alteration of the refractive index of the phase-change meta-atoms. Consequently, written laser pulses directly define the device pattern, eliminating the need for top-down lithography processes. This not only significantly enhances the flexibility of the silicon photonic device but also reduces fabrication errors and phase noise caused by lithography and etching. Collectively, these advantages demonstrate that our proposed AOMLE paves the way for more energy-efficient and flexible optical artificial intelligence processors.

References

1. Shen, Y., Harris, N., Skirlo, S., et al.: Deep learning with coherent nanophotonic circuits. Nat. Photonics **11**(7), 441–446 (2017)
2. Lin, X., Rivenson, Y., Yardimci, N.T., et al.: All-optical machine learning using diffrac-tive deep neural networks. Science **361**(6406), 1004–1008 (2018)
3. Feldmann, J., Youngblood, N., Wright, C.D., et al.: All-optical spiking neurosynaptic networks with self-learning capabilities. Nature **569**(7755), 208–214 (2019)
4. Xu, X., Tan, M., Corcoran, B., et al.: 11 TOPS photonic convolutional accelerator for optical neural networks. Nature **589**(7840), 44–51 (2021)
5. Feldmann, J., Youngblood, N., Karpov, M., et al.: Parallel convolutional processing using an integrated photonic tensor core. Nature **589**(7840), 52–58 (2021)
6. Zhang, H., Gu, M., Jiang, X.D., et al.: An optical neural chip for implementing complex-valued neural network. Nat. Commun. **12**(1), 457 (2021)
7. Ashtiani, F., Geers, A.J., Aflatouni, F.: An on-chip photonic deep neural network for image classification. Nature **606**(7914), 501–506 (2022)
8. Liu, W., Li, M., Guzzon, R., et al.: A fully reconfigurable photonic integrated signal processor. Nat. Photonics **10**(3), 190–195 (2016)
9. Zhou, T., Lin, X., Wu, J., et al.: Large-scale neuromorphic optoelectronic computing with a reconfigurable diffractive processing unit. Nat. Photonics **15**(5), 367–373 (2021)
10. Xu, Z., Tang, B., Zhang, X., et al.: Reconfigurable nonlinear photonic activation function for photonic neural network based on non-volatile opto-resistive RAM switch. Light Sci. Appl. **11**(1), 288 (2022)
11. Zhou, W., Dong, B., Farmakidis, N., et al.: In-memory photonic dot-product engine with electrically programmable weight banks. Nat. Commun. **14**(1), 2887 (2023)
12. Wu, T., Menarini, M., Gao, Z., et al.: Lithography-free reconfigurable integrated photonic processor. Nat. Photonics (2023)
13. Wang, Q., Rogers, E., Gholipour, B., et al.: Optically reconfigurable metasurfaces and photonic devices based on phase change materials. Nat. Photonics **10**(1), 60–65 (2016)
14. Zhang, Y., Fowler, C., Liang, J., et al.: Electrically reconfigurable non-volatile metasurface using low-loss optical phase-change material. Nat. Nanotechnol. **16**(8), 661–666 (2021)
15. Fang, Z., Chen, R., Zheng, J., et al.: Ultra-low-energy programmable non-volatile silicon photonics based on phase-change materials with graphene heaters. Nat. Nanotechnol. **17**(9), 842–848 (2022)
16. Chen, R., Fang, Z., Perez, C., et al.: Non-volatile electrically programmable integrated photonics with a 5-bit operation. Nat. Commun. **14**(1), 3465 (2023)
17. Wuttig, M., Bhaskaran, H., Taubner, T., et al.: Phase-change materials for non-volatile photonic applications. Nat. Photonics **11**(8), 465–476 (2017)
18. Feldmann, J., Stegmaier, M., Gruhler, N., et al.: Calculating with light using a chip-scale all-optical abacus. Nat. Commun. **8**(1), 1256 (2017)
19. Ríos, C., Youngblood, N., Cheng Z., et al.: In-memory computing on a photonic platform. Sci. Adv. **5**(11), eaau5759 (2019)
20. Wu, C., Yu, H., Lee, S., et al.: Programmable phase-change metasurfaces on waveguides for multimode photonic convolutional neural network. Nat. Commun. **12**(1), 96 (2021)
21. Hughes, T.W., Williamson, I.A.D., Minkov, M., et al.: Wave physics as an analog recurrent neural network. Sci. Adv. **5**(12), eaay6946 (2019)
22. Backer, A.S.: Computational inverse design for cascaded systems of metasurface optics. Opt. Express **27**(21), 30308–30331 (2019)
23. Hillenbrand, J., Getty, L.A., Clark, M.J., et al.: Acoustic characteristics of American English vowels. J. Acoust. Soc. Am. **97**, 3099–3111 (1995)

Brainstem Functional Parcellation Based on Spatial Connectivity Features Using Functional Magnetic Resonance Imaging

Meiyi Wang, Zuyang Liang, Cong Zhang, Yuhan Zheng, Chunqi Chang, and Jiayue Cai[✉]

School of Biomedical Engineering, Shenzhen University Medical School, Shenzhen University, Shenzhen, China
jiayuec@szu.edu.cn

Abstract. The brainstem controls almost all normal functions in the life, such as breathing, memory, movement, and is closely related to many neurological diseases. Despite the importance of the brainstem, the delineation of its functional sub-regions remains largely unexplored. In this study, we aim to explore functional parcellation of the brainstem using functional magnetic resonance imaging (fMRI), and propose a novel framework by combining spatial functional connectivity features of the brainstem and NCut spectral clustering. Firstly, functional connectivity between the brainstem and other cortical and sub-cortical brain regions is estimated using fMRI data. Secondly, the estimated spatial functional connectivity features are used to detect functional sub-regions of the brainstem using NCut spectral clustering. Finally, the Dice coefficient was used to evaluate the reproducibility of brainstem functional parcellation. The results show that the Dice coefficient obtained by the proposed method was 0.74, which is higher than that of the parcellation using temporal features of the brainstem (Dice coefficient of 0.32). In addition, NCut spectral clustering outperformed other clustering methods regarding the reproducibility of brainstem functional parcellation. The proposed method explores the potentials of spatial functional connectivity features for brainstem functional parcellation. It may serve as a promising tool for studying the functions and dysfunctions of the brainstem.

Keywords: functional parcellation · functional magnetic resonance imaging · functional connectivity · brainstem

1 Introduction

The brainstem, as an indispensable component of the central nervous system, plays a crucial role in maintaining vital functions in individuals. Many critical physiological functions, including but not limited to heartbeat, respiration, and digestion, are closely associated with the brainstem [1]. In the human body, it can be regarded as one of the most crucial organs. Current research indicates that many neurological disorders are related to brainstem dysfunctions [2]. For instance, the neuropathological changes associated with

Parkinson's disease primarily occur in the brainstem [3]. The delineation of functional sub-regions of the brainstem contributes to the understanding of functional organizations of the brainstem and its relationship with brain diseases.

While there is a collection of functional parcellation studies for the cortex and subcortex of the brain, functional parcellation of the brainstem remains largely unexplored, in spite of its importance in the understanding of brain functions and dysfunctions [4]. Currently, the sub-regions of the brainstem are mostly delineated by using anatomical principles [5, 6]. These methods rely on the morphological features and spatial information of different sub-structures [7]. One study conducted in vivo segmentation of the brainstem by using a semi-supervised approach, resulting in eleven sub-regions of the brainstem [8]. In another study, a data-driven approach was utilized to extract functional parcellations of the brainstem [9]. This approach adopted a modularity-based criterion to generate functional sub-regions of the brainstem using temporal brainstem voxel signals. While the brainstem voxel signals provide important information regarding the temporal features of the brainstem, the spatial connectivity pattern between the brainstem and other cortical and sub-cortical brain regions is another key feature to characterize the brainstem. However, such spatial features have not been taken into account for functional parcellation of the brainstem.

In this study, a novel framework was proposed for brainstem functional parcellation using fMRI, which incorporates spatial functional connectivity features. The proposed method combines the spatial functional connectivity features of the brainstem with NCut spectral clustering [10]. The functional connectivity patterns between the brainstem and other cortical and subcortical regions of the brain characterize the spatial features of the brainstem. NCut spectral clustering is used to partition the brainstem into functional sub-regions based on the similarity between spatial patterns of brainstem voxels. To the best of our knowledge, this study is the first to utilize spatial functional connectivity features for the functional parcellation of the brainstem.

2   Materials and Methods

This study proposes a new framework for brainstem functional parcellation by combining spatial functional connectivity features of the brainstem and NCut spectral clustering. Figure 1 shows the overall workflow of the proposed method. Firstly, using fMRI data, the functional connectivity between brainstem voxels and cortical as well as subcortical brain regions is estimated, thereby capturing the spatial features of the brainstem. Secondly, the correlation between spatial functional connectivity features of brainstem voxels is calculated to generate a similarity matrix between brainstem voxels. Lastly, NCut spectral clustering [10], an unsupervised machine learning algorithm, is employed to partition brainstem voxels into functional sub-regions based on the generated similarity matrix. To assess the efficacy of the proposed method, the reproducibility of functional sub-region delineation is evaluated using the Dice coefficient [11].

2.1   fMRI Dataset and Preprocessing

The data for this study was obtained from the publicly available Human Connectome Project (HCP) dataset [12]. The participants with substantial head motion parameters and

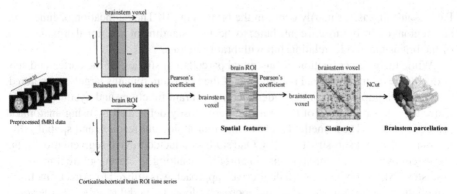

Fig. 1. The overall workflow of the proposed method for brainstem functional parcellation.

those who did not undergo repeated scanning sessions were excluded from the analysis due to concerns regarding data quality and reliability. As a result, the resting-state fMRI data from 170 healthy adult participants were included in this study.

The fMRI data was collected using a 3T magnetic field strength and a gradient-echo EPI sequence. The scanning parameters included a repetition time (TR) of 720ms, an echo time (TE) of 33.1ms, a flip angle (FA) of 52°, a field of view (FOV) of 208 × 180mm (RO × PE), a matrix size of 64 × 64, a slice thickness of 2.0mm, 72 slices, and an isotropic voxel size of 2.0mm. The data was acquired with a multiband factor of 8, an echo spacing of 0.58ms, and a bandwidth (BW) of 2290Hz/Px. Each participant underwent two fMRI scans, with each scan consisting of 1200 volumes. The total scan time for each participant was approximately 14 min [13]. In this study, the two scans are referred to as REST1 and REST2, respectively.

The preprocessing pipeline for HCP data consists of spatial and temporal processing procedures. Spatial preprocessing involves a series of steps, including spatial artifact removal, surface generation, cross-modal registration, and alignment to a standard space. On the other hand, temporal preprocessing involves the application of MELODIC ICA, where artifact and motion-related time courses are regressed out from both volumetric and grayordinate data [12]. Moreover, this approach also acts as a noise reduction technique for the signals originating from the brainstem region.

A template derived based on multimodal data [14] was used to define the regions of interest (ROIs), including both brainstem and brain cortex/subcortex. The average time series within each cortical and subcortical ROIs, as well as the voxel time series within the brainstem, were extracted for the subsequent analyses.

2.2 Spatial Functional Connectivity Features of Brainstem Voxels

In this study, we evaluated the similarity between voxels of the brainstem using their spatial features (i.e., functional connectivity patterns), rather than commonly used temporal features (i.e., original fMRI time series) [9]. Specifically, spatial functional connectivity features were assessed by correlating the signals from brainstem voxels with those from brain cortical/subcortical ROIs. These functional connectivity features captured the spatial patterns of interaction between the brainstem and other cortical and subcortical brain regions. Subsequently, the similarity matrix between the voxels of the brainstem was constructed by computing the correlation between spatial functional connectivity features of brainstem voxels.

2.3 NCut Spectral Clustering for Brainstem Functional Parcellation

2.3.1 Clustering Methods

To extract functional sub-regions of the brainstem from the similarity matrix obtained from brainstem spatial functional connectivity features, we employed NCut spectral clustering method [15]. Spectral clustering projects the data into a lower-dimensional space and then applies a clustering algorithm, such as K-means, to group the data points. By leveraging the spectral properties of the similarity matrix, this method can effectively capture the underlying structure and identify functional sub-regions of the brainstem [16].

Given an input sample set $D = (x_1, x_2, \ldots, x_n)$, the similarity matrix generation method, and the number of clusters k, the similarity matrix S is constructed based on the provided similarity matrix generation method. Subsequently, the similarity matrix undergoes an absolute value transformation. Furthermore, only the similarity values corresponding to adjacent voxels are retained, while those values between non-adjacent voxels is set to zero. This leads to the construction of the adjacency matrix W and the degree matrix D. Subsequently, the Laplacian matrix L is computed. The graph partitioning requires constructing a normalized Laplacian matrix $L_n = D^{-\frac{1}{2}} L D^{-\frac{1}{2}}$. Eigenvalues of L_n are calculated. The computed eigenvalues are arranged in ascending order, with the first k eigenvalues being selected to construct the eigenvector f. The eigenvector f is then normalized to form the $n * k$ dimensional feature matrix H. Each row in H is treated as a k-dimensional sample, yielding a total of n samples. A clustering method is then applied, typically utilizing the K-means clustering algorithm. The final outcome of this process is the cluster partition $C = (c_1, c_2, \ldots, c_k)$ [17].

2.3.2 Group-Level Analysis

After obtaining individual-level brainstem functional parcellations using NCut spectral clustering, group-level analysis is performed to derive a population-level brainstem template. Group-level analysis is a method that combines individual brainstem functional parcellations to obtain a representative brainstem template that captures shared characteristics of the group.

We employed a two-level analysis to obtain the group-level parcellation [18]. It is accomplished by constructing an adjacency matrix A of size $N * N$, where N represents the number of vertices in the brainstem. The edges in the stability graph are weighted based on the frequency of occurrence of the same parcel assignments for two vertices, v_i and v_j, across all individual subject parcellations. Once the stability graph is constructed, it is further subdivided into different regions using a graph partitioning algorithm, such as spectral clustering with normalized cuts [19], resulting in a group-level parcellation.

2.3.3 Optimal Parcellations

Determining the optimal number of functional sub-regions is an important issue. Eigen-map is one of the methods that can be used for estimating the number of clusters [21]. Eigengap is computed by analyzing the eigenvalues obtained from the spectral decomposition of a given matrix. Specifically, the eigengap is computed by calculating the absolute difference between consecutive eigenvalues. Larger eigengap typically indicates significant differences between distinct clusters and serves as a basis for selecting the appropriate number of clusters. A common approach is to identify the peaks in the eigenvalue gaps as the optimal number of clusters [22].

2.3.4 Commonly used Clustering Methods

To evaluate the performance of the proposed framework, we compared the performance of NCut spectral clustering with other commonly clustering methods, including K-means clustering, modularity and RatioCut spectral clustering. K-means is a widely used clustering algorithm that aims to partition data into k clusters by minimizing the within-cluster sum of squares [18, 23]. Modularity is a popular community detection method, which measures the density of connections within communities and compares it to a random network, aiming to maximize a modularity quality function [24]. RatioCut can be regarded as a simplification of the NCut clustering method. It disregards considerations of node degrees and connectivity and instead focuses on minimizing the number of cut edges [25].

2.4 Reproducibility of Brainstem Functional Parcellation

The Dice coefficient was employed to assess the reproducibility of brainstem functional parcellation [26, 27].

The Dice coefficient measures the overlap between two parcellation results, where a higher value indicates a greater similarity between the parcellations [28]. The calculation of the Dice coefficient is as follows:

$$Dice(A, B) = \frac{2|A \cap B|}{|A| + |B|}$$

The numerator of the Dice coefficient is twice the intersection of sets A and B, and the denominator is the sum of the lengths of sets A and B. Therefore, the Dice coefficient ranges from 0 to 1, where a higher value indicates a greater overlap and similarity between two parcellation results.

3 Experiments

We utilized this framework to extract functional subregions of the brainstem from 170 healthy subjects. Firstly, we built an individual similarity matrix for each subject using spatial functional connectivity features of the brainstem. Subsequently, NCut spectral clustering was applied to the matrix of each subject, resulting in an individual brainstem functional parcellation for each participant. Finally, a two-level group analysis method was employed to derive the group-level parcellation by integrating the individual brain-stem functional parcellation results. The eigengap method was used to select the optimal number of clusters, and the range of the number of clusters was set to 15–25 by refer-ring to the previous literature [20]. To evaluate the proposed method, we used the Dice coefficient between the two scans, REST1 and REST2, to measure the reproducibility of brainstem functional parcellation. Higher reproducibility indicates better parcellation performance.

4 Results

We computed the eigen-gap values within the range of 15 to 25 subregions. The eigen-gap indicates significant transitions in the eigenvalues. As shown in Fig. 2, there is a substantial jump at 21 clusters, leading to the selection of 21 sub-regions as the optimal division for functional parcellation of the brainstem.

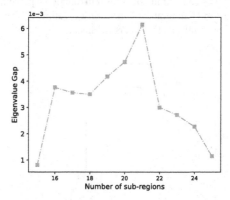

Fig. 2. Eigen-gap analysis for functional parcellation of the brainstem.

To validate the proposed method, we compared different clustering methods and features with the number of clusters ranging from 15 to 25. Specifically, we first compared the reproducibility between the proposed spatial functional connectivity features and traditional temporal features, and then compared different clustering methods, including NCut, RatioCut, K-means and modularity.

Figure 3 shows the Dice coefficient of brainstem functional parcellation using the proposed method with spatial functional connectivity features, compared to that using temporal fMRI time series features. The boxplot was drawn based on the Dice coefficients across different number of clusters. It can be seen that the brainstem parcellation based on spatial connectivity features outperformed that based on temporal features in terms of the Dice coefficient. This trend was consistent throughout the number of clusters ranging from15 to 25.

For the optimal parcellation with 21 brainstem subregions, the Dice coefficient using the proposed method with spatial functional connectivity features was 0.74. Such per-formance was higher than that obtained using the fMRI temporal time series features, which yielded a Dice coefficient of 0.32.

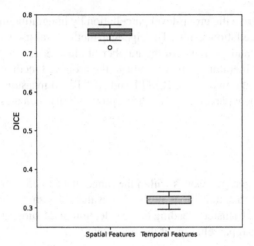

Fig. 3. Comparisons of the reproducibility of brainstem functional parcellation using spatial functional connectivity features and temporal fMRI time series features. The boxplot was drawn based on the Dice coefficients across different numbers of clusters.

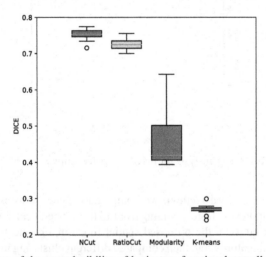

Fig. 4. Comparisons of the reproducibility of brainstem functional parcellation using different clustering methods. The boxplot was drawn based on the Dice coefficients across different numbers of clusters.

The reproducibility of different clustering methods using spatial functional connectivity features of the brainstem were compared in Fig. 4. The boxplot was drawn based on the Dice coefficients across different numbers of clusters. The results showed that the NCut method achieved higher Dice coefficients compared to other methods. The consistent trend was observed across different numbers of clusters ranging from 15 to 25.

When the brainstem was divided into 21 subregions, the results showed that NCut spectral clustering exhibited the best reproducibility with a Dice coefficient of 0.74, outperforming that of RatioCut (a Dice coefficient of 0.69), modularity (a Dice coefficient of 0.40), and K-means (a Dice coefficient of 0.27).

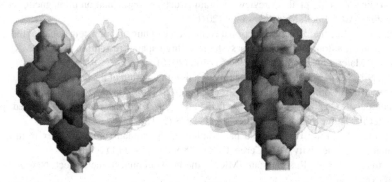

Fig. 5. Visualization of functional sub-regions of the brainstem obtained using the proposed method.

Finally, the group-level brainstem functional parcellation obtained using the proposed method is shown in Fig. 5. The brainstem is partitioned into 21 functional sub-regions.

5 Conclusions

This study proposed a novel framework for brainstem functional parcellation by combining the spatial functional connectivity features of the brainstem and NCut spectral clustering. The proposed method exhibited the best reproducibility in terms of Dice coefficient, outperforming that achieved by using temporal fMRI time series features and other clustering methods. The results demonstrated the advantage of spatial functional connectivity features of the brainstem over temporal fMRI time series features and the superiority of NCut spectral clustering. The current study provides the delineation of functional sub-regions of the brainstem, benefiting the exploration of functional organizations of the brainstem and brainstem pathology of brain diseases.

References

1. Arber, S., Costa, R.M.: Networking brainstem and basal ganglia circuits for movement. Nat. Rev. Neurosci. **23**(6), 342–360 (2022)
2. Benghanem, S., et al.: Brainstem dysfunction in critically ill patients. Crit. Care **24**, 1–14 (2020)
3. Grinberg, L.T., et al.: Brainstem pathology and non-motor symptoms in PD. J. Neurol. Sci. **289**(1–2), 81–88 (2010)
4. Bijsterbosch, J., et al.: Challenges and future directions for representations of functional brain organization. Nat. Neurosci. **23**(12), 1484–1495 (2020)

5. Iglesias, J.E., et al.: Bayesian segmentation of brainstem structures in MRI. Neuroimage **113**, 184–195 (2015)
6. Sander, L., et al.: Accurate, rapid and reliable, fully automated MRI brainstem segmentation for application in multiple sclerosis and neurodegenerative diseases. Hum. Brain Mapp. **40**(14), 4091–4104 (2019)
7. González-Villà, S., et al.: A review on brain structures segmentation in magnetic resonance imaging. Artif. Intell. Med. **73**, 45–69 (2016)
8. Bianciardi, M., et al.: In vivo functional connectome of human brainstem nuclei of the ascending arousal, autonomic, and motor systems by high spatial resolution 7-Tesla fMRI. Magn. Reson. Mater. Phys., Biol. Med. **29**, 451–462 (2016)
9. Haq, N.F., et al.: Connectivity based functional segmentation of the brainstem. In: 2021 IEEE International Conference on Image Processing (ICIP), pp. 51–55 (2021)
10. Craddock, R.C., et al.: A whole brain fMRI atlas generated via spatially constrained spectral clustering. Hum. Brain Mapp. **33**(8), 1914–1928 (2012)
11. Schaefer, A., et al.: Local-global parcellation of the human cerebral cortex from intrinsic functional connectivity MRI. Cereb. Cortex **28**(9), 3095–3114 (2018)
12. Smith, S.M., et al.: Resting-state fMRI in the human connectome project. Neuroimage **80**, 144–168 (2013)
13. Glasser, M.F., et al.: The minimal preprocessing pipelines for the human connectome project. Neuroimage **80**, 105–124 (2013)
14. Glasser, M.F., et al.: A multi-modal parcellation of human cerebral cortex. Nature **536**(7615), 171–178 (2016)
15. Zhang, C., et al.: Image segmentation based on multiscale fast spectral clustering. Multimedia Tools Appl. **80**, 24969–24994 (2021)
16. Shen, X., Papademetris, X., Constable, R.T.: Graph-theory based parcellation of functional subunits in the brain from resting-state fMRI data. Neuroimage **50**(3), 1027–1035 (2010)
17. Shi, J., Malik, J.: Normalized cuts and image segmentation. IEEE Trans. Pattern Anal. Mach. Intell. **22**(8), 888–905 (2000)
18. Arslan, S., et al.: Human brain mapping: A systematic comparison of parcellation methods for the human cerebral cortex. Neuroimage **170**, 5–30 (2018)
19. Van Den Heuvel, M., Mandl, R., Hulshoff Pol, H.: Normalized cut group clustering of resting-state FMRI data. PloS one **3**(4), e2001 (2008)
20. Tang, Y., et al.: A probabilistic atlas of human brainstem pathways based on connectome imaging data. Neuroimage **169**, 227–239 (2018)
21. Dalmaijer, E.S., Nord, C.L., Astle, D.E.: Statistical power for cluster analysis. BMC Bioinform. **23**(1), 1–28 (2022)
22. Tibshirani, R., Walther, G., Hastie, T.: Estimating the number of clusters in a data set via the gap statistic. J. Royal Stat. Soc.: Ser. B (Stat. Methodol.) **63**(2), 411–423 (2001)
23. Zhou, G., et al.: Characterizing functional pathways of the human olfactory system. Elife **8**, e47177 (2019)
24. Zelditch, M.L., Goswami, A.: What does modularity mean? Evol. Dev. **23**(5), 377–403 (2021)
25. Wang, S., Siskind, J.M.: Image segmentation with ratio cut. IEEE Trans. Pattern Anal. Mach. Intell. **25**(6), 675–690 (2003)
26. Moghimi, P., et al.: Evaluation of functional MRI-based human brain parcellation: a review. J. Neurophysiol. **128**(1), 197–217 (2022)
27. Liu, Y., et al.: Understanding of internal clustering validation measures. In: 2010 IEEE International Conference on Data Mining. IEEE (2010)
28. Garcia-Garcia, M., et al.: Detecting stable individual differences in the functional organization of the human basal ganglia. Neuroimage **170**, 68–82 (2018)

Other AI Related Topics

A Reinforcement Learning Approach for Personalized Diversity in Feeds Recommendation

Li He[1], Kangqi Luo[2], Zhuoye Ding[2], Hang Shao[3], and Bing Bai[1(✉)]

[1] Tsinghua University, Beijing, China
baibing12321@163.com
[2] JD.com, Beijing, China
[3] Zhejiang Future Technology Institute, Jiaxing, China

Abstract. Feeds recommendation has been widely used in various applications, such as e-commerce site, where users can constantly browse products generated by never-ending feeds. It's important to not only consider instant metrics but also pay more attention to long-term user engagement. In this paper, we focus on optimizing user browsing depth, which represents users' willingness to stay within the e-commerce feed streams. By analyzing the ranking and re-ranking stages, we find that the re-ranking stage is a suitable phase for maximizing user browsing depth. First, we evaluate the current status of our used re-ranking module and identify that the fixed diversity rule neglects unique propensity to the degree of diversity in each user request. Hence there is a need to personalize diversity in the granularity of user requests. Then, we note that the personalized diversity process of user request granularity can be modelled as a Markov decision process (MDP). Finally, by solving three issues of MDP elements design, acquisition of interaction data, off-policy learning and policy selection, we propose a *Personalized Diversity Reranking Model in the granularity of user request (PDRM-request)* based on reinforcement learning. We conduct offline experiments and deploy the PDRM-request model in a live e-commerce site to perform A/B testing. The results show that the our approach achieves deeper user browsing depth and more diversified recommended lists than the existing baseline.

Keywords: Reinforcement learning · Personalized diversity · E-commerce recommendations

1 Introduction

Users can browse endless products if they want in e-commerce feeds recommendation. Apart from instant feedback, such as clicks and purchases, it is important

Supplementary Information The online version contains supplementary material available at https://doi.org/10.1007/978-981-99-9119-8_42.

for feeds streams to consider long-term user engagement [17], including user browsing depth, revisits, and dwell time. This work focuses on user browsing depth, which is the number of products that a user browses in a single day and it reflects a user's tendency to kill time and hang out on the e-commerce site.

In order to optimize the user browsing depth, we review the ranking and re-ranking modules. The ranking stage aims at improving relevance through supervised learning, such as click-through rate or conversion rate [18,22], which is not suitable for optimizing long-term user engagement. Conversely, the re-ranking stage is usually the last part of a recommender system and is the closest to the users. By considering the mutual influences between the recommended items, the re-ranking module controls the trade-off between relevance and diversity of the final recommended list, which directly affects the long-term user engagement. Thus, we analyze our current re-ranking module and observe that it employs a greedy search method to select products incrementally one by one in each user request. To maintain the diversity of recommendations and circumvent bad user experiences, a fixed diversity rule in the sense of product category is adopted to ensure that similar products are not too close to each other in the final list. However, we note that this fixed diversity rule does not account for the fact that each user's request may have a unique tendency towards diversity. Motivated by this observation, we argue that it is more reasonable to personalize the relevance-diversity trade-off in the granularity of user requests.

In essence, the personalized diversity process of a user request can be formulated as a MDP [24]. We treat the user's request as a state and the diversity rule as an action, where the immediate reward is the number of products the user browses in each request. In this way, we can optimize the policy of selecting diversity rule, so as to maximize the cumulative user browsing reward within a session. Therefore, we not only possess the ability of personalized diversity in the granularity of user requests, which can meet the personalized and real-time diversity demands of users, but also improve the long-term user participation, i.e., user browsing depth. Nevertheless, achieving this personalized diversity policy presents three challenges. First, designing the MDP elements for the personalized diversity process is difficult. Second, interaction data acquisition for policy learning should be solved under the current fixed diversity rule situation. Third, learning and evaluating a policy model is also not easy.

To address these challenges, we first describe our approach in Sect. 3.2 to the MDP element design. For data collection, since the diversity rule in our greedy-based re-ranking is fixed, which always selects the same diversity rule for all user requests, the data generated by it cannot be used for policy learning. We use a policy which follows a uniform distribution over the action space as a behavior policy. Our online performance during data collection on a small portion of real traffic doesn't degrade, thanks to designing balanced diversity rules and adopting a uniform distribution behavior policy. To handle the third challenge, we use off-policy REINFORCE algorithm [9] for off-policy learning and importance sampling technique for off-policy evaluation [25]. We conduct offline experiment to learn and select a good personalized diversity policy network.

We call the *Personalized Diversity Re-ranking Model in the granularity of user request* as *PDRM-request* model. Online A/B testing show significant improvements in both the user browsing depth averaged by users and the number of distinct categories of browsed items averaged by users over the baseline.

2 Background and Motivation

In the context of feed streaming, it is crucial to prioritize long-term user engagement [36], which is exemplified by metrics such as user browsing depth and dwell time. These metrics demonstrate not only that users can locate items of interest, but also that they are willing to return to and spend time on the feed stream. Our work focuses specifically on the user browsing depth, which refers to the number of products browsed by a user within a single day. A higher browsing depth indicates greater user satisfaction, as it suggests that the user enjoys scrolling through and exploring the e-commerce portal.

2.1 Greedy-Based Re-ranking

We study the configurations of our existing ranking and re-ranking stages, and try to find a breakthrough point for maximizing user browsing depth. The ranking stage focuses on instant metrics, such as click-through rate or conversion rate, and predicts the relevance of each user-item pair based on a scoring function that is learned from labeled data. It is well-known that it may be sub-optimal to directly display top items to users as the scoring function applies to each item individually and overlooks mutual influences between items. Instead, we aim to provide users with relevant and diverse results that offer more choices and improve their experience. The re-ranking stage, which controls the relevance-diversity trade-off of the recommended list, is an appropriate phase to achieve the maximization of user browsing depth. Thus, we analyze the current status of our re-ranking module, which takes n items with point-wise relevance scores as inputs and generates a recommended list of k items for each user request. Please refer to the supplementary[1] for more detailed descriptions.

2.2 Personalized Diversity in User Request Granularity

As mentioned above, the given diversity rule is a critical component for avoiding bad user experience. However, the diversity rule we applied is fixed, which does not account for the fact that each user's request may have a unique propensity for diversity. On one hand, diversity for recommender system should be user-specific, depending on the user's long-term preferences [27]. Users with narrow taste may be more tolerant of having more similar products in the recommended list, while those with wider interests may expect more diverse products. On the other hand, even for the same user, his (her) intentions may differ in different requests. For

---

[1] https://github.com/heli223/PDRM-request-appendix/.

example, when a user clicks many products in the same category, it may indicate that (s)he is very interested in this category at this time. Hence, it is fine to show similar products in the recommended list. When (s)he clicks on a wide range of products, (s)he may not have obvious purchasing intention, and the products in the recommended list can be more diverse. Inspired from this observation, it is more reasonable to apply the personalized diversity in the granularity of user requests. Assigning an appropriate diversity rule to each request will make the person who wants to browse be willing to browse more and make the person who wants to click continue to click. Nevertheless, industrial recommender system is a very complicated framework. Considering the convenience and compatibility of the implementation, we still follow the greedy-based re-ranking approach while achieving the personalized diversity in the granularity of user request. Although the list retrieval system (i.e., list generation and list evaluation) [26, 28] is popular recently, we do not apply it as it is not compatible with our used components. Back to the personalized diversity and our focus (i.e., user browsing depth), we found that the personalized diversity process of user request granularity can be modelled as a Markov decision process (MDP) to help optimize user browsing depth. Therefore, the personalized diversity policy can choose the diversity rule with a small immediate reward but make big contribution to the rewards for future user requests. In this way, we not only own the ability of personalized diversity in the granularity of user requests, but also improve the user browsing depth.

3   The Proposed Approach

There are indeed challenges in the implementation of personalized diversity policy based on reinforcement learning. The first challenge is to design appropriate reinforcement learning elements for the personalized diversity process. The second is related to the fixed diversity rule, which makes it difficult to collect interaction data for policy learning. The third challenge is off-policy learning and evaluation, which is not straightforward.

3.1   Overview of the Framework

The personalized diversity re-ranking model (*PDRM-request*) framework in Fig. 1 consists of three main parts: a point-wise learning to rank (LTR) model, a personalized diversity policy network which works in parallel with LTR, and a re-rank module. The framework works as follows. When a user request comes, the LTR model assigns a relevance score to each candidate item generated by the recall phase. Meanwhile, the personalized diversity policy network selects a diversity rule from a pre-defined set of rules based on the information extracted from the user request. Once the above both steps are completed, the re-rank module applies the selected diversity rule to the candidate items with ranking scores and outputs the recommended list for the user request.

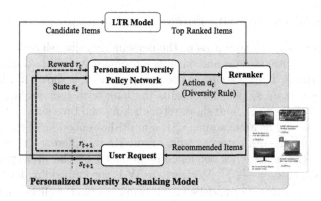

Fig. 1. The framework of the personalized diversity re-ranking model.

3.2 MDP Formulation of Personalized Diversity Process

The personalized diversity policy network observes information from the user request, takes a diversity rule from the pre-defined diversity rule set, and receives immediate feedback from the users. To formalize the sequential interactions between the user requests and the personalized diversity re-ranking model for e-commerce feeds recommendation, we represent it as a Markov decision process (MDP). A MDP is usually defined by a tuple $(\mathcal{S}, \mathcal{A}, R, \mathbf{P}, \gamma)$, where $\mathcal{S}$ is state space, $\mathcal{A}$ is action space, $R : \mathcal{S} \times \mathcal{A} \to \mathbb{R}$ is the mean reward function which takes values in the real number space $\mathbb{R}$, and $\mathbf{P} : \mathcal{S} \times \mathcal{A} \times \mathcal{S} \to \mathbb{R}$ is the state transition probability. The discount factor $\gamma \in [0, 1)$ is used to measure the present value of future reward. The policy is defined as $\pi : \mathcal{S} \times \mathcal{A} \to [0, 1]$, which represents the distribution over actions for any state $s \in \mathcal{S}$. $a \in \mathcal{A}$ has a probability $\pi(a|s)$.

For our personalized diversity re-ranking model, we devise the MDP elements as follows. **State space $\mathcal{S}$**: A set of states represented by a tuple at time step t, denoted as $s_t = (u, req_t, c_t, e_t)$. Here u denotes the user information, req_t denotes contextual information of the user request, c_t represents items those the user clicks and e_t represents the items previously exposed to the user before step t. **Action space $\mathcal{A}$**: A finite set of pre-defined diversity rules, where an action represents a candidate diversity rule available in the re-ranking stage. **Reward R**: The mean reward function which takes values in the real number space $\mathbb{R}$. The immediate reward obtained by taking action a at state s is denoted as $r(s, a)$ and it is measured by the number of products that a user browses in each user request to optimize the cumulative users browsing depth. **Transition probability $\mathbf{P}$**: The probability of the next state s_{t+1} given the current state s_t and action a_t is represented as $p(s_{t+1}|s_t, a_t)$. The objective of the agent is to seek a policy that maximizes the expected discounted cumulative rewards $\max_\pi J(\pi) = \max_\pi E_{\tau \sim \pi}[R(\tau)]$, where $R(\tau) = \sum_{t=0}^{|\tau|} \gamma^t r(s_t, a_t)$. The expectation is taken over the trajectories τ obtained by following the policy: $s_0 \sim \rho_0$ is the initial state, $a_t \sim \pi(\cdot|s_t)$, and $s_{t+1} \sim p(\cdot|s_t, a_t)$. For specific elements design, please refer to appendix.

3.3   Personalized Diversity Policy

We use a policy network to represent the personalized diversity policy $\pi_\theta(a_t|s_t)$, which models the user state s_t and outputs a probability distribution over actions in the action space. The structure of the policy network is shown in Fig. 2. It first models the user state $\mathbf{s}_t$, which contains the long-term user profile, contextual information of the request, and short-term clicking and browsing history. Taking the click sequence c_t as an example, $\mathbf{c}_t^i$ is a multi-hot encoding of the i-th clicked item, which includes its category and brand ID in different granularity. The embedding of the whole click sequence is modelled by the average embedding over all items, i.e., $\mathbf{c}_t = \frac{1}{|c_t|} \sum_i \mathbf{W}\mathbf{c}_t^i$, where $\mathbf{W}$ is the embedding lookup matrix. The exposure sequence e_t is modelled in the same way as the click sequence. Then the user state embedding is formed by concatenating the user profiles, contextual information, and short-term behavior features: $\mathbf{s}_t = \mathbf{u} \oplus \mathbf{req}_t \oplus \mathbf{c}_t \oplus \mathbf{e}_t$, where $\oplus$ denotes a vector concatenation. Given the state embedding $\mathbf{s}$, we implement a two layer perceptrons to learn the high-order feature representation $\mathbf{h}$. Finally, the policy $\pi_\theta(a|s)$ is modelled by a softmax layer over all actions.

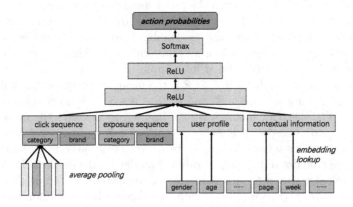

Fig. 2. The structure of the policy network.

3.4   Off-Policy Learning

Given the explicit representation of $\pi_\theta(a|s)$, policy optimization methods are available to optimize the long-term rewards $J(\pi_\theta)$. We focus on REINFORCE algorithm [30], which optimizes the parameters θ directly by using the gradient ascent algorithm. Thanks to the log-trick, the gradient of $J(\pi_\theta)$ with respect to the policy parameters can be approximated with the formula [9,28]: $\nabla_\theta J(\pi_\theta) = \mathbb{E}_{s_t \sim d_t^\pi(\cdot), a_t \sim \pi_\theta(\cdot|s_t)} [R(s_t, a_t) \nabla_\theta \log \pi_\theta(a_t|s_t)]$, where $d_t^\pi(\cdot)$ denotes the state distribution of π_θ at step t, and $R(s_t, a_t) = \sum_{t'=t}^{|\tau|} \gamma^{t'-t} r(s_{t'}, a_{t'})$ denotes the discounted future reward from the t-th step. The vanilla REINFORCE algorithm is almost performed on-policy, which means that π_θ is deployed and interacts with users in real time and each policy update only utilizes user feedback collected while acting according to the most recent version of the policy. However, training the policy online and generating trajectories immediately is not desirable in recommender systems due to cost, risk, and infrastructure constraints, as unsatisfactory policies may harm online performance. Therefore, it is more crucial to apply off-policy learning in e-commerce instead. An alternative approach is to train the policy network using logged feedback data collected by a behavior policy β, which is different from the target policy π_θ. Since the model is trained offline, the online performance is not affected by the trial-and-error online learning process.

Selections of Behavior Policy and Action Space. To achieve off-policy learning in the personalized diversity re-ranking task, the first step is to collect logged feedback data using a proper behavior policy with a proper action space. However, in our applied greedy-based re-ranking approach, the diversity rule is fixed and always selects the same diversity rule for all requests. Hence, the data generated by it cannot be used for off-policy learning. To address this, we use an uniform distribution as the behavior policy β, where actions are equally selected from a pre-defined action space $\mathcal{A}$ for every user request. To construct candidate action spaces, we follow the guidelines in Sect. 3.2 and create several candidate rule sets as action space candidates, denoted by $\mathcal{A}^{(1)}, \mathcal{A}^{(2)}$, and so on. The main differences between these candidates lie in the number of actions, and the highest (or lowest) degree of the diversity rule in each action space. The selection of action space $\mathcal{A}$ is crucial because a better action space with its uniform distribution policy leads to a better starting point of policy optimization. Thus, we deploy different diversity rule sets along with their uniform distribution policies on multiple small slices of traffic for A/B testing. The diversity rule set and its uniform distribution policy which has the best online A/B testing performance is selected as the final action space and the final behavior policy for data collection. Thanks to the benefits of constructing a balanced candidate diversity rule set and utilizing an uniform distribution policy, the optimal action space $\mathcal{A}$ and its uniform distribution policy β show no significant A/B testing performance degradation compared to the greedy-based re-ranking approach described in Sect. 2.1. With the selected action space and behavior policy, we can safely collect logged feedback data for a long period of time to conduct off-policy learning.

Off-Policy Policy Gradient. With the provided behavior policy, we can obtain an approximation of the policy gradient $\nabla_\theta J(\pi_\theta)$ based on the logged data collected by policy β. The intuition behind off-policy policy gradient estimation is to circumvent the distribution mismatch between the target policy π_θ and the behavior policy β. To do this, we use importance weighting [20, 23] to derive an approximation of the off-policy REINFORCE gradient. Please refer to the supplementary for detailed derivation.

4 Experimental Results

4.1 Offline Experiments

To answer the first research question of how to learn and select a good personalized diversity policy network, we conduct offline experiments.

Data Set. We design several diversity rule sets as action space candidates. We deploy those candidate diversity rule sets by using their uniform distribution policies on several small portions of traffic. Then, the diversity rule set together with its uniform distribution policy which has superior online A/B testing result is used as the action space and the behavior policy to collect the interaction data. The logged data consists of many user browsing sessions, where each session contains a complete interaction history $\{(s_1, a_1, r_1), \cdots, (s_T, a_T, r_T)\}$, starting from the first user request until the user leaves the recommender system at step T. When a user's request arrives, the online service collects the user state s_t, and performs the personalized diversity re-ranking according to action a_t sampled from the behavior policy, i.e., $a_t \sim \beta(\cdot|s_t)$. We also record the user's immediate reward, which is the number of browsed items of the request. For our offline experiment, we collect 16 consecutive days of user interaction data from a real-world e-commerce platform. To make the experimental results be reliable, the training data and testing data are separated in time order. The first 12 days are used for training, and the last 4 days are for testing.

Baseline. We denote our personalized diversity re-ranking model in the granularity of user request as the **PDRM-request** model. We compare its performance with an offline baseline called **BASE-uniform**, which is based on the greedy-based re-ranking approach described in Sect. 2.1. BASE-uniform uses the uniform distribution behavior policy β and the selected action space, which is the one that performs best during online A/B testing illustrated in Sect. 4.1.

Metrics. To perform policy model selection, we have to evaluate the performance of the policy model $\pi_\theta(a_t|s_t)$ when it is deployed online. In the reinforcement learning setting, we only have access to the log data from the behavior policy β, but we have to exploit them to give a evaluation of π_θ. Following [11,25], we use importance sampling technique to correct the distribution mismatch between the models β and π_θ. For the t-th user request, we denote its discounted future exposure reward under the behavior policy β as $R^\beta(s_t, a_t) = \sum_{t'=t}^{|\tau|} \gamma^{t'-t} r^\beta(s_{t'}, a_{t'})$. We use $R^{\pi_\theta}(s_t, a_t) = \frac{\pi_\theta(a_t|s_t)}{\beta(a_t|s_t)} R^\beta(s_t, a_t)$ to estimate the performance of $\pi_\theta(a_t|s_t)$ if it is deployed for the t-th user request instead. Then, we sum the discounted future exposure rewards of all user requests in the test data set for policy π_θ and behavior policy β, respectively. We denote the two sums as R^{π_θ} and R^β, respectively. We then calculate their relative improvement percentage, denoted as *ExposureGain*, namely,

$ExposureGain = \frac{R^{\pi_\theta}}{R^\beta} - 1$, which is used as the offline evaluation metric. A higher value of $ExposureGain$ indicates better model performance. We select the policy network with the highest $ExposureGain$ value as the optimal policy network.

Experimental Settings. The policy is implemented as a multi-layer neural network with an embedding dimension of 5 and two fully connected layers with 256 and 128 hidden units, respectively. As mentioned in Sect. 3.4, we use the off-policy REINFORCE algorithm based on the work in [9]. During training, we use the primitive function of the REINFORCE gradient as the objective function. To update the policy, we use the Adam [16] optimizer with a batch size of 10000 and learning rate of 0.001. The discount factor γ is set to 0.7 for experiment. All experiments are implemented on TensorFlow[2].

Offline Results. We report the best offline evaluation result of the testing data in Table 1, where $BASE - uniform$ is set as the base. We can observe that $PDRM - request$ outperforms the baseline $BASE - uniform$. The result suggests that our personalized diversity policy model is effective, as it enables personalized recommendations for users, leading to improved performance.

Table 1. The results of offline evaluation.

Model	Exposure Gain
$BASE - uniform$	+0.0%
$PDRM - request$	+0.0648%

4.2 Online Experiments

We focus on the second research question: how well our proposed personalized diversity re-ranking model performs when it is deployed online.

Baseline. We deploy our model $PDRM\text{-}request$ at a real-world e-commerce site to verify its effectiveness. We point that in order to ensure the online performance, we keep the KL-divergence between the deployed policy π_θ and the uniform distribution β under control. Therefore, we adopt a smoothing operation over the final action distribution, i.e., softmax operation[3]. For online comparison, we compare it with the **BASE-fixed** approach, which is the greedy-based re-ranking described in Sect. 2.1 and applies the given fixed diversity rule.

---

[2] https://www.tensorflow.org/.

[3] Other operations, such as epsilon-greedy or PPO algorithm, can be used to control the distance, which is not the focus of this paper, so we do not discuss it in detail.

Online Metrics. We utilize three online metrics, including the user browsing depth averaged by users (UBD, for short), the number of distinct categories of browsed items averaged by users (UNOC), and the number of clicks averaged by users (UCTR, for short). These metrics assess the willingness of users to click and browse, as well as the diversity of the recommended items. A higher value of the above metric indicates a better recommendation performance.

Online Results. We present the online experimental results in Table 2, where we only report the relative improvement of *PDRM-request* over the *BASE-fixed* model, in consideration of commercial concerns. The online A/B testing was conducted for half a month, and the results show that the *PDRM-request* model has statistically significant improvements over the *BASE-fixed* model, with a +0.50% increase in the UBD metric and a +2.52% improvement in the UNOC metric. Despite a slight decrease in the UCTR metric, this decrease is not statistically significant. Overall, our proposed model enhances long-term user engagement with no significant loss of clicks, and produces a more diversified recommendation result that encourages exploration and alleviates the Matthew Effect in recommender systems. For more details of an attempt on reward design, limitations, and future work related to our paper, please refer to the supplementary material due to space constraints.

Table 2. The results of online A/B testing.

Model	UBD	UNOC	UCTR
BASE − fixed	+0.0%	+0.0%	+0.0%
PDRM − request	**+0.50% (p = 0.0)**	**+2.52% (p = 0.0)**	−0.22% (p = 0.21)

5   Related Work

The first category related to this paper is reinforcement learning (RL) for recommendations. Value-based approaches like Q-learning [19], and policy-based ones such as policy gradients [30] are two classical approaches to solve RL problems [24]. Applying RL on recommendation and searching tasks has been a hot research topic recently. Some examples [6,9,15,31–34,36] include DQN-based RL framework for online news recommendation and policy gradient-based top-K recommender system for video recommendation.

The second category is re-ranking methods. One class of re-ranking approaches use multiple point-wise ranked items as inputs, and output the refined scores, by modelling the complex dependencies between items in different ways [3,4,7,35]. These include ideas such as the listwise context model [3] and groupwise scoring functions [4], etc. To solve the challenges in re-ranking, there is another popular class of re-ranking approaches, *i.e.*, list generation and list evaluation, which is also called as list recall and list ranking [26,28]. Diversity management also can be treated as a re-ranking model. More and more studies have tried to balance instant metrics and diversity in result lists to enhance user satisfaction [1,2,5,8,10,13,14,29]. For example, the work [1] describes their journey in tackling the problem of diversity for Airbnb search. Moreover, some researchers proposed to utilize users' behaviors for personalized diversified recommendation [12,21,27].

6   Conclusion

We propose a personalized diversity re-ranking model called *PDRM-request*, which is designed to improve user browsing depth in feeds recommendation. It can be easily deployed as a follow-up component after any ranking module. Unlike existing models that consider user or user clusters as the granularity for personalized diversity, this model focuses on the granularity of user requests, taking into account the fact that different requests have different tendencies towards diversity. The personalized diversity process is formulated as a Markov decision process, and an off-policy reinforcement learning approach is employed to learn the optimal personalized diversity policy that maximizes long-term user engagement. Offline experiments are carried out on real-world data collected from an e-commerce site to learn and select a good policy. The PDRM-request model is deployed in a live e-commerce portal. Online A/B testing results show that our model is effective in improving user browsing depth and UNOC metrics in feeds recommendation, demonstrating its usefulness in practical applications.

Acknowledgements. This work is supported in part by Provincial Key R&D Program of Zhejiang under contract No. 2021C01016, in part by Young Elite Scientists Sponsorship Program by CAST under contract No. 2022QNRC001.

References

1. Abdool, M., et al.: Managing diversity in Airbnb search. In: SIGKDD, pp. 2952–2960. ACM (2020)
2. Adomavicius, G., Kwon, Y.: Improving aggregate recommendation diversity using ranking-based techniques. IEEE Trans. Knowl. Data Eng. **24**(5), 896–911 (2012)
3. Ai, Q., Bi, K., Guo, J., Croft, W.B.: Learning a deep listwise context model for ranking refinement. In: SIGIR, pp. 135–144. ACM (2018)
4. Ai, Q., Wang, X., Bruch, S., Golbandi, N., Bendersky, M., Najork, M.: Learning groupwise multivariate scoring functions using deep neural networks. In: SIGIR, pp. 85–92. ACM (2019)
5. Ashkan, A., Kveton, B., Berkovsky, S., Wen, Z.: Optimal greedy diversity for recommendation. In: IJCAI, pp. 1742–1748 (2015)
6. Bai, X., Guan, J., Wang, H.: A model-based reinforcement learning with adversarial training for online recommendation. In: NeurIPS, pp. 10734–10745 (2019)
7. Bello, I., et al.: Seq2Slate: re-ranking and slate optimization with RNNs. CoRR abs/1810.02019 (2018)
8. Chen, L., Zhang, G., Zhou, E.: Fast greedy MAP inference for determinantal point process to improve recommendation diversity. In: NeurIPS, pp. 5627–5638 (2018)

9. Chen, M., Beutel, A., Covington, P., Jain, S., Belletti, F., Chi, E.H.: Top-K off-policy correction for a REINFORCE recommender system. In: WSDM, pp. 456–464. ACM (2019)

10. Cheng, P., Wang, S., Ma, J., Sun, J., Xiong, H.: Learning to recommend accurate and diverse items. In: WWW, pp. 183–192. ACM (2017)

11. Cortes, C., Mansour, Y., Mohri, M.: Learning bounds for importance weighting. In: NeurIPS, pp. 442–450. Curran Associates, Inc. (2010)

12. Eskandanian, F., Mobasher, B., Burke, R.: A clustering approach for personalizing diversity in collaborative recommender systems. In: UMAP, pp. 280–284. ACM (2017)

13. Gelada, C., Kumar, S., Buckman, J., Nachum, O., Bellemare, M.G.: DeepMDP: learning continuous latent space models for representation learning. In: ICML, vol. 97, pp. 2170–2179. PMLR (2019)

14. Gogna, A., Majumdar, A.: Balancing accuracy and diversity in recommendations using matrix completion framework. Knowl. Based Syst. **125**, 83–95 (2017)

15. Gong, Y., et al.: Exact-K recommendation via maximal clique optimization. In: SIGKDD, pp. 617–626. ACM (2019)

16. Kingma, D.P., Ba, J.: Adam: a method for stochastic optimization. In: ICLR (2015)

17. Lalmas, M., O'Brien, H., Yom-Tov, E.: Measuring User Engagement. Synthesis Lectures on Information Concepts, Retrieval, and Services. Morgan & Claypool Publishers (2014)

18. Li, J., Ren, P., Chen, Z., Ren, Z., Lian, T., Ma, J.: Neural attentive session-based recommendation. In: CIKM, pp. 1419–1428. ACM (2017)

19. Mnih, V., et al.: Playing atari with deep reinforcement learning. CoRR abs/1312.5602 (2013)

20. Munos, R., Stepleton, T., Harutyunyan, A., Bellemare, M.G.: Safe and efficient off-policy reinforcement learning. arXiv preprint arXiv:1606.02647 (2016)

21. Noia, T.D., Ostuni, V.C., Rosati, J., Tomeo, P., Sciascio, E.D.: An analysis of users' propensity toward diversity in recommendations. In: RecSys, pp. 285–288. ACM (2014)

22. Pradel, B., et al.: A case study in a recommender system based on purchase data. In: SIGKDD, pp. 377–385. ACM (2011)

23. Precup, D., Sutton, R.S., Dasgupta, S.: Off-policy temporal-difference learning with function approximation. In: ICML, pp. 417–424 (2001)

24. Sutton, R.S., Barto, A.G.: Reinforcement Learning - An Introduction. Adaptive Computation and Machine Learning. MIT Press, Cambridge (1998)

25. Swaminathan, A., Joachims, T.: Batch learning from logged bandit feedback through counterfactual risk minimization. J. Mach. Learn. Res. **16**, 1731–1755 (2015)

26. Wang, F., et al.: Sequential evaluation and generation framework for combinatorial recommender system. CoRR abs/1902.00245 (2019)

27. Wang, Y., et al.: Personalized re-ranking for improving diversity in live recommender systems. CoRR abs/2004.06390 (2020)

28. Wei, J., Zeng, A., Wu, Y., Guo, P., Hua, Q., Cai, Q.: Generator and critic: a deep reinforcement learning approach for slate re-ranking in e-commerce. CoRR abs/2005.12206 (2020)

29. Wilhelm, M., Ramanathan, A., Bonomo, A., Jain, S., Chi, E.H., Gillenwater, J.: Practical diversified recommendations on Youtube with determinantal point processes. In: CIKM, pp. 2165–2173. ACM (2018)

30. Williams, R.J.: Simple statistical gradient-following algorithms for connectionist reinforcement learning. Mach. Learn. **8**, 229–256 (1992)

31. Zhao, X., Xia, L., Tang, J., Yin, D.: Deep reinforcement learning for search, recommendation, and online advertising: a survey. SIGWEB Newsl. **2019**(Spring), 4:1–4:15 (2019)
32. Zhao, X., Xia, L., Zhang, L., Ding, Z., Yin, D., Tang, J.: Deep reinforcement learning for page-wise recommendations. In: RecSys, pp. 95–103. ACM (2018)
33. Zhao, X., Zhang, L., Ding, Z., Yin, D., Zhao, Y., Tang, J.: Deep reinforcement learning for list-wise recommendations. CoRR abs/1801.00209 (2018)
34. Zheng, G., et al.: DRN: a deep reinforcement learning framework for news recommendation. In: WWW, pp. 167–176. ACM (2018)
35. Zhuang, T., Ou, W., Wang, Z.: Globally optimized mutual influence aware ranking in e-commerce search. In: IJCAI, pp. 3725–3731 (2018)
36. Zou, L., Xia, L., Ding, Z., Song, J., Liu, W., Yin, D.: Reinforcement learning to optimize long-term user engagement in recommender systems. In: SIGKDD, pp. 2810–2818. ACM (2019)

Domain Incremental Learning
for EEG-Based Seizure Prediction

Zhiwei Deng[1], Tingting Mao[1], Chenghao Shao[1], Chang Li[1(✉)],
and Xun Chen[2]

[1] Hefei University of Technology, Hefei 230009, Anhui, China
{zhiweideng,Tingtingmao,Chenghaoshao,changli}@mail.hfut.edu.cn
[2] University of Science and Technology of China, Hefei 230001, Anhui, China
xunchen@ustc.edu.cn

Abstract. When building seizure prediction systems, the typical research scenario is patient-specific. In this scenario, the model is limited to performing well for individual patients and cannot acquire knowledge transferable to new patients to learn a set of universal parameters applicable to all patients. To this end, we investigate a new task scenario, domain incremental (DI) learning, which aims to build a unified epilepsy prediction system that performs well across patients by incrementally learning new patients. However, the neural network is susceptible to the problem of catastrophic forgetting (CF) during incremental training, which quickly forgets the knowledge learned from past tasks due to differences in domain distributions. To address this problem, we introduce an experience replay (ER) method, which stores a few samples from previous patients and then replays them in new patient training to review past knowledge. In addition, we propose a novel ER-based centroid matching method (ER-CM) that computes the class centroid in the feature space using subsets stored in the memory buffer. The ER-CM regularizes incremental training by matching the distance between sample embeddings and class centroid, providing additional guidance for parameter updates. Experimental results demonstrate that the ER approach substantially reduces CF and significantly improves performance when combined with CM.

Keywords: Electroencephalogram (EEG) · seizure prediction · domain incremental learning · catastrophic forgetting · experience replay · centroid matching

1 Introduction

Machine learning algorithms [8,10,13,22,23] have been successfully applied in seizure prediction with excellent performance. However, this success remains limited due to most state-of-the-art solutions to seizure prediction being developed

This work is supported by the National Natural Science Foundation of China (Grants 41901350, 32271431).

for patient-specific settings. In this scenario, the algorithms cannot consistently acquire and transfer knowledge across different domains (patients). Specifically, an algorithm trained in the patient-specific setting can only perform well for one patient at a time, which is not capable of performing well for all patients simultaneously. In a more open and real-world scenario, it is critical to learn a set of parameters that behaves well for all patients to develop a unified and parameters-shared system. In that perspective, the model could learn more general and cross-subject discriminative features and thus acquire superior generalization capabilities.

Multi-task learning (joint learning) can learn a set of shared weight parameters from domain-biased distributions across patients, showing outstanding performance on each patient. However, the cost of memory storage and training is expensive. As the number of patients increases, it becomes infeasible to store and retrain the entire data for all patients. In addition, multi-task learning is inefficient since knowledge learned from past tasks cannot be accumulated and all tasks must be trained from scratch when a new task arrives. Therefore, an artificial agent must be able to incrementally learn new patients from the task stream while preserving knowledge of past patients.

Significant distribution differences in the EEG signals across patients have been observed [11]. During incremental learning, abrupt domain shifts (switching between patients) and over-biased learning of new data distributions often lead to forgetting past knowledge in modern neural networks, which is well-known as catastrophic forgetting (CF) [17]. The modern neural networks quickly forget what is learned from past tasks, preventing them from learning progressively on new patients (tasks) [6,15]. In this research, we explore domain incremental learning, also called lifelong learning or continuous learning. It aims to address the setting where samples from different tasks (patients) are learned sequentially in task streams, with each task ideally encountered only once. In this case, the model is only trained on new patients rather than all patients' data at once.

Rehearsal-based methods [1,4,7,16,19] in incremental learning have gained attention recently due to their simplicity and effectiveness, are commonly used in text and image classification. In this family of methods, a few examples of previous tasks or embedding features are stored directly in a fixed-size buffer or compressed in a generative model and then replayed to mitigate CF when trained on a new patient. In simple terms, rehearsal-based methods use previous task experiences stored in the memory buffer to help adjust parameters for new tasks. For instance, experience replay (ER) [4] has shown impressive results in computer vision by jointly training stacked examples retrieved from the memory buffer and the current task. We revisit the role of the memory buffer and consider how to gain additional guidance on updating parameters from it.

Motivated by prototype networks [21], prototypes exist in the feature space, and all embedding points of each class are clustered around the prototype of that class. In this paper, we present empirical replay with centroid matching (ER-CM), a novel method for incremental learning that leverages a few samples in memory to learn generic class centroids over the time step. The ER-CM regularizes the model by minimizing the distance between the feature embeddings and the corresponding class centroids, which steers the model optimization. Specifically, ER-CM computes the class centroid from mini-batches retrieved from the memory buffer and matches stacked samples to the class centroid to obtain sample predictions. We calculate the cross-entropy error between these predictions and labels and combine the classification loss to constrain the parameters for updating. In addition, we propose the sliding average update (SAU) strategy to mitigate centroid shifts caused by underrepresented samples in each replay subset (B_m) from the memory buffer, which is due to the random sampling nature of ER and the limited size of B_m. Our SAU strategy continuously calculates and updates the class centroids, providing additional information to preserve the performance of original tasks.

The main contributions of this work are as follows:

1. We focus on the challenge of incremental learning across patients in the EEG domain for seizure prediction for the first time. To address this, we introduce the ER method to alleviate catastrophic forgetting of previously seen tasks.
2. We further propose a novel centroid matching method based on the ER, termed ER-CM, which assists the model in acquiring broader and more discriminative embedding properties by implicitly limiting the gap between each embedding vector and the class centroid. To address the centroid shift due to the limited size and random sampling nature of each replay subset (B_m), we propose the sliding average update (SAU) strategy, which continuously updates the class centroids based on each replay subset.
3. Extensive experiments demonstrate that the ER-CM approach consistently outperforms the ER approach and other baseline approaches in the Children's Hospital Boston and the Massachusetts Institute of Technology (CHB-MIT) [20] and the American Epilepsy Society Prediction Challenge (Kaggle) [3] databases.

2   Related Work

The Definition of Domain Incremental Learning: Domain incremental learning for epilepsy prediction studies the problem of progressively learning from the task stream to classify pre-ictal and inter-ictal signals, in which significant distribution shifts are frequently noticed among patients. We consider the problem of supervised EEG classification over the task stream, where the system receives a set of EEG segments (obtained via the sliding window) and labels from the current task distribution D_t at time step t. Formally, we define a task stream with unknown distribution as $D = \{D_1, ..., D_N\}$, where each task consists of input EEG segments $X = \{x_1, ..., x_n\}$ and the associated labels $Y = \{y_1, ..., y_n\}$. Given a primitive model (*e.g.*, the convolutional neural network) with two components: the encoder maps the input EEG samples to d-dimensional embedding vectors, denoted as $f : X \rightarrow R^d$, with parameters θ_f; the classifier $g : R^d \rightarrow R^c$ that maps the d-dimensional embedding vectors to output predictions of the c

categories, parameterized as θ_g. A domain incremental algorithm A^{DI} is defined as:

$$A_t^{DI} : \langle (f,g)_{t-1}, (X_B, Y_B)_t, M_{t-1} \rangle \to \langle (f,g)_t, M_t \rangle, \tag{1}$$

where $(f,g)_t$ is the encoder and classifier at incremental step t, $(X_B, Y_B)_t$ denotes a mini-batch of size B that the model receives from task D_t, M_t is an external memory for storing a subset of previously seen samples or a model from the previous time step M_{t-1}. The model parameters θ_f and θ_g at time step t are updated based on the mini-batch $(X_B, Y_B)_t$ and the memory buffer M_{t-1} during the domain incremental learning.

Evaluation Metrics: To measure the degrees of the model forgetting, we define the average forgetting F_t to assess how much of the acquired knowledge is lost from the AUC perspective, which can be formulated as:

$$
\begin{aligned}
f_{i,j} &= \max_{k \in \{1,\dots,i-1\}} a_{k,j} - a_{i,j}, \\
F_t &= \frac{1}{t-1} \sum_{j=1}^{t-1} f_{t,j}, \forall j < t,
\end{aligned}
\tag{2}
$$

where $a_{t,j}$ represents the AUC evaluated on the test set of task j after training the network from task 1 to t. $f_{i,j}$ represents the degree of model forgetting for task j after being trained on tasks 1 to i. Generally, forgetting occurs when the average AUC decreases with incremental learning new tasks, and the low average forgetting values indicate less forgetting.

3   Methods

Experience replay (ER) [4] is one of the most competitive methods that has been widely utilized in various fields. It deploys a memory buffer to store a subset of data from previous tasks for replay. There are two crucial components are involved: memory retrieval and memory update. In memory retrieval, ER uses random sampling to replay samples. For memory update, ER employs reservoir sampling [24] to ensure that every example in tasks has an equal chance of being stored in the memory buffer. At incremental step 1, when receiving a mini-batch (X_B, Y_B), the model optimizes parameters θ_f and θ_g using the standard cross-entropy loss (see time step 1 in Fig. 1). For incremental steps greater than 1, a mini-batch of size B_m is retrieved from the buffer, and trained jointly with a mini-batch of size B obtained from the current task. The classifier generates prediction scores for stacked samples, and the model parameters are optimized by minimizing the cross-entropy loss between these scores and the actual labels. The classification loss ℓ_{ce} for time steps greater than 1 can be defined as:

$$\ell_{ce} = -\frac{1}{B_m + B} \sum_{i=1}^{B_m + B} y_i \log \hat{y}_i, \tag{3}$$

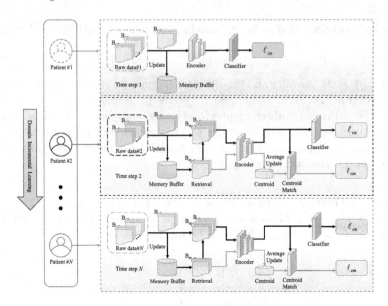

Fig. 1. The overall framework of ER-CM.

where y_i is the label corresponding to the ith example, $\hat{y}_i$ denotes the prediction scores for the ith sample (obtained by the softmax function).

The ER method enables the acquisition of knowledge from prior tasks by simply replaying a few previously seen samples during the new task training. To obtain additional guidance on updating parameters from these replay samples, we further propose the ER-based centroid matching (ER-CM) method. ER-CM guides the model to learn more common semantic features or latent patterns that existed in both past and current tasks by minimizing the distance between stacked batches and class centroids in feature space. An overview of ER-CM is illustrated in Fig. 1. For the incremental step of 1, the standard cross-entropy loss is used as in ER. For incremental steps greater than 1, both the classification loss and the centroid matching loss ℓ_{cm} jointly guide the updating of parameters.

The mini-batch (B_m) retrieved from the memory buffer is assumed to contain a positive and b negative samples $(a + b = B_m)$. The encoder f generates embedding vectors of the replayed subsets, which are used to compute the class centroid c for positive and negative samples, respectively. The formula is as follows:

$$c = \begin{cases} \frac{1}{a}\sum_{i=1}^{a} f(x_i), & \text{positive}, \\ \frac{1}{b}\sum_{i=1}^{b} f(x_i), & \text{negative}. \end{cases} \tag{4}$$

Instead of using the stacked samples of size $(B + B_m)$, the class centroids are calculated from the representation of the replayed small batch of samples (X_{B_m}, Y_{B_m}). It avoids an overly biased estimation of the class centroid as most samples that are stacked are taken from the current task (size B). Moreover, the limited size of each replay subset (B_m) extracted from the memory buffer may result in centroid bias due to inadequate sample representation. To address this issue, we propose the sliding average update (SAU) strategy, which continuously updates the class centroid based on each replay subset. Specifically, suppose an observed sequence of class centroid as $\{c_1, c_2, ..., c_n\}$, where c_i is the class centroid calculated for the ith replay subset in the task stream. The ith class centroid after the SAU strategy takes the following form:

$$f_{cm}^i = \frac{c_{i-1} + c_i}{2}, \tag{5}$$

where c_0 is initialized by the standard normal distribution. In cases where only one class is present in the replay subset, the calculation of class centroid would not be feasible. In such cases, the class centroid will be preserved as the previously updated value.

The cosine similarity is adopted to determine the distance of each example to the class centroid in the latent space, which we use as the prediction scores y_i' after softmax. Our centroid matching loss is defined as:

$$y_i' = \sigma(\text{cosine}(f(x_i), f_{cm})),$$
$$\ell_{cm} = -\frac{1}{B + B_m} \sum_{i=1}^{B+B_m} y_i \log y_i', \tag{6}$$

where σ denotes the softmax function.

The ER-CM incorporates the classification loss (ℓ_{ce}) and the centroid matching loss (ℓ_{cm}) to optimize the final model. The overall loss function can be expressed as:

$$\ell = \ell_{ce} + \lambda \ell_{cm}, \tag{7}$$

where λ is the loss balance weight, we set 1 in our experiments.

4 Experiments

In this section, we first describe our experimental setup in the DI scenario, including the benchmark datasets, preprocessing, baselines, and implementation details. Following that, we evaluate the effectiveness of the proposed methods and analyze the obtained results.

4.1 Experiment Setup

Preprocessing: We conduct experiments on the CHB-MIT [20] and Kaggle [3]. The CHB-MIT scalp EEG database collected EEG recordings from 22 pediatric subjects with intractable epilepsy. A total of 182 seizures were recorded and annotated by experts. These recordings contain 23 cases, each consisting of 9 to 42 consecutive .edf files. We consider the commonly used definition of EEG activity periods in works of literature [18,23] of seizure prediction, *i.e.*, inter-ictal, pre-ictal, seizure prediction horizon (SPH), and ictal. The same setup as research in [9,12,18] is followed for the rest of the EEG activity period. We set SPH to 1 min, pre-ictal to 30 min. The inter-ictal indicates the signals between at least 4 h before the onset and 4 h after the end of the seizure. The seizure onset period (SOP) refers to the period during which a seizure is expected to occur after the system is alerted. In cases where the interval between two seizures is less than 28 min, we regard only one leading seizure. Channel inconsistency across patients leads to misaligned input shapes of EEG signals stored in the memory buffer, hindering the network from performing incremental learning across domains. Consequently, only raw EEG data from 18 channels common to all patients are loaded. Please refer to [2] for details on specific channels.

The Kaggle records intracranial EEG from 5 dogs and 2 subjects, where the iEEG signals from the dogs are sampled from 16 electrodes at 400 Hz using an ambulatory monitoring system (15 electrodes in dog 5), and the iEEG data from 2 patients are sampled at 5000 Hz. The EEG signals are organized into 10-min EEG segments, where each segment is labeled 'Pre-ictal' or 'Inter-ictal' and then stored in the .mat file. The organizer defines the SPH as 5 min, and we similarly set the SOP as 30 min [10,23]. To align the data shape with those of other dogs, the iEEG signals sampled from 15 electrodes in dog 5 are zero-padded to 16 channels. As in literature [9,23], we resample the intracranial EEG signals of dogs to 200 Hz.

Long-range EEG signals from two publicly available datasets were analyzed using sliding window analysis with a window length of 15 s [9]. For most sub-jects, there were significantly more inter-ictal signals than pre-ictal signals. The overlap oversampling strategy with a 15 s sliding window is used to obtain extra preictal data for training [23,26]. Due to the discrepancy in the EEG electrode locations of the recorded signals across the two datasets, incremental learning is not performed jointly for the two datasets but separately for both. In domain incremental learning, each subject's EEG data in two datasets was considered as a separate task, with two classes of EEG samples (positive for pre-ictal and neg-ative for inter-ictal) available for each task. Subjects in both datasets had 2 to 14 leading seizures. In the patient-specific setting, the leave-one-out cross-validation strategy is used to assess all leading seizures of each patient separately. How-ever, the network is trained and evaluated in a domain incremental fashion. The inconsistent leading seizures across patients makes the above evaluation strategy infeasible. To make the results more statistically significant and reduce accidental evaluation errors, we define the last 5-fold leave-one-out cross-validation strat-egy. Specifically, we temporally load the signals of pre-ictal and inter-ictal of each seizure to compose an N-fold training set (Both pre-ictal and inter-ictal signals divided into N parts). We conduct five iterations, each of which selects one of the last five folds as test data and the remaining $N-1$ parts as training data. We report the average metrics for each patient across all five iterations. Subjects with fewer than five leading seizures are not evaluated. To prevent over-fitting during training, we take the first 75% of samples from each fold of the

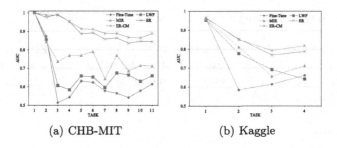

(a) CHB-MIT   (b) Kaggle

Fig. 2. Average AUC at the end of each task observed on CHB-MIT and Kaggle with replay batch size of 64 and memory (M) of 5k in rehearsal-based methods. Each task represents a patient. Upon completion of training for the ith task at time step i, the model's average performance is evaluated on tasks 1 to i. The ER-CM consistently outperforms these compared approaches by significant margins.

training data for training and the remaining 25% for validation. Considering these above definitions and limitations, we evaluated 60 seizures from 11 patients in CHB-MIT and 38 seizures of 4 dogs from Kaggle.

Architectures and Training Settings: We use RepNet-MMCD [9] as the architecture for extensive validation in our experiments. RepNet-MMCD is a lightweight CNN architecture that uses deep separable convolutions to reduce computational burden during training. In addition, re-parameterization is used during testing to further mitigate deployment costs. A modified Monte Carlo dropout strategy is also used to improve model reliability. All baselines use the same model architecture for a given task stream, which is optimized using the AdamW [14] optimizer with a mini-batch of size 128. The network consists of a feature extractor (encoder) and a classifier, trained with learning rates of 0.004 and 0.0003, respectively. We train the model from scratch with 40 epochs and set 10 patience in the early stopping technique to prevent overfitting.

We compare our proposed ER-CM against the following reference baselines:

- **Joint Learning:** The method trains the model offline for all patients over multiple epochs, where mini-batches of each epoch are sampled i.i.d. It is not an incremental learning method and is typically considered to approximate the upper bound of the incremental learning task. We set 40 epochs for training and the mini-batch is 128.
- **Fine-Tune:** Fine-Tune refers to a model that is incrementally trained without any measures to avoid forgetting, where model parameters on the new task are initialized from the previous task parameter vector.
- **LwF:** Learning without forgetting (LWF) is a regularization method that utilizes knowledge distillation to preserve experience from past seen task [5]. The student model is trained with the current task, while the teacher model is trained after learning the last task.
- **ER:** Experience replay (ER) is a rehearsal-based method that stores replay samples in a restricted-size memory buffer. It applies random sampling in

memory retrieval to replay samples and uses reservoir sampling to update memory.
- **MIR:** Maximally interfered retrieval [1] is a variant of ER that focus on the memory update strategy. The MIR algorithm utilizes the principle of maximum increase in loss after a virtual update to retrieve memory samples from a larger subset.

4.2   Results

We compare ER-CM with several strong baselines described in Sect. 4.1. From the Fig. 2, we can make several observations. First, the ER-CM consistently outperforms these baseline approaches by significant margins over the entire task streams of two public datasets. With incremental learning of tasks, ER-CM becomes more prominent against catastrophic forgetting. Moreover, all rehearsal-based methods (MIR, ER, ER-CM) show comparable performance and greatly outperform the regularization-based method (LWF).

Table 1. Average performance of all methods on CHB-MIT and Kaggle at the end of training. The average AUC, average S_n, average recall, average FPR, and average forgetting are reported for the memory buffer with sizes 3k, 5k, and 10k. The experiments are executed more than three times and the best performance is marked in bold. The ↑ indicates that a higher value is considered to be better performance.

Method		CHB-MIT				Kaggle			
		AUC ↑	S_n ↑	Recall ↑	Forgetting ↓	AUC ↑	S_n ↑	Recall ↑	Forgetting ↓
Joint Learning		0.905 ± 0.019	0.776 ± 0.018	0.913 ± 0.019	–	0.814 ± 0.001	0.526 ± 0.010	0.893 ± 0.001	–
Fine-Tune		0.609 ± 0.005	0.464 ± 0.027	0.687 ± 0.027	0.315 ± 0.025	0.641 ± 0.020	0.289 ± 0.072	0.872 ± 0.061	0.165 ± 0.028
LWF		0.674 ± 0.014	0.667 ± 0.114	0.520 ± 0.104	0.236 ± 0.022	0.675 ± 0.033	0.174 ± 0.103	0.845 ± 0.079	0.196 ± 0.083
M=3k	MIR	0.688 ± 0.001	0.562 ± 0.005	0.700 ± 0.010	0.300 ± 0.005	0.710 ± 0.002	0.515 ± 0.018	0.766 ± 0.009	0.186 ± 0.004
	ER	0.847 ± 0.012	0.738 ± 0.002	0.863 ± 0.006	0.116 ± 0.018	0.788 ± 0.001	0.525 ± 0.008	**0.859 ± 0.005**	0.065 ± 0.009
	ER-CM	**0.865 ± 0.001**	**0.752 ± 0.013**	**0.886 ± 0.008**	**0.098 ± 0.003**	**0.803 ± 0.002**	**0.601 ± 0.088**	0.851 ± 0.039	**0.041 ± 0.004**
M=5k	MIR	0.695 ± 0.018	0.599 ± 0.023	0.707 ± 0.050	0.290 ± 0.015	0.690 ± 0.023	0.499 ± 0.004	0.740 ± 0.023	0.201 ± 0.026
	ER	0.856 ± 0.012	0.736 ± 0.002	0.887 ± 0.016	0.108 ± 0.015	0.786 ± 0.002	0.485 ± 0.014	**0.892 ± 0.015**	0.074 ± 0.006
	ER-CM	**0.887 ± 0.002**	**0.764 ± 0.036**	**0.909 ± 0.005**	**0.083 ± 0.008**	**0.807 ± 0.011**	**0.518 ± 0.038**	0.888 ± 0.005	**0.039 ± 0.012**
M=10k	MIR	0.661 ± 0.006	0.516 ± 0.009	0.731 ± 0.005	0.330 ± 0.001	0.701 ± 0.016	0.535 ± 0.004	0.744 ± 0.010	0.184 ± 0.008
	ER	0.858 ± 0.011	0.724 ± 0.018	0.913 ± 0.011	0.103 ± 0.016	0.798 ± 0.006	0.541 ± 0.020	0.880 ± 0.006	0.070 ± 0.011
	ER-CM	**0.889 ± 0.001**	**0.753 ± 0.001**	**0.914 ± 0.001**	**0.080 ± 0.005**	**0.815 ± 0.006**	**0.567 ± 0.030**	0.880 ± 0.006	**0.036 ± 0.015**

To thoroughly assess the robustness of the ER-CM approach and its ability to generalize to the DI scenario, we conduct comprehensive experiments on the CHB-MIT and Kaggle datasets, respectively. As shown in Table 1, the ER and its variant (ER-CM) are substantially superior to the Fine-Tune method, which lacks any regularization or episodic memory during the training. Besides, ER-CM noticeably improves upon the baseline ER for the memory buffer sizes of 3k, 5k, and 10k. For example, in the CHB-MIT dataset, CM helps ER with 5k memory achieve gains of 3.6% (0.856 → 0.887), 3.8%, and 2.6% in average AUC, average sensitivity and average recall, respectively, and effectively reduce the average forgetting by 19.5%. Interestingly, across all memory buffer settings shown in Table 1, we can observe that CM mostly preserves knowledge of positive samples from past tasks (see S_n), while barely forgetting knowledge of negative samples (see Recall).

Table 2. Impact of memory buffer size on sliding average update strategy. At the end of the training, the average performance with and without the SAU strategy is evaluated for memory buffers of 3k, 5k, and 10k, respectively. The memory batch size is 64 and the best performance is marked in bold.

Method		CHB-MIT				Kaggle			
		AUC ↑	S_n ↑	Recall ↑	Forgetting ↓	AUC ↑	S_n ↑	Recall ↑	Forgetting ↓
M=3k	Without SAU	0.860 ± 0.007	0.738 ± 0.026	**0.894 ± 0.002**	0.109 ± 0.001	0.779 ± 0.004	0.467 ± 0.000	**0.885 ± 0.014**	0.081 ± 0.001
	With SAU	**0.865 ± 0.001**	**0.752 ± 0.013**	0.886 ± 0.008	**0.098 ± 0.003**	**0.803 ± 0.002**	**0.601 ± 0.088**	0.851 ± 0.039	**0.041 ± 0.004**
M=5k	Without SAU	0.871 ± 0.005	0.716 ± 0.002	0.909 ± 0.012	0.089 ± 0.001	0.796 ± 0.003	0.493 ± 0.029	**0.898 ± 0.016**	**0.037 ± 0.011**
	With SAU	**0.887 ± 0.002**	**0.764 ± 0.036**	**0.909 ± 0.005**	**0.083 ± 0.008**	**0.807 ± 0.011**	**0.518 ± 0.038**	0.888 ± 0.005	0.039 ± 0.012
M=10k	Without SAU	0.850 ± 0.009	0.714 ± 0.014	0.903 ± 0.003	0.106 ± 0.009	0.795 ± 0.026	0.523 ± 0.030	0.880 ± 0.021	0.057 ± 0.023
	With SAU	**0.889 ± 0.001**	**0.753 ± 0.001**	**0.914 ± 0.001**	**0.080 ± 0.005**	**0.815 ± 0.006**	**0.567 ± 0.030**	**0.880 ± 0.006**	**0.036 ± 0.015**

4.3 Ablation Studies

We use ER-CM with a replay subset of 64 and a memory buffer of 5k as study cases to analyze the impacts of memory buffer sizes and the SAU strategy, respectively. Table 2 shows the effect of memory buffer size with and without SAU strategy. It can be seen that ER-CM with the SAU strategy consistently outperforms ER-CM without the SAU strategy. Moreover, when M is increased to 10k, the performance of ER-CM without the SAU strategy is significantly lower or even inferior to the baseline ER. For instance, the average AUC of CHB-MIT is 0.850, while the average AUC of ER is 0.858. This can be attributed to the insufficient representations of samples in the fixed-size replay subset as the memory size increases, leading to an overly biased estimation of the class centroid during training. In contrast, the centroid matching with the SAU strategy continually adjusts and updates the estimate of the new class centroid using information computed from past replay subsets.

Table 3. Average performance of CNN and AdderNet on CHB-MIT and Kaggle at the end of training.

Architectures		CHB-MIT				Kaggle			
		AUC ↑	S_n ↑	Recall ↑	Forgetting ↓	AUC ↑	S_n ↑	Recall ↑	Forgetting ↓
CNN [25]	Joint Learning	0.815 ± 0.025	**0.692 ± 0.041**	0.820 ± 0.028	–	**0.806 ± 0.008**	0.491 ± 0.004	0.899 ± 0.017	–
	Fine-Tune	0.599 ± 0.009	0.231 ± 0.032	0.872 ± 0.042	0.314 ± 0.017	0.612 ± 0.018	0.493 ± 0.113	0.646 ± 0.141	0.303 ± 0.018
	LWF	0.633 ± 0.040	0.500 ± 0.005	0.683 ± 0.044	0.145 ± 0.009	0.564 ± 0.006	0.047 ± 0.019	**0.966 ± 0.033**	0.261 ± 0.033
	MIR	0.683 ± 0.039	0.450 ± 0.015	0.821 ± 0.062	0.214 ± 0.011	0.668 ± 0.050	0.497 ± 0.047	0.733 ± 0.023	0.212 ± 0.044
	ER	0.800 ± 0.011	0.621 ± 0.017	0.871 ± 0.018	0.113 ± 0.001	0.742 ± 0.041	0.580 ± 0.115	0.735 ± 0.156	0.107 ± 0.036
	ER-CM	**0.834 ± 0.011**	0.639 ± 0.031	**0.892 ± 0.009**	**0.098 ± 0.008**	0.800 ± 0.004	0.493 ± 0.043	0.882 ± 0.012	**0.057 ± 0.009**
AdderNet [26]	Joint Learning	0.872 ± 0.007	**0.717 ± 0.034**	0.896 ± 0.021	–	0.794 ± 0.028	0.482 ± 0.047	0.919 ± 0.004	–
	Fine-Tune	0.608 ± 0.006	0.356 ± 0.019	0.888 ± 0.019	0.328 ± 0.017	0.601 ± 0.015	0.366 ± 0.043	0.774 ± 0.065	0.395 ± 0.022
	LWF	0.621 ± 0.019	0.486 ± 0.081	0.638 ± 0.044	0.216 ± 0.034	0.600 ± 0.002	0.340 ± 0.110	0.762 ± 0.128	0.271 ± 0.052
	MIR	0.744 ± 0.036	0.588 ± 0.053	0.746 ± 0.045	0.223 ± 0.034	0.748 ± 0.032	0.504 ± 0.071	0.786 ± 0.027	0.131 ± 0.019
	ER	0.859 ± 0.015	0.684 ± 0.036	0.902 ± 0.014	0.104 ± 0.017	0.825 ± 0.014	**0.572 ± 0.027**	0.875 ± 0.041	0.056 ± 0.028
	ER-CM	**0.880 ± 0.008**	0.695 ± 0.022	**0.913 ± 0.011**	**0.093 ± 0.011**	**0.847 ± 0.007**	0.542 ± 0.018	**0.919 ± 0.008**	**0.034 ± 0.012**

4.4 Performance on Other Architectures

We extend these approaches to other representative models such as AdderNet [26] and CNN [25]. The CNN is a classical architecture stacked by standard convolutional layers, ReLU activation functions, and max pooling layers. In contrast, the AdderNet replaces multiplication with additive operations in convolutional computation to reduce the computational overhead. We follow the training setup presented in the original paper. As shown in Table 3, the proposed ER-CM method consistently outperforms other baseline methods on both networks, which further validates the generalizability of ER-CM.

5   Conclusion

We investigate the problem of catastrophic forgetting in supervised domain incremental scenarios for seizure prediction tasks and present a simple yet effective solution using the ER method. In addition, we further propose a novel centroid matching (CM) technique that leverages sample embeddings stored in an episodic memory buffer to compute class centroids and guides the parameter update by aligning samples with the corresponding class centroid in the feature space. Our method also implements the SAU strategy that continually adjusts and updates the current class centroid estimate using prior centroid information. The empirical analysis of two benchmark datasets demonstrates that CM can effectively regularize the model, reduce catastrophic forgetting in the DI setting. We hope that it can inspire other multiclassification problems and expect future research to further explore its potential and limitations.

References

1. Aljundi, R., et al.: Online continual learning with maximally interfered retrieval. arXiv:1908.04742 (2019)
2. Bhattacharya, A., Baweja, T., Karri, S.: Epileptic seizure prediction using deep transformer model. Int. J. Neural Syst. **32**(02), 2150058 (2022)
3. Brinkmann, B.H., et al.: Crowdsourcing reproducible seizure forecasting in human and canine epilepsy. Brain **139**(6), 1713–1722 (2016)
4. Chaudhry, A., et al.: On tiny episodic memories in continual learning. arXiv preprint arXiv:1902.10486 (2019)
5. Hinton, G., Vinyals, O., Dean, J.: Distilling the knowledge in a neural network. arXiv preprint arXiv:1503.02531 (2015)
6. Hou, S., Pan, X., Loy, C.C., Wang, Z., Lin, D.: Learning a unified classifier incrementally via rebalancing. In: Proceedings of the IEEE/CVF Conference on Computer Vision and Pattern Recognition, pp. 831–839 (2019)
7. Iscen, A., Zhang, J., Lazebnik, S., Schmid, C.: Memory-efficient incremental learning through feature adaptation. In: Vedaldi, A., Bischof, H., Brox, T., Frahm, J.-M. (eds.) ECCV 2020. LNCS, vol. 12361, pp. 699–715. Springer, Cham (2020). https://doi.org/10.1007/978-3-030-58517-4_41
8. Lahmiri, S., Shmuel, A.: Accurate classification of seizure and seizure-free intervals of intracranial EEG signals from epileptic patients. IEEE Trans. Instrum. Meas. **68**(3), 791–796 (2018)
9. Li, C., Deng, Z., Song, R., Liu, X., Qian, R., Chen, X.: EEG-based seizure prediction via model uncertainty learning. IEEE Trans. Neural Syst. Rehabil. Eng. **31**, 180–191 (2022)

10. Li, C., Huang, X., Song, R., Qian, R., Liu, X., Chen, X.: EEG-based seizure prediction via transformer guided CNN. Measurement **203**, 111948 (2022)
11. Li, S., Zhou, W., Yuan, Q., Liu, Y.: Seizure prediction using spike rate of intracranial EEG. IEEE Trans. Neural Syst. Rehabil. Eng. **21**(6), 880–886 (2013)
12. Li, Y., Liu, Y., Guo, Y.Z., Liao, X.F., Hu, B., Yu, T.: Spatio-temporal-spectral hierarchical graph convolutional network with semisupervised active learning for patient-specific seizure prediction. IEEE Trans. Cybern. **52**(11), 12189–12204 (2021)
13. Liang, D., Liu, A., Gao, Y., Li, C., Qian, R., Chen, X.: Semi-supervised domain-adaptive seizure prediction via feature alignment and consistency regularization. IEEE Trans. Instrum. Meas. **72**, 1–12 (2023). https://doi.org/10.1109/TIM.2023.3261919
14. Loshchilov, I., Hutter, F.: Decoupled weight decay regularization. In: International Conference on Learning Representations (2017)
15. Mai, Z., Li, R., Jeong, J., Quispe, D., Kim, H., Sanner, S.: Online continual learning in image classification: an empirical survey. Neurocomputing **469**, 28–51 (2022)
16. Mai, Z., Li, R., Kim, H., Sanner, S.: Supervised contrastive replay: revisiting the nearest class mean classifier in online class-incremental continual learning. In: Proceedings of the IEEE/CVF Conference on Computer Vision and Pattern Recognition, pp. 3589–3599 (2021)
17. McCloskey, M., Cohen, N.J.: Catastrophic interference in connectionist networks: the sequential learning problem. In: Psychology of Learning and Motivation, vol. 24, pp. 109–165. Elsevier (1989)
18. Ozcan, A.R., Erturk, S.: Seizure prediction in scalp EEG using 3D convolutional neural networks with an image-based approach. IEEE Trans. Neural Syst. Rehabil. Eng. **27**(11), 2284–2293 (2019)
19. Rebuffi, S.A., Kolesnikov, A., Sperl, G., Lampert, C.H.: iCaRL: incremental classifier and representation learning. In: 2017 IEEE Conference on Computer Vision and Pattern Recognition (CVPR), pp. 5533–5542 (2016)
20. Shoeb, A.H.: Application of machine learning to epileptic seizure onset detection and treatment, Ph. D. thesis, Massachusetts Institute of Technology (2009)
21. Snell, J., Swersky, K., Zemel, R.: Prototypical networks for few-shot learning. In: Advances in neural information processing systems, vol. 30 (2017)
22. Tawhid, M.N.A., Siuly, S., Li, T.: A convolutional long short-term memory-based neural network for epilepsy detection from EEG. IEEE Trans. Instrum. Meas. **71**, 1–11 (2022)
23. Truong, N.D., et al.: Convolutional neural networks for seizure prediction using intracranial and scalp electroencephalogram. Neural Netw. **105**, 104–111 (2018)
24. Vitter, J.S.: Random sampling with a reservoir. ACM Trans. Math. Softw. **11**(1), 37–57 (1985)
25. Xu, Y., Yang, J., Zhao, S., Wu, H., Sawan, M.: An end-to-end deep learning approach for epileptic seizure prediction. In: 2020 2nd IEEE International Conference on Artificial Intelligence Circuits and Systems (AICAS), pp. 266–270. IEEE (2020)
26. Zhao, Y., Li, C., Liu, X., Qian, R., Song, R., Chen, X.: Patient-specific seizure prediction via adder network and supervised contrastive learning. IEEE Trans. Neural Syst. Rehabil. Eng. **30**, 1536–1547 (2022)

Multitask Learning-Based Early MTT Partition Decision for Versatile Video Coding

Wu Liu, Yue Li, and Mingxing Nie$^{(\boxtimes)}$

University of South China, Hengyang 421009, China
niemx@usc.edu.cn

Abstract. Versatile Video Coding (VVC) introduces a new block partition structure called Multi-Type Tree (MTT), which includes four partitioning modes: horizontal-vertical binary tree partitioning, horizontal-vertical ternary tree partitioning. This new block partition structure significantly improves compression performance, but at the same time greatly increases the computational complexity of VVC. To reduce the computational complexity of MTT in VVC inter-frame coding, a Multitask learning-Based early MTT partition decision for Versatile Video Coding is proposed. Firstly, for each Coding Unit (CU), two types of features related to the optimal MTT partitioning are extracted, namely encoding parameter features and encoding intermediate information features. Secondly, to reduce the number of neural network parameters, the horizontal or vertical partitioning in MTT is jointly learned, and lightweight neural networks are constructed to decide whether to skip the horizontal or vertical partitioning of binary or ternary trees. Experimental results show that under the Random Access (RA) configuration, the proposed method can reduce the VVC inter-frame computational complexity by an average of 27.79%, while only increasing the Bjontegaard delta bit rate (BDBR) by 1.14%.

Keywords: Versatile video coding · Multi-type tree · Multi-task learning · Block partition

1 Introduction

With the rapid development of information acquisition technology, new video formats continue to emerge, such as 4K/8K and 360° panoramic video. Although the new video formats can give viewers a better visual experience, their data volume is very large, which brings new serious challenges to the field of video compression. In order to store and transmit video data more efficiently, in July 2020, Joint Video Explore Team (JVET) launched the new generation video

This research was supported in part by the National Natural Science Foundation of China under Grant 62001209; in part by the Research Foundation of Education Bureau of Hunan Province under Grant No. 21B0424.

compression standard H.266/VVC [1]. Compared with the previous generation of video compression standard H.265/High Efficiency Video Coding (HEVC), the coding efficiency has been improved by about 40% while maintaining the same subjective video quality [2].

Similar to the coding structure of HEVC, VVC is also encoded based on hybrid coding framework. In order to further improve coding efficiency, VVC introduced many new coding techniques [3,4]. For example, in order to support more flexible block partitioning shapes, VVC used a nested multi-type tree based on Quadtree with nested multi-type tree (QTMT), which increases the number of partition modes for each CU to six [5]: Non-partition (NT), Quadtree partition (QT), Horizontal binary tree partition (H_BT), Vertical binary tree partition (V_BT), Horizontal ternary tree partition (H_TT) and Vertical ternary tree partition (V_TT), as shown in Fig. 1. Under the RA configuration, the QTMT partitioning structure can reduce the coding rate by 8.5% [6], but it leads to about 1.7 times more computational complexity for VVC than HEVC [7]. Currently, the high complexity has become a major obstacle to deploying VVC in real-time applications on devices that require low power consumption, such as smartphones and unmanned aerial vehicles. Therefore, it is necessary to study fast QTMT decision method to reduce the complexity of VVC.

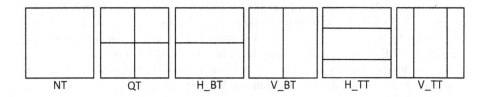

Fig. 1. Six partition modes.

In this paper, a multi-task learning-based early MTT partitioning decision method for VVC is proposed, which cleverly combines multi-task learning with the MTT module of VVC for the first time, and effectively solves the problems of a large number of model parameters and low prediction accuracy. The main contribution of this paper is as follows:

(1) Some new features related to MTT partitioning have been proposed, experimental results show that these features have good prediction effect, and the proposed method can effectively reduce the computational complexity.

(2) A lightweight neural network based on multi-task learning is proposed to reduce the computational complexity of MTT, the lightweight neural network model has fewer parameters and low training difficulty.

2   Related Work

2.1   Fast Algorithm in HEVC

The QTMT module of VVC is extended from the Quadtree module of HEVC. The existing fast algorithms in HEVC can be mainly divided into two categories:

the methods based on Machine Learning (ML) [8–12] and the methods based on encoding intermediate information [13–15]. For example, Bouaafia *et al.* [8] proposed two fast CU partitioning methods based on ML. The first is the online Support Vector Machine (SVM) fast algorithm. Another method is to design a deep convolutional neural network to predict the optimal size of each CU. Lee *et al.* [11] used characteristic information based on Sobel operator and rate distortion to determine the optimal size of each CU in advance. In the method based on intermediate information. For example, Tan *et al.* [13] predicted residual error through statistical analysis and designed a residual threshold to determine whether the CU needs further division.

2.2   Fast Algorithm in VVC

Since QTMT partitioning in VVC is more complex and flexible than QT partitioning in HEVC, the above method cannot be used directly in VVC. Fast methods in VVC also fall into two categories: the methods based on ML [16–22] and the methods based on intermediate information [23–26]. In the method based on ML, methods [16–20] is used for RA configuration inter-frame coding. For example, Pan *et al.* [16] designed a Multi-information Convolutional Neural Network (MF-CNN) model, which jointly uses multi-domain information to terminate the CU partitioning process in advance. Methods [21,22] are used for All Intra (AI) configuration intra-frame coding. For example, Tissier *et al.* [21] proposed a two-stage learning method is proposed to reduce the computational complexity of CUs in VVC encoders, including CNN and Decision Tree.

In the method using intermediate coding information, methods [23,24] is used for inter-frame coding. For example, Won *et al.* [23] proposed a fast partitioning algorithm of binary and ternary trees based on Mean Absolute Error (MAE) function, using the MAE value to compare with a threshold value to determine whether to further partition. Methods [25,26] is used for intra-frame encoding. For example, Peng *et al.* [26] sets adaptive threshold to classify CUs into simple, ordinary and complex types according to texture features, and skips the calculation of all partition modes of simple CU.

3   Background and Motivation

In the VVC encoding process, the current frame is first divided into multiple Coding Tree Units (CTUs) of the same size. Then, each CTU is divided into CU leaf nodes, and then CUs is recursively divided. Due to the addition of a variety of partitioning modes and partitioning rules, the partitioning results of a frame image become diverse. In order to obtain the best result of the current frame partitioning, it is necessary to traverse all possible partitioning cases for each CU and calculate the Rate-Distortion cost (RDcost) for each CU partitioning mode. Finally, the mode with the lowest RDcost is selected as the best CU partitioning mode. The RDcost is calculated as follows:

$$RDcost = D + \lambda \times K_m \tag{1}$$

where D is the distortion, K_m represents the number of bits of mode m, m includes the six partition modes show in Fig. 1, and λ is the Lagrange multiplier.

Although the exhaustive search method in VVC can obtain the optimal partition mode of CUs, it increases the RDcost calculation several times, which brings a sharp increase in computational complexity. Figure 2 shows an example of optimal CU partitioning in a frame of BQSquare sequence in RA configuration, where the left subgraph is a 128×128 CU partition, Only one partition mode is selected as the optimal mode for a CU. Therefore, if we can accurately predict the optimal partition mode of CUs in advance and skip the RDcost calculation of the remaining partition modes, the complexity will be reduced effectively.

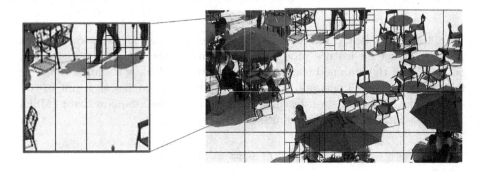

Fig. 2. A Partition Example.

4 Proposed Approach

4.1 Multi-task Learning Model

Multi-task Learning (MTL) can combine datasets from multiple tasks, and thereby alleviating the problem of data sparsity by utilizing useful information from other related learning tasks. In addition, when multiple tasks learn together, the unrelated parts of the tasks act as trace noise, and adding trace noise can improve the generalization ability of the model.

In VVC, since the binary tree partitioning of CUs in the same direction is closely related to the ternary tree partitioning, MTL can be applied to the MTT module of VVC based on this feature. Therefore, in this paper, the binary tree horizontal partitioning skip and ternary tree horizontal partitioning skip of the same CU are combined into a multi-task problem, while the vertical orientation constitutes another multi-task problem. Then, two types of multitask learning models are constructed: Horizontal Multitask Model (HMTL) and Vertical Multitask Model (VMTL). In order to reduce the number of parameters in the model, this paper employs lightweight neural network to build multi-task learning model. The specific structure of the model is shown in Fig. 3. At the input

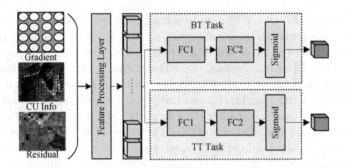

Fig. 3. Network Model.

layer, the residuals, CU information and gradients of two single tasks are input into the model for feature processing according to the calculation method in Sect. 4.2, and the obtained features are then processed using a simple two-layer fully connected (FC) network. The final output is the prediction result of two single tasks. The multi-task learning model utilizes Mean Square Error (MSE) as the loss function, which is defined as follows:

$$MSE(y, y') = \frac{\sum_{i=1}^{n}(y_i - y_i')^2}{n} \qquad (2)$$

where y' is the predicted value, y is the actual value, and n is the dimension.

4.2   Feature Analysis

In order to obtain the features most relevant to the optimal partitioning mode, the coding information of each CU and the corresponding optimal partitioning mode are extracted as data sets in the original VVC encoding process. In this paper, eight types of coding information are selected for correlation research. Then, according to the correlation from high to low, six kinds of encoded information are chosen as the input features for the model, Fig. 4 illustrates the analysis of thermal map characteristics. The dataset is obtained by encoding the BlowingBubbles sequence, Although the data set is extracted from only one sequence, the experimental results demonstrate that the model also exhibits good prediction performance on other sequences, which also proves that the method proposed in this paper has good generalization. The following is a detailed explanation of the selected features:

1) Maximum subblock residual variance (Max_res): In inter-frame coding, the residual value represents the changes of pixel value. However, block partition tends to divide pixels with similar changes into the same block, so the partitioning mode becomes more necessary when the variance value of subblock residuals obtained after partitioning is smaller. The residual value of pixel points is calculated as follows:

$$R_{i,j} = |P_{i,j} - O_{i,j}| \qquad (3)$$

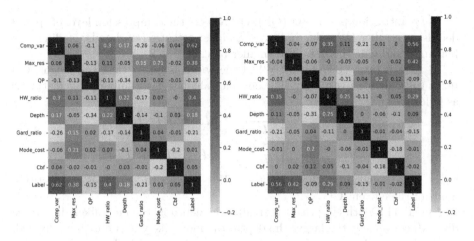

Fig. 4. The analysis of thermal map characteristics. The left is the HMTL data set and the right is the VMTL data set.

where $R_{i,j}$ represents the residual value of point (i,j) in the subblock, $P_{i,j}$ represents the predicted luma value, and $O_{i,j}$ represents the original luma value. In order to ensure that the subblocks of variance calculation are of the same size, the binary tree partitioning mode is considered to have two subblocks of equal size, and the ternary tree partitioning mode is considered to have four subblocks of equal size. Finally, the residual variance values of all CU subblocks are calculated based on the current partitioning mode, and the maximum value is normalized as a feature. The variance calculation is as follows:

$$Var = \frac{\sum_{i=0}^{H-1} \sum_{j=0}^{W-1} (R_{i,j} - \bar{R})^2}{H \times W} \tag{4}$$

where Var is the residual variance value of subblock, H is the height of subblock, W is the width, $R_{i,j}$ is the residual value of subblock point (i, j), $\bar{R}$ is the average residual value of subblock.

2) Comparison value of variance of subblock residuals (Comp_var): Judging from only one direction will result in significant prediction errors. Therefore, within the same partition tree, we can compare the partition modes in two different directions to obtain the maximum residual variance value of subblock, and then skip the partition mode with large residual variance value of subblock through comparison. This feature is calculated as follows:

$$S = \begin{cases} 1 & Var\_H > Var\_V \\ 0 & Var\_H < Var\_V \end{cases} \quad D = \begin{cases} 1 & Var\_V > Var\_H \\ 0 & Var\_V < Var\_H \end{cases} \tag{5}$$

where Var_H represents the maximum residual variance value of current CU horizontal subblock, Var_V represents the maximum residual variance value of vertical subblock, S represents the horizontal binary tree and ternary tree features, and D represents the vertical binary tree and ternary tree features.

3) Quantization parameter (QP): QP reflects the compression level of spatial details. A smaller QP value indicates a higher retention of details, leading to a tendency to divide the data into smaller blocks.

4) Aspect ratio (HW_ratio): When the width is larger than the height, CU tends to be divided vertically. The specific calculation formula of this feature is as follows:

$$HW\_ratio = \begin{cases} \dfrac{H}{H+W} & M\epsilon\{H\_BT, H\_TT\} \\ \dfrac{W}{H+W} & M\epsilon\{V\_BT, V\_TT\} \end{cases} \tag{6}$$

where H is the height of the current CU, W is the width, and M is the partition mode of the current CU.

5) QTMT Depth (Depth): the smaller the depth is, the more likely it is to be divided, conversely, the larger the depth, the more likely it is not to be divided.

6) Horizontal and vertical gradient ratio (Gard_ratio): The gradient value can effectively represent the motion in a specific direction. In this paper, HMTL model uses G_h/G_v, VMTL model uses G_v/G_h, and the specific calculation formula of gradient is as follows:

$$G_h = \sum_{i=0}^{H-1}\sum_{j=0}^{W-1}|R_{i,j+1} - R_{i,j}| \quad G_v = \sum_{i=0}^{H-1}\sum_{j=0}^{W-1}|R_{i+1,j} - R_{i,j}| \tag{7}$$

where G_h and G_v respectively represent horizontal and vertical gradients. H is the height of the current CU, W is the width, and $R_{i,j}$ represents the residual value of the point (i, j).

4.3 Model Training

In order to obtain the lightweight neural network structure with the best performance, we tested five different fully connected network structures. The test results are shown in Table 1, "Quantity" represents the number of parameters and "Accuracy" represents the model accuracy. The structure $6\times20\times20$ achieved the highest prediction accuracy, and both tasks use the same structure.

Table 1. Model Architecture Testing.

Structure	$6\times10\times50$	6×30	6×20	$6\times30\times30$	$6\times20\times20$
Quantity	671	241	161	1171	581
Accuracy	80.14%	78.43%	78.13%	81.48%	83.32%

After data cleaning and redundancy removal, a total of 284,497 data sets were used to train the previously constructed multi-task model, including 110,589 data sets for the HMTL model and 173,908 data sets for the VMTL model. Train the model precision convergence about 500 times, and both models achieved an accuracy of over 80%. Figure 5 illustrates the training process.

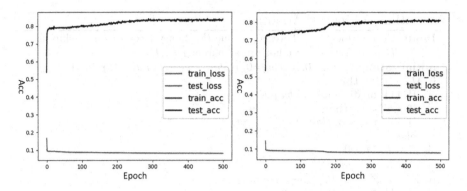

Fig. 5. Train Process. The left is the HMTL model and the right is the VMTL model, "Acc" represents the accuracy and "Epoch" represents the number of iterations

4.4 The Whole Algorithm Proposed

In this paper, the objective is to identify unnecessary partitioning modes using the algorithm, skip the calculation of RDcost, and reduce the computational complexity in the search process for the optimal CU partitioning mode. Additionally, skip flags are introduced to minimize the impact of incorrect predictions. For the binary tree or ternary tree partitioning of the same CU, if the HMTL model predicts the horizontal direction and skips it, the VMTL model will not make predictions for the vertical direction. The overall flow of the proposed algorithm is presented in Algorithm 1, Where "Skip" indicates that the calculation of the current mode is skipped ahead of time.

5 Experimental Results and Discussion

5.1 Experimental Conditions

In order to evaluate the performance of the proposed method, the latest test software VTM19.2 and the test software VTM6.0 of VVC were tested respectively with the original VTM as the anchor point. The experiment employed a total of 21 recommended general test videos, ranging from Class A1 to Class E, with RA configuration and QPs of 22, 27, 32, and 37. To mitigate the impact of incorrect predictions, the decision to skip the partitioning mode was based on a confidence level exceeding 95% in the model prediction. Therefore, the threshold (th) is set to 0.05. Encoding performance was evaluated using encoding time saving TS and BDBR [27]. Typically, better performance is indicated by greater encoding time reduction and smaller BDBR increase. To quantify coding performance, we use a performance metric similar to what is called a "Factor" in [28]. A higher Factor value denotes superior performance, The formulas are defined as follows:

$$TS = \frac{Time_{org} - Time_{pre}}{Time_{org}} \tag{8}$$

Algorithm 1: Proposed Algorithm

Input: Current mode M, Threshold value th, Binary tree horizontal skip flag
      B_flag, Ternary tree horizontal skip flag T_flag

initialization:$B\_flag=0$, $T\_flag=0$, $p\_bh=1$, $p\_bv=1$, $p\_th=1$, $p\_tv=1$

if $M==H\_BT$ **then**
  | HMTL prediction—>p_bh,p_th;
  | **if** $p\_bh < th$ **then**
  | | Skip and $B\_flag=1$
  | **else**
  | | $B\_flag=0$

if $M==V\_BT$ && $B\_flag=0$ **then**
  | VMTL prediction—>p_bv,p_tv;
  | **if** $p\_bv < th$ **then**
  | | Skip

if $M==H\_TT$ && $p\_th < th$ **then**
  | Skip and $T\_flag=1$;
if $M==V\_TT$ && $p\_tv < th$ && $T\_flag=0$ **then**
  | Skip;
end

$$Factor = \frac{TS}{BDBR} \qquad (9)$$

where $Time_{org}$ represents the total encoding time of the original VTM encoder, and $Time_{pre}$ represents the total encoding time with the proposed algorithm added. The computer configuration for the experiment is: "11th Gen Intel(R) Core(TM) i7-11700F @ 2.50GHz, 16GB-RAM"

5.2   Coding Performance Evaluation

Table 2 shows the overall performance of the proposed method. The fast MTT partitioning method proposed in VTM19.2 can save 13.92%-41.63% encoding time, with an average saving of 27.79%. The corresponding BDBR increases by 0.56%-1.79%, with an average increase of only 1.14%. To better demonstrate the effectiveness of the algorithm proposed in this paper, a comparison is made with the methods proposed by Pan [16] and Li [24]. In order to make the experimental comparison fair, the same test platform version is used. The algorithm proposed in this paper is implemented on VTM6.0, and the comparison data with Pan's method is obtained. Similarly, Li's algorithm implemented on VTM19.2 is compared with the experimental results of the algorithm proposed in this paper. On VTM6.0, Pan's method achieves an average time reduction of 25.42% with an average BDBR increase of 2.53%. In contrast, the proposed method achieves an average time reduction of 26.68% with an average BDBR increase of 0.98%. On VTM19.2, Li's algorithm saves an average of 23.95% of time, and BDBR increases an average of 1.21%. The results indicate that, on average, the proposed method outperforms both Pan's and Li's algorithms in terms of TS and

BDBR. In other words, the method in this paper achieves a greater reduction in coding time with a smaller increase in BDBR, and get a higher Factor value.

Table 2. Experimental Result.

Class	Sequence	Pan[16]		Li[24]		Proposed(V6.0)		Proposed(V19.2)	
		BDBR	TS	BDBR	TS	BDBR	TS	BDBR	TS
A1	Campfire	2.80	30.08	1.84	33.35	1.24	30.23	1.31	30.37
	FoodMarket4	1.59	42.90	0.70	25.33	0.92	35.82	0.80	41.63
	Tango2	3.68	34.05	0.87	21.19	1.39	30.90	1.73	35.66
A2	CatRobot1	5.59	30.62	0.85	16.28	0.61	25.21	0.79	29.11
	DaylightRoad2	4.43	29.20	0.91	17.00	1.13	27.21	1.60	31.90
	ParkRunning3	1.61	21.30	0.87	27.35	0.73	28.50	0.87	29.54
B	MarketPlace	3.22	36.47	1.20	21.54	1.35	30.47	1.38	30.50
	RitualDance	2.97	31.23	1.89	29.53	1.52	30.71	1.79	26.86
	BasketballDrive	2.96	32.39	1.29	27.14	1.40	30.96	1.59	31.23
	BQTerrace	0.98	13.80	0.86	26.45	0.22	23.42	0.56	26.42
	Cactus	5.20	25.42	1.12	25.41	0.94	26.19	0.75	27.54
C	BasketballDrill	1.59	24.38	1.60	32.98	1.25	28.60	1.54	26.14
	PartyScene	1.84	14.94	1.36	33.65	0.69	24.94	0.86	25.33
	RaceHorsesC	2.23	22.55	1.92	32.63	1.05	25.87	1.29	26.02
D	BasketballPass	1.56	21.18	1.49	22.71	0.75	22.58	0.86	20.58
	BlowingBubbles	2.29	16.97	1.44	22.94	0.79	23.63	1.05	23.91
	BQSquare	0.84	9.69	1.04	18.78	0.35	15.30	0.65	13.92
	RaceHorses	2.24	20.33	1.96	26.83	1.40	25.36	1.38	23.80
E	FourPeople	1.76	25.26	0.93	15.63	0.88	23.04	0.97	26.11
	Johnny	1.69	24.92	0.63	12.65	0.63	25.77	1.20	28.95
	KristenAndSara	2.11	26.21	0.65	13.53	0.92	25.53	0.93	28.13
	Average	2.53	25.42	1.21	23.95	0.98	26.68	1.14	27.79
	Factor	10.45		19.79		27.22		24.38	

5.3   Model Performance Evaluation

In order to provide a clearer analysis of the number of model parameters, a comparison is made between the network structure in this paper and Pan's [16] as shown in Table 3. The number of model parameters used in this paper is only 1162, which is far less than Pan's model with 25.6M. In addition, the additional consumption brought by the model is tested under four different QPS. The result is to take the average of three sequences (BasketballDrill, BlowingBubbles and FourPeople). The additional time added in this paper is only 0.98% on average, while the additional time added by the Pan's model is 5.21%. Combined with the experimental test results, it is shown that the neural network model constructed in this paper can bring better prediction effect with fewer parameters.

Table 3. Model Parameter Quantity.

	Structure	Quantity	Size	QP22	QP27	QP32	QP37	Average
Proposed	Full-6*20*20	1162	4.528KB	0.73%	0.78%	1.06%	1.38%	0.98%
Pan [16]	ResNet-50	25.6M	102.4MB	4.23%	5.18%	5.91%	5.53%	5.21%

6   Conclusion

In order to reduce the computational complexity of VVC inter-frame coding, this paper proposes a Multitask learning-Based early MTT partition decision for VVC inter-frame coding. The proposed multi-task learning model is simple in structure, easy to be integrated into VVC test software, and can effectively reduce the complexity of coding computation. Experimental results show that the proposed method can achieve good coding performance on different versions of the test platform. In the latest test platform VTM19.2, the average BDBR increase is only 1.14%, and the encoding time can be reduced by 27.79%.

References

1. Bross, B.: Versatile video coding (draft 1), jvet-j1001 (2018)
2. Zhang, Q., Wang, Y., Huang, L., Jiang, B.: Fast CU partition and intra mode decision method for H. 266/VVC. IEEE Access **8**, 117539–117550 (2020)
3. An, J., Huang, H., Zhang, K., Huang, Y., Lei, S.: Quadtree plus binary tree structure integration with JEM tools. JVET-B0023, Joint Video Exploration Team (JVET) (2016)
4. Saldanha, M., Sanchez, G., Marcon, C., Agostini, L.: VVC intra-frame prediction. In: Versatile Video Coding (VVC) Machine Learning and Heuristics, pp. 23–33. Springer, Cham (2022). https://doi.org/10.1007/978-3-031-11640-7_3
5. Bross, B., Chen, J., Liu, S., Wang, Y.K.: Versatile video coding (draft 5). Joint Video Experts Team (JVET) of Itu-T Sg 16, 3–12 (2019)
6. François, E., et al.: VVC per-tool performance evaluation compared to HEVC. In: IBC (2020)
7. Siqueira, Í., Correa, G., Grellert, M.: Rate-distortion and complexity comparison of HEVC and VVC video encoders. In: 2020 IEEE 11th Latin American Symposium on Circuits & Systems (LASCAS), pp. 1–4. IEEE (2020)
8. Bouaafia, S., Khemiri, R., Sayadi, F.E., Atri, M.: Fast CU partition-based machine learning approach for reducing HEVC complexity. J. Real-Time Image Proc. **17**, 185–196 (2020)
9. Duanmu, F., Ma, Z., Wang, Y.: Fast mode and partition decision using machine learning for intra-frame coding in HEVC screen content coding extension. IEEE J. Emerg. Sel. Top. Circuits Syst. **6**(4), 517–531 (2016)

10. Momcilovic, S., Roma, N., Sousa, L., Milentijevic, I.: Run-time machine learning for HEVC/H. 265 fast partitioning decision. In: 2015 IEEE International Symposium on Multimedia (ISM), pp. 347–350. IEEE (2015)
11. Lee, D., Jeong, J.: Fast CU size decision algorithm using machine learning for HEVC intra coding. Signal Proc. Image Commun. **62**, 33–41 (2018)
12. Bakkouri, S., Elyousfi, A.: Machine learning-based fast CU size decision algorithm for 3D-HEVC inter-coding. J. Real-Time Image Proc. **18**, 983–995 (2021)
13. Tan, H.L., Ko, C.C., Rahardja, S.: Fast coding quad-tree decisions using prediction residuals statistics for high efficiency video coding (HEVC). IEEE Trans. Broadcast. **62**(1), 128–133 (2016)
14. Shen, L., Zhang, Z., An, P.: Fast CU size decision and mode decision algorithm for HEVC intra coding. IEEE Trans. Consum. Electron. **59**(1), 207–213 (2013)
15. Huade, S., Fan, L., Huanbang, C.: A fast CU size decision algorithm based on adaptive depth selection for HEVC encoder. In: 2014 International Conference on Audio, Language and Image Processing, pp. 143–146. IEEE (2014)
16. Pan, Z., Zhang, P., Peng, B., Ling, N., Lei, J.: A CNN-based fast inter coding method for VVC. IEEE Signal Process. Lett. **28**, 1260–1264 (2021)
17. Li, J., Zhang, S., Yang, F.: Random forest accelerated CU partition for inter prediction in H. 266/VVC. In: 2022 IEEE International Conference on Multimedia and Expo (ICME), pp. 01–06. IEEE (2022)
18. Liu, Y., Abdoli, M., Guionnet, T., Guillemot, C., Roumy, A.: Light-weight CNN-based VVC inter partitioning acceleration. In: 2022 IEEE 14th Image, Video, and Multidimensional Signal Processing Workshop (IVMSP), pp. 1–5. IEEE (2022)
19. Amestoy, T., Mercat, A., Hamidouche, W., Menard, D., Bergeron, C.: Tunable VVC frame partitioning based on lightweight machine learning. IEEE Trans. Image Process. **29**, 1313–1328 (2019)
20. Yeo, W.H., Kim, B.G., et al.: CNN-based fast split mode decision algorithm for versatile video coding (VVC) inter prediction. J. Multimedia Inf. Syst. **8**(3), 147–158 (2021)
21. Tissier, A., Hamidouche, W., Mdalsi, S.B.D., Vanne, J., Galpin, F., Menard, D.: Machine learning based efficient QT-MTT partitioning scheme for VVC intra encoders. IEEE Trans. Circuits Syst. Video Technol. **33**, 4279–4293 (2023)
22. Li, T., Xu, M., Tang, R., Chen, Y., Xing, Q.: DeepQTMT: a deep learning approach for fast QTMT-based CU partition of intra-mode VVC. IEEE Trans. Image Process. **30**, 5377–5390 (2021)
23. Won, D.J., Moon, J.H.: Fast inter CU partitioning algorithm using MAE-based prediction accuracy functions for VVC. J. Broadcast Eng. **27**(3), 361–368 (2022)
24. Li, Y., Luo, F., Zhu, Y.: Temporal prediction model-based fast inter CU partition for versatile video coding. Sensors **22**(20), 7741 (2022)
25. Fu, T., Zhang, H., Mu, F., Chen, H.: Fast CU partitioning algorithm for H. 266/VVC intra-frame coding. In: 2019 IEEE International Conference on Multimedia and Expo (ICME), pp. 55–60. IEEE (2019)
26. Peng, S., Peng, Z., Ren, Y., Chen, F.: Fast intra-frame coding algorithm for versatile video coding based on texture feature. In: 2019 IEEE International Conference on Real-time Computing and Robotics (RCAR), pp. 65–68. IEEE (2019)
27. Barman, N., Martini, M.G., Reznik, Y.: Revisiting Bjontegaard delta bitrate (BD-BR) computation for codec compression efficiency comparison. In: Proceedings of the 1st Mile-High Video Conference, pp. 113–114 (2022)
28. Kuang, W., Chan, Y.L., Tsang, S.H., Siu, W.C.: Fast HEVC to SCC transcoder by early CU partitioning termination and decision tree-based flexible mode decision for intra-frame coding. IEEE Access **7**, 8773–8788 (2019)

EEG Extended Source Imaging with Variation Sparsity and L_p-Norm Constraint

Shu Peng[1], Feifei Qi[3], Hong Yu[1,2], and Ke Liu[1,2(✉)] [iD]

[1] School of Computer Science and Technology, Chongqing University of Posts and Telecommunications, Chongqing 400065, China
liuke@cqupt.edu.cn
[2] Key Laboratory of Big Data Intelligent Computing, Chongqing University of Posts and Telecommunications, Chongqing, China
[3] School of Internet Finance and Information Engineering, Guangdong University of Finance, Guangzhou 510521, China

Abstract. Accurately reconstructing the location and extent of cortical sources is crucial for cognitive research and clinical applications. Regularization methods that use the L_1-norm in the spatial variation domain effectively estimate cortical extended sources. However, in the variation domain, employing L_1-norm constraint tends to overestimate the extent of sources. Hence, to achieve more precise estimations of both the location and extent of sources, further sparseness-enforced regularizations are required. In this work, we develop a robust EEG source imaging method, VSSI-L_p, to estimate extended cortical sources. VSSI-L_p employs the L_p-norm ($0 < p < 1$) in the variation domain to promote sparsity. Using alternating direction method of multipliers (ADMM) and generalized soft-thresholding (GST) algorithm, we can efficiently derive the solution of VSSI-L_p. According to numerical simulations plus real data analysis, VSSI-L_p outperforms both traditional L_2 and L_1-norm-based methods, and the L_1-norm-based method in the variation domain for reconstructing extended sources, validating the outstanding performance of L_p-norm and variation constraint.

Keywords: EEG source imaging · L_p-norm · Variation sparsity · generalized soft-thresholding

1 Introduction

As a non-invasive tool, Electroencephalography (EEG) is used extensively in neuroscience research because of its excellent millisecond-level time resolution. EEG source imaging (ESI) aims to reconstruct cortical activities from EEG signals, essential in neuroscience research and clinical diagnosis (e.g., epileptic seizure area localization). Moreover, ESI can also provide higher spatial resolution in BCIs [6,8], obtaining more precise outcomes.

© The Author(s), under exclusive license to Springer Nature Singapore Pte Ltd. 2024
L. Fang et al. (Eds.): CICAI 2023, LNAI 14474, pp. 500–511, 2024.
https://doi.org/10.1007/978-981-99-9119-8_45

To handle this ESI task, the current density model uses triangles to represent sources and divides the cortex into a fixed triangular mesh [12]. With Maxwell's equations, the EEG signal is somehow a linear combination of the source amplitudes [9]. Then ESI estimates the potential source activities by solving this linear inverse problem, which is to find a source configuration that best suits the scalp EEG measurement. However, the inverse problem is fully underdetermined due to the candidate sources (typically more than 5000) vastly outnumbered the scalp EEG electrodes (tens to hundreds) [7,16]. To obtain a unique source configuration, employing appropriate constraints on the source spaces is therefore necessary.

The most commonly employed constraint is the L_2-norm regularization, like the minimum norm estimate (MNE) [7], which obtains the target source configuration with the minimum energy. However, the solutions of MNE are biased towards superficial sources because the fields generated by scalp sources are stronger than the deep sources with less energy [7]. One way to compensate for this bias is to weight the regularization term with the lead-field matrix, which is referred to as the weighted MNE (wMNE) [13]. Furthermore, in considering the dependencies between adjacent sources, the low-resolution electromagnetic tomography (LORETA) approach was proposed. LORETA minimizes the L_2-norm of the second-order spatial derivative in source space, so as to derive smoothness and local spatial coherent solutions. In general, these L_2-norm-based methods are welcomed due to their computational efficiency, but they limit spatial resolution as they produce diffused estimations, though.

Sparse methods with L_0-norm provide better spatial resolution than the L_2-norm-based approaches, but L_0-norm optimization is computationally infeasible with large-scale data. To approximate the L_0-norm, L_1-norm constraints are commonly used [14]. However, the sparse constraint on the original source space only produces some point sources, providing little information on the size of cortical activities [7,11]. In contrast, employing L_1-norm regularization in the transform domain, such as variation transform, will provide more accurate estimations of extended sources [4,12]. Nonetheless, as suggested in [2], in the transform domain, methods based on L_1-norm tend to overestimate the extent of sources, especially for small-sized sources. Therefore, more sparseness-enforced constraints are necessary to achieve more accurate estimations [3].

To better approximate the solution of L_0-norm with sufficient sparsity, several studies have adopted the L_p-norm $(0 < p < 1)$. L_p-norm offers flexible recovery by controlling the value of p. Moreover, L_p-norm-based methods require fewer measurements to achieve reliable reconstruction [3]. Therefore we propose a new ESI algorithm in this work, to accurately estimate locations and extents of sources, named Variation Sparse Source Imaging based on L_p-norm (VSSI-L_p). Specifically, we utilize the L_p-norm regularization for spatial variation sources to obtain sparse and robust solutions in the variation domain. The value of p is alterable to fit sparsity and noise flexibly, enabling more reliable estimations. Moreover, we employ the Alternating Direction Method of Multipliers (ADMM) algorithm [20], in order to solve the optimization problem efficiently.

The structure of this paper is outlined as follows. In Sect. 2, we introduce the details of VSSI-L_p. In Sect. 3, we present the simulation design and evaluation metrics. In Sect. 4, we compare the performance of VSSI-L_p with the benchmark algorithms, followed by a brief discussion and conclusion in Sect. 5.

2   Method

We can use the following formula to describe the linear relationship between potential sources and EEG [2,10]

$$b = Ls + \varepsilon \tag{1}$$

in which $b \in \mathbb{R}^{m \times 1}$ is the scalp EEG measurement from m sensors. $s \in \mathbb{R}^{n \times 1}$ denotes the current sources of n sources. $L \in \mathbb{R}^{m \times n}$ is so-called the lead-field matrix, describing the conductivity from potential sources to scalp electrodes. ε is the measurement noise typically assumed to follow a Gaussian distribution [17].

The goal of ESI is to characterize the location and extents information of potential source s with a giving EEG data b. Unfortunately, the number of potential sources n is much bigger than the number of EEG electrodes m, and numerous source configurations are suitable for the scalp measurements. Therefore, narrowing the solution space with constraints is needed for the EEG inverse problem.

$$s = \arg\min_{s} \|b - Ls\|_2^2 + f(s) \tag{2}$$

where the former term is the data fitting term, and the latter term is the regularization term which imposes the constraints.

Evidence has revealed that EEG signals largely arise from synchronized neural electrical activity and the cortical activation is compact [1]. Based on this, we assume the sources have the attributes that are locally smooth and globally clustered [10]. To achieve this, we impose sparsity on the variation domain of sources and penalize the differences in amplitude between adjacent dipoles [4]. Specifically, we introduce the variation operator V, which is defined as

$$V = \begin{bmatrix} v_{11} & v_{12} & \cdots & v_{1n} \\ v_{21} & v_{22} & \cdots & v_{2n} \\ \vdots & \vdots & \ddots & \vdots \\ v_{P1} & v_{P2} & \cdots & v_{Pn} \end{bmatrix} \begin{cases} v_{pi} = 1, v_{pj} = -1, i < j; & \text{if source i,j share edge p} \\ v_{pi} = 0; & \text{otherwise} \end{cases} \tag{3}$$

Here, P represents the number of edges of all triangular grids in source model. Each row of matrix V refers to the corresponding triangle edge. The values 1 and -1 in the pth row characterize a pair of adjacent sources over the pth edge. Then, each non-zero element in the variation source $u = Vs \in \mathbb{R}^{P \times 1}$ denotes the difference of amplitude between the two adjacent sources. To reconstruct locally smooth and globally clustered cortical activities, we assume that the variation source, Vs, is sparse.

Compared to the L_2-norm and L_1-norm regularization, previous studies have revealed that the L_p-norm $(0 < p < 1)$ can provide more accurate solutions with less measurement data [3]. In this work, we employ the L_p-norm regularization term to develop a precise and robust ESI method, VSSI-Lp, to reconstruct extended sources with variation sparsity. The VSSI-Lp algorithm intends to solve the following non-convex optimization problem

$$s = \arg\min_s \|b - Ls\|_2^2 + \lambda\|Vs\|_p^p \tag{4}$$

where $\| \cdot \|_p^p = (\sum_i |s_i|^p)$ with $0 < p < 1$, and $\lambda > 0$ is the regularization parameter. In this work, the value of p of the L_p-norm is empirically selected within the result of simulations.

Equation (4) can be rewritten as

$$s = \arg\min_s \|b - Ls\|_2^2 + \lambda\|u\|_p^p \quad s.t., u = Vs \tag{5}$$

which can be efficiently solved using the ADMM algorithm. Hence, the augmented Lagrangian function is derived as

$$\mathcal{L}(s,u,z) = \|b - Ls\|_2^2 + \lambda\|u\|_p^p + z^\top(Vs - u) + \frac{\rho}{2}\|Vs - u\|_2^2 \tag{6}$$

where $\rho > 0$ is the Lagrangian penalty parameter and $z \in \mathbb{R}^{P \times 1}$ is the Lagrangian multiplier. The variables s, u, z can be updated by alternately minimizing the augmented Lagrangian function $\mathcal{L}$. In the kth iteration, these variables are updated as

$$s^{k+1} = (2L^\top L + \rho V^\top V)^{-1}[2L^\top b + V^\top(\rho u^k - z^k)]$$
$$u^{k+1} = \arg\min_u \lambda\|u\|_p^p + \frac{\rho}{2}\|Vs^{k+1} - u + \frac{1}{\rho}z^k\|_2^2 \tag{7}$$
$$z^{k+1} = z^k + \rho(Vs^{k+1} - u^{k+1})$$

Letting $y = Vs^{k+1} + \frac{1}{\rho}z^k$, u^k is optimized as

$$u^k = \arg\min_u \frac{1}{2}\|y - u\|_2^2 + \frac{\lambda}{\rho}\|u\|_p^p \tag{8}$$

which can be solved using the generalized soft-thresholding (GST) function [21].

In each iteration, we alternately update the variables s, u, z. Generally, the iteration is terminated by reaching the maximum number of iterations or when the relative change of the estimated source s reaches the tolerance.

As for application details, the proposed method VSSI-L_p was conducted on a standard PC (Corei9-10980XE CPU 3 GHz and 128 GB RAM). The algorithm will converge after 500 ADMM iterations, which takes about 35 s, for the given simulation configurations in Sect. 3. For reproducibility purposes, the code for the proposed method is available at https://github.com/Mashirops/VSSI-Lp.git.

3   Simulation Design and Performance Metrics

VSSI-Lp is compared with two conventional L_2-norm constraint ESI methods: (1) wMNE [13], (2) LORETA [15], and two sparse constraint methods implemented in the original source domain: (3) L_1-norm regularization [19] which solves

$$s_{L_1} = \arg \min_s \|b - Ls\|_2^2 + \lambda\|s\|_1, \tag{9}$$

(4) L_p-norm regularization (in this work, we set $p = 0.8$ only for the following formula) [21] which solves

$$s_{L_p} = \arg \min_s \|b - Ls\|_2^2 + \lambda\|s\|_p^p, \tag{10}$$

and (5) VB-SCCD [4].

3.1   Numerical Simulation

Given the absence of ground truth, several Monte Carlo numerical simulations were conducted with Brainstorm [18], using the default ICBM 152 head structure, to validate the performance of those ESI algorithms. The cortex surface was downsampled into 6004 triangular meshes and each triangular stood for a dipole source perpendicular to the cortical surface. We calculated the lead-field matrix L through BEM models based on the 64-channel Neuroscan Quik-cap sensor system.

On the cortex, we randomly selected a seed triangle and added adjacent triangle grids one by one till the whole area reaches a specified value, so as to construct an extended source. Then we applied an amplitude on the constructed source to obtain the ground truth s_{real}. By multiplying it with the lead-field matrix L, we obtained the clean EEG signals. To further simulate actual EEG signals for experiments, we added Gaussian white noise on the clean EEG data. By changing the signal-to-noise ratio (SNR), the noise level is controllable. Here, SNR is defined as $10\log_{10}\left[\frac{\sigma^2(Ls)}{\sigma^2(\varepsilon)}\right]$, where $\sigma^2(\cdot)$ denotes the variance. Monte Carlo numerical simulations in the following scenarios are conducted:

1) Various SNRs - we made use of four levels of SNR (-5, 0, 5 and 10 dB) with only one patch source around $6\,\mathrm{cm}^2$ to evaluate the robustness of our proposed ESI method to noise levels;
2) Various number of channels - we considered using data with varying numbers of channels with SNR = 5 dB, including 100% (62 channels), 75% (47 channels), 50% (31 channels), and 25% (16 channels) to evaluate the robustness of our proposed ESI method to different amounts of data. For each simulation, channels of all non-complete cases were randomly selected.

For each case, we conducted 50 Monte Carlo simulations.

3.2   Performance Metrics

To fully evaluate the performance of ESI algorithms, we carry our four performance metrics. (1) The area under the receiver operating characteristic (ROC) curve (AUC) [4,11], describing the sensitivity and specificity of the reconstructed sources. (2) Spatial dispersion (SD) [11,12], measuring the spatial blurring of the reconstructed sources w.r.t. the ground truth. (3) The distance of localization error (DLE) [9,12], measuring the localization error of the reconstructed sources w.r.t. the ground truth. (4) The normalized relative mean square error (nRMSE) [10], measuring the relative squared error between the normalized reconstructed sources and the normalized ground truth.

Details of four performance metrics can be found in [12]. In general, higher AUC values with lower SD, DLE and nRMSE values imply better performance of the ESI methods. The significance is assessed using the Kruskal-Wallis test. Suppose that the statistic from the Kruskal-Wallis test is significant, we will further conduct Wilcoxon rank sum tests to determine whether VSSI-L_p yields significantly superior estimations against each benchmark algorithm. The Otsu's threshold is employed to visualize the imaging results [10,12].

4   Results

4.1   Simulation Results Analysis

Effect of Different p-values. For the L_p-norm-based methods, the value of p primarily affects the sparsity of the solutions. Here, we compared the performance of VSSI-L_p with various p-values from 0.1 to 0.9 with one patch source under the SNR = 5 dB, to test the influence of the p-value. Figure 1 depicts the performance metrics under different values of p. For $p < 0.5$, the L_p-norm enforced the sparsity too aggressively, leading to a high error rate in source estimation, evidenced by the low AUC, and large DLE, SD, and nRMSE values.

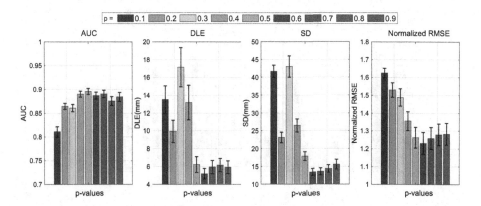

Fig. 1. Performance metrics under various values of p. The figure shows the Mean $\pm$ SEM (standard error of the mean) of the results for 50 Monte-Carlo simulations.

Conversely, the performance of VSSI-L_p remains stable when $p \geq 0.6$. Consequently, for VSSI-L_p, the p-value in the later simulations are empirically selected as 0.6.

Effect of SNRs. Figure 2 presents the performance metrics under various SNRs. As the SNR increases, all algorithms show improved performance, indicated by the increased AUC ($p < 0.05$), decreased DLE ($p < 0.05$), SD ($p < 0.05$) and nRMSE ($p < 0.05$) values. Because L_1 and L_p-norm-based methods enforce sparsity on the original domain, they always produce point estimations, resulting in the lowest SD values at all SNR levels. However, the L_1 and L_p-norm regularizations produce many false estimations, indicated by the large DLE values, and provide little information about source extents, indicated by the lowest AUC values. VSSI-L_p outperforms VB-SCCD, wMNE and LORETA, indicated by the largest AUC, lowest SD and nRMSE values.

Figure 3 provides an imaging example under different SNRs. As expected, wMNE and LORETA produce too diffused estimations, while the L_1-norm and L_p constraint in the original source domain obtained several point sources around

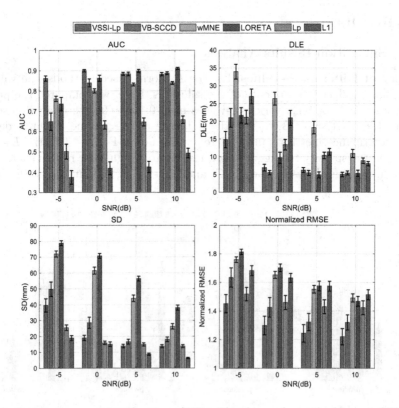

Fig. 2. Performance metrics of various SNRs. This figure shows the Mean ± SEM of the results for 50 Monte-Carlo simulations.

or in the ground truth. VB-SCCD shows better estimations than the other benchmark methods, although it provides some spurious sources around the actual activities. Among all the ESI methods, the reconstructions by the proposed VSSI-L_p are the most accurate in matching the ground truth.

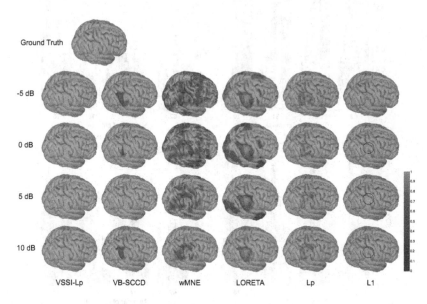

Fig. 3. Imaging example under different SNR levels. The thresholds of the estimated maps are obtained using Ostu's method.

Effect of the Number of Channels. Figure 4 presents the performance metrics under different numbers of EEG channels. As the number of channels decreases, the performance of all methods declines due to the loss of measurement information. VSSI-L_p exhibits good robustness and provides more accurate information on extended sources than other methods even with some missing data, indicated by larger AUC ($p < 0.05$) and lower DLE ($p < 0.05$), SD ($p < 0.05$, except for L_1-norm) and nRMSE ($p < 0.05$) values.

4.2   Real Data Result Analysis

In this subsection, we utilized the public EEG dataset to further assess the practical efficacy of VSSI-L_p, which is the epilepsy EEG data from Brainstorm. Detailed descriptions can be found at https://neuroimage.usc.edu/brainstorm/ DatasetEpilepsy. In this work, we followed the tutorial of Brainstorm to derive the head model, lead-field matrix and EEG data for source localization. The EEG data is presented in Fig. 5(a), which is an average of 58 tails, and the data at the peak (0 ms) is used for source imaging. Figure 5(b) presents the

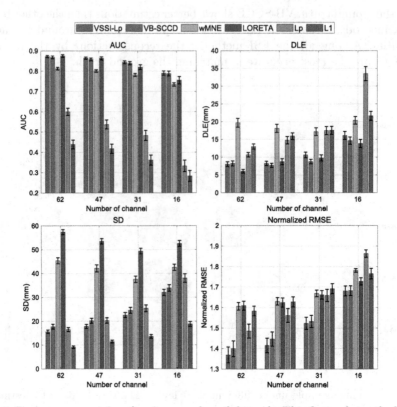

Fig. 4. Performance metrics of various number of channels. This figure shows the Mean ± SEM of the results for 50 Monte Carlo simulations.

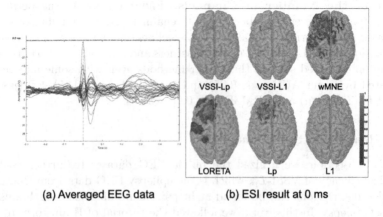

(a) Averaged EEG data   (b) ESI result at 0 ms

Fig. 5. Estimated sources of epilepsy data. (a) is the waveforms of averaged EEG data; (b) is the result of each algorithm at 0 ms.

imaging results. wMNE and LORETA provide diffused estimations, while the result of LORETA is smoother. L_1-norm regularization obtains 2 point sources, and L_p-norm regularization yields several incoherent point sources around the left frontal lobe. Results of VSSI-L_p and VB-SCCD provide clear information about the location and extent of potential sources, which conform to clinical findings in [5].

5   Discussion and Conclusion

Here in this work, we proposed a new ESI method, VSSI-L_p, to reconstruct the location and extent of brain activity. VSSI-L_p method enforces the sparsity of potential brain sources using the L_p-norm regularization in the variation domain. By utilizing the ADMM and GST algorithms, the solution of VSSI-L_p can be efficiently obtained. Numerical simulations and real data analysis reveal the superior performance of VSSI-L_p.

Due to the highly under-determined nature of ESI, such methods are difficult to work out. Even worse, it is essential for neuroscience and neurology applications to infer the spatial distribution of potential brain sources from limited measurements. Methods based on L_2-norm constraint, such as wMNE and LORETA, produce too blurred and diffused results, indicated by the large DLE and SD values in Fig. 2. Methods based on L_1-norm and L_p-norm improve the spatial resolution of their estimations by enforcing sparsity in the original source domain. However, these conventional sparse constrained methods provide little information on the extent of brain activity because they miss the most active sources on the cortex.

To estimate the localization and extent of potential extended sources, VB-SCCD [4] employed L_1-norm sparse constraint in the spatial variation domain which significantly improved the reconstructions of extended sources. However, as suggested in [2], VB-SCCD over-estimates the extent of cortical activities, especially for sources with small extents. This may be because the mathematical properties of L_1-norm make it not sparse enough [3]. To enforce sparsity more aggressively, we proposed VSSI-L_p, which employed the L_p-norm ($0 < p < 1$) instead of the L_1-norm regularization in the variation domain. Results of Monte Carlo simulations demonstrate the superiority of VSSI-L_p over VB-SCCD with higher AUC values and lower DLE, SD, and nRMSE values in most cases.

In this work, the regularization parameter, λ, was selected using cross-validation. In our future work, we plan to investigate using the Bayesian probability framework to model VSSI-L_p and allow the model to infer the parameters automatically. Additionally, we will also apply the proposed method for brain disease diagnosis, cortical network analysis and fine motor imagery decoding.

Acknowledgments. This work was supported in part by the National Natural Science Foundation of China under Grants 62136002, and the Natural Science Foundation of Chongqing under Grant CSTB2022NSCQ-MSX0291 and cstc2022ycjh-bgzxm0004.

References

1. Bai, X., Towle, V.L., He, E.J., He, B.: Evaluation of cortical current density imaging methods using intracranial electrocorticograms and functional MRI. Neuroimage **35**(2), 598–608 (2007)
2. Becker, H., Albera, L., Comon, P., Gribonval, R., Wendling, F., Merlet, I.: Brain-source imaging: from sparse to tensor models. IEEE Signal Process. Mag. **32**(6), 100–112 (2015)
3. Chartrand, R., Staneva, V.: Restricted isometry properties and nonconvex compressive sensing. Inverse Prob. **24**(3), 035020 (2008)
4. Ding, L.: Reconstructing cortical current density by exploring sparseness in the transform domain. Phys. Med. Biol. **54**(9), 2683 (2009)
5. Dümpelmann, M., Ball, T., Schulze-Bonhage, A.: sLORETA allows reliable distributed source reconstruction based on subdural strip and grid recordings. Hum. Brain Mapp. **33**(5), 1172–1188 (2012)
6. Fang, T., et al.: Decoding motor imagery tasks using ESI and hybrid feature CNN. J. Neural Eng. **19**(1), 016022 (2022)
7. He, B., Sohrabpour, A., Brown, E., Liu, Z.: Electrophysiological source imaging: a noninvasive window to brain dynamics. Annu. Rev. Biomed. Eng. **20**, 171–196 (2018)
8. Hou, Y., Zhou, L., Jia, S., Lun, X.: A novel approach of decoding EEG four-class motor imagery tasks via scout ESI and CNN. J. Neural Eng. **17**(1), 016048 (2020)
9. Liu, K., Wang, Z., Yu, Z., Xiao, B., Yu, H., Wu, W.: WRA-MTSI: a robust extended source imaging algorithm based on multi-trial EEG. IEEE Trans. Biomed. Eng. **70**(10), 2809–2821 (2023)
10. Liu, K., Yu, Z.L., Wu, W., Gu, Z., Li, Y.: Imaging brain extended sources from EEG/MEG based on variation sparsity using automatic relevance determination. Neurocomputing **389**, 132–145 (2020)
11. Liu, K., Yu, Z.L., Wu, W., Gu, Z., Li, Y., Nagarajan, S.: Bayesian electromagnetic spatio-temporal imaging of extended sources with Markov random field and temporal basis expansion. Neuroimage **139**, 385–404 (2016)
12. Liu, K., Yu, Z.L., Wu, W., Gu, Z., Li, Y., Nagarajan, S.: Variation sparse source imaging based on conditional mean for electromagnetic extended sources. Neurocomputing **313**, 96–110 (2018)
13. Lucka, F., Pursiainen, S., Burger, M., Wolters, C.H.: Hierarchical Bayesian inference for the EEG inverse problem using realistic FE head models: depth localization and source separation for focal primary currents. Neuroimage **61**(4), 1364–1382 (2012)
14. Ou, W., Hämäläinen, M.S., Golland, P.: A distributed spatio-temporal EEG/MEG inverse solver. Neuroimage **44**(3), 932–946 (2009)
15. Pascual-Marqui, R.D., Michel, C.M., Lehmann, D.: Low resolution electromagnetic tomography: a new method for localizing electrical activity in the brain. Int. J. Psychophysiol. **18**(1), 49–65 (1994)
16. Sohrabpour, A., He, B.: Exploring the extent of source imaging: recent advances in noninvasive electromagnetic brain imaging. Curr. Opin. Biomed. Eng. **18**, 100277 (2021)
17. Sohrabpour, A., Lu, Y., Worrell, G., He, B.: Imaging brain source extent from EEG/MEG by means of an iteratively reweighted edge sparsity minimization (IRES) strategy. Neuroimage **142**, 27–42 (2016)

18. Tadel, F., Baillet, S., Mosher, J.C., Pantazis, D., Leahy, R.M.: Brainstorm: a user-friendly application for MEG/EEG analysis. Comput. Intell. Neurosci. **2011**, 1–13 (2011)
19. Tibshirani, R.: Regression shrinkage and selection via the Lasso. J. Roy. Stat. Soc. Ser. B (Methodol.) **58**(1), 267–288 (1996)
20. Wang, Y., Yin, W., Zeng, J.: Global convergence of ADMM in nonconvex nonsmooth optimization. J. Sci. Comput. **78**, 29–63 (2019)
21. Zuo, W., Meng, D., Zhang, L., Feng, X., Zhang, D.: A generalized iterated shrinkage algorithm for non-convex sparse coding. In: Proceedings of the IEEE International Conference on Computer Vision, pp. 217–224 (2013)

Perceptual Quality Assessment
of Omnidirectional Audio-Visual Signals

Xilei Zhu, Huiyu Duan(ID), Yuqin Cao, Yuxin Zhu, Yucheng Zhu, Jing Liu,
Li Chen, Xiongkuo Min, and Guangtao Zhai(✉)

Institute of Image Communication and Network Engineering, Shanghai Jiao Tong
University, Shanghai, China
{xilei_zhu,huiyuduan,caoyuqin,rye2000,zyc420,hilichen,
minxiongkuo,zhaiguangtao}@sjtu.edu.cn, jliu_tju@tju.edu.cn

Abstract. Omnidirectional videos (ODVs) play an increasingly important role in the application fields of medical, education, advertising,
tourism, *etc*. Assessing the quality of ODVs is significant for service-
providers to improve the user's Quality of Experience (QoE). However,
most existing quality assessment studies for ODVs only focus on the
visual distortions of videos, while ignoring that the overall QoE also
depends on the accompanying audio signals. In this paper, we first
establish a large-scale audio-visual quality assessment dataset for omnidirectional videos, which includes 375 distorted omnidirectional audio-
visual (A/V) sequences generated from 15 high-quality pristine omnidirectional A/V contents, and the corresponding perceptual audio-visual
quality scores. Then, we design three baseline methods for full-reference
omnidirectional audio-visual quality assessment (OAVQA), which combine existing state-of-the-art single-mode audio and video QA models via
multimodal fusion strategies. We validate the effectiveness of the A/V
multimodal fusion method for OAVQA on our dataset, which provides
a new benchmark for omnidirectional QoE evaluation. Our dataset is
available at https://github.com/iamazxl/OAVQA.

Keywords: Audio-visual Quality · Omnidirectional videos · Quality
assessment · Dataset

1 Introduction

Virtual Reality (VR) has attracted substantial attention from industry and
research communities due to its ability to provide users with a stereoscopic
and immersive experience through Head-Mounted Displays (HMDs) [6,8]. Omnidirectional Videos (ODVs), *a.k.a*, 360° videos, panoramic videos or spherical
videos, have emerged as a significant form of VR content. By using VR HMDs
and adjusting their head orientation, users can explore the audio-visual content
in any direction. This immersive experience of simulating real-world scenes has
contributed to the popularity of ODVs in various application fields, including
medical, education, advertising, tourism, *etc*.

L. Fang et al. (Eds.): CICAI 2023, LNAI 14474, pp. 512–525, 2024.
https://doi.org/10.1007/978-981-99-9119-8_46

Compared to traditional videos, ultra high-definition ODVs contain more scene information and multi-channel audio information, which results in a doubling of ODV data volume. Due to the huge amount of data, playback stucking and quality switching caused by network delays and fluctuations usually occur during video transmission, which leads to the degradation of ODVs quality, and further affects the QoE of ODVs. Moreover, ODVs may also suffer from the distortions introduced during the process of capturing or displaying, which further decreases the QoE. Therefore, to provide users with a smooth viewing experience, it is important to monitor the quality of ODVs during the procedure of shooting, codec, transmission, *etc.*, and perform optimization accordingly.

In the past few decades, many objective quality assessment methods have been proposed for traditional plane videos [19,23], and some recent works have also explored the problem of audio-visual video quality assessment [21]. Recently, with the popularity of VR, many studies have explored the problem of omnidirectional image quality assessment [3,24] and omnidirectional video quality assessment [13]. However, most omnidirectional video quality assessment research only focuses on the single-mode signal, *i.e.*, visual information, few works have investigated the multimodal quality assessment of ODVs incorporating audio information. As an important part of ODVs, spatial audio may strongly influence the human perceptual quality, thus it is necessary to conduct in-depth research on the audio-visual quality assessment of the omnidirectional videos.

In this paper, we make three contributions to the omnidirectional audio-visual quality assessment (OAVQA) field. Firstly, we construct a large-scale omnidirectional audio-visual quality assessment dataset to solve the poverty problem of the corresponding dataset. We first collected 15 high-quality reference omnidirectional audio-visual (A/V) content, and generated 375 distorted ODVs degraded from them. Subsequently, 22 subjects were recruited to participate in the subjective quality assessment experiment, and the audio-visual quality ratings of the reference and distorted videos were collected. Secondly, we design three baseline methods for full-reference omnidirectional AVQA. The baseline models first utilize the existing state-of-the-art audio and video single-mode quality assessment methods to predict the audio quality and video quality of ODVs, respectively, then utilize different multimodal fusion strategies to fuse A/V prediction results and obtain the overall quality results of the ODVs. Thirdly, we compare and analyze the prediction performance of these models on our dataset, and establish a new benchmark for future studies on OAVQA.

2 Related Work

2.1 Omnidirectional Video Quality Assessment Dataset

Table 1 provides an overview of several existing omnidirectional video quality assessment datasets. It can be observed that most of the existing ODV quality assessment datasets lack spatial audio information, and mainly focus on visual distortions, while audio distortions are rarely been considered.

2.2 Quality Assessment Models

Omnidirectional Video Quality Assessment. As a common storage format of ODVs, ERP projection has severe mapping stretches near the poles. In order to solve this problem, Yu *et al.* [31] proposed a spherical PSNR scheme (S-PSNR), which is based on a set of uniform sampling points on the spherical surface, the corresponding position on the mapping plane is calculated by different mapping formulas. Sun *et al.* proposed the Weighted to Spherically uniform PSNR (WS-PSNR) [25], which is directly performed in the original format and combined with different stretching weights according to different mapping methods. Anwar *et al.* [1] established an ODVs quality assessment model using the Bayesian inference method, and evaluated the impact of buffering on users' perceptual quality at different bitrates. Fan *et al.* [10] established an ODVs dataset that contains various distortions such as compression distortion and quality switching, and then used machine learning methods to establish VQA models.

Table 1. An overview of omnidirectional video quality assessment datasets. "Mute" means mute audio and "ambisonics" indicates spatial audio. SI and TI represent spatial information and temporal information respectively. QP indicates quantization parameter and CRF means constant rate factor, which is used to control the video bitrate.

Dataset	Video Num	Audio	Distortion Type	QoE
Schatz *et al.* [22]	10	Mute	Stalling	MOS(1~5)
Meng *et al.* [20]	774	Mute	Frame size, Frame rate, Quantization stepsize, Resolutions	MOS(1~10)
Fei *et al.* [11]	468	Mute	Bandwidth, Packet loss, Latency, Presence	MOS(1~5)
Anwar *et al.* [1]	208	Mute	Bitrate, Stalling	MOS
Fan *et al.* [10]	48	Mute	Bitrate, Gender, Presence, TI, SI	MOS(0~9)
IVQAD [9]	150	Mute	Bitrate, Frame rate, Resolution	MOS(1~5)
VQA-ODV [18]	600	Mute	QP, Projection format	DMOS(0 ~ 60)
Fela *et al.* [12]	576	Ambisonics	QP, Resolution, Audio bitrate	MOS(0 ~ 100)
Ours	**390**	**Ambisonics**	**Audio bitrate, CRF, Resolution, Noise, Blur, Stucking**	**MOS(1~10)**

Omnidirectional Audio-Visual Quality Assessment. As an important part of ODVs, the influence of spatial audio on perceptual quality has rarely been studied. Zhang *et al.* [33] presented a quality assessment methodology for audio-visual multimedia in virtual reality environment. They presented a panoramic audio-visual dataset and the quality factors which represent different distortions were applied as the input to neural network. Fela *et al.* [14] utilized PSNR and its variants designed for ODVs, *i.e.*, WS-PSNR, CPP-PSNR and S-PSNR [25,31,32], as the quality scores and studied the perceptual audio-visual quality prediction based on the fusion of these scores [13]. Four machine learning models including multiple linear regression, decision tree, random forest, and support vector machine (SVM), were tested.

3 Omnidirectional Audio-Visual Quality Assessment Dataset (OAVQAD)

3.1 Reference and Distorted Contents

We first captured 162 different ODVs with different scenes using a professional VR camera Insta360 Pro2. Then, we selected 15 high-quality ODVs from the collected ODVs as the reference videos in our OAVQAD. We utilized FFmpeg to clip the duration of the selected ODVs to 6 s. Each ODV has a resolution of 8K (7680 × 3840) in equirectangular projection (ERP) format with a frame rate of 29.97 fps. All ODVs contain first order ambisonics (FOA) with 48,000 Hz audio sampling rate and four audio channels. The audio and video formats are shown in Table 2. The ODV contents include acappella chorus, shopping, guitar playing, restaurant ordering, *etc.* Figure 1 shows the ERP format previews of the selected 15 reference ODVs.

We utilized advanced audio coding (AAC) as the audio encoding method provided by FFmpeg 4.4, and used constant bitrate (CBR) mode to set the audio bitrate to 64Kbps, 32Kbps and 16Kbps, respectively, thereby generating three levels of perceptually well-separated audio compression distortion. Then, we chose HEVC as the video encoding method provided by FFmpeg libx265 encoder, and for each source video we applied 3 different compression levels, *i.e.,* 32, 37 and 42 in constant rate factor (CRF) mode. Besides, we also set the video resolution to three levels including 4K (3840×1920), 2K (1920×960), 1K (1080×540). Moreover, in order to adapt to a wider range of application scenarios, we further introduced more abundant distortion types and added three

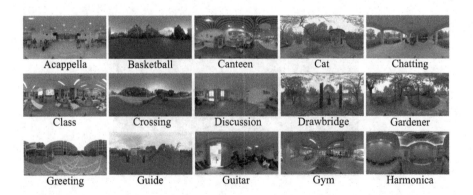

Fig. 1. EPR format previews of 15 reference ODVs used in our OAVQAD.

Table 2. Omnidirectional audio and video format parameters.

	Resolution	Frame rate	Bitrate	Format	Bit depth	Duration	Encoding
Video	8K	29.97fps	144Mbps	YUV420	8bit	6s	H.265
Audio	-	-	3072Kbps	FOA	16bit	6s	AAC-LC

types of distortions [5,7] including noise, blur, and stucking, and generated distorted ODVs with various levels of these distortions. To summarize, we applied 25 distortion conditions to 15 reference ODVs, resulting in a total of 375 (15 × 25) distorted ODVs.

3.2   Subjective Experiment Methodology

Experiment Apparatus. Since the subjective experiment was needed to be conducted in a VR immersive environment, we used HTC Vive Pro Eye as the HMD to demonstrate ODVs and collect subjective quality ratings. The subjective experiment platform used to play 8K ODVs and perform scoring interaction was build based on Unity 1.1.0 as shown in Fig. 2.

Experiment Procedure. The subjective experiment was conducted in a subjective study room in a university. A total of 22 subjects (14 males and 8 females) were invited to participate in the subjective experiment. The subjects were between 20 and 28 years old (mean 22.62, variance 5.23) and were all graduate and undergraduate students. All subjects had normal or corrected-to-normal vision and hearing. In the experiment, subjects firstly received the guidance on the use of VR equipment, including HMD and controllers. Then a training session was performed for the subjects, making them be familiarized with the user interface as well as the general range and types of distortions. In the testing session, subjects watched 390 ODVs and gave perceptual scores of the overall A/V quality. The order of the test videos was random for each subject to avoid bias.

Fig. 2. Demonstration of the subjective experiment interface based on the Unity platform.

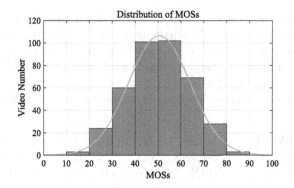

Fig. 3. Histogram of MOS distribution in the database.

3.3 Subjective Data Processing and Analysis

We followed the subjective data processing method recommended by ITU [2,4] to perform the outlier detection and subject rejection. None of the 22 subjects was identified as an outlier and eliminated. We normalized the raw scores of subjects to Z-scores ranging between 0 and 100 and calculated the mean of Z-scores to obtain the final mean opinion scores (MOSs), which are formulated as follows:

$$z_{ij} = \frac{r_{ij} - \mu_i}{\sigma_i}, \quad z'_{ij} = \frac{100\left(z_{ij} + 3\right)}{6}, \tag{1}$$

$$\mathrm{MOS}_j = \frac{1}{N} \sum_{i=1}^{N} z'_{ij}, \tag{2}$$

where r_{ij} is the original score of the i-th subject on the j-th sequence, μ_i and σ_i are the mean rating and the standard deviation given by subject i, N is the total number of subjects. Figure 3 draws the histogram of MOS distribution over the entire database, indicating that the perceptual quality scores are widely distributed in the $[0, 100]$ interval, basically covering every score segment, and generally showing a normal distribution. It also manifests that the perceptual quality distribution conforms to our expectations and the distortions setting is quite reasonable.

4 Objective Omnidirectional Audio-Visual Quality Assessment

4.1 Single-Mode Models

Many video and audio quality assessment methods have been proposed separately in previous studies. These quality assessment algorithms, only predict quality of single-modal audio or video signals, can be called as single-mode quality assessment methods. We first utilize the existing state-of-the-art single-mode quality assessment methods to predict the omnidirectional video and audio quality, respectively. Since both the single-mode AQA and VQA prediction scores can characterize one aspect of the distortion severity of the distorted video, it is reasonable to directly use the single-mode models to predict the overall audio-visual quality score of the ODVs.

The well-known single-mode assessment models adopted in this paper are introduced as follows:

- **Video**: VMAF [19], SSIM [28], MS-SSIM [29], VIFP [23], FSIM [34], GMSD [30], WS-PSNR [25], CPP-PSNR [32], S-PSNR [31].
- **Audio**: PEAQ [27], STOI [26], VISQOL [16], LLR [17], SNR [17], segSNR [15].

4.2 Weighted-Product Fusion

A single-mode audio/visual quality assessment metric can only characterize one quality aspect thus cannot fully represent the overall subjective perceptual quality of an ODV. Therefore, it is important to use appropriate multimodal feature fusion method to predict the A/V quality of ODVs. The simplest fusion method is to directly multiply the quality scores of a VQA model and an AQA model as the overall quality score of ODVs.

However, for human audio-visual perception, video and audio quality often occupy different importance in ODVs, and people may pay more attention to visual quality. The weighted product can balance the influence of different modalities by assigning different weights to each of them, so the weighted product is a better choice for score fusion compared to the direct multiplication method. The weighted product can be formulated as

$$Q_{av} = \hat{Q}_v^w \cdot \hat{Q}_a^{1-w}, \tag{3}$$

where $\hat{Q}_a$ and $\hat{Q}_v$ are normalized score of the audio and video, w and $1 - w$ represent the weights of video and audio quality respectively, $0 \leq w \leq 1$. $\hat{Q}_a$ and $\hat{Q}_v$ are calculated by $\hat{Q}_a = \frac{Q_a - Q_{a_{\min}}}{Q_{a_{\max}} - Q_{a_{\min}}}$ and $\hat{Q}_v = \frac{Q_v - Q_{v_{\min}}}{Q_{v_{\max}} - Q_{v_{\min}}}$, where $Q_{a_{\min}}$, $Q_{a_{\max}}$, $Q_{v_{\min}}$ and $Q_{v_{\max}}$ bound Q_a and Q_v respectively. The optimal weights depend on the used single-mode A/V quality evaluation models and we vary the weight from 0 to 1 with 0.05 step increment to find the optimal weight w. Since the score ranges of the video and audio quality assessment models may be different, the multiplication method can only be performed after they are appropriately scaled or normalized.

Table 3. Video and audio quality prediction algorithms and their corresponding feature types.

Category	Models	Feature	Decomposed features.
Video	VMAF [19]	6	4 scales of VIF, detail loss, motion
	SSIM [28]	2	Luminance similarity, contrast and structural similarity
	MS-SSIM [29]	6	Luminance similarity, 5 scales of contrast and structural similarity
	VIFP [23]	4	4 scales of VIFP features
	FSIM [34]	3	Phase congruency, gradient magnitude, and chrominance similarity
	GMSD [30]	2	Mean and standard deviation of gradient magnitude similarity
	WS-PSNR [25]	3	PSNR of Y, U, V components
	CPP-PSNR [32]	3	PSNR of Y, U, V components
	S-PSNR [31]	3	PSNR of Y, U, V components
Audio	PEAQ [27]	11	11 model output variables before the neural network
	STOI [26]	1	The complete algorithm
	VISQOL [16]	3	Narrowband, wideband, fullband versions of VISOOL
	LLR [17]	1	The complete algorithm
	SNR [17]	1	The complete algorithm
	seg-SNR [15]	1	The complete algorithm

4.3   Support Vector Regression Fusion

Since Support Vector Regression (SVR) is a commonly used machine learning algorithm for establishing nonlinear relationships between inputs and outputs, we also utilize the SVR method to integrate the quality prediction scores of single-mode models

$$Q_{av} = SVR(Q_v, Q_a), \qquad (4)$$

where Q_v and Q_a represent the quality prediction scores of video and audio, respectively, and Q_{av} denotes the fused A/V quality scores. In this case, SVR uses the single-mode quality scores predicted by traditional AQA and VQA algorithms respectively as the inputs, and the quality score (*i.e.*, MOS) as the labels for regression function training.

The performance of SVR fusion methods can be further improved by substituting scores with quality-aware feature vectors $\mathbf{f}_v$ and $\mathbf{f}_a$, which can be either hand-crafted features or extracted features from existing popular AQA and VQA models. In this way, we can better fuse video and audio quality prediction results by fully utilizing the quality features of audio and video, thereby improving the performance of the entire model. This feature-based fusion method can be expressed as:

$$Q_{av} = SVR(\mathbf{f}_v, \mathbf{f}_a). \qquad (5)$$

The video and audio quality-aware feature vectors used here are extracted from some existing AQA and VQA models, which are summarized in Table 3.

5   Experiment Validation

5.1   Evaluation of Single-Mode Models

We test different single-mode quality assessment models (6 audio models and 9 video models) on our omnidirectional AVQA dataset to analyze the effectiveness of single-mode quality models. Experimental results are illustrated in Fig. 4.

For AQA models, STOI, VISQOL, SNR, and segSNR yield relatively good performances on our database, in which STOI achieves the both highest SRCC and PLCC performance. Most of the VQA models show similar performance, and all of them are not able to predict A/V quality effectively with SRCC and PLCC below 0.6. The above analysis shows that most single-mode quality assessment models have a poor performance on our OAVQAD, indicating the necessity of fusing single-mode quality prediction results for more accurate OAVQA.

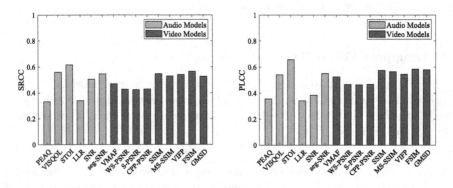

Fig. 4. Performances of single-mode models on overall audio-visual quality prediction.

5.2   Evaluation of Weighted-Product Fusion

For weighted-product fusion methods, we randomly divide the dataset into 80% training set and 20% test set. All distorted ODVs from the same reference ODVs are placed in the same set to ensure that the video content of the two set are completely separated.

In the weighted-product fusion, a total of 54 (9 video models × 6 audio models) weighted product quality fusion models are generated. In order to normalize the prediction scores of the single-mode quality prediction models, the following normalization functions are used: $Q'_{VMAF} = \frac{1}{100}Q_{VMAF}$, $Q'_{WS\text{-}PSNR} = \frac{1}{29}(Q_{WS\text{-}PSNR} - 23)$, $Q'_{S\text{-}PSNR} = \frac{1}{29}(Q_{S\text{-}PSNR} - 23)$, $Q'_{CPP\text{-}PSNR} = \frac{1}{29}(Q_{CPP\text{-}PSNR} - 23)$, $Q'_{GMSD} = 1 - \frac{1}{0.26}Q_{GMSD}$, $Q'_{PEAQ} = 1 + \frac{1}{3.5}(Q_{PEAQ} - 0.21)$, $Q'_{LLR} = 1 - \frac{1}{1.2-0.7}(|Q_{LLR}| - 0.7)$, $Q'_{SNR} = \frac{1}{20}Q_{SNR}$, $Q'_{segSNR} = \frac{1}{35+2}(Q_{segSNR} + 2)$. The prediction scores of other models are already bounded in $[0, 1]$, no further normalization is needed.

Table 4 shows the performance of weighted product fusion models. Among these methods, the models fused by VQA algorithms VMAF, MS-SSIM, GMSD, and the AQA algorithms STOI, VISQOL, SNR show relatively better performances. The model combining GMSD and STOI achieves the best performance in terms of SRCC. In addition, with the same AQA components, the performance of fusion models using different VQA components has little difference, which manifests that different AQA components have larger impact on the performance of fusion models. Moreover, the mean optimal weight for visual modality of 54 weighted product models is 0.7231, suggesting that visual modality has a greater impact on QoE than audio modality.

5.3   Evaluation of SVR Fusion

SVR fusion includes two methods including the score-based fusion and the feature-based fusion. A total of 108 (9 video models × 6 audio models × 2 SVR conditions) models are tested and the normalization process is no longer required. In SVR fusion models, the radial basis function (RBF) is selected as the kernel function, the parameter γ of the kernel function is 0.05, and the penalty factor C is 1024. Table 5 shows the performance of SVR fusion models.

Table 4. Performances of weighted-product fusion-based A/V quality models. The top 3 models are in bold.

Criteria	Video	Weighted Product					
	Model	PEAQ	STOI	VISQOL	LLR	SNR	segSNR
SRCC	VMAF	0.5783	**0.7790**	0.7157	0.5745	0.7432	0.6660
	WS-PSNR	0.5252	0.7348	0.6911	0.5507	0.7124	0.6658
	S-PSNR	0.5182	0.7292	0.6886	0.5460	0.7068	0.6576
	CPP-PSNR	0.5246	0.7333	0.6914	0.5499	0.7121	0.6652
	SSIM	0.5605	0.7717	0.7289	0.5123	0.6783	0.6372
	MS-SSIM	0.6131	**0.7998**	0.7511	0.6161	0.7596	0.6942
	VIFP	0.5916	0.7746	0.7332	0.5978	0.7499	0.7017
	FSIM	0.5386	0.7563	0.7259	0.5638	0.6632	0.6188
	GMSD	0.6151	**0.8044**	0.7358	0.6246	0.7530	0.6844
PLCC	VMAF	0.6124	0.7885	0.7265	0.6324	0.7442	0.6484
	WS-PSNR	0.5595	0.7576	0.7407	0.6020	0.7351	0.5960
	S-PSNR	0.5558	0.7530	0.7404	0.5966	0.7287	0.5984
	CPP-PSNR	0.5594	0.7567	0.7401	0.6001	0.7340	0.5931
	SSIM	0.5984	0.7917	0.7604	0.5601	0.6917	0.6886
	MS-SSIM	0.6405	**0.8124**	0.7792	0.6561	0.7710	0.7270
	VIFP	0.6188	**0.8057**	0.7294	0.6415	0.7522	0.6758
	FSIM	0.5806	0.7743	0.7682	0.6015	0.6693	0.6650
	GMSD	0.6357	**0.8112**	0.7518	0.6557	0.7587	0.6894

Table 5. Performances of SVR fusion-based A/V quality models. The top 3 models in terms of each metric are in bold.

Criteria	Video Model	SVR (Quality Score)						SVR (Quality Feature)					
		PEAQ	STOI	VISQOL	LLR	SNR	segSNR	PEAQ	STOI	VISQOL	LLR	SNR	segSNR
SRCC	VMAF	0.5481	0.7855	0.7141	0.5676	0.5688	0.6391	0.8343	0.8428	0.8566	0.6052	0.6119	0.6818
	WS-PSNR	0.5306	0.7625	0.6974	0.5506	0.5453	0.6269	0.8035	0.7787	0.8171	0.5612	0.5582	0.6346
	S-PSNR	0.5221	0.7593	0.6966	0.5418	0.5365	0.6202	0.8030	0.7764	0.8123	0.5550	0.5476	0.6263
	CPP-PSNR	0.5301	0.7626	0.6982	0.5495	0.5452	0.6272	0.8039	0.7806	0.8174	0.5612	0.5584	0.6356
	SSIM	0.5023	0.7246	0.6734	0.4651	0.5636	0.6222	0.7385	0.7475	0.7654	0.5314	0.5643	0.6492
	MS-SSIM	0.5809	0.7984	0.7407	0.5963	0.6020	0.6727	0.8201	0.8342	0.8654	0.6136	0.6103	0.6752
	VIFP	0.5983	**0.8412**	**0.8149**	0.6149	0.6043	0.6887	**0.8751**	**0.8726**	**0.8881**	0.6545	0.6464	0.7311
	FSIM	0.5046	0.7290	0.6775	0.4604	0.5727	0.6227	0.7485	0.7413	0.7646	0.5275	0.5603	0.6357
	GMSD	0.5749	**0.8048**	0.7450	0.6178	0.6020	0.6669	0.8426	0.7982	0.8459	0.6084	0.5940	0.6599
PLCC	VMAF	0.5845	0.8111	0.7808	0.6275	0.6075	0.6832	0.8440	0.8543	0.8619	0.6527	0.6552	0.7303
	WS-PSNR	0.5789	0.7831	0.7521	0.6125	0.5978	0.6704	0.8113	0.8012	0.8286	0.6312	0.6118	0.6919
	S-PSNR	0.5703	0.7802	0.7472	0.6044	0.5895	0.6599	0.8109	0.7975	0.8229	0.6268	0.6030	0.6800
	CPP-PSNR	0.5770	0.7832	0.7514	0.6115	0.5974	0.6712	0.8118	0.8026	0.8286	0.6313	0.6122	0.6930
	SSIM	0.4297	0.7340	0.7125	0.4892	0.4234	0.5577	0.7656	0.7729	0.7721	0.5641	0.5674	0.6404
	MS-SSIM	0.6187	0.8168	0.7874	0.6542	0.6476	0.7075	0.8350	0.8508	0.8697	0.6630	0.6632	0.7191
	VIFP	0.6358	**0.8565**	**0.8374**	0.6752	0.6591	0.7382	**0.8779**	**0.8828**	**0.8941**	0.6950	0.6862	0.7748
	FSIM	0.4275	0.7330	0.7102	0.4827	0.4298	0.5427	0.7647	0.7647	0.7676	0.5556	0.5595	0.6286
	GMSD	0.6065	**0.8249**	0.7986	0.6564	0.6495	0.6964	0.8473	0.8170	0.8539	0.6488	0.6409	0.6889

It can be observed that quality score-based SVR fusion models achieve similar performance compared with the weighted-product fusion models, while quality feature-based SVR fusion models achieve much better performance compared to above two methods. The models combining the AQA components, PEAQ, STOI and VISQOL, and the VQA components VIFP and GMSD have relatively better performance.

Figure 5 demonstrates the performance improvement obtained by each single-mode AQA and VQA model, which further confirms the above phenomenon. The performance improvement of each single-mode model is calculated by averaging the SRCC improvements of all combinations of this model with the models from another perceptual mode. It can be observed that only VISQOL and VIFP models gain performance improvement by replacing weighted-product with SVR, suggesting that the weighted-product fusion is generally a more feasible method. Futhermore, Fig. 5 also illustrates that it is more efficient to decompose the single-mode VQA and AQA scores into ODVs' quality features. It can be observed that the feature-based regression models achieve different degrees of performance improvement for different VQA and AQA fusion, among which PEAQ achieved a significant improvement with nearly 50%. Some of these models, *e.g.*, STOI, LLR, SNR and segSNR, have a small performance progress caused by feature extraction, we reasonably speculate that these algorithm models are not easy to decompose.

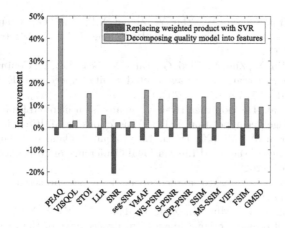

Fig. 5. Performance improvements in terms of SRCC introduced by replacing weighted-product fusion with quality score-based SVR fusion, and decomposing quality models into features during SVR fusion.

6   Conclusion

In this work, we construct an informative omnidirectional audio-visual quality assessment dataset, which involves 390 omnidirectional videos with ambisonics and the corresponding perceptual scores collected from 22 participants under immersive environment. Based on our dataset, we design three types of baseline AVQA models which combine AQA and VQA models via two multimodal fusion methods to predict quality scores of ODVs. Moreover, quantitative analyses for the performance of these models are conducted to evaluate the predictive effect of different objective models. The experiment results on our dataset show that SVR fusion based on quality-aware features have the best performance. Our dataset, objective baseline methods and established benchmark can great facilitate the further research of dataset design and algorithm improvement for OAVQA.

Acknowledgement. This work is supported by National Key R&D Project of China (2021YFE0206700), NSFC (61831015, 62101325, 62101326, 62271312, 62225112), Shanghai Pujiang Program (22PJ1407400), Shanghai Municipal Science and Technology Major Project (2021SHZDZX0102), STCSM (22DZ2229005).

References

1. Anwar, M.S., Wang, J., Ullah, A., Khan, W., Ahmad, S., Fei, Z.: Measuring quality of experience for 360-degree videos in virtual reality. SCIENCE CHINA Inf. Sci. **63**, 1–15 (2020)
2. Duan, H., Guo, L., Sun, W., Min, X., Chen, L., Zhai, G.: Augmented reality image quality assessment based on visual confusion theory. In: Proceedings of the IEEE International Symposium on Broadband Multimedia Systems and Broadcasting (BMSB). pp. 1–6. IEEE (2022)

3. Duan, H., Min, X., Sun, W., Zhu, Y., Zhang, X.P., Zhai, G.: Attentive deep image quality assessment for omnidirectional stitching. IEEE J. Selected Topics Signal Process. (JSTSP) (2023)
4. Duan, H., Min, X., Zhu, Y., Zhai, G., Yang, X., Le Callet, P.: Confusing image quality assessment: toward better augmented reality experience. IEEE Trans. Image Process. (TIP) **31**, 7206–7221 (2022)
5. Duan, H., et al.: Develop then rival: a human vision-inspired framework for superimposed image decomposition. IEEE Transactions on Multimedia (TMM) (2022)
6. Duan, H., Shen, W., Min, X., Tu, D., Li, J., Zhai, G.: Saliency in augmented reality. In: Proceedings of the ACM International Conference on Multimedia (ACM MM), pp. 6549–6558 (2022)
7. Duan, H., et al.: Masked autoencoders as image processors. arXiv preprint arXiv:2303.17316 (2023)
8. Duan, H., Zhai, G., Min, X., Zhu, Y., Fang, Y., Yang, X.: Perceptual quality assessment of omnidirectional images. In: Proceedings of the IEEE International Symposium on Circuits and Systems (ISCAS), pp. 1–5 (2018)
9. Duan, H., Zhai, G., Yang, X., Li, D., Zhu, W.: Ivqad 2017: an immersive video quality assessment database. In: Proceedings of the IEEE International Conference on Systems, Signals and Image Processing (IWSSIP), pp. 1–5. IEEE (2017)
10. Fan, C.L., Hung, T.H., Hsu, C.H.: Modeling the user experience of watching 360 videos with head-mounted displays. ACM Trans. Multimed. Comput., Commun. Appl. (TOMM) **18**(1), 1–23 (2022)
11. Fei, Z., Wang, F., Wang, J., Xie, X.: Qoe evaluation methods for 360-degree vr video transmission. IEEE J. Selected Topics Signal Process. (JSTSP) **14**(1), 78–88 (2019)
12. Fela, R.F., Pastor, A., Le Callet, P., Zacharov, N., Vigier, T., Forchhammer, S.: Perceptual evaluation on audio-visual dataset of 360 content. In: Proceedings of the IEEE International Conference on Multimedia and Expo Workshops (ICMEW), pp. 1–6. IEEE (2022)
13. Fela, R.F., Zacharov, N., Forchhammer, S.: Perceptual evaluation of 360 audiovisual quality and machine learning predictions. In: 2021 IEEE 23rd International Workshop on Multimedia Signal Processing (MMSP), pp. 1–6. IEEE (2021)
14. Fela, R.F., Zacharov, N., et al.: Towards a perceived audiovisual quality model for immersive content. In: 2020 Twelfth International Conference on Quality of Multimedia Experience (QoMEX), pp. 1–6. IEEE (2020)
15. Hansen, J.H., Pellom, B.L.: An effective quality evaluation protocol for speech enhancement algorithms. In: Proceedings of the Fifth International Conference on Spoken Language Processing (1998)
16. Hines, A., Gillen, E., Kelly, D., Skoglund, J., Kokaram, A., Harte, N.: Visqolaudio: An objective audio quality metric for low bitrate codecs. J. Acoust. Soc. America **137**(6), EL449–EL455 (2015)
17. Hu, Y., Loizou, P.C.: Evaluation of objective quality measures for speech enhancement. IEEE Trans. Audio Speech Lang. Process. **16**(1), 229–238 (2007)
18. Li, C., Xu, M., Du, X., Wang, Z.: Bridge the gap between vqa and human behavior on omnidirectional video: A large-scale dataset and a deep learning model. In: Proceedings of the ACM International Conference on Multimedia, pp. 932–940 (2018)
19. Li, Z., Aaron, A., Katsavounidis, I., Moorthy, A., Manohara, M.: Toward a practical perceptual video quality metric. The Netflix Tech Blog **6**(2), 2 (2016)

20. Meng, Y., Ma, Z.: Viewport-based omnidirectional video quality assessment: database, modeling and inference. IEEE Trans. Circuits Syst. Video Technol. **32**(1), 120–134 (2022)

21. Min, X., Zhai, G., Zhou, J., Farias, M.C., Bovik, A.C.: Study of subjective and objective quality assessment of audio-visual signals. IEEE Trans. Image Process. (TIP) **29**, 6054–6068 (2020)

22. Schatz, R., Sackl, A., Timmerer, C., Gardlo, B.: Towards subjective quality of experience assessment for omnidirectional video streaming. In: 2017 Ninth International Conference on Quality of Multimedia Experience (QoMEX), pp. 1–6 (2017)

23. Sheikh, H.R., Bovik, A.C.: Image information and visual quality. IEEE Trans. Image Process. (TIP) **15**(2), 430–444 (2006)

24. Sun, W., Min, X., Zhai, G., Gu, K., Duan, H., Ma, S.: Mc360iqa: A multi-channel cnn for blind 360-degree image quality assessment. IEEE J. Selected Topics Signal Process. (JSTSP) **14**(1), 64–77 (2019)

25. Sun, Y., Lu, A., Yu, L.: Weighted-to-spherically-uniform quality evaluation for omnidirectional video. IEEE Signal Process. Lett. **24**(9), 1408–1412 (2017)

26. Taal, C.H., Hendriks, R.C., Heusdens, R., Jensen, J.: An algorithm for intelligibility prediction of time-frequency weighted noisy speech. IEEE Trans. Audio Speech Lang. Process. **19**(7), 2125–2136 (2011)

27. Thiede, T., et al.: Peaq-the itu standard for objective measurement of perceived audio quality. J. Audio Eng. Society **48**(1/2), 3–29 (2000)

28. Wang, Z., Bovik, A.C., Sheikh, H.R., Simoncelli, E.P.: Image quality assessment: from error visibility to structural similarity. IEEE Trans. Image Processing (TIP) **13**(4), 600–612 (2004)

29. Wang, Z., Simoncelli, E.P., Bovik, A.C.: Multiscale structural similarity for image quality assessment. In: Proceedings of the Thrity-Seventh Asilomar Conference on Signals, Systems & Computers, 2003. vol. 2, pp. 1398–1402. IEEE (2003)

30. Xue, W., Zhang, L., Mou, X., Bovik, A.C.: Gradient magnitude similarity deviation: a highly efficient perceptual image quality index. IEEE Trans. Image Process. (TIP) **23**(2), 684–695 (2013)

31. Yu, M., Lakshman, H., Girod, B.: A framework to evaluate omnidirectional video coding schemes. In: Proceedings of the IEEE International Symposium on Mixed and Augmented Reality, pp. 31–36. IEEE (2015)

32. Zakharchenko, V., Choi, K.P., Park, J.: Quality metric for spherical panoramic video. In: Optical Engineering + Applications (2016)

33. Zhang, B., Yan, Z., Wang, J., Luo, Y., Yang, S., Fei, Z.: An audio-visual quality assessment methodology in virtual reality environment. In: 2018 IEEE International Conference on Multimedia & Expo Workshops (ICMEW), pp. 1–6 (2018)

34. Zhang, L., Zhang, L., Mou, X., Zhang, D.: FSIM: a feature similarity index for image quality assessment. IEEE Trans. Image Process. (TIP) **20**(8), 2378–2386 (2011)

Research on Tongue Muscle Strength Measurement and Recovery System

Xiaotian Pan[1], Ling Kang[1], Qia Zhang[1], Hang Liu[1], Jin Ai[1], Jianxiong Zou[1], Donghong Qiao[1], Menghan Hu[1], Yue Wu[2], and Jian Zhang[1(✉)]

[1] East China Normal University, Shanghai, China
51215904089@stu.ecnu.edu.cn, jzhang@cee.ecnu.edu.cn
[2] Ninth People's Hospital Affiliated to Shanghai Jiao Tong University School of Medicine, Shanghai, China

Abstract. Dysphagia is caused by movement disorders such as muscular systems or neurological diseases that participate in speech movement. Speech difficulty sufferers as the main victim of Dysphagia often have problems with speech accuracy and difficulty communicating, which greatly affect their daily life. The correction and treatment of patients with Dysphagia have become a hot topic in current research. Clinically, tongue pressure (TP) is used as an indicator to evaluate the function of the tongue muscle, thus reflect the status of Dysphagia. In this paper, we designed and produced a sensor that can accurately measure tongue pressure and developed corresponding hardware and software systems. Based on physiotherapy, we have developed a WeChat mini-program that not only visually displays muscle strength values, but also includes rehabilitation games. Users can use tongue muscle compression sensor to complete game tasks that increase the motivation of users to perform physiotherapeutic exercises and improve rehabilitation effectiveness.The demo video of the proposed system is available at:
https://figshare.com/articles/media/sensor_demo_show_mp4/23578689

Keywords: Dysphagia · Pressure Sensor · Micro Air Bags · Rehabilitative Training Game

1 Introduction

The tongue is an important muscle organ in the human body, and its interaction with the hard palate forms the basis of speech and swallowing. Muscular neurological disorders can cause tongue weakness, leading to speech and swallowing disorders. Dysphagia is a clinical manifestation of inability to effectively transport food to the stomach due to damage or weakness of tongue muscles and other organs. Clinically, tongue pressure (T P) is an important evaluation reference indicator for dysphagia and speech function [1]. By studying the tongue muscle and lip muscle pressure of children with speech disorders, we can provide help for language correction, significantly reducing the pain and suffering of the patient and their family. Currently, medical professionals generally use TP measurement to examine the tongue function, because it has quantifiable strength, endurance, and training capabilities [2]. TP refers to the force with which the tongue contacts the hard palate.

L. Fang et al. (Eds.): CICAI 2023, LNAI 14474, pp. 526–531, 2024.
https://doi.org/10.1007/978-981-99-9119-8_47

A variety of devices for TP detection have been developed, many based on the principle of strain gage manometry. These devices use strain gauges [3], load measurement [4, 5], force resistors [6, 7] and bulb pressure sensors [8–11]. The instrument can also be used for dysphagia testing and treatment [2, 12–16]. But they all have problems such as inconvenient measurement and single function. To address this issue, we designed and developed a bladder-type sensor and developed corresponding circuits that can accurately reflect TP. We also designed a mobile terminal rehabilitation game to enhance the motivation for therapeutic exercise.

2 Methods

2.1 Sensor Design and Production

The sensor structure shown in Fig. 1 (a) consists of a silicone layer, a bladder layer, and a circuit layer, which jointly function to form a TP measurement sensor and achieve feedback of TP. We designed and printed silicone molds, and used the silicone impression method to produce the silicone layer. The circuit layer consists of a pressure sensing chip and a flexible circuit. We used food-grade silicone adhesive to bond the silicone layer to the circuit layer. Figure 1 (b) is an example of an experimental diagram of the sensor in the oral cavity.

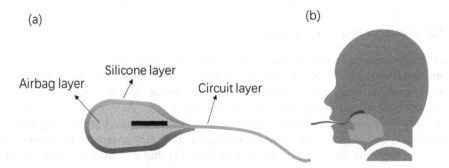

Fig. 1. (a) Sensor structure (b) Schematic diagram of the test scenario

Before being used for clinical human testing, the performance of the sensor will be tested using a pressure machine as a pressure applying device. In the group of adults without dysphagia or language problems, the average maximum pressure is approximately 60 kPa, with a range of 40–80 kPa. Tongue force intensity in the group with dysphagia difficulties is significantly lower than in the normal population. Considering the actual contact area, the range of TP is approximately 4–10 N.

To evaluate the sensor performance, this experiment used a pressure machine as the pressure generating device. Figure 2(a) shows the relationship between the sensor output and pressure. It can be seen from the figure that the sensor's measurement range is 0–15 N, and when the pressure exceeds 15 N, the sensor output does not increase further. A linear regression analysis was performed on the tongue pressure range (4–10 N).

The corresponding function relationship was $y = 4.348x + 89.815$, with a correlation coefficient $R^2 = 0.991$. The linear relationship is strong, and the sensitivity reaches 4.131 kPa/N.

Figure 2(b) shows the status of the sensor under different pressures generated by the pressure machine. It can be seen that as the pressure increases, the sensor deformation becomes more obvious, and the cavity volume decreases continuously.

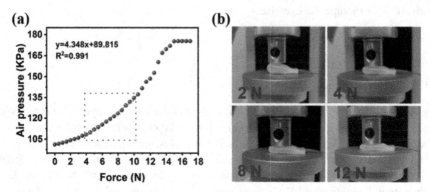

Fig. 2. (a) Sensor pressure-air pressure curve (b) Sensor status under different pressures

2.2 Sensor Design and Production

The hardware circuit is mainly composed of an Arduino Uno micro-controller, a BMP280 pressure sensor, a LCD1602 display screen, a CC2540 low-power Bluetooth chip, and a power module. Figure 3 is the hardware framework of the system. The pressure sensor collects internal air pressure data from the air bag, and the ADC module converts the pressure value into a digital signal that is written to a register. Through the SPI protocol, the Uno micro-controller reads the pressure value from the corresponding register. On the one hand, the Uno communicates with the LCD display screen through the I2C protocol to transmit pressure data, and the pressure data is displayed real-time on the LCD display screen. The Uno sends the data through the CC2540 module to the mobile terminal for communication. The sampling period of the entire process is 0.05 s.

Traditional tongue and lip muscle physiotherapy often involves using a tongue and lip pressing device to perform resistive exercises to train the strength of the tongue and lip muscles. However, this exercise can be boring. This research developed a WeChat mini-program, which is connected to the sensor hardware through Bluetooth. The WeChat mini-program adds a rehabilitative training game training mode. Users can operate the game character by pressing the air bag sensor of the tongue and lip muscle to complete game tasks. Meanwhile, the rehabilitative training game acquires the Bluetooth transmission value through the backend and compares it in real time with the set value. It is judged that the air bag has been pressed and the bird flies upward. Otherwise, the bird flies downward.

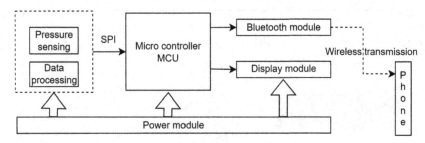

Fig. 3. System diagram

During vertical motion, the bird also flies forward at a preset value. A specified distance is generated to create obstacles, and the lowest height of the obstacle passage and the width of the passage through which it can pass are randomly generated within a specified range. Figure 4(a) shows the process of obstacle generation. Figure 4(b) is a hardware circuit diagram showing a real-time display on the LCD screen.

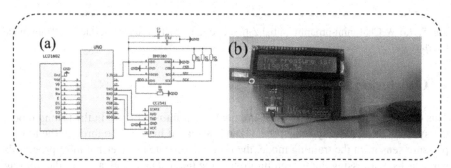

Fig. 4. (a) Circuit diagram (b) Physical circuit diagram

3   Demonstration Setup

After completing the connection of the hardware circuit, it is necessary to open the switch of the Bluetooth adapter on the WeChat mini-program side, search for nearby Bluetooth devices, and select the target Bluetooth device to complete the connection. Depending on needs, the user can enter the test mode or training mode. Figure 5(a) is the interface of the WeChat mini-program initialization, where the user can connect to the Bluetooth adapter and enter the practice mode from this interface, or directly enter the test mode.

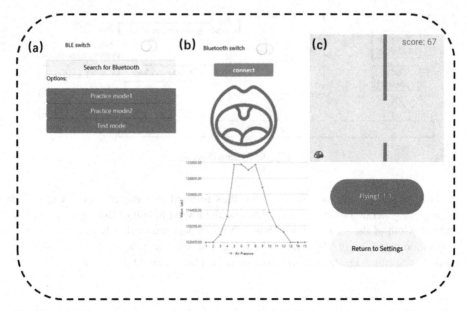

Fig. 5. (a) WeChat mini-program initialization interface (b) Test mode interface (c) Train mode interface

4   Conclusion

This demo paper proposes a TP measurement and training system. In the testing mode, the system can accurately reflect the TP and display it in real time through the tongue pressure sensor. In the training mode, the user can operate the WeChat mini-program's game tasks by pressing the sensor, improving the interest and motivation of the training exercise. The system has obvious advantages in evaluating and training the tongue and lip muscles of patients with speech and dysphagia.

References

1. Baxter, S.L., Marks, C., Kuo, T.T., Ohno-Machado, L., Weinreb, R.N.: Machine learning-based predictive modeling of surgical intervention in glaucoma using systemic data from electronic health records. Am. J. Ophthalmol.Ophthalmol. **208**, 30–40 (2019). https://doi.org/10.1016/j.ajo.2019.07.005
2. Oh, J.-C.: Effects of tongue strength training and detraining on tongue pressures in healthy adults. Dysphagia **30**(3), 315–320 (2015). https://doi.org/10.1007/s00455-015-9601-x
3. Yokoyama, S., et al.: Tongue pressure modulation for initial gel consistency in a different oral strategy. PloS one **9**(3), e91920(2014). https://doi.org/10.1371/journal.pone.0091920
4. Shaker, R., Cook, I.J., Dodds, W.J., Hogan, W.J.: Pressure-flow dynamics of the oral phase of swallowing. Dysphagia **3**, 79–84 (1988). https://doi.org/10.1007/BF02412424
5. Nagao, K., Kitaoka, N., Kawano, F., Komoda, J., Ichikawa, T.: Influence of changes in occlusal vertical dimension on tongue pressure to palate during swallowing. Prosthodont. Res. Pract. **1**, 16–23 (2002). https://doi.org/10.2186/prp.1.16

6. Miller, J.L., Watkin, K.L.: The influence of bolus volume and viscosity on anterior lingual force during the oral stage of swallowing. Dysphagia **11**, 117–124 (1996). https://doi.org/10.1007/BF00417901

7. Tsuga, K., Hayashi, R., Sato, Y., Akagawa, Y.: Handy measurement for tongue motion and coordination with laryngeal elevation at swallowing. J. Oral Rehabil.Rehabil. **30**(10), 985–989 (2003). https://doi.org/10.1046/j.1365-2842.2003.01077.x

8. Pouderoux, P., Kahrilas, P.J.: Deglutitive tongue force modulation by volition, volume, and viscosity in humans. Gastroenterology **108**, 1418–1426 (1995). https://doi.org/10.1016/0016-5085(95)90690-8

9. Crow, H.C., Ship, J.A.: Tongue strength and endurance in different aged individuals. J. Gerontol. Ser. A Biol. Sci. Med. Sci. **51**, M247–M250 (1996). https://doi.org/10.1093/gerona/51a.5.m247

10. Utanohara, Y., Hayashi, R., Yoshikawa, M., Yoshida, M., Tsuga, K., Akagawa, Y.: Standard values of maximum tongue pressure taken using newly developed disposable tongue pressure measurement device. Dysphagia **23**(3), 286–290 (2008). https://doi.org/10.1007/s00455-007-9142-z

11. Hayashi, R., Tsuga, K., Hosokawa, R., Yoshida, M., Sato, Y., Akagawa, Y.: A novel handy probe for tongue pressure measurement. Int. J. Prosthodont.Prosthodont. **15**(4), 385–388 (2002)

12. Yoshida, M., Kikutani, T., Tsuga, K., Utanohara, Y., Hayashi, R., Akagawa, Y.: Decreased tongue pressure reflects symptom of dysphagia. Dysphagia **21**, 61–65 (2006). https://doi.org/10.1007/s00455-005-9011-6

13. Maeda, K., Akagi, J.: Decreased tongue pressure is associated with sarcopenia and sarcopenic dysphagia in the elderly. Dysphagia **30**, 80–87 (2015). https://doi.org/10.1007/s00455-014-9577-y

14. Hewitt, A., et al.: Standardized instrument for lingual pressure measurement. Dysphagia **23**, 16–25 (2008). https://doi.org/10.1007/s00455-007-9089-0

15. McKenna, V.S., Zhang, B., Haines, M.B., Kelchner, L.N.: A systematic review of isometric lingual strength-training programs in adults with and without dysphagia. Am. J. Speech Lang. Pathol.Pathol. **26**, 524–539 (2017). https://doi.org/10.1044/2016_AJSLP-15-0051

16. Van den Steen, L., De Bodt, M., Guns, C., Elen, R., Vanderwegen, J., Van Nuffelen, G.: Tongue-strengthening exercises in healthy older adults: effect of exercise frequency—a randomized trial. Folia Phoniatr. Logop., 1–8 (2020). https://doi.org/10.1159/000505153

Updates and Experiences of VenusAI Platform

Meng Wan[1], Rongqiang Cao[1,2]($\boxtimes$), Kai Li[1], Xiaoguang Wang[1], Zongguo Wang[1,2], Jue Wang[1,2], and Yangang Wang[1,2]

[1] Computer Network Information Center, Chinese Academy of Sciences, Beijing 100083, China
caorq@sccas.cn

[2] School of Computer Science and Technology, Chinese Academy of Sciences, Beijing 100049, China

Abstract. **[Objective]** This paper presents an overview introduction of the VenusAI platform, focusing on its technical updates and sharing the experiences gained since its deployment. The objective is to highlight the platform's advancements, challenges, and valuable insights for other researcher and engineer in the field of AI platform. **[Coverage]** This paper uses keywords search and citation secondary search to collect papers and information from international computer journals, conferences and open source code warehouse. **[Methods]** The workflow engine tailored to different processes is designed for streamlining the AI development and enhancing operational efficiency. Then the phased optimization strategy is proposed to address unexpected events and ensure smooth resource allocation operations. Additionally, the disk repair mechanism is utilized to handle disk errors and maintain data integrity. **[Experience]** We provide detail of the challenges and experiences encountered on the VenusAI platform, aiming to share valuable insights and best practices for AI development. VenusAI's user-friendly toolsets, advanced functionalities, and up-to-date datasets and models make it a leading platform for AI research and development, catering to the diverse needs of both non-IT professionals and advanced researchers. **[Conclusion]** In conclusion, the VenusAI platform has undergone significant technical updates to improve its performance, efficiency, and stability. The implementation of the workflow engine, operation optimization strategies, and disk repair mechanism has enhanced the platform's capabilities and user experience. The experiences gained from managing VenusAI provide valuable insights for the operation of high-performance computing cluster-based AI platforms. The experience learned and challenges overcome contribute to the continuous improvement and innovation of VenusAI and similar platforms in the future.

Keywords: AI platform · Workflow engine · Research efficiency · Resource allocation · SLURM framework

1 Introduction

With the rapid development of artificial intelligence (AI) technologies, an increasing number of research institutes require large-scale computational resources to support their AI project development. However, traditional computing resources often fail to meet the

© The Author(s), under exclusive license to Springer Nature Singapore Pte Ltd. 2024
L. Fang et al. (Eds.): CICAI 2023, LNAI 14474, pp. 532–538, 2024.
https://doi.org/10.1007/978-981-99-9119-8_48

demands of flexible, complex calculations [1]. In this context, the VenusAI [2, 3] platform based on high-performance computing clusters has emerged. Currently, VenusAI consists of three core modules: data, models, and computing power. It offers over 200 online open datasets, state of the art(SOTA) models, and algorithm services. Researchers can engage in one-stop AI development activities within an integrated development environment (IDE) accessible through a web interface. This significantly enhances the efficiency and accuracy of AI projects, thereby fostering rapid technological advancements.

VenusAI was officially launched and operationalized in 2022. During its practical implementation, we identified certain design and technical limitations and undertook optimization and updates to enhance users' AI development activities on the platform. The key technical updates include:

- Workflow engine for different processes: We expanded a workflow engine that encompasses data processing, model development, and permission review processes. The automation of workflow enhances the overall operational efficiency of the platform.
- Operation Optimization: We implemented optimization techniques during startup, runtime, and stop stage. These measures prevent resource wastage by monitoring user activities and ensuring efficient resource allocation.
- Automated repair for disk errors: We incorporated automated repair mechanisms to address disk errors and ensure the integrity of data storage under faulty disk conditions.

During the operation and maintenance process, we gained insights into architecture design, functional design requirements, and operational methods from various perspectives. The main areas of focus included:

- Toolsets: We evaluated and refined the tools available on VenusAI to facilitate user interactions and streamline AI development processes.
- Usage: We provide diversified services for different users to conduct artificial intelligence research and development.
- Content: We focused on iterative updates of datasets and models, implementing efficient methods for data management, model versioning, and tracking changes.

By addressing these areas of improvement and incorporating user feedback, we have strived to create a more robust and user-friendly AI platform on VenusAI.

2 Related Work

This section compares VenusAI, a high-performance computing cluster-based AI platform utilizing the SLURM framework, with three popular cloud-based platforms: Google Colab [4], IBM Watson [5], and Alibaba Cloud Machine Learning Platform for AI [6].

Google Colab is a widely used cloud-based platform offering GPU and TPU resources for AI development. IBM Watson provides AI services and integration with data sources but may lack the same level of control and performance. Alibaba Cloud Machine Learning Platform for AI offers AI tools and integration with their ecosystem but relies on Kubernetes for resource management [7–10].

In contrast, VenusAI's advantage lies in its high-performance computing clusters and SLURM framework, which provides powerful computing resources, efficient allocation, and scalability for large-scale AI projects on HPC.

3   Methods & Updates

In the Methods & Updates section, we introduce a workflow engine designed to enhance operational efficiency and streamline the AI development process. To address unexpected challenges, we implemented a phased optimization strategy for startup and shutdown processes, ensuring smoother resource allocation operations and enhancing the platform's resilience. An important update in VenusAI is the implementation of a disk repair mechanism using the fsck.ext3 command, which proactively identifies and repairs disk bad sectors, ensuring data integrity and platform reliability.

3.1   Workflow Engine

To improve operational efficiency and reduce network communication and IO overhead, we introduced a workflow engine tailored to different processes within VenusAI. This engine encompasses automated workflows for data processing, model development, and review. By utilizing this workflow engine, users can navigate seamlessly through the various stages of AI development, from data preparation to model training and evaluation, while streamlining the review and approval process. This automation significantly minimizes manual effort, enhances overall efficiency, and optimizes the utilization of computational resources. Additionally, the standardized framework provided by the workflow engine ensures consistent and reliable task execution. Overall, the integration of the workflow engine in VenusAI has streamlined the AI development process, resulting in improved operational efficiency and reduced network communication and IO overhead (Fig. 1).

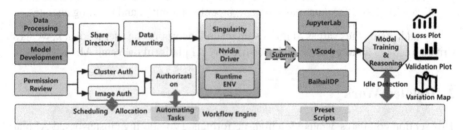

Fig. 1. Workflow Engine Process. We have built three automated workflows to improve execution efficiency.

3.2   Operation Optimization

During the actual deployment of VenusAI, we encountered unexpected situations such as resource scarcity, node failures, network initialization errors, and dependency startup conflicts. To address these challenges, we implemented a phased optimization strategy for startup and shutdown processes. This strategy ensures the proper sequential and parallel execution of resource allocation stages, effectively mitigating potential issues.

By adopting this phased optimization approach, we can proactively manage and mitigate the impact of unexpected events, ensuring smoother resource allocation operations. It enhances the platform's resilience and minimizes disruptions caused by resource scarcity or operational challenges. This optimized approach guarantees a more robust and reliable resource allocation process, contributing to the overall efficiency and stability of VenusAI.

3.3 Bad-Roads Repair

In addition to the aforementioned updates, we also addressed the issue of disk errors, specifically related to bad-roads disk. During the shutdown process, if errors occur or network instability occurs during large file read/write operations, it may lead to disk bad sectors, primarily due to the ext3 file system. To resolve this issue, we implemented a automation script for disk repair using the fsck.ext3 mechanism. When encountering disk bad sectors, VenusAI automatically initiates the disk repair process by invoking the fsck.ext3 command [11]. This command scans the disk for errors, identifies and repairs any bad sectors, and ensures the integrity of data storage. By proactively addressing disk errors, we minimize the risk of data corruption and improve the overall reliability and performance of the platform.

This update significantly enhances the robustness and stability of VenusAI, ensuring that disk errors are promptly identified and repaired. By incorporating this disk repair mechanism, we provide a more reliable platform for AI development, reducing the impact of disk-related issues on user workflows and data integrity.

4 Guidance & Experience

In this section, we aim to share the challenges and experience of the VenusAI platform, which serves as a valuable resource for users seeking guidance, inspiration, and best practices in their AI development journey (Fig. 2).

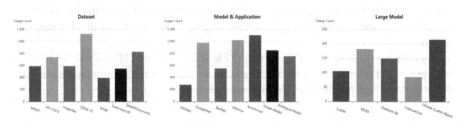

Fig. 2. Comparison of user usage (different from page view) of various products and services.

4.1 Toolsets

Statistical analysis of platform usage data over the past 2 years reveals that non-IT professionals comprise a significant proportion, accounting for a staggering 62%. Consequently, simplifying workflows, enhancing computing flexibility, and automating environment configurations have become paramount objectives for AI platform.

To address the challenges faced by non-IT users in computing environment configurations and repetitive setup processes, an AI platform must offer domain-specific foundational models tailored to different academic disciplines. Additionally, incorporating pre-built environments and commonly used libraries such as PyTorch, TensorFlow, Anaconda, and CUDA is essential. Moreover, the integration of flexible data processing tools like file transfer plugins and error correction utilities greatly enhances users' efficiency in model development. Furthermore, efficient data visualization tools empower users to intuitively observe and analyze model results, errors, and other relevant data. Lastly, the provision of cloud-based Integrated Development Environments (IDEs) like JupyterLab [12] and VSCode [13] proves instrumental in facilitating remote work, collaborative efforts, and addressing challenges related to model environment migration and mobile demonstrations.

The experience gained from these user-friendly tools demonstrates VenusAI's commitment to providing accessible and efficient solutions, supporting the diverse needs of non-IT professionals in their AI development endeavors.

4.2 Flexible Usage Methods

For advanced AI research and development users, the platform goes beyond the interface-based operations and offers more sophisticated and advanced functionalities. To cater to tasks that require distributed computing across multiple nodes, the platform incorporates an MPI (Message Passing Interface) program [14], enabling users to write custom parallel computing scripts seamlessly. Moreover, by modifying the SLURM job management system [15], users can leverage the cluster environment through command-line interfaces for complex tasks such as distributed training and large-scale model inference.

It is important to note that research projects in different domains exhibit varying demands for computational resources. Therefore, the computing cluster needs to have a scheduling control mechanism that allows for elastic scaling, expanding during peak times and contracting during idle periods. This ensures optimal utilization of resources and efficient allocation based on the ever-changing requirements of diverse scientific endeavors.

By offering these advanced features and capabilities, VenusAI caters to the needs of experienced AI researchers, empowering them to tackle complex computational tasks and unlocking new possibilities for innovation and discovery.

4.3 Datasets

In addition to the platform's objective functionalities, the content provided plays a crucial role in enhancing the user experience. Ensuring regular updates of data and models is of paramount importance for the platform's continuous development and user satisfaction.

To meet the evolving needs of users, the platform incorporates popular open-source datasets, including CIFAR-10, MS-COCO and ImageNet in the computer vision domain, Weather and Speech Command in the time series domain, and Sentiment140 in the NLP domain [16–20]. Furthermore, a diverse range of state-of-the-art artificial intelligence models such as Autoformer, Informer, MOSS (Massive OpenAI Scaling System), and LLaMA are made available to users [20–22].

Notably, the introduction of the alphafold2 model service on the VenusAI platform has resulted in a significant influx of new users and the return of existing users. This exemplifies the platform's commitment to staying abreast of advancements in the field and catering to users' demands for cutting-edge models. The availability of up-to-date and sought-after models greatly influences user engagement and satisfaction, further cementing VenusAI's position as a leading platform in the AI research and development landscape.

5 Conclusion

In conclusion, the VenusAI platform has successfully addressed the needs of both non-IT professionals and advanced AI researchers. By simplifying workflows, enhancing computational flexibility, and automating environment configurations, the platform has empowered non-IT professionals to overcome barriers and improve research efficiency. For advanced AI researchers, the platform offers advanced functionalities, including cross-node computing and distributed training capabilities. Regular updates of datasets and models have ensured relevance and attracted a growing user base. Overall, VenusAI's user-friendly tools and continuous improvements position it as a valuable platform driving innovation in artificial intelligence research.

Our future plans at VenusAI involve continuous improvement and innovation. We aim to expand our model library, embrace emerging technologies, and enhance the scalability of our computational clusters. We value user feedback and seek collaborations to drive advancements in AI research. Our goal is to establish VenusAI as a leading platform, empowering researchers and contributing to the field of artificial intelligence.

References

1. Sevgi, U.T., Erol, G., Doğruel, Y., et al.: The role of an open artificial intelligence platform in modern neurosurgical education: a preliminary study. Neurosurg. Rev. **46**(1), 86 (2023)
2. Yao, T., Wang, J., Wan, M., et al.: VenusAI: an artificial intelligence platform for scientific discovery on supercomputers. J. Syst. Architect. **128**, 102550 (2022)
3. Wan, M., Cao, R., Wang, Y., et al.: OpenVenus: an open service interface for HPC environment based on SLURM. In: International Conference on Smart Computing and Communication, pp. 131–141. Springer Nature Switzerland, Cham (2022). https://doi.org/10.1007/978-3-031-28124-2_13
4. Bisong, E.: Google colaboratory. In: Bisong, E. (ed.) Building Machine Learning and Deep Learning Models on Google Cloud Platform: A Comprehensive Guide for Beginners, pp. 59–64. Apress, Berkeley, CA (2019). https://doi.org/10.1007/978-1-4842-4470-8_7
5. Strickland, E.: IBM Watson, heal thyself: How IBM overpromised and underdelivered on AI health care. IEEE Spectr. **56**(4), 24–31 (2019)

6. Tao, Y., Zhang, Y., Cui, C., et al.: Automated spectral classification of galaxies using machine learning approach on Alibaba cloud AI platform (PAI). arXiv preprint arXiv:1801.04839 (2018)

7. Liao, W., Xie, L., Xi, J., et al.: Intelligent parking lot control system based on Alibaba Cloud platform and machine learning. In: 2021 6th International Conference on Intelligent Computing and Signal Processing (ICSP), pp. 908–911. IEEE (2021)

8. Jeffery, A., Howard, H., Mortier, R.: Rearchitecting kubernetes for the edge. In: Proceedings of the 4th International Workshop on Edge Systems, Analytics and Networking, pp. 7–12 (2021)

9. Canesche, M., Bragança, L., Neto, O.P.V., et al.: Google Colab CAD4U: hands-on cloud laboratories for digital design. In: 2021 IEEE International Symposium on Circuits and Systems (ISCAS), pp. 1–5. IEEE (2021)

10. Lee, K.Y., Kim, J.: Artificial intelligence technology trends and IBM Watson references in the medical field. Korean Med. Educ. Rev. **18**(2), 51–57 (2016)

11. Both D Both D Files, Directories, and Links. Using and Administering Linux: Volume 1: Zero to SysAdmin: Getting Started, pp. 513–547 (2020)

12. Kumar, M.: HPC/AI Deep dive training experiences with containers and JupyterLab. In: 2022 International Conference on Knowledge Engineering and Communication Systems (ICKES), pp. 1–5. IEEE (2022)

13. Taş, R., Tanrıöver, Ö.Ö.: Building a decentralized application on the ethereum blockchain. In: 2019 3rd International Symposium on Multidisciplinary Studies and Innovative Technologies (ISMSIT), pp. 1–4. IEEE (2019)

14. Ragunthar, T., Ashok, P., Gopinath, N., et al.: A strong reinforcement parallel implementation of k-means algorithm using message passing interface. Mater. Today: Proc. **46**, 3799–3802 (2021)

15. Chadha, M., John, J., Gerndt, M.: Extending SLURM for dynamic resource-aware adaptive batch scheduling. In: 2020 IEEE 27th International Conference on High Performance Computing, Data, and Analytics (HiPC), pp. 223–232. IEEE (2020)

16. Jinliang, N.: Cifar10 image classification based on ResNet. Системный анализ в проектировании и управлении **23**(1), 412–415 (2019)

17. Tong, K., Wu, Y.: Rethinking PASCAL-VOC and MS-COCO dataset for small object detection. J. Vis. Commun. Image Represent. **93**, 103830 (2023)

18. Beyer, L., Hénaff, O.J., Kolesnikov, A., et al.: Are we done with imagenet?. arXiv preprint arXiv:2006.07159 (2020)

19. Vadillo, J., Santana, R.: Universal adversarial examples in speech command classification. arXiv preprint arXiv:1911.10182 (2019)

20. Chen, M., Peng, H., Fu, J., et al.: AutoFormer: searching transformers for visual recognition. In: Proceedings of the IEEE/CVF International Conference on Computer Vision, pp. 12270–12280 (2021)

21. Zhou, H., et al.: Informer: beyond efficient transformer for long sequence time-series forecasting. Proc. AAAI Conf. Artif. Intell. **35**(12), 11106–11115 (2021). https://doi.org/10.1609/aaai.v35i12.17325

22. Touvron, H., Lavril, T., Izacard, G., et al.: LLaMA: open and efficient foundation language models. arXiv preprint arXiv:2302.13971 (2023)

SyncRec: A Synchronized Online Learning Recommendation System

Yixuan Zhang, Wenkang Zhang, Linqi Liu, Xuxue Sun$^{(\boxtimes)}$, Hao Zeng$^{(\boxtimes)}$,
Weijuan Zhao, Youbing Zhao, and Weifan Chen

Communication University of Zhejiang, Hangzhou 310018, China
{sunxuxue,hao.zeng}@cuz.edu.cn

Abstract. With the rapid development of digitalization and big data technology, numerous online learning materials have become available for self-regulated online learning. However, there is still a lack of a practical recommendation platform that can achieve synchronization between massive online learning materials and multiple users at different stages. To fill the gaps, we present a synchronized online learning recommendation system (SyncRec). The multi-source heterogeneous information fusion module integrates online learning materials from different digital platforms. The dynamic knowledge status tracing module tracks the real-time knowledge status and learning progress of online learners via dynamic mapping to a set of knowledge trees. Furthermore, the personalized recommendation module achieves adaptive recommendation of digital learning materials for each self-regulated online learner based on current knowledge status and learning needs as well as preferences. The demonstrated system helps improve learning outcomes and user experiences. An illustration video could be found here (https://github.com/Edith-xuan/video/blob/main/demo.mp4).

Keywords: Multi-source Heterogeneous Information Fusion ·
Knowledge Status Tracing · Synchronized Recommendation · Online
Learning

1 Introduction

1.1 Background

Online learning platforms have gained increasing attention in the past decade due to the rich multimodal learning materials, such as the MOOC platform [1]. However, existing platforms in real practice are not feasible for self-regulated online learners at different learning stages due to issues of confusion and information overload in the e-learning process. Moreover, learning outcomes may vary among different users, which also presents challenges [2]. There is a need for a practical system that can synchronize the real-time knowledge status of online learners at different stages with massive heterogeneous digital learning materials, and further provide adaptive recommendations for online learning. Such a system can reduce the cognitive overload of self-regulated online learners and improve both the learning outcomes as well as the learning experiences.

© The Author(s), under exclusive license to Springer Nature Singapore Pte Ltd. 2024
L. Fang et al. (Eds.): CICAI 2023, LNAI 14474, pp. 539–544, 2024.
https://doi.org/10.1007/978-981-99-9119-8_49

1.2 Literature Review

Existing literature addresses the topic of online learning recommendations from different perspectives. In some studies, online learning content is recommended at the group level, such as the multilayer bucket recommendation method for similar online learners [3], and the clustering strategy-based method for the recommendation of similar learning content [4]. Others focus on individual-level online learning recommendations, such as personalized recommendations based on learner preferences [5], and reinforcement learning-based dynamic recommendations [6]. Moreover, some used a hybrid method with collaborative filtering and association rules for online learning recommendation [7]. However, there are few practical online learning recommendation systems that can effectively synchronize massive online learning materials with real-time user knowledge status, as well as user needs and preferences.

To fill the above gaps, we propose a new online learning recommendation system that effectively synchronizes heterogeneous online learning materials with varying user knowledge status and personal needs as well as preferences.

2 Methodology

With fused online learning materials and traced user knowledge status, the system achieves adaptive online learning recommendations, as shown in Fig. 1.

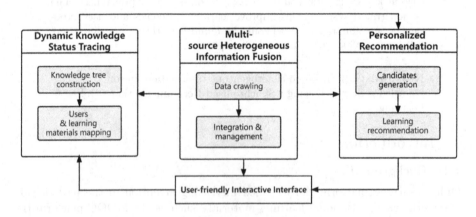

Fig. 1. Framework illustration of SynRec system

2.1 Multi-source Heterogeneous Information Fusion

We utilize the Selenium and Python toolkit with robust anti-scraping capability to facilitate information retrieval from online courses and videos over three digital platforms (e.g., MOOC, NetEase Cloud Classroom, and Bilibili). Online learning materials vary significantly among different platforms. In order to address the consistency issue, the developed system fuses and integrates multi-source information. To improve overall performance, the Supabase hosting platform [8] is used for data storage due to its self-hosting feature and database scalability, as well as support for frequent requests.

2.2 Dynamic Knowledge Status Tracing

In order to achieve dynamic tracking of the status of user knowledge, we use chapters from online courses or sequences of e-learning videos as tree nodes. The developed system utilizes a knowledge graph to capture the dependency between knowledge points and further provides query functionality. We employ the Neo4j graph database [9] to create and store the knowledge tree. When a user selects a learning plan, the learning process starts at the root node of the corresponding knowledge tree. The system keeps track of changes in user knowledge through interactive interfaces. After accomplishing the learning on a specific tree node based on recommended online resources, the user can alter the knowledge status via interactive self-assessment and automatically enroll into the next stage.

2.3 Personalized Recommendation

Based on the above modules, we further synchronize the fused massive online learning materials with the tracked real-time user knowledge status, as illustrated in Fig. 2. We use course/video tags from the fusion module and user knowledge status from the tracing module to generate candidates via automatic matching. Since the deep learning-based recommendation method could effectively discover the underlying patterns [10], we further employ a hybrid recommendation

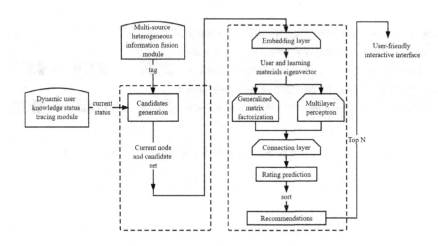

Fig. 2. Personalized adaptive recommendation

method with traditional matrix factorization and multilayer perceptron (MLP), which is capable of extracting both low-dimensional and high-dimensional features simultaneously. Unlike existing methods, the system generates embedding vectors based on multiple features, including the user ID, the status of user knowledge, the online learning material ID, and the ratings. The generalized matrix factorization (GMF) layer can learn the interactions between users and online learning materials. The MLP layer can retain the effective components in the high-dimensional sparse features and transform others into low-dimensional representations. Nonlinearity is further learned by concatenating multiple fully connected layers. Finally, the results of the GMF layer and the MLP layer are combined to generate a predicted rating for recommendations.

3   Application Scenarios

In this demo, information is collected from more than 16000 digital learning materials and further integrated into the Supabase platform, as shown in Fig. 3(a). We use course category, number of learners, and teacher information from MOOC and NetEase platforms to represent the information about tags, video publishers, and viewer counts respectively. To further illustrate the tracking of knowledge status, we use a triplet list that stores the dependency of the nodes to build a knowledge tree for each learning topic, as shown in Fig. 3(b). When a user sets up a learning goal via an interactive interface, the system automatically maps the knowledge status to the root node of the corresponding knowledge tree. When the user finishes the learning process of the current node, the system dynamically alters the user knowledge status via a self-assessment method. Moreover, to improve the fidelity of the recommendations, the system synchronizes the status of user knowledge with massive learning materials and displays the recommendations via a user-friendly interface, as shown in Fig. 4.

Fig. 3. (a) multi-source information integration, (b) knowledge tree

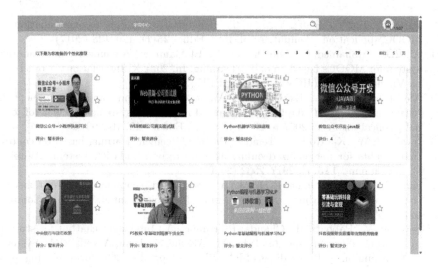

Fig. 4. Recommendation results

4 Conclusion

The proposed system synchronizes the dynamically evolving knowledge status with heterogeneous multi-source online learning materials. The synchronization mechanism enables effective recommendation of learning materials for multiple online learners at different stages. The proposed system with a user-friendly interface addresses the issues of confusion and cognitive overload during the self-regulated online learning process and improves learning outcomes as well as user experiences.

In future work, more interactive functionalities can be introduced to explicitly and implicitly retrieve online learner behaviors. The data could then be augmented with more online resources and user characteristics to improve modeling fidelity. In addition, comprehensive assessments, such as quizzes and tests, can be incorporated to improve the rationality of the tracing of knowledge status.

Acknowledgements. This work was supported in part by the Teaching Reform Project from Communication University of Zhejiang: "Research on Contextualized Teaching Mode for the New Generation of Engineering Students Based on Convergence Media" and the Key Laboratory of Film and TV Media Technology of Zhejiang Province (No. 2020E10015).

References

1. Deng, R., Benckendorff, P., Gannaway, D.: Progress and new directions for teaching and learning in MOOCs. Comput. Educ. **129**, 48–60 (2019)
2. Zhang, H., Shen, X., Yi, B., Wang, W., Feng, Y.: KGAN: knowledge grouping aggregation network for course recommendation in MOOCs. Expert Syst. Appl. **211**, 118344 (2023)

3. Pang, Y., Jin, Y., Zhang, Y., Zhu, T.: Collaborative filtering recommendation for MOOC application. Comput. Appl. Eng. Educ. **25**(1), 120–128 (2017)
4. Ali, H.A., Mohamed, C., Abdelhamid, B., El Alami, T.: A course recommendation system for MOOCs based on online learning. In: 2021 XI International Conference on Virtual Campus (JICV), pp. 1–3. IEEE (2021)
5. Xie, H., Chu, H.C., Hwang, G.J., Wang, C.C.: Trends and development in technology-enhanced adaptive/personalized learning: a systematic review of journal publications from 2007 to 2017. Comput. Educ. **140**, 103599 (2019)
6. Intayoad, W., Kamyod, C., Temdee, P.: Reinforcement learning based on contextual bandits for personalized online learning recommendation systems. Wireless Pers. Commun. **115**(4), 2917–2932 (2020)
7. Xiao, J., Wang, M., Jiang, B., Li, J.: A personalized recommendation system with combinational algorithm for online learning. J. Ambient. Intell. Humaniz. Comput. **9**, 667–677 (2018)
8. Sevagen, V., Pabbati, H., Chanda, P., Kumar, A.: Intelligent chatbot for student monitoring and mentoring. In: Tuba, M., Akashe, S., Joshi, A. (eds.) ICT Systems and Sustainability: Proceedings of ICT4SD 2022, pp. 393–399. Springer, Singapore (2022)
9. Šestak, M., Heričko, M., Družovec, T.W., Turkanović, M.: Applying k-vertex cardinality constraints on a Neo4j graph database. Futur. Gener. Comput. Syst. **115**, 459–474 (2021)
10. Chen, W., Cai, F., Chen, H., Rijke, M.D.: Joint neural collaborative filtering for recommender systems. ACM Trans. Inform. Syst. (TOIS) **37**(4), 1–30 (2019)

Interpretability-Based Cross-Silo Federated Learning

Wenjie Zhou[1], Zhaoyang Han[1], Chuan Ma[2], Zhe Liu[1,2], and Piji Li[1(✉)]

[1] College of Computer Science and Technology, Nanjing University of Aeronautics
and Astronautics, Nanjing, China
lipiji.pz@gmail.com
[2] Research Institute of Basic Theories, Zhejiang Lab, Hangzhou, China

Abstract. The severe challenge encountered in cross-silo federated learning (FL) is the performance degradation caused by data heterogeneity. To overcome it, we propose two methods, FedGDI and FedCI, identifying these clients with unbalanced categories based on an interpretability mechanism. We firstly iteratively generate feature maps of last global model and local client models selected, then these unbalanced local models are identified by comparing the feature maps. For clients with unbalanced categories, these local update parameters are further adjusted by minimizing the gradient distance between the global model and clients' models, so as to reduce the adverse impact on the performance of FL model. We adopt different client-filtering strategies to filter clients. FedGDI filters clients by taking advantage of the cosine similarity between the gradient of these local client models and last aggregation global model; FedCI samples clients by clustering clients based on gradient distance. We evaluate the effectiveness of FedGDI and FedCI through multiple datasets, and from experimental results it can be concluded that our methods outperform these state-of-the-art (SOTA) schemes.

Keywords: Federated Learning · Interpretability · Data Heterogeneity

1 Introduction

The emergence of federated learning (FL) [1] makes it possible for clients to jointly train a model without sharing data. However, due to FL settings for data privacy, different clients do not share data information, such as their data distribution, which causes data imbalance and slows convergence of the global model. For example, multiple e-commerce platforms aim to jointly train a recommendation model, while user behavior data from different source generally follows different distributions. Thus, solving adverse impact of data heterogeneity [2] is urgent and beneficial to the application of FL.

To the best of our knowledge, existing researches on solving data heterogeneity focuses on outputs of FL model, then resolves the problem directly through devised loss function or privacy-violated knowledge. That is, the FL model is treated as a black-box [3] model. The operation on a black-box model is not interpretable.

Converting a black-box model into an interpretable model requires the utilization of interpretability techniques. Current works that generate visual explanations fall into

L. Fang et al. (Eds.): CICAI 2023, LNAI 14474, pp. 545–556, 2024.
https://doi.org/10.1007/978-981-99-9119-8_50

four categories: gradient-based methods [4], activation-based methods [5], region-based methods [6] and perturbation-based methods [7].

In this paper, we work on exploring the feasibility of interpretability in alleviating adverse effects caused by data imbalance. Our main goal is to tune the FL optimization process in an explainable way. The adjustment for data heterogeneity depends only on the input of the FL model, not its output. Thus, our solution can be widely applied to different FL models. Empirical research indicates that client models with unbalanced categories do mainly learn features of the majority classes. Therefore, we adopt one activation-based method: Gradient-weighted Class Activation Mapping (GradCAM) [5], which is capable of characterizing features related to these classification results. GradCAM helps capture changes in interpretable results and then identify clients with unbalanced categories (called unbalanced clients).

Moreover, inspired by the work of detecting malicious clients [8], we design a gradient distance-based interpretable federated learning method (named FedGDI). We consider that clients with a larger cosine distance between their own gradient and the last global aggregation gradient are more likely to have data heterogeneity. So FedGDI firstly calculate and rank the cosine distance between client gradient and global gradient to identify clients with greater risk on affecting global model's performance. The likelihood of a client data imbalance is then measured by calculating the similarity of the interpretable results between the client and the last aggregated global model. Then we narrow the gradient distance difference between the unbalanced client model and the last global aggregation model. In order to achieve unbiased sampling, we propose another solution, a clustering-based interpretable federated learning method, named FedCI. FedCI includes a client-selecting strategy, hierarchical clustering modified from [9]. We cluster clients based on the similarity between client models, and then extract clients from the clusters to participate in each FL round. The similarity of the interpretability results between clients selected and last global aggregation model is calculated to evaluate unbalanced clients. Then the imbalance client's parameters are adjusted like in FedGDI.

Our main contributions are summarized as follows:

- To explore the power of interpretability in alleviating adverse effects caused by the data imbalance, we designed two interpretable solutions, FedGDI, and FedCI, both of which not only improved FL model performance but also made the improvement interpretable.
- We can take advantage of the interpretability to infer the general situation of data heterogeneity in FL model frameworks.
- We conducted evaluations on four datasets: MNIST [10], CIFAR-10 [11], CIFAR-100 [11], CINIC-10 [12]. Experimental results show that our solutions outperform these SOTA solutions. With these encouraging results, we confirm that, employing the interpretability in cross-silo FL is promising to overcome the influence of data heterogeneity.

2   Related Work

At present, existing methods [13–21] solving data imbalance are the utilization of encrypted data privacy information or disclosure of privacy to adjust unbalanced data.

Table 1. Solutions on data imbalance in cross-silo FL

Technique	Proposed Methods	Trait
Cryptography	Dubhe [13], FedeAMC [15]	Expensive overhead
Loss Function	Monitor [16], Fed-Focal Loss [17]	Application limitation
Devised Framework	Astraea [18], FedMA [14], CCVR [19], FedSens [23], FedPer++ [20]	Privacy violation
Interpretability	FedGDI, FedCI	Interpretable

Cryptographic Techniques. Zhang et al. [13] introduced Dubhe based on homomorphic encryption. Each client decides whether to participate in each round based on a self-determined probability. And the data distribution is transferred between the client and the server through an encryption vector to guide the client's decision. Thus, the additional encryption operation incurs more computational overhead. Wang et al. [15] proposed a FL-based AMC (FedeAMC) under the condition of class imbalance and noise varying. However, either homomorphic encryption or other encryption computation suffers from heavy computational and communication overheads, and this kind of solutions are too expensive to be implemented in complex FL models.

Loss Function. Wang et al. [16] designed a monitor mechanism, where the monitor detects the imbalanced composition of training data in each round by using auxiliary data and the server loads the self-defined Ratio Loss to alleviate data imbalance. Sarkar et al. [17] proposed a new loss function called Fed-Focal Loss, which addresses the class imbalance by reshaping cross-entropy loss. However, the kind of solutions are not suitable to solve the problem of data imbalance in FL frameworks where the degree of data imbalance is large. Moreover, the kind of solutions only design the loss function according to these outputs of FL models, thus it is not practical.

Devised Framework. Duan et al. [18] developed Astraea, a self-balancing data sampling scheme that is dedicated to addressing the class imbalance problem by the designed mediator to improve of the model accuracy in FL. Lou et al. [19] proposed the Classifier Calibration with Virtual Representation (CCVR) to alleviate the adverse effects of data heterogeneity by adjusting the classifier using virtual representation. Wang et al. [14] designed the Federated matched averaging (FedMA) algorithm, which constructs the shared global model in a layer-wise manner by matching and averaging hidden elements with similar feature extraction signatures. Xu et al. [20] performed a linear combination based classifier collaboration method for achieving better FL model performance, with a feature-regularized training strategy. However, the server in the kind of solutions has the global knowledge of class distributions of all participants or other data information, which violate these strict privacy requirements of participants.

To improve FL performance without expensive overhead, we explored to explore the feasibility of interpretability in mitigating performance degradation practically (Table 1).

3   Preliminaries

3.1   System Model

To formulate cross-silo FL model, we hypothesize that there are n clients, a central server (CS) and an interpreter, as shown in Fig. 1. Each client j, $j \in \{1,2,3,...,n\}$, has a local dataset D_j, following a dirichlet distribution. In each local round, every client performs Q iterations of model training. The CS collects local model updates from selected clients, aggregates local updates to get the global optimization after operations conducted by the interpreter, and then distributes the updated global model to every clients. During the local training, clients utilize stochastic gradient descent (SGD) to update their own model weights. Let η denote the learning rate, and W represent model weights. The local update W_j^h of j-th client in h-th iteration can be calculated as:

$$W_j^h = W_j^{h-1} - \eta \dot{G}_j^h, \tag{1}$$

where G_j^h is the gradient of the h-th iteration of j-th client.

3.2   Related Notions

The Federated Averaging Algorithm (FedAvg). FedAvg [24] is a distributed framework that allows multiple users to train a machine learning model simultaneously. Local clients are responsible for training local data to get their own local models, and the central server is responsible for weighted aggregation to get the global model: First, local clients download model parameters from the server, update local model parameters, and conduct local training. Secondly, the local models are constantly updated through SGD. When the pre-determined local training times is reached, the updated local model parameters are uploaded to the server. Then the server sample m clients randomly from all the clients n. Then the server averages the gradient updates of these m clients to form a global update. Finally, the aggregated model parameters are passed back to client devices. Repeat the above steps until the number of communications reaches the set. Client sampling formula of FedAvg is as follows:

$$\theta^{t+1} = \sum_{i \in S_t} \frac{n_i}{n} \theta_i^{t+1} + \sum_{i \notin S_t} \frac{n_i}{n} \theta^t, \tag{2}$$

where S_t is the set of clients selected for t-th aggregation, θ^t represents the current global model parameters.

Dirichlet Distribution. Dirichlet distribution is a probability distribution, which is determined by concentration parameter and base distribution. In fact, dirichlet distribution is the conjugate prior of the categorical distribution and multinomial distribution. Let $Dir(\cdot)$ denote dirichlet distribution, and α denote its concentration parameter. The larger the value of α, the closer the data distribution is to uniform distribution; the smaller the value of α, the more concentrated the data distribution will be. Due to the distinct characteristics of the dirichlet distribution (i.e. the summation

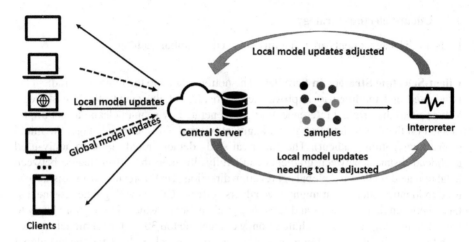

Fig. 1. The System Model of Interpretability-based FL (FedGDI and FedCI)

equal to 1), the Dirichlet distribution is widely used to deal with categorical data with a source of Gamma-distributed random variates (set gamma distribution as the base distribution), we can easily sample a random vector $x = (x_1, \ldots, x_K)$ from the K-dimensional Dirichlet distribution with parameters $(\alpha_1, \ldots, \alpha_K)$ (that is, $x = (x_1, \ldots, x_K) \sim \text{Dir}(\alpha_1, \ldots, \alpha_K)$). First, draw K independent random samples $y_1, \ldots, y_K$ from Gamma distributions each with density $Gamma(\alpha_i, 1) = \frac{y_i^{\alpha_i - 1} e^{-y_i}}{\Gamma(\alpha_i)}$, and then set $x_i = \frac{y_i}{\sum_{j=1}^{K} y_j}$.

The Interpretability. The interpretability [25,26] is used to provide explanations in a human-comprehensible manner that relate to predicted results provided by models. For the goal of accountability, these methods related to the interpretability can be divided into two categories: intrinsic methods and post-hoc methods. Intrinsic methods essentially rely on models that are interpretable by design, often referred as white-box models or transparent models. That is, intrinsic methods tend to have a strong task dependency, which limits their practicality. The post-hoc methods are model-agnostic, and they can be applied to a kind of models (e.g. neural networks) instead of specific models. Thus, the inherent ability of these post-hoc methods is to explain the model after training, which are often much more adaptable. In this paper, a post-hoc method is adopted to make the overall optimization interpretable and practical.

4 Interpretability-Based FL Methods

As shown in Fig. 1, we introduce our solutions in details here: 1. Select clients for local updates and the corresponding interpretability verification; 2. Perform interpretability verification on these clients selected; 3. Adjust model parameters.

4.1 Client-Selecting Strategy

In the initial stage, clients are randomly selected for initial training.

Client-Selecting Strategy in FedGDI. The notion related to the reference gradient [8] is adopted and modified to improve the practicality of the client-selecting strategy. We modified the original reference gradient ,generated by a small clean dataset implemented on the server, into last global aggregation gradient (the initial stage: the initial global aggregation gradient). The empirical study demonstrated that commonly-used euclidean distance between gradients can't fully illustrate the offset degree between local update direction and global aggregation direction. Therefore, the cosine distance is used to identify clients that might have adverse effects. Clients with greater cosine distance between their gradients and reference gradient are considered to be more likely to have data heterogeneity. In each iteration, we select the top 30% of local model updates with lower cosine distance. Let g_i, g_0 represent the i-th local gradient and the last global gradient, respectively. The i-th client's cosine similarity can be calculated as:

$$S_i = \frac{< g_i, g_0 >}{\|g_i\| \cdot \|g_0\|}. \tag{3}$$

Client-Selecting Strategy in FedCI. We redefine a representative gradient shown in [9] as the difference between a local model and last global model, instead of the difference between local models. Representative gradient G_{rep} is treated as the reference in subsequent hierarchical clustering. The i-th client's representative gradient is computed as:

$$W_{ori} - \eta \cdot G_{rep} = W_{update} \rightarrow G_{rep} = p_i - p_0 \tag{4}$$

where p_i and p_0 represent the i-th client's local update and the last global update respectively. According to the similarity of clients' representative gradients, we can divide them into c clusters by utilizing the ward's method [28], and then sample m clients with cluster sampling for local updates and interpretability verification. Based on empirical results, when the number of clusters reaches m or the distance between the two nearest clusters exceeds the set threshold, the clustering stops and c is set. Every γ iterations, we perform the hierarchical clustering operation, where γ is set as 10 in our experiments.

4.2 Interpretable Verification

These clients selected for interpretability verification are named as validation client. To assess the learning ability of each validation client model, a single set of samples extracted from each category of the dataset is set as the validation set S_{val}, the input of each validation client model for interpretability results. As described in Algorithm 1, the interpreter interprets each validation client's classification results on S_{val}: These features affecting the final classification results are highlighted, where the average gradient values on the feature map can be calculated as:

$$\xi_C^\beta = \frac{1}{Z} \sum_i \sum_j \frac{\partial y^\beta}{\partial A_{ij}^C}, \tag{5}$$

Algorithm 1. Similarity calculation of interpretability results

Input: Model structure M, Model parameters P_m, Validation set S_{val}, and Reference images img_{ref}

Output: Similarity Sim

1: Load parameters P_m to model structure $M \rightarrow$ model M_m
2: Input S_{val} into M_m: obtain the weights of feature map $W = [w_1, w_2, ..., w_n]$
3: Calculate the mean values of the gradient on the feature map with Eq. 5
4: Keep only the positive areas on the validation category β: $L_{Grad}^{\beta} = RELU\left(\sum_C \xi_C^{\beta} A^C\right)$
5: Save the above interpretability results img_{comp}
6: Calculate the similarity between img_{comp} and img_{ref} with Eq. 6: $Sim = [Sim^1 ... Sim^{\varepsilon} ... Sim^N]$
7: **return** Sim

Algorithm 2. Dynamic Parameter Adjustment

Input: Model M, Local parameters of j-th client P_l^j, Last global parameters P_g, Validation data of j-th client D_v^j, and FL aggregation times T

Output: Adjusted parameters P_{adj}^j

1: Load the dataset based on the pre-defined class of dataset
2: Load parameters P_l^j and P_g^j to last global model M_g and j-th client model M_l^j, respectively
3: Determine number of external cycles T_{ec} and internal cycles T_{ic} with Eq. 7
4: **for** $i = 0$ to T_{ec} **do**
5:     Input D_v^j to M_g^j: generated gradients G_g
6:     **for** $i = 0$ to T_{ic} **do**
7:         Input D_v^j to M_l^j: generated gradients G_{loc}
8:         Calculate gradient distance and backward update for parameters P_{local}^j with Eq. 8
9: **return** P_{adj}^j

---

where A, C, (i,j), and Z represent the feature map, its channels, its location and its size respectively, y^{β} represents its logits and β represents the category. These highlighted feature maps are saved as interpretability results img_{comp}. We then can infer those clients with imbalanced data: These interpretability results of last aggregated global model img_{ref} are taken as the measurement standard. Then the similarity between each validation client's N-sized interpretability results img_{comp} and the referenced results img_{ref} is calculated as follows:

$$Sim^{\varepsilon} = \frac{2\sum_{i=1}^N |cof_{x,i}^{\varepsilon}||cof_{y,i}^{\varepsilon}| + C}{\sum_{i=1}^N |cof_{x,i}^{\varepsilon}|^2 + \sum_{i=1}^N |cof_{y,i}^{\varepsilon}|^2 + C} \cdot \frac{2|\sum_{i=1}^N cof_{x,i}^{\varepsilon} cof_{y,i}^{*\varepsilon}| + C}{2\sum_{i=1}^N |cof_{x,i}^{\varepsilon} cof_{y,i}^{*\varepsilon}| + C}, \quad (6)$$

where $cof_{x/y,i}^{\varepsilon}$ and $cof_{x/y,i}^{*\varepsilon}$ represent the coefficients of ε-th image x (resp. y) of img_{ref} (resp. img_{comp}), the complex conjugate of $cof_{x/y,i}$, respectively, and C denotes a small positive constant for the robustness. A category with the most variation in the similarity is considered as its imbalanced category for the validation client.

Table 2. Comparison results on MNIST and CIFAR-10 (VS. FedMA)

Dataset	Method			
	BASELINE	FedMA	FedGDI	FedCI
MNIST	0.93	**0.99**	**0.99**	**0.99**
CIFAR-10(VGG-9)	0.64	0.74	**0.7409**	**0.7517**

4.3 Dynamic Parameter Tuning

As shown in Algorithm 2, these imbalanced client model's parameters are dynamically adjusted by minimizing distance between gradients of j-th client model's parameters P_l and last global model's parameters P_g. The number of external cycles T_{ec} and internal cycles T_{ic} in Dynamic Parameter Adjustment are dynamically determined by FL aggregation times T. The determined equation can be expressed as:

$$T_{ec} = min\{\lfloor \phi T + R \rfloor, 20\}, T_{ic} = min\{\lfloor \psi T + O \rfloor, 30\}, \tag{7}$$

where hyper-parameters of iteration variation ϕ and ψ are set to 0.5 and 1.5, respectively, then R and O are two positive constant 5 and 10, respectively. The 20 and 30 are added in Eq. 7 for the initial stabilization. Then the adjustment can correct these adverse parameters by the gradients converging. The distance between gradients can be computed as the following:

$$d = \sum_{k=1}^{n} \|g_k{}' - g_0^{k}{}'\|_2^2 \tag{8}$$

where $g_k{}'$ and $g_0^{k}{}'$ are gradients generated by the local client model and last global model on the k-th sample of the validation set S_{val}, respectively. To mitigate negative impact caused by the data heterogeneity, we consider minimizing the above gradient distance to adjust parameters P_{adj}^{j}. However, the initial phase of FL is unstable and easily affected. Thus, the dynamic adjustment amplitude is extremely necessary. We flexibly adjust these above parameters (T_{ec} and T_{ic}) based on FL aggregation times T. Repeat the above steps for the FL aggregation times until the global model M_g converges.

5   Experiment

Our schemes are classified as the category of devised framework and for improving the global FL optimization. Thus, we compare our work with FedAvg, FedMA [14] and CCVR [19], where the FedAvg is the BASELINE, FedMA and CCVR are the SOTA methods in the category of devised framework. To avoid accidental results, the following experimental results are the average results of repeated experiments.

5.1   Experiment Settings

The MNIST dataset is a handwritten digits dataset of 10 categories. Both CIFAR-10 [11] and CIFAR-100 [11] are composed of 60,000 colour images. The CINIC-10 is a dataset constructed from two different sources: ImageNet and CIFAR-10, which contains 270,000 images. These proposed methods are implemented on a A5000 GPU.

We adopt these same datasets and network architectures used in comparisons. That is, a 4-layer CNN network with a 2-layer MLP projection head like CCVR for CIFAR-10. For CIFAR-100 and CINIC-10, we adopt MobileNetV2 [29]. When comparing with FedMA, LeNet-5 [30] is adopted on MNIST [10] and VGG-9 [31] on CIFAR-10. 132 clients are set up in the experiment. In each round, 16 clients are selected for local updates. To simulate practical FL scenarios, we sample $p_i \sim Dir_K(\alpha)$ and assign a $p_{i,k}$ proportion of samples from class i to client k, making the data obey the Dirichlet distribution. Incidentally, we set the parameter α in the Dirichlet distribution to 0.5 unless otherwise specified.

5.2 Results

We evaluate the effectiveness of our designed solutions, compared with the optimal experimental values of CCVR [19], FedMA [14] and BASELINE.

We compare FedGDI (resp. FedCI) with FedMA (resp. CCVR) on different datasets. From Table 2 and Table 3, we know that both solutions outperform the SOTA solutions.

FedGDI and FedCI VS. FedMA. From experiment results shown in Table 2, we have these following observations: 1) In MNIST, all solutions perform the best and the BASELINE has only a minor accuracy loss. 2) For CIFAR-10, the accuracy of the FedGDI and FedCI is higher than that of the FedMA and BASELINE. 3) All solutions have a downward trend in the FL model accuracy with the data imbalance incurred, but FedGDI and FedCI are more robust on different datasets. The MNIST structure is simple, so the improvement effect of all solutions is significant, but for the more complex CIFAR-10 data set, our methods are more prominent.

FedGDI and FedCI VS. CCVR. As shown in Table 3, it's apparent that FedGDI and FedCI achieved higher accuracy than CCVR. Figure 2 shows the accuracy of CIFAR-10 with varying degrees of disequilibrium α. By comparing these results shown in Fig. 2, we can observe that as the value of α decreases (the greater the degree of imbalance), our solutions become more competitive.

The main reason of the above trend is that FedGDI and FedCI make targeted adjustments on these negatively affected model parameters based on interpretability results instead of these FL models' outputs. The degree of data imbalance does not have a linear change on the model output, so the method of relying on the model output to adjust has limitations. Thus, our methods are more robust to the degree of data imbalance that other solutions. And these solutions like CCVR can only alleviate the performance degradation of the FL models caused by data imbalance within a certain range.

The Effects of Client-Selecting Strategies. As seen in these above charts and figures, we can observe that FedCI outperforms FedGDI on different benchmarks. To investigate reasons for the above phenomenon, we conducted related ablation experiments: the effect of client-selecting strategy alone on the FL performance. From Table 4, we can infer that the interpretability works better in FedCI. Therefore, we can infer that the

Table 3. Comparison results on CIFAR-10, CIFAR-100 and CINIC-10 (VS. CCVR)

Dataset	Method			
	BASELINE	CCVR	FedGDI	FedCI
CIFAR-10(CNN+MLP)	0.6939	0.7143	**0.7411**	**0.7504**
CIFAR-10(α=0.1)	0.5953	0.6322	**0.6418**	**0.6671**
CIFAR-10(α=0.05)	0.5276	0.5556	**0.5716**	**0.5911**
CIFAR-100	0.6679	0.6723	**0.6746**	**0.6975**
CINIC-10	0.6224	**0.7053**	0.6923	**0.7205**

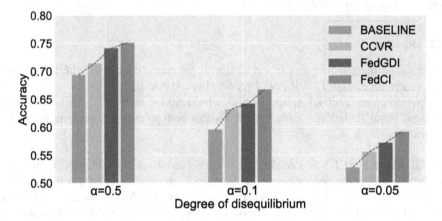

Fig. 2. The robustness to different degrees of disequilibrium (α=0.5, 0.1, 0.05.)

Table 4. Ablation experiments on the effect of interpretability (α=0.5)

Dataset	Method			
	FedGDI	*FedGDI**	FedCI	*FedCI**
CIFAR-10	0.7411	0.7298	0.7504	0.7306
CIFAR-100	0.6746	0.6645	0.6975	0.6647
CINIC-10	0.6923	0.6425	0.7205	0.6488

*FedGDI** (resp. *FedCI**) denotes FedGDI (resp. FedCI) without interpretability.

upper limit of performance improvement of *Dynamic Parameter Adjustment* depends on the characteristic of client-selecting strategy (unbiased sampling or biased sampling) as well. The reason is that the expected value of the client aggregation derived from unbiased sampling is closer to the global aggregation derived from considering all clients. The unbiased sampling strategy take advantage of client data in FL, that is, some clients with unique data distribution can be selected.

The Effectiveness of the Interpretability. From Table 4, we can infer that the performance gain of FedGDI and FedCI is due to the adoption of the interpretability. These client-selecting strategies are not main factors for FL model improvement. Removing the interpretability part is harmful for improving FL model performance. For instance, ablation experiments on CINIC-10 show that FL model accuracy decreases by 4.98% when using FedGDI's selecting strategy alone, and decreases by 7.17% when using FedCI's selecting strategy alone.

6 Conclusion

This paper aims to explore the power of interpretability in solving data heterogeneity in FL frameworks. To this end, we propose FedGDI and FedCI, both of which employ the interpretability to alleviate adverse effects caused by the data imbalance. FedGDI and FedCI successfully improved FL model performance by combining client-filtering strategies with interpretability. We conducted experiments on different datasets, and encouraging experiment results demonstrate that our solutions outperform other SOTA solutions. Furthermore, we also conducted ablation experiments on client-filtering strategies and interpretability verification. In future, we will further explore the power of the interpretability in FL frameworks, including exploring how to mitigate model performance degradation in cross-silo FL with stricter privacy protection.

Acknowledgment. This research is supported by the National Natural Science Foundation of China (No.62106105), the CCF-Tencent Open Research Fund (No.RAGR20220122), the CCF-Zhipu AI Large Model Fund (No.CCF-Zhipu202315), the Scientific Research Starting Foundation of Nanjing University of Aeronautics and Astronautics (No.YQR21022), and the High Performance Computing Platform of Nanjing University of Aeronautics and Astronautics.

References

1. Smith, V., Chiang, C.-K., Sanjabi, M., Talwalkar, A.S: Federated multi-task learning. In: Proceedings of NeurIPS, vol. 30 (2017)
2. Zawad, S., et al.: Curse or redemption? How data heterogeneity affects the robustness of federated learning. In: Proceedings of AAAI, pp. 10807–10814 (2021)
3. Dezfouli, A., Bonilla, E.V.: Scalable inference for gaussian process models with black-box likelihoods. In: Proceedings of NeurIPS (2015)
4. Adebayo, J., Gilmer, J., Goodfellow, I., Kim, B.: Local explanation methods for deep neural networks lack sensitivity to parameter values. arXiv preprint arXiv:1810.03307 (2018)
5. Selvaraju, R.R., Cogswell, M., Das, A., Vedantam, R., Parikh, D., Batra, D.: Grad-CAM: visual explanations from deep networks via gradient-based localization. In: Proceedings of ICCV (2017)
6. Petsiuk, V., Das, A., Saenko, K.: RISE: randomized input sampling for explanation of black-box models. In: British Machine Vision Conference (2018)
7. Ribeiro, M., Singh, S., Guestrin, C.: "why should I trust you?": explaining the predictions of any classifier. In: Proceedings of KDD (2016)
8. Cao, X., Fang, M., Liu, J., Gong, N.Z.: FLTrust: Byzantine-robust federated learning via trust bootstrapping. In: Proceedings of NDSS (2021)
9. Sattler, F., Müller, K.-R., Samek, W.: Clustered federated learning: model-agnostic distributed multitask optimization under privacy constraints. IEEE Trans. Neural Netw. Learn. Syst. **32**(8), 3710–3722 (2021)

10. Lecun, Y., et al.: Learning algorithms for classification: a comparison on handwritten digit recognition. Stat. Mech. Perspect. (2000)
11. Krizhevsky, A., Hinton, G.: Learning multiple layers of features from tiny images. In: Handbook of Systemic Autoimmune Diseases, vol. 1, no. 4 (2009)
12. Darlow, L.N., Crowley, E.J., Antoniou, A., Storkey, Amos J.: CINIC-10 is not ImageNet or CIFAR-10 (2018)
13. Zhang, S., Li, Z., Chen, Q., Zheng, W., Leng, J., Guo, M.: Dubhe: towards data unbiasedness with homomorphic encryption in federated learning client selection. In: Proceedings of ICPP, pp. 1–10 (2021)
14. Wang, H., Yurochkin, M., Sun, Y., Papailiopoulos, D., Khazaeni, Y.: Federated learning with matched averaging. In: Proceedings of ICLR (2020)
15. Wang, Y., Gui, G., Gacanin, H., Adebisi, B., Sari, H., Adachi, F.: Federated learning for automatic modulation classification under class imbalance and varying noise condition. IEEE Trans. Cogn. Commun. Netw. 8(1), 86–96 (2022)
16. Wang, L., Xu, S., Wang, X., Zhu, Q.: Addressing class imbalance in federated learning. In: Proceedings of AAAI, vol. 35, pp. 10165–10173 (2021)
17. Sarkar, D., Narang, A., Rai, S.: Fed-focal loss for imbalanced data classification in federated learning. CoRR, vol. abs/2011.06283 (2020)
18. Duan, M., Liu, D., Chen, X., Liu, R., Tan, Y., Liang, L.: Self-balancing federated learning with global imbalanced data in mobile systems. IEEE Trans. Parallel Distrib. Syst. 32(1), 59–71 (2021)
19. Luo, M., Chen, F., Dapeng, H., Zhang, Y., Liang, J., Feng, J.: No fear of heterogeneity: classifier calibration for federated learning with Non-IID data. In Proceedings of NeurIPS, vol. 34, pp. 5972–5984 (2021)
20. Xu, J., Yan, Y., Huang, S.-L.: FedPer++: toward improved personalized federated learning on heterogeneous and imbalanced data. In: Proceedings of IJCNN, pp. 01–08 (2022)
21. Yang, M., Wang, X., Zhu, H., Wang, H., Qian, H.: Federated learning with class imbalance reduction. In: Proceedings of EUSIPCO, pp. 2174–2178 (2021)
22. Kairouz, P., et al.: Advances and open problems in federated learning. Found. Trends® Mach. Learn. 14, 1–210 (2021)
23. Zhang, D.Y., Kou, Z., Wang, D.: FedSens: a federated learning approach for smart health sensing with class imbalance in resource constrained edge computing. In Proceedings of IEEE INFOCOM, pp. 1–10 (2021)
24. McMahan, B., Moore, E., Ramage, D., Hampson, S., y Arcas, B.A.: Communication-efficient learning of deep networks from decentralized data. In: Proceedings of AISTATS (2017)
25. Madsen, A., Reddy, S., Chandar, S.: Post-hoc interpretability for neural NLP: a survey. ACM Comput. Surv. 55(8), 155:1–155:42 (2023)
26. Akhilan Boopathy, et al.: Proper network interpretability helps adversarial robustness in classification. In: Proceedings of PMLR, pp. 1014–1023 (2020)
27. Lai, F., Zhu, X., Madhyastha, H.V., Chowdhury, M.: Oort: efficient federated learning via guided participant selection. In: Proceedings of OSDI, pp. 19–35 (2021)
28. Ward, J.H., Jr.: Hierarchical grouping to optimize an objective function. J. Am. Stat. Assoc. 58, 236–244 (1963)
29. Sandler, M., Howard, A., Zhu, M., Zhmoginov, A., Chen, L.-C.: MobileNetV2: Inverted residuals and linear bottlenecks. In: Proceedings of CVPR (2018)
30. Lecun, Y., Bottou, L., Bengio, Y., Haffner, P.: Gradient-based learning applied to document recognition. In: Proceedings of the IEEE, vol. 86, no. 11, pp. 2278–2324 (1998)
31. Simonyan, K., Zisserman, A.: Very deep convolutional networks for large-scale image recognition. Comput. Sci. (2014)

Nature-Inspired and Nature-Relevant UI Models: A Demo from Exploring Bioluminescence Crowdsourced Data with a Design-Driven Approach

Yancheng Cao, Keyi Gu, Ke Ma, and Francesca Valsecchi[✉]

College of Design and Innovation, Tongji University, Shanghai, China
francesca@tongji.edu.cn

Abstract. This demo project is an interactive design demo with a nature-centric perspective, focusing on bioluminescence in the natural world. We obtained a dataset related to bioluminescence from iNaturalist and performed data visualization to glean insights into the scientific classification, geographical distribution, and temporal patterns of bioluminescence. Based on the insights above, we designed a Bioluminescence World Clock, which demonstrates the flow of time by dynamically changing rhythmic processes of bioluminescence. Moreover, this interface is part of a public data collection service inspired by the Citizen Science model which contributes to the original data set through new data gathering and sharing.

Keywords: nature-centred perspective · bioluminescence · data-driven design

1 Introduction

Bioluminescence is the entry point of this design research that uses data-driven design as a method and focuses on the investigation of nature-based datasets. Although bioluminescence is not common in most organisms and the phenomenon is not readily observable, it usually occurs in many marine vertebrates and invertebrates, as well as some fungi, microorganisms, bacteria, and terrestrial arthropods such as fireflies. Bioluminescence is functionally essential for some of these organisms [1], so researchers are still exploring various aspects of bioluminescence, such as the luminescence mechanisms of different organisms, the differences in energy transfer processes, and the causes of varied luminescent colours [2]. In addition, the macroscopic aspects of bioluminescence are also worth investigating to understand lighting process. [3] Different Bioluminescence use luminescencing to achieve diverse physiological meanings, including courtship [4], attracting communicators, hunting, warning, and defence against predators (such as cuttlefish) [5].

iNaturalist is a nature-related platform where users can upload their observations of organisms in the natural world. Therefore, on this platform, we were able to obtain data on bioluminescence observed by users, including the identified species, the latitude and longitude of the observation locations, and the time of the organism's discovery. After conducting data analysis, it was discovered that bioluminescence exhibit certain patterns in terms of species, geographic distribution, and occurrence time which serve as the inspiration for our demo research results.

Inspired by the findings above, we have created a demo which includes a world clock feature along with related alarm clock and stopwatch functions, as well as the discovering function in Apple Watch, which are inherently connected to time and geography. We harness the dynamic shifts in various bioluminescent phenomena to depict time, enabling individuals to alter their perception of time and cast the magical bioluminescence of the natural world as the central character.

2   Data Visualization

We used open-source software RAW Graphs[1] and Tableau[2] for the dataset analysis witch can be seem in Fig.1. They can support a large number of data formats, thus allowing us to explore multiple dimensions and configurations of the data quickly.

Firstly, we looked at the variety and diversity of species that showcase bioluminescence, and we visualised them in the sunburst map. The diagram shows that at the level of Kingdom, among bioluminescent organisms, Fungi is a unique type, followed by Animalia. In terms of geographical distribution, we employed Tableau to visualize the latitude and longitude data related to bioluminescence. Our data visualization revealed that bioluminescent phenomena are predominantly concentrated in forested regions, coastal areas, and near rivers.

Shifting our focus to the temporal distribution, we conducted an analysis of bioluminescence over time. This examination allowed us to discern that various bioluminescent species display unique temporal patterns and rhythms.

3   Design Output

Building upon the insights derived from our data visualization research, we gained a comprehensive understanding of the scientific classification, geographical distribution, and temporal distribution of bioluminescence. These findings were subsequently integrated into our data-driven user interface design, facilitating a more informed and data-centric approach to our interface development.

---

[1] https://www.rawgraphs.io/.
[2] https://www.tableau.com/.

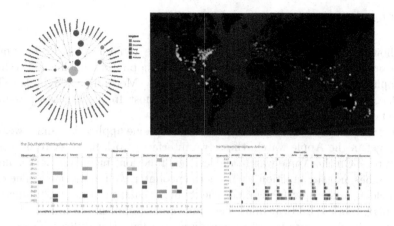

Fig. 1. The result of the visualization of the bioluminescence

3.1 Concept

Our intuition was to transform the data visualisation insights into a bioluminescence world clock that can work as a flexible and cross-platform widget integrated and embodied in daily-use devices such as the Apple Watch and mobile phones.

The data-driven UI design concept is represented by the Bioluminescence World Clock, which visually portrays the dynamic changes in bioluminescence to depict the passage of time. Our platform includes a user interface for locating natural bioluminescent events and uploading data to the cloud, thereby influencing the imagery displayed on the Bioluminescence World Clock.

3.2 Application Field

We employed Houdini as our prototyping software and specifically utilized the "particle system" function for visualizing dynamic bioluminescent imagery. The system in Houdini is defined by spatial position data and various adjustable parameters, while the carrier continually generates new particles. Within the interface, we have the capability to fine-tune the speed of particle generation and dissipation. Designers can manipulate particle movement by applying additional force around the carrier to influence the output's shape.

3.3 Visual Style

Since the luminescent effect of particles is a chemical reaction related to the function of luminophores in their bodies, we choose to use the visual effects of dynamic particles to reflect the repeated chemical reactions of bioluminescence.

By the particle system we mentioned above in Houdini, we can create visual style bioluminescence made up of different particles. Their movement can be shaped by particle parameters, external force system, external model, and node system in the Houdini system, which can reflect their endless vitality.

4   User Experience

As shown in Fig. 2, we finally generated the dynamics of 12 bioluminescence organisms expressing the time flow. Cyanophyta, Panellus, Polychaeta, Ctenophora, Omphalotus, Mycena, Lampyridae, Merulinidae, Sylliidae, Tubiporidae, Pocillopora and Alcyonacea are the most frequent categories in our datasets.

Our output includes a UI widget in a smartphone application and a wearable UI widget in the Apple watch: these two interfaces work together to implement our system. On the Apple watch, we designed functions including clocks, alarms, stopwatches and the function for users to encounter bioluminescence. The clock, alarm clock, and stopwatch will use the dynamic changes of the bioluminescence to display, remind, and calculate the time. The equipment can notice the user's closest bioluminescence to help them discover them. Hopefully, we can inspire more users to discover bioluminescence in this way.

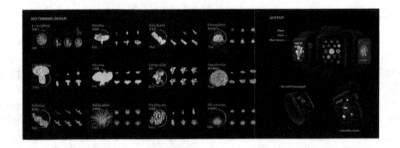

Fig. 2. The output of the interaction design

5   Limitations and Future Work

5.1   More Comprehensive Dataset

We focused on accessible bioluminescent areas but encountered limitations in regions like Asia and Africa where our platform is underutilized. This led to a dearth of data in less-explored regions. Collaborating with other datasets presents an opportunity for data enrichment.

5.2   Remote Interactive Graphics

Ideally, the bioluminescence types and particle movement in the World Clock could adapt based on continuously observed and uploaded data. As a future enhancement, we envision a seamless connection between the data used for World Clock outputs and the cloud-based data from platforms like iNaturalist, allowing real-time updates to influence the UI design. Houdini's node-based approach facilitates the incorporation of external data, making this a technically viable prospect. However, we acknowledge that implementing this may require additional project time.

References

1. Wilson, T., Hastings, J.W.: Bioluminescence. Annu. Rev. Cell Dev. Biol. **14**(1), 197–230 (1998)
2. Hastings, J.W.: Chemistries and colors of bioluminescent reactions: a review. Gene **173**(1), 5–11 (1996)
3. Hastings, J.W.: Biological diversity, chemical mechanisms, and the evolutionary origins of bioluminescent systems. J. Mol. Evol. **19**, 309–321 (1983)
4. Buck, J.: Synchronous rhythmic flashing of fireflies. II. Q. Rev. Biol. **63**(3), 265–289 (1988)
5. Bioluminescence (2022) National Geographic Education. https://education.nationalgeographic.org/resource/bioluminescence

Lightening and Growing Your Stromatolites: Draw and Dive into Nature

Yancheng Cao, Xinghui Chen, Siheng Feng, and Jing Liang[✉]

College of Design and Innovation, Tongji University, Shanghai, China
12046@tongji.edu.cn

Abstract. This demo is centered on a nature-centered perspective, providing individuals with the opportunity to explore the evolution of stromatolite phenomena in the natural world. The installation allows users to manipulate two key natural conditions: water dynamics and light exposure, using acrylic panels, which can control the morphological transformation of simulated image. When user interact with the device, the unique image generated. The experience is further enriched through external projection onto a large screen, creating an immersive simulation of natural evolution.

Keywords: nature-centric perspective · stromatolite · interactive experience

1 Introduction

This project is a design project based on the phenomenon of stromatolites in nature. Stromatolites are laminated biosedimentary structures usually attributed to the trapping and binding as well as chemical action of non-skeletal algae in shallow-water environments [4]. The device aims to create an interactive experience where people can control the growth of stromatolites by interacting with the environmental conditions. The fucntion allows users to witness the process of stromatolites formation, which is influenced by sunlight and water flow, resulting in different patterns. Stromatolites is a layered growth structure formed by the combination or precipitation of ancient microbial mats or biofilms (mainly composed of cyanobacteria) with sediment, possibly accompanied by non-biological surface precipitation [1]. Stromatolites records the interaction between microorganisms, sediment, and flowing water throughout Earth's history, provides only a small fraction of the structures preserved in Ordovician and Paleozoic carbonates [2], potentially shedding light on the long-term history of life and the environment.

2 System Design

The growth of stromatolite is closely related to the environmental conditions in which it exists, with key morphological factors including light exposure, water

L. Fang et al. (Eds.): CICAI 2023, LNAI 14474, pp. 562–566, 2024.
https://doi.org/10.1007/978-981-99-9119-8_52

dynamics, and surface microbial presence. [3] Under different environmental conditions, stromatolite exhibits distinct growth patterns. In this demo, we primarily focus on the impact of light exposure and water dynamics on the growth of stromatolite. By using different lighting ways and predefined water dynamics, the interactive process generates stromatolites models with different shapes. Our system involves three acrylic panels, light-dependent resistors, Processing for graphics generation, and a projection device, showed in Fig. 1.

Fig. 1. The hardware setup of the entire interactive system consists of several components.

2.1 Water Dynamics Control - Three Acrylic Panels

The research team has created three acrylic panels using laser cutting, with graphical cues on the panels indicating the magnitude of water dynamics which can be seem in Fig. 2. Users can select an acrylic panel, which then links to the algorithm program associated with their chosen water dynamics condition. These water dynamics conditions are broadly grouped into three levels, each representing a unique longitudinal cross-sectional shape of stromatolites. And the algorithm generates shapes based on abstract representations of these natural stromatolite forms.

2.2 Light Exposure Control - Light-Dependent Resistors

Since the growth of stromatolite depends on light exposure, the position, direction, and intensity of light, the driving force behind stromatolite growth in this system is generated by user-controlled lighting. We have set up four light-dependent resistors at specific positions on the acrylic panel to sense the intensity, position and the duration of light.

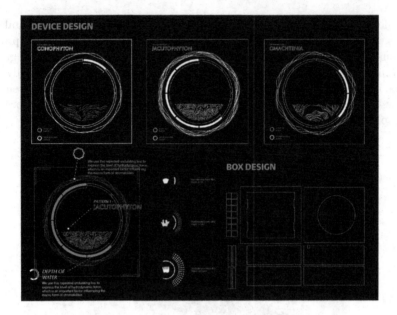

Fig. 2. Hardware device design sketches.

2.3 System Implementation - Arduino and Projection

Various acrylic panels can trigger distinct circuit connections, and these circuits from multiple panels run in parallel. These signals are then directed to an Arduino board, which subsequently transmits them to the Processing software on the computer. Based on the input parameters, diverse stromatolite graphics are generated. Ultimately, the visuals created in Processing are projected onto a large screen for display.

3 Algorithm Implementation

The algorithmic graphics for this project are developed in Processing, utilizing the Java programming language. In the algorithm, we leverage data from four light-dependent resistors to establish a coordinate system and ascertain the light source's position from the flashlight. These visuals are generated through an agent class, which draws line-shaped graphics with different colors, thickness, and movement patterns along predefined trajectories.

4 User Experience

The foremost value of this project lies in its ability to transition individuals from a human-centric perspective to a nature-centric one, highlighting the transformative processes that nature has undergone over millennia. Human actions and societal interactions are no longer central to this project; instead, it emphasizes the presentation of natural evolution that has unfolded across millennia, achieved through the interplay of software and hardware components.

In this project, users act as controllers, influencing the environmental factors that impact stromatolite growth. Through their engagement with the interactive device, users can observe the gradual vertical development of stromatolites driven by light exposure. They also witness the alteration of growth patterns and rules when transitioning between water dynamics conditions, as illustrated in Fig. 3. This immersion allows users to genuinely experience the enchanting and large-scale transformations inherent in nature's long-term evolution. Our aspiration is for individuals to find inspiration in this interactive demonstration and cultivate a deeper respect and awe for the natural world.

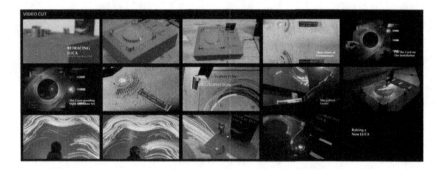

Fig. 3. Hardware device design sketches.

5 Limitations and Future Work

5.1 Incorporating the Influence of Microbial Conditions

At present, our interactive device primarily focuses on factors like light exposure and water dynamics, yet we acknowledge that the presence of microbes on layered rock plays a pivotal role in influencing growth. In our future endeavors, we aspire to integrate microbial conditions into the growth dynamics of the entire interactive device.

5.2 Improved Diversity of Stromatolites Growth Patterns

To enhance the distinguishability of stromatolites growth patterns, we aim to introduce greater diversity in the shapes of stromatolite growth under different water dynamics conditions.

References

1. Grotzinger, J.P., Rothman, D.H.: An abiotic model for stromatolite morphogenesis. Nature **383**(6599), 423–425 (1996)
2. Bosak, T., Knoll, A.H., Petroff, A.P.: The meaning of stromatolites. Ann. Rev. Earth Planet. Sci. **41**, 21–44 (2013)
3. Seong-Joo, L., Browne, K.M., Golubic, S.: On stromatolite lamination. In: Microbial Sediments, pp. 16–24. Springer, Heidelberg (2000). https://doi.org/10.1007/978-3-662-04036-2_3
4. Hofmann, H.J.: Stromatolites: characteristics and utility. Earth Sci. Rev. **9**(4), 339–373 (1973)

Robotics

A CNN-Based Real-Time Dense Stereo SLAM System on Embedded FPGA

Qian Huang, Yu Zhang, Jianing Zheng, Gaoxing Shang, and Gang Chen[✉]

Sun Yat-sen University, Guangzhou, China
cheng83@mail.sysu.edu.cn

Abstract. Simultaneous localization and mapping (SLAM) is the task to estimate agent's ego-motion in the map and reconstruct the 3D geometric of an unknown environment in parallel. Although many SLAM algorithms have been proposed in the past decades, few efforts have been devoted to conducting accurate real-time dense SLAM on resource- and computation-constrained platforms. In this paper, we leverage a shared binary neural network (BNN) architecture to learn robust feature descriptors for depth estimation and pose estimation modules simultaneously, which not only improves the system's accuracy, but also reduces the computation cost. Also, we propose several optimization strategies targeting feature extraction, feature aggregation as well as feature matching, and to accelerate them on embedded platform. Experimental results demonstrate that our design maintains accurate real-time pose estimation while yielding high-quality dense 3D maps. Our demo video is available at https://github.com/CICAIsubmission/CICAI2023.

Keywords: Simultaneous localization and mapping · Hardware accelerator · Real time · Dense map reconstruction

1 Introduction

Among the existing SLAM approaches, feature representation, map density, running platform are all key factors that are directly related to the application scenarios. Therefore, we investigate the state-of-the-art algorithms. Recent researches [5,7,8] propose SLAM systems that run on embedded SoC, but relying on handcrafted feature descriptors result in limited ability to cope with the highly dynamic environments. As an improvement, the work in [6,9] explore feature representation based on CNN on FPGA to enhance the robustness of SLAM, but did not implement a complete SLAM system on the SoC. Ling, et al. [4]implements a dense SLAM that can provide a lot of useful information for advanced applications such as automatic obstacle avoidance, but it can only run on CPU/GPU devices which consume a lot of power. It can be seen that no efforts have been devoted to developing dense stereo SLAM with CNN-based features on the embedded platforms.

In this paper, we propose a robust hardware-software co-design accelerator for SLAM task and implement it on embedded platforms. A light weight binary

© The Author(s), under exclusive license to Springer Nature Singapore Pte Ltd. 2024
L. Fang et al. (Eds.): CICAI 2023, LNAI 14474, pp. 569–574, 2024.
https://doi.org/10.1007/978-981-99-9119-8_53

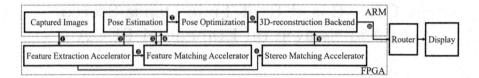

Fig. 1. The architecture of the demo system. ❶ denotes original image left&right.
❷,❸ denote key-points. **❹** denotes the feature of origin image. **❺** denotes the keypoints
of last frame or global map points. **❻** denotes the results of matching. **❼** denotes the
estimated pose of the camera. **❽** denotes optimized keyframes. **❾** denotes depth map
of the current camera perspective. **❿** denotes the global dense map. Finally, the dense
map is transmitted to the display through a router.

neural network(BNN) is applied in our architecture to generate robust feature
descriptors, which is useful for accurate feature matching which is required in
slam system sub-tasks like stereo matching and pose-estimation. To reduce the
hardware resource consumption, we proposeseveral optimization strategies for
hardware implemented algorithms like non-maximal suppression (NMS), heap-
sorting and feature matching. We also optimize the methods for mapping sub-
task while maintaining real-time dense map reconstruction. The experimental
results demonstrate that our accelerator provides more accurate localization and
high-quality dense mapping solution when compared to state-of-the-art SLAM
systems. What's more, we provide a demo video to explain how the accelerator
runs on the embedded device.

2 Proposed Slam Accelerator

As shown in Fig. 1, we develop an FPGA-based hardware accelerator to alleviate
the computational load on the ARM processor in order to enhance the performance of dense stereo SLAM. This hardware accelerator is specifically designed
to target the most time-consuming tasks, namely feature extraction, feature
matching, and stereo matching. Furthermore, we achieve significant improvements in the efficiency of the localization backend and 3D reconstruction processes by employing algorithmic optimizations. The implementation details can
be found in [3].

Feature Extraction Accelerator. The feature extraction accelerator is dedicated to extracting features from the original images. To obtain the image features necessary for key-point association and stereo matching, we integrate a
light-weighted BNN following the work in [1]. To obtain high-quality key-points
for pose estimation, we employ the FAST corner point detection algorithm to
identify candidate key-points. Subsequently, we apply NMS to mitigate the over-
concentration of key-points in texture-rich regions, followed by heap-sorting to
select a fixed number of key-points. The selected key-points are then transmitted
to the ARM processor for further utilization in the localization backend module.

These components work concurrently within a pipelined structure to maximize overall throughput.

Feature Matching Accelerator. In order to achieve accurate pose estimation, it is necessary to match the key-points extracted from each frame with global map points based on the learned binary descriptors. However, low-power ARM processors cannot handle the computational complexity of feature matching, which takes a long time to complete. To address this challenge, we have parallelized the matching process by leveraging the capabilities of FPGAs. Our proposed matching accelerator can efficiently execute the matching task for 1024 key-points in less than $5ms$, while without the need for intervention from the ARM processor.

Stereo Matching Accelerator. Stereo matching accelerator estimates dense depth maps from stereo images. We follow the StereoEngine accelerator designed in [1] and integrate it as the stereo matching accelerator. Then, through steps such as cost aggregation and post-processing, a high-quality dense depth map can be obtained. The depth map contains accurate distance information for each pixel, allowing for precise localization of objects and accurate reconstruction of the 3D environment.

Localization Backend. To reduce redundancy and optimize computational efficiency in our SLAM system, we implement a strategy to discard keyframes and sparse features that fall outside the optimization sliding window which do not contribute to the bundle adjustment (BA) optimization process. This selective approach allows us to focus computational resources on relevant data points, improving the overall efficiency of the system without compromising accuracy.

3D-reconstruction. We observed that many voxels were empty or no need to update. Motivated by this, we propose a coarse-to-fine TSDF calculation method to expedite the map integration process. It initially searches for chunks (containing a block of voxels) that are likely to represent item surface at a coarse level and skips blanks regions. We further subdivide it into smaller chunks, and keep performing the same surface finding strategy until reaching the minimum voxel level. By minimizing redundant evaluations and voxel updates, our proposed method reduces the map integration time from 127 ms to 32 ms.

3   Experiment

Experiment Setup. For the purpose of live demonstration, we deploy our proposed SLAM accelerator on an embedded development board with Xilinx ultra-scale+ SoC, which is a heterogeneous platform that consists of programmable logic as well as a quad-core ARM Cortex-A53 CPU. The stereo images are transmitted to the embedded development board via the TCP-IP protocol. Subsequently, all visual SLAM computation is executed on the embedded development board. The output of the system including Disparity Map, Camera Trajectory and 3D-reconstruction are then transmitted back to a display via the TCP-IP protocol to showcase the running effect. We develop the localization sub-task using Verilog hardware description language and deploy it on the programmable logic. The mapping sub-task is developed in C++ and implemented on the ARM Cortex-A53 CPU. To evaluate the performance of the proposed SLAM accelerator in terms of accuracy, real-time responsiveness and 3D map's density, we make use of KITTI dataset [2] for both quantitative and qualitative analyses. The qualitative results of localization and mapping is illustrated in Fig. 2.

Accuracy Evaluation of Localization. The test results on the dataset for visual SLAM indicate that the accuracy of our method is comparable with CB-SLAM [4], even though CB-SLAM [4] has worked on a high-performance platform. The accuracy improvement mainly attributes to the learned-features used for stereo matching and feature matching procedures. Meanwhile, the feature aggregation strategy also improves the robustness of the feature descriptors.

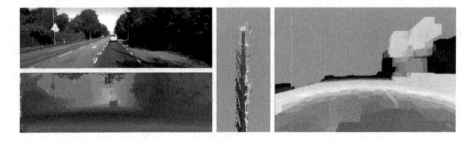

Fig. 2. Qualitative mapping results on the testing. From top to bottom, column 1: Input Image, Disparity Map obtained through stereo matching; column 2: Camera Trajectory; column 3: Effect of 3D-reconstruction.

Table 1. Accuracy Evaluation of Online Mapping.

ta/m	0.1	0.2	0.3	0.4	0.5	0.6	0.7	0.8	0.9	1
Ours	**0.398**	**0.563**	**0.667**	**0.737**	**0.784**	**0.818**	**0.842**	**0.860**	**0.873**	**0.885**
CB-SLAM [4]	0.342	0.514	0.634	0.715	0.769	0.806	0.833	0.853	0.868	0.881
Ours-fast	**0.242**	**0.382**	**0.496**	**0.583**	**0.648**	**0.697**	**0.735**	**0.765**	**0.789**	**0.809**
CB-SLAM-fast [4]	0.219	0.350	0.464	0.555	0.626	0.680	0.723	0.756	0.783	0.804

Performance Evaluation of Online Mapping. Table 1 illustrates the dense mapping accuracy evaluation results on *sequence 05*, in which both the implementations with and without coarse-to-fine (i.e., fast in Table 1) strategies are taken into consideration. According to Table 1, we can see that our online mapping sub-task achieves high-quality 3D mapping than that of CB-SLAM [4]. This is because the learned-features can generate more accurate depth maps for 3D reconstruction. In addition, although the coarse-to-fine strategy generates less accurate dense map, it significantly speeds up ($4\times$) the 3D reconstruction procedure, providing alternative solution for online mapping.

Real-Time Performance and Energy Consumption. The proposed accelerator runs at 13.1 frames per second (fps) on the target embedded device with energy consumption of only 6.201 W. The corresponding energy efficiency is 1.01×10^8 *J/pixel*. Compared to [4] (fps: 5.4, power: 36 W) that runs on Intel i7, our accelerator achieves $2.42\times$ speed up and $14.2\times$ power efficiency.

4   Conclusion

This paper presents an accurate and real-time dense SLAM accelerator, which can be efficiently deployed on embedded devices. To improve the performance of our accelerator, we make several optimizations targeting both software and hardware architecture. The accelerator achieves better performance in terms of absolute trajectory error, online mapping, real-time responsiveness and energy efficiency when compared with the state-of-the-art SLAM implementations.

References

1. Chen, G., et al.: Stereoengine: an FPGA-based accelerator for real-time high-quality stereo estimation with binary neural network. IEEE Trans. Comput.-Aided Des. Integr. Circ. Syst. **39**, 4179–4190 (2020)
2. Geiger, A., et al.: Vision meets robotics: the kitti dataset. Int. J. Rob. Res. **32**, 1231–1237 (2013)
3. Huang, Q., et al.: EDS-SLAM: an energy-efficient accelerator for real-time dense stereo slam with learned feature matching. In: Proceedings of IEEE/ACM International Conference On Computer Aided Design (ICCAD) (2023)
4. Ling, Y., et al.: Real-time dense mapping for online processing and navigation. J. Field Rob. **35**, 1004–1036 (2019)
5. Liu, R., et al.: ESLAM: an energy-efficient accelerator for real-time orb-slam on FPGA platform. In: Proceedings of the 56th Annual Design Automation Conference 2019, pp. 1–6 (2019)
6. Liu, Y., et al.: MobileSP: an FPGA-based real-time keypoint extraction hardware accelerator for mobile VSLAM. IEEE Trans. Circuits Syst. I Regul. Pap. **69**(12), 4919–4929 (2022)
7. Vemulapati, V., et al.: FSLAM: an efficient and accurate slam accelerator on SoC FPGAs. In: Proceedings of International Conference on Field-Programmable Technology (ICFPT) (2022)

8. Wang, et al.: ac^2 slam: FPGA accelerated high-accuracy slam with heapsort and parallel keypoint extractor. In: Proceedings of International Conference on Field-Programmable Technology (ICFPT) (2021)
9. Xu, Z., et al.: CNN-based feature-point extraction for real-time visual slam on embedded FPGA. In: Proceedings of IEEE 28th Annual International Symposium on Field-Programmable Custom Computing Machines (FCCM) (2020)

Land-Air Amphibious Robots: A Survey

Bo Hu[1,2], Zhiyan Dong[1,2(✉)], and Lihua Zhang[1,2]

[1] Academy for Engineering and Technology, Fudan University, Shanghai, China
22210860040@m.fudan.edu.cn
[2] Institute of AI and Robotics, Fudan University, Shanghai, China
{dongzhiyan,lihuazhang}@fudan.edu.cn

Abstract. This paper explores the development of land-air amphibious robots that combine the advantages of ground robots and aerial robots. The land-air amphibious robot is a multi-modal robot that can move and work both on land and in the air. They have the ability to perform tasks in different environments, have a wide range of application scenarios and high research value, and have aroused the interest of scientific researchers and research institutions. Based on all the articles we investigated, this paper first summarizes the research status of land-air amphibious robots and divides land-air amphibious mobile robots into three categories: wheeled, tracked, and legged, according to different motion structures in land travel mode. The article analyzes the advantages and disadvantages of different land-air amphibious robots and discusses the technical problems and research difficulties in the research of land-air amphibious robots. In addition, the article also puts forward the vision for the future development of land-air amphibious robots and the further work that needs to be done.

Keywords: Land-Air amphibious robots · Structural design · Bionic mechanisms · Motion modes

1 Introduction

Most mobile robots are usually designed for a single type of motion mode, such as swimming [1], land walking [2–4] or flying [5], which limits their ability to move in complex and varied scenes. There are also some deficiencies in their environmental adaptability. To improve the environmental adaptability of mobile robots, researchers drew inspiration from amphibians and set out to develop amphibious robots [6].

The research and development of robotics provide powerful automated tools for exploring the natural world and improving productivity. In the research and design of modern mobile robots, to cope with the increasingly complex task requirements, the working space of mobile robots is no longer limited to a single environment, and its scope of activity is no longer limited to a single spatial dimension such as land or air, but is developing in a diversified direction [7]. A multiphibious mobile robot refers to the use of a variety of different motion mechanisms to build a complete robot system to achieve a variety of different forms

L. Fang et al. (Eds.): CICAI 2023, LNAI 14474, pp. 575–586, 2024.
https://doi.org/10.1007/978-981-99-9119-8_54

of motion. At present, multiphibious mobile robots mainly include water-land amphibious robots, land-air amphibious robots, water-air amphibious robots, triphibious robots [8], and so on. Among them, the multiphibious feature refers to the robot's ability to adopt multiple motion modes such as wheeled, legged, crawling, and flying, including the variability of the motion mode of a single robot and the combined deformation of multiple robots [9].

To successfully perform various tasks in complex scenarios and flexibly respond to changing environmental conditions, land-air amphibious robots have emerged as the times require, which combine the advantages of aircraft and ground mobile robots. Compared with robots with a single motion mode, land-air amphibious robots make up for their shortcomings, such as single scene application function, limited task execution, low flexibility, and poor adaptability, and have broad application prospects [10,11]. The significance of land-air amphibious robots research lies in the ability to play the respective advantages of aerial robots and land robots at the same time, and the two complement each other. Land-air amphibious robots can not only perform ground tasks independently but also can switch to flight mode when the ground task execution conditions are not met, so they have excellent environmental adaptability.

This paper first summarizes the research status of land-air amphibious robots, divides the existing land-air amphibious robots into three different types: wheel-based, track-based, and leg-based, and introduces these three different types of land-air amphibious robots, and analyzes their advantages and disadvantages. Then, the existing technical problems and research difficulties in the study of land-air amphibious robots are analyzed. Finally, based on the existing land-air amphibious robots, the idea and future work are put forward, and the future development of land-air amphibious robots is prospected.

2   Land-Air Amphibious Robots

2.1   Land-Air Amphibious Robots Based on Wheeled Motion

Early land-air amphibious robots usually used wheeled mechanisms as their ground motion schemes [12]. However, with the deepening of research and the increasing need to make robots meet the needs of different application scenarios, Meiri N and Zarrouk D proposed a reconfigurable composite flying robot, "Flying STAR" (FSTAR) [13]. The robot is equipped with a stretching mechanism and propellers, allowing it to fly over obstacles or move inside pipes. When the extension mechanism of the robot is in operation, the arm of the robot is extended downward so that the free wheel at the end of the arm touches the ground, thus entering the land travel mode. The robot can reduce its width while deforming to adapt to crawling in limited spaces or under obstacles. The team tested the robot on different outdoor surfaces, recording its fastest ground speed of 2.6 m/s and a movement speed of 2.21 m/s in a deformation mode at a 15° angle. Beyond this speed, the robot will partially or completely lift off the ground and enter a flight state. Experiments have shown that the robot has high maneuverability and a very smooth transition between flight and land-walking modes. However, the robot can only move on the ground with a fixed attitude because the wheeled mechanism cannot guarantee motion stability under rough terrain.

Dudley C J, Woods A C and Leang K K of the University of Utah in the United States designed a miniature rolling flying robot called "ATR", with a total mass of 35 g and a payload of 10 g [14]. The robot's design revolves around a tiny quadcopter encased in a lightweight spherical exoskeleton that can rotate around the quadcopter. The spherical exoskeleton provides flexible ground motion capabilities while maintaining essential aerial robotic properties. "ATR" can fly in the air or roll on the ground and can enter and move in narrow spaces such as wind tunnels. However, the robot has some limitations, one of which is that the spherical shell is easy to get stuck in the terrain, while it is prone to deformation when moving, which affects the performance of ground motion.

To provide a robotic platform capable of implementing multiple motion mechanisms to perform flexible and challenging tasks, the team of Kuswadi S, Tamara M N, and Sahanas D A proposed an adaptive morphology-based flying robot "PENS-FlyCrawl" [15]. The robot achieves flight by using a dual-rotor mechanism and adds a simple lever with a tilting landing gear function for crawling. The lever can be rotated 360° to simulate the movement effect of a wheeled robot. The tilt-rotor mechanism enables the dual-rotor aircraft to achieve vertical take-off, efficient landing, and high-speed flight. However, due to the consideration of energy efficiency, the flight mechanism chooses dual rotors, which makes stability the core issue that needs to be solved urgently. At the same time, using a lever as a ground motion mechanism will cause severe vibration of the overall body of the robot, which will have a certain impact on the motion stability of the robot.

Fig. 1. Unmanned flying truck "Black Knight Transformers" [16].

In the military field, Advanced Tactics Inc., California, USA has developed a land-air amphibious aircraft named "Black Knight" [17]. The aircraft combines the characteristics of an off-road truck and a drone. It is similar to an ordinary truck when driving on the ground. It uses eight rotor blades when flying, and the speed in the air can reach 150 miles per hour (about 240 km per hour). The aircraft is capable of off-road driving on the ground, vertical take-off and landing, and autonomous flight, with a total weight of about 2000 kg and a load capacity of more than 450 kg. When driving on the ground, even on rough terrain, its speed can still reach 112 km/h; in flight mode, the flight speed can reach 240 km/h. The rotors used by the "Black Knight" can be interchanged, and the cargo compartment can also be quickly converted into a crew compartment. When taking off, the eight-rotor engines extend outward, and the rotors are tilted forward at a certain angle in the level flight state to increase the level flight speed. In land travel mode, all rotor engines are folded in to reduce vehicle width and facilitate traversing tight terrain. The two motion modes of the "Black Knight" are shown in Fig. 1.

Latscha S, Kofron M, and Stroffolino A from the University of Pennsylvania combined a snake-like transmission mechanism, a wheeled robot, and a quadrotor to design a land-air amphibious robot named "H.E.R.A.L.D" [18] for urban search and rescue missions. The flight function of H.E.R.A.L.D is mainly realized by quadrotor aircraft, and the task load during flight is increased by using large-size propellers. In the ground motion mode, the robot is composed of two serpentine transmission structures with seven degrees of freedom. The quadrotor and the snake mechanism are connected by crossed magnetic elements. The magnetic connection system enables the two to form a passive joint connection relationship, which reduces energy consumption to a certain extent. In addition, H.E.R.A.L.D can also separate the quadrotor from the wheeled robot. After separation, the ground moving mechanism becomes two independent snake-like robots. Each snake-like robot has seven degrees of freedom and is equipped with a free wheel at the end, so it can change the direction of movement flexibly and freely and has strong ground exploration and obstacle-crossing ability. However, the design of this robot has multiple connecting parts, and there are some safety risks in practical applications.

2.2 Land-Air Amphibious Robots Based on Tracked Motion

To solve the inherent defects of the poor adaptability of land-air amphibious robots based on wheeled motion in complex terrain, and considering the advantages of the track mechanism with strong off-road ability, climbing ability, and not easy to slip on wet, muddy, or soft ground, etc., and is less affected by the terrain and able to cope with more complex working conditions, engineers and researchers began to try to use track motion schemes to replace wheeled schemes.

On January 3, 2020, robotics technology supplier Robotic Research LLC launched a new deformable drone Pegasus-Mini [19], at the 2020 International Consumer Electronics Show (CES 2020). The drone can travel on land as well as move in flight. The two sides of the robot are equipped with two "T-shaped" landing gears, and land tracks are installed on the landing gears so that it can complete ground tasks like a tank when it lands. The Pegasus-Mini has a compact structure measuring approximately 16 × 8 in., weighs only 4.2 pounds, and has a payload of up to 2 pounds. It can operate for up to 30 min in flight mode and up to two hours in ground operation mode. The robot can switch motion modes by controlling the rotation of the motors on both sides. However, one of its obvious disadvantages is that it is susceptible to external interference when the motion mode is switched, which affects the normal progress of the mode conversion.

To expand the working space range of quadrotor drones and improve endurance, and avoid the waste of electric energy caused by long-term flight, Polish designer Witold Mielniczek developed a new type of land-air amphibious drone called "X-Tankcopter" [20], as shown in Fig. 2. The appearance design of the UAV is based on military tanks. It is equipped with land tracks on both sides, and four propellers with a diameter of 4.5 in. are placed in the middle of the tracks, making the quadrotor UAV have the function of land and air amphibious movement. However, since the robot needs an additional power source for walking on land, there are some disadvantages, such as the extra power battery increases the load, reduces the endurance, the mechanical structure is relatively complicated, and mechanical failures are prone to occur.

Fig. 2. X-Tankcopter [20].

In 2021, a technology company called Phractyl demonstrated the land-air amphibious vehicle they are developing-"Macrobat". The aircraft is designed to operate in environments that lack level ground and ground transportation infrastructure. The Macrobat is a single-seat manned vehicle with a range of 93 miles, a payload of 330 pounds, and a top speed of approximately 112 MPH. When the Macrobat takes off, the track mechanism will assume part of the function of the landing gear, lifting the body and tilting it by 45° to create a proper take-off angle. The propulsion system will then activate and generate enough lift to get it off the ground. Once the aircraft is fully airborne, the track mechanism automatically retracts under the empennage to minimize drag for a stable and fast cruising flight. The unique tracked landing gear structure enables Macrobat to land safely on rough terrain. However, this also means that the landing must be performed at a very low speed. Otherwise, the excessive momentum could cause the craft to tip over.

In 2022, five EU countries, Poland, Finland, Spain, the Czech Republic, and Austria, jointly participated in the development of a composite land-air amphibious flight platform called "HUUVER". The platform adopts a combination of a track mechanism and a multi-rotor lift mechanism. As shown in Fig. 3, HUUVER completely abandons the car body of the traditional land-air amphibious robot in appearance but designs a narrow rectangular body structure in the middle of the platform for installing various sensor devices. This design makes the HUU-VER robot platform have good ground obstacle-crossing ability. In the middle of the track, there are four brushless motors on the left and right, and when switching to flight motion, eight horizontal propellers start to run. When performing ground tasks, HUUVER mainly uses connected tracks to complete the movement. However, the disadvantages of this robotic platform are low energy utilization during track movement and relatively poor maneuvering ability.

Fig. 3. HUUVER UAV first prototype [21].

2.3   Land-Air Amphibious Robot Based on Leg-Foot Motion

With the deepening of people's understanding of the formation mechanism and functional characteristics of biological systems, the development of modern bionics provides a new perspective for mobile robots. The design of these robots is based on the concept of bionics, fusing biological performance with electromechanical systems. They have many advantages, including the ability to adapt to different environments, a wide range of work spaces, good at avoiding danger, high survival ability, and excellent spatial mobility. As a result, they can substitute humans in carrying out various activities in unpredictable environments [6].

Traditional robots can only inspect steel bridge structures from the outside because it is difficult for them to navigate inside the complex steel bridge structure without colliding. To meet the automation needs of long-term inspection and maintenance of bridge safety, Photchara Ratsamee's team at Osaka University proposed a composite robot for steel bridge inspection and repair [22]. Modeled on reptilian creatures, the robot has six legs, each with three joints. At the second joint, the researchers connected the propeller and motor drive together. When the robot is in flight mode, its six legs will unfold to maintain overall balance and then take off and land vertically through the lift provided by the propellers. To keep the robot stable in three-dimensional space, the team also proposed a control scheme based on the vibration of the vibrator. The core idea of this scheme is to compensate for the vibration generated by the joint actuator during flight. However, the disadvantage of this robot is that the leg parts are too dense and prone to mutual interference.

Stanford University proposed a composite flying robot called The Stanford Climbing and Aerial Maneuvering Platform (SCAMP) [23]. Based on the perching and climbing behavior of various animals, including flightless birds, the robot can fly, perch and climb outdoor surfaces. In addition to the most basic quadrotor components, the composite robot is equipped with two elastic barbs for climbing and two rear feet for support and climbing. Among them, the rear feet can not only provide a certain amount of traction but also can be used to change the angle of the robot's torso, thereby enhancing its ability to adapt to unstructured terrain. Since the robot platform combines two mechanisms of flying and climbing at the same time, which not only increases the weight but also has a certain strong coupling effect.

However, most current land-air amphibious robots lack dexterous object manipulation capabilities beyond basic functions. With the deepening of the research, it is hoped that the land-air amphibious robot can not only fly like a bird but also can freely switch the motion mode and can grab objects. Although flying robots have limited payload capacity, they can use ground contact to grab objects through specific rotations. Therefore, researchers Fan Shi and Moju Zhao simulated the behavior of birds grabbing prey based on mechanical analysis and external contact constraints and proposed a new type of composite flying robot named "HYDRUS" [24]. The researchers focused on a transformable multi-joint flying robot, which consists of four rotors and a four-degree-of-freedom "snake" body, and the bottom is equipped with multiple triangular brackets to act as "legs and feet". In the ground mode, the robot first reaches the heavy object through the foot movement, then fixes the object through the bending of the body, and finally starts the rotor to lift the object. The most obvious disadvantage of this robot is that the actuators may interfere with each other during grabbing, which affects the stability of the overall motion.

In the journal SCIENCE ROBOTICS in 2021, Kyunam Kim and Patrick Spieler proposed a multi-modal mobile robot platform called Legs onboard drone (LEONARDO) [25]. Through the synchronous control of two sets of distributed motors and a pair of joints with multiple degrees of freedom, the robot makes up for the defect between the two different motion mechanisms of flight and walking. By combining these two different motion mechanisms, LEONARDO has achieved some complex maneuvers and can maintain a delicate balance under complex extreme conditions, such as walking on steel wires and sliding skateboards. The robot chooses to fly or walk depending on whether the ground is detected by the plantar contact sensor, take-off and landing are accomplished by generating appropriate walking and flight trajectories and monitoring the plantar contact state to switch between the two modes. Using LEONARDO, the experimental team also demonstrated that the robot can use its redundant joints to achieve agile walking movements, staggered flight maneuvers, and use propellers and leg joints to descend stairs and cross obstacles. Through the combination of mechanical structure design and synchronous control strategy, LEONARDO has achieved unique multi-modal mobility capabilities, enabling it to complete tasks and operations that are difficult for single-modal motion robots. However, the biped mechanical structure adopted by the robot is not stable enough compared with the quadruped structure, and its control is also difficult. Table 1 summarizes the advantages and disadvantages of land-air amphibious robots based on different motion structures in the land travel mode.

Table 1. Comparison of land-air amphibious robots based on different motion structures in land travel mode.

Motion structures in land travel mode	Advantages	Disadvantages
Wheeled Based	Simple, high speed and high efficiency	Movement is unstable under rough ground
Tracked Based	Strong ability to overcome obstacles on the ground	High energy consumption, poor mobility
Legged Based	Strong adaptability to complex terrain	Complex structure and low movement efficiency

3   Existing Problems of Land-Air Amphibious Robots

The land-air amphibious robots can work in the air as well as on land. Although this robot has certain potential and application prospects, there are also some problems and challenges.

1) Complexity: The land-air amphibious robots need to have the ability to move on land and in the air at the same time, and the robot needs to take into account the different movement and control requirements in the two environments, which will lead to the more complex structure and mechanical design of the robot, increasing the difficulty of research and development and manufacturing.
2) Power and energy: Movement on land and in the air requires the use of different power and energy systems. For example, a robot may use a wheeled drive system on land, but a thruster or rotor system in the air, which means that the robot may need to use two different energy systems, which will also bring a greater load to the robot.
3) Control and navigation: It is a complex and difficult task to realize the autonomous control and navigation of land-air amphibious robots because the robots need to have the ability to perceive and navigate in different environments (land or air) to perform accurate positioning, path planning, and obstacle avoidance [26]. At the same time, land-air amphibious robots also need to be able to switch between land and air motion modes autonomously.
4) Structure and weight: To achieve movement on land and in the air, land-air amphibious robots may need to have more complex structures and mechanical components, which may increase the weight of the robot. Greater weight may have a negative impact on the robot's performance, such as stability in flight and walking and the robot's endurance.

5) Environmental adaptability: Due to the large differences in environmental conditions such as climate, temperature, and air density between land and air, the performance and adaptability of robots in these two environments may be different, so they need to have the ability to adapt to different environmental conditions.

Although these problems still exist in the current research of land-air amphibious robots, and there are not enough mature products yet, but with the continuous development and progress of science and technology, land-air amphibious robots will continue to innovate and improve, and more and more solutions to existing problems will emerge. Solutions to problems to meet various challenges and play an important role in military, rescue, exploration, and other fields.

4   Future Prospects

At present, the research on land-air amphibious robots is still in its infancy. Based on the existing land-air amphibious robots, we can provide the following prospects for the design and development of land-air amphibious robots in the future:

1) Mechanical structure optimization: By constantly exploring and innovating new mechanical structure designs, researchers can develop more efficient, lighter, and more compact land-air amphibious robots. This includes the use of new materials and the optimization of the overall structural design of the robot to improve the motion performance and stability of the robot.

2) Autonomous control and intelligent decision-making: By designing more advanced autonomous control algorithms and intelligent decision-making systems, the autonomous perception, decision-making, and action capabilities of land-air amphibious robots in complex environments can be realized so that robots can adapt to different tasks and environmental requirements.

3) Multi-modal perception and navigation: Further improve the perception and navigation capabilities of land-air amphibious robots so that they can more accurately identify the environment on land and in the air. This involves using multiple sensors, including infrared cameras, LiDAR, barometers, and more, and fusing the information collected by the sensors to model the surrounding environment for more accurate positioning navigation, path planning, and obstacle avoidance.

4) Energy efficiency and endurance: Researchers can work to improve the energy efficiency and endurance of land-air amphibious robots. This includes the development of more efficient power systems, new energy technologies, and intelligent energy management strategies to extend the working time of robots in different modes.

5) Multi-modal switching ability: During the movement of land-air amphibious robots, it involves switching between two different motion modes: land and air. In the future, we can further study how to make the robot perform fast and stable transition and switch in these two different working states.

6) Application expansion and safety: Explore the potential of land-air amphibious robots in various practical application fields, and study how to improve the safety and reliability of robots.

7) Triphibious robots: Most of the existing multiphibious robots only involve the ability to work in two kinds of motion scenarios, including land-water amphibious, land-air amphibious, and air-water amphibious, and there are few experiments and research on triphibious robots, compared with amphibious robots, triphibious robots have stronger environmental adaptability, stronger mobility and better functionality, which makes triphibious robots have a wider range of application scenarios, and also brings greater development difficulty [27].

Through continuous research and innovation in the above aspects, the performance and application prospects of land-air amphibious robots will be further improved, providing more possibilities for future technological development and practical applications.

5   Conclusion

The research and development of land-air amphibious robots require the integration of knowledge from multiple disciplines, including machinery, control, materials, communications, energy and power, physics, robotics, power electronics technology, and computer science. For now, the research on land-air amphibious robots is still in the experimental stage, the technology of land-air amphibious robots is still immature, and the application practice is far from meeting people's expectations. However, with the continuous development and progress of science and technology, the research on land-air amphibious robots will also be continuously improved to overcome existing bottlenecks and existing problems to create robot products that are more in line with actual needs and have more application value.

References

1. Yuh, J.: Design and control of autonomous underwater robots: a survey. Auton. Robot. **8**, 7–24 (2000)
2. Raibert, M., Blankespoor, K., Nelson, G., Playter, R.: Bigdog, the rough-terrain quadruped robot. IFAC Proc. Vol. **41**(2), 10822–10825 (2008)
3. Ortigoza, R.S., et al.: Wheeled mobile robots: a review. IEEE Lat. Am. Trans. **10**(6), 2209–2217 (2012)
4. Moskvin, I., Lavrenov, R., Magid, E., Svinin, M.: Modelling a crawler robot using wheels as pseudo-tracks: model complexity vs performance. In: 2020 IEEE 7th International Conference on Industrial Engineering and Applications (ICIEA), pp. 1–5. IEEE (2020)
5. Fahlstrom, P.G., Gleason, T.J., Sadraey, M.H.: Introduction to UAV Systems. John Wiley & Sons, Hoboken (2022)
6. Ren, K., Yu, J.: Research status of bionic amphibious robots: a review. Ocean Eng. **227**, 108862 (2021)

7. Xu, X., Liu, B., Pan, D., Lu, X.: Research and application of amphibious bionic robots. J. Unmanned Undersea Syst. **31**, 143–151 (2023)
8. Guo, Z., Li, T., Wang, M.: A survey on amphibious robots. In: 2018 37th Chinese Control Conference (CCC), pp. 5299–5304. IEEE (2018)
9. Parra-Vega, V., Sanchez, A., Izaguirre, C., Garcia, O., Ruiz-Sanchez, F.: Toward aerial grasping and manipulation with multiple UAVs. J. Intell. Rob. Syst. **70**(1–4), 575–593 (2013)
10. Liu, Z., Song, M., Liu, Y., Bai, B.: Design, modeling and simulation of a reconfigurable land-air amphibious robot. In: 2021 IEEE 11th Annual International Conference on CYBER Technology in Automation, Control, and Intelligent Systems (CYBER), pp. 346–352. IEEE (2021)
11. Li, K., Han, B., Zhao, Y., Zhu, C.: Motion planning and simulation of combined land-air amphibious robot. In: IOP Conference Series: Materials Science and Engineering, vol. 428, p. 012057. IOP Publishing (2018)
12. Kossett, A., D'Sa, R., Purvey, J., Papanikolopoulos, N.: Design of an improved land/air miniature robot. In: 2010 IEEE International Conference on Robotics and Automation, pp. 632–637. IEEE (2010)
13. Meiri, N., Zarrouk, D.: Flying star, a hybrid crawling and flying sprawl tuned robot. In: 2019 International Conference on Robotics and Automation (ICRA), pp. 5302–5308. IEEE (2019)
14. Dudley, C.J., Woods, A.C., Leang, K.K.: A micro spherical rolling and flying robot. In: 2015 IEEE/RSJ International Conference on Intelligent Robots and Systems (IROS), pp. 5863–5869. IEEE (2015)
15. Kuswadi, S., Tamara, M.N., Sahanas, D.A., Islami, G.I., Nugroho, S.: Adaptive morphology-based design of multi-locomotion flying and crawling robot "pens-flycrawl". In: 2016 International Conference on Knowledge Creation and Intelligent Computing (KCIC), pp. 80–87. IEEE (2016)
16. McCluskey, B.: U.s. military to use 'black night transformer' drone to evacuate wounded warriors (video + pics) (2014). https://www.guns.com/news/2014/01/13/u-s-military-use-black-night-transformer-drone-evacuate-wounded-warriors-video
17. Jian, M.: Flight System Design of an Air-ground Vehicle. Master's thesis, Beijing Institute of Technology (2015)
18. Latscha, S., et al.: Design of a hybrid exploration robot for air and land deployment (herald) for urban search and rescue applications. In: 2014 IEEE/RSJ International Conference on Intelligent Robots and Systems, pp. 1868–1873. IEEE (2014)
19. Choi, C.: Pegasus robots can fly and drive (2021). https://insideunmannedsystems.com/pegasus-robots-can-fly-and-drive/
20. Kesteloo, H.: It's a drone. it's a tank. it's the x-tankcopter! (2019). https://dronedj.com/2019/02/28/x-tankcopter/
21. European GNSS Agency: D3.1 a fully assembled hardware platform (2022). https://huuver.eu/wp-content/uploads/2022/01/D3.1_A-fully-assembled-hardware-platform.pdf
22. Ratsamee, P., et al.: A hybrid flying and walking robot for steel bridge inspection. In: 2016 IEEE International Symposium on Safety, Security, and Rescue Robotics (SSRR), pp. 62–67. IEEE (2016)
23. Pope, M.T., et al.: A multimodal robot for perching and climbing on vertical outdoor surfaces. IEEE Trans. Rob. **33**(1), 38–48 (2016)
24. Shi, F., Zhao, M., Murooka, M., Okada, K., Inaba, M.: Aerial regrasping: pivoting with transformable multilink aerial robot. In: 2020 IEEE International Conference on Robotics and Automation (ICRA), pp. 200–207. IEEE (2020)

25. Kim, K., Spieler, P., Lupu, E.S., Ramezani, A., Chung, S.J.: A bipedal walking robot that can fly, slackline, and skateboard. Sci. Rob. **6**(59), eabf8136 (2021)
26. Zhang, E., Sun, R., Pang, Z., Liu, S.: Obstacle capability of an air-ground amphibious reconnaissance robot with a planetary wheel-leg type structure. Appl. Bionics Biomech. **2021**, 1–12 (2021)
27. Zhu, Y., Guo, Z., Li, T., Wang, M.: Implementation and performance assessment of triphibious robot. In: 2019 IEEE International Conference on Mechatronics and Automation (ICMA), pp. 1514–1519 (2019). https://doi.org/10.1109/ICMA.2019. 8816394

Author Index

Printed in the United States
by Baker & Taylor Publisher Services